Field to Palette

Dialogues on Soil and Art in the Anthropocene

Alexandra R. Toland · Jay Stratton Noller · Gerd Wessolek

CRC Press
Taylor & Francis Group
Boca Raton London New York

CRC Press is an imprint of the
Taylor & Francis Group, an **informa** business

Cover Image: Lara Almarcegui, Aushub aus Basel (Excavation materials from Basel), Kunsthaus Baselland, 2015.

Photo: Serge Hasenböhler.

Image reprinted with permission of the artist.

CRC Press
Taylor & Francis Group
6000 Broken Sound Parkway NW, Suite 300
Boca Raton, FL 33487-2742

© 2019 by Taylor & Francis Group, LLC
CRC Press is an imprint of Taylor & Francis Group, an Informa business

No claim to original U.S. Government works

Printed on acid-free paper

International Standard Book Number-13: 978-1-138-29745-6 (Paperback)
978-1-138-58509-6 (Hardback)

Visit the Taylor & Francis Web site at
http://www.taylorandfrancis.com

and the CRC Press Web site at
http://www.crcpress.com

To Tilia and Taavi, and the future generation
of soil stewards the world over

Foreword

People are intimately connected to the soil. The study of the soil would easily be the most vibrant and well-funded scientific discipline in the universe—if only all humans realized the importance of that intimate connection. That is not the case. Most people of the planet will never see a soil profile, be enthralled by the beauty of argillans, or watch stagnating water over a textural discontinuity, nor will they partake a deep understanding of the importance of soil for our daily livelihood. There is great beauty in the soil and although humankind may not overwhelmingly realize the importance of soil, there is potential to enhance its appreciation through the arts. See here the purpose of this book: an eclectic collection of unexpected enlightening, a global journey balancing art and science.

The first depictions of soil profiles were made long before soil science as a scientific discipline was established. In many art galleries across the world, there are paintings of landscapes, usually from the seventeenth century onward. They illustrate how artists viewed the landscape but also how the naturalists' view and the countryside have changed over time. Landscape painting was particularly popular in Europe. Hans Jenny (1899–1992) was a dedicated visitor of art galleries and to him soils were highly aesthetic. In the late 1960s he wrote an article on the image of soil in landscape art from medieval times to the mid-1900s. In nineteen paintings he discussed medieval rocks, Renaissance paintings, landscapes of the noble moods, trends toward naturalism, Mediterranean painters, red soils, and the abstract landscape. The old landscape painters saw things that most other humans failed to see. They painted soil features that we now recognize as podzols (Jan van Goyen), paleosols (Jacob van Ruysdael), oxisols (Paul Gauguin), or vertisols (George Lambert). At the time they were painted, these soils had no name and no description, and soil science had yet to be established. We can look at these early soil depictions and note that there is an element of great aesthetics. Perhaps, there was the hidden invitation to study what was seen and the arts may have opened the eyes for the science to follow.

This book follows in that belief. The book contains a series of dialogues between artists and soil scientists about the cultural meaning and value of soil and the way it is studied for practical purposes or simply for the need to understand our natural world. There is deep reflection on the aesthetic value of the soil and its importance for humankind. If one thing becomes clear reading this book it is that the pursuit for discovery is a common thread among artists and soil scientists alike. Hats off to the editors and all contributors for a unique book that connects different people on a natural resource on which so much depends: the soil! Let us hope that the dialogues presented in this book may open the eyes of a new generation, now and in the future. May they take up a shovel, an auger, as well as the pen and palette. May they wander the fields and imagine all we do not know yet.

Alfred E. Hartemink
Department of Soil Science, FD Hole Soils Lab
University of Wisconsin–Madison
Madison, Wisconsin
E-mail: hartemink@wisc.edu

Introduction

Alexandra R. Toland, Gerd Wessolek,
Jay Stratton Noller

Field to Palette – Dialogues on Soil and Art in the Anthropocene is an investigation of the cultural meanings, representations, and values of soil in a time of planetary change. It is a critical reflection on soil-related issues of the Anthropocene, including land take, groundwater pollution, desertification, and biodiversity loss. It is also a celebration of resilience in the face of such challenges and a call to action. The title of the book is on the one hand a nod to grassroots social organizing and locally controlled food production methods developed by "field to plate" movements worldwide. On the other hand, it is a call to the field of soil science for increased interdisciplinary engagement with the arts and humanities. With contributions from over one hundred internationally renowned artists, curators, and leading environmental scientists, *Field to Palette* presents a set of visual methodologies and worldviews that expand our understanding of soil.

Inspired by the rich biological and pedological diversity of soil types found in terrestrial landscapes the world over, we used diversity as a guiding principal in the selection of book contributions. Special attention was given to ensure gender and cultural diversity as well as representation from different geographic locations. Scientists included in the book represent a wide range of disciplinary interests, from agronomy and crop science to geomorphology, soil hydrology, microbiology, physical geography, and environmental engineering. Artistic positions were similarly chosen to represent a broad range of creative formats, with examples from the visual and performing arts, architecture, landscape design, product design, textile design, culinary arts, and film. For the sake of diversity, the authors do not merely visualize the physical, aesthetic properties of soil, such as color or texture, but explore a wide range of cultrual articulations, moving between attraction and disgust, dependence and exploitation, reverence and loss, use and degradation.

The chapters in this book are framed as dialogues among different disciplines and individuals that come together as a chorus of lively voices. The slogan "Give Soil a Voice" often comes up in soil awareness and education discussions. For this book, we wanted to hear the voices of those who give soil a voice. We wanted to hear stories of the past and visions of the future. We wanted to hear expert opinions in the form of research narratives, critical questioning, and sometimes disagreement. We wanted to hear the voices of prominent scientists as well as those who do not usually attend scientific meetings but have something important to say about soil. We wanted to facilitate dialogue because we believe that dialogue is a fundamental process of change and is often overlooked in soil protection contexts that value data over discourse and policy statements over human experience.

The result of that process is a rich volume of authentic exchanges about the material properties, cultural histories, environmental functions, and existential threats of the soil in a range of different practices, places and cultural traditions. It is a collection of conversations in different formats and time frames, some carried out over many months and even years, others as fleeting exchanges via email. While the chapter topics are relatively straightforward, the style and personal tenor of the writing fluctuates from solemn to humorous, fictional to factual, poetic to prosaic, and objectively distanced to deeply personal.

Based on variables of research interests and geographic proximity (which in some cases only meant living on the same continent) we conducted an interdisciplinary "matchmaking" experiment to connect people who otherwise would not likely have come in contact with one another. In some cases this was a success. In other cases, disciplinary differences, human chemistry, time and geographical restraints, or factors unbeknownst to us truncated the process. Over the course of four years, the list of contributors shifted multiple times, as did the publishing relationship, the title, and the structural focus of the book. By no means an exhaustive overview, this book represents instead an iterative and relational process of bringing scientists and artists together to share perspectives on human-soil relationships and their importance for the existence of life on the planet.

The perspectives shared here are as heterogeneous as the interests of the participating authors and point to an important aspect of the interdisciplinary experiment itself. In many cases disciplinary boundaries are diffuse, revealing a complex web of methodological and epistemological impulses that challenge assumptions about artistic and scientific practice. About half of the artists in the book have some degree of training in different scientific fields and view their work as research. Meanwhile, a good number of contributing scientists openly discuss the aesthetic aspects of their work, and a handful actively paint, photograph, write poetry, or pursue other artistic endeavors.

To honor the complexity of interdisciplinarity and avoid a repetition of well-intentioned but generalized exchanges, we needed to focus efforts around specific ideas. In transdisciplinary stake-holder processes *boundary objects* are often used to do just this. Boundary objects, according to sociologists and Science Technology Society (STS) scholars Susan Leigh Star and James Griesemer (1989),[1] are conceptual entities that bridge different understandings of information by different user groups. Boundary objects are interpreted differently depending on the group, but contain enough content to allow members of different disciplines and social groups to talk and work together. Boundary objects, which can be material or theoretical in nature, must be specific enough to keep discussions focused and avoid superficiality, but general enough to allow new ideas and possibly new boudary objects to emerge and conversation to remain open. Boundary objects both "inhabit several communities of practice and satisfy the informational requirements of each…" and "are both plastic enough to adapt to local needs and constraints of the several parties employing them, yet robust enough to maintain a common identity across sites" (Star, 2015).[2]

For a long time, we debated what these could be. Current soil science research topics? Artistic genres? Land use conflicts of the Anthropocene? We initially looked at the internal organizational structure of the International Union of Soil Sciences for clues. Established concepts of soil genesis, soil organic matter (SOM), and soil security served as initial boundary objects at the outset of the book. Based on earlier research, however, we finally decided to focus on the concept of *soil functions* as boundary objects for the dialogue process of the book.[3] Whether we speak of structural functions, environmental functions, political functions, bodily functions, or aesthetic functions, we can agree that "function" is a term that is widely used and accepted by various groups. Philippe Baveye, citing Kurt Jax (2005) describes four common uses of the term, before unpacking its meaning for soil scientific inquiry. Function, according to Baveye, is understood as:

1. a state change in time (more or less synonymous to "process"),
2. as a shorthand notation for "functioning" (referring to some state or trajectory of a given system, and to the sum of the processes that sustain the system),

3. as the specific role of parts of the system in the different processes they are engaged in, and, finally,
4. as a "service" provided to humans and possibly other living beings (plants or animals)."[4]

Soil functions embody a number of technological, ecological, and social facets that have been variously described by Winfried Blum and adopted by the European Union and other organizations as a way of explaining what soils do and how we humans are utterly dependent upon what they do. Having been in use for over fifty years,[5] the soil functions concept is robust enough to provide a structure for the book that readers of all disciplines and non-disciplines can relate to, and flexible enough to enable authentic dialogue among thinkers of very different backgrounds. Using soil functions as boundary objects for interdisciplinary dialogue, we can begin thinking about the cultural meanings and hidden interactions of particular phenomena in the landscape and what these could mean to a wider public.

Seeing soil functions through the lens of artistic practice, we can poetically translate the soil function of *biomass production* into terms of *sustenance* and *nourishment*. *Habitat and gene pool* can be reinterpreted as *home*. The *storage function* of the soil, for example in carbon and water sequestration processes is read as the capacity to be a *vessel*, a *repository* or a *receptacle*. The ionic *buffering and filtering* function of the soil is seen in terms of *transformation*. The conflicted *platform* function of the soil is understood as not only a literal platform for buildings and roads but as a social *stabilizer* and *bioinfrastructure* for human and more-than-human communities.[6] Using the six main soil functions as a conceptual guide, a wealth of imagery and ideas begin to emerge beyond the common tropes of Mother Earth and cupped hands holding a seedling. By focusing on soil functions the dialogue becomes enriched with new ways of thinking about soil and our relationships to it. At the same time, our own functions as human beings are called upon to strengthen the capacity of our species for soil stewardship, social responsibility, and ethical living on and with the Earth.

Endnotes

1. Star, Susan Leigh; Griesemer, James (1989). "Institutional Ecology, 'Translations' and Boundary Objects: Amateurs and Professionals in Berkeley's Museum of Vertebrate Zoology, 1907-39". *Social Studies of Science*. 19 (3): 387–420.

2. Star, S. L. (2015) page 157. "Misplaced Concretism and Concrete Situations" in Boundary Objects and Beyond – Working with Leigh Star, Eds. Geoffrey C. Bowker, Stefan Timmermans, Adele E. Clarke, and Ellen Balka (Cambridge: MIT Press), reprinted from Feminism, Method and Information Technology (1994) Aarhus University: Feminist Research Network, Gender-Nature-Culture.

3. For a comprehensive discussion on soil functions and art, see Toland, A. (2015) Soil Art – Transdisciplinary Approaches to Soil Protection. Doctoral thesis at the TU Berlin, Faculty VI Planning, Building, Environment; Institute for Ecology; Dept. of Soil Protection; For a theoretical overview of soil functions, see for example: Blum, W. E. H. (1993). "Soil protection concept of the Council of Europe and integrated soil research," in *Soil and Environment, Integrated Soil and Sediment Research: A Basis for Proper Protection*, Vol. 1, Eds. H. J. P. Eijsackers and T. Hamers (Dordrecht: Kluwer Academic Publisher), 37–47; Blum, W. E. H. (2005). Functions of soil for society and the environment. *Reviews in Environmental Science and Bio/Technology* 4, 75–79. doi: 10.1007/s11157-005-2236-x; and Blum, W. E. H., Warkentin, B. P., and Frossard, E. (2006). "Soil, human society and the environment," in *Functions of Soils for Human Societies and the Environment*, Vol. 266, Eds. E. Frossard, W. E. H. Blum, and B. P. Warkentin (London: The Geological Society of London).

4. Baveye, P. C., Baveye, J., and Gowdy, J. (2016). "Soil 'Ecosystem' Services and Natural Capital: Critical Appraisal of Research on Uncertain Ground". *Frontiers in Environmental Science*. 4 (41). doi: 10.3389/fenvs.2016.00041. p. 11.

5. Ibid.

6. For more on the idea of Soil as Bioinfrastucture, see: Puig de la Bellacasa, M. (2015). "Ecological Thinking, Material Spirituality, and the Poetics of Infrastructure," in *Boundary Objects and Beyond – Working with Leigh Star*, Eds. Geoffrey C. Bowker, Stefan Timmermans, Adele E. Clarke, and Ellen Balka (Cambridge: MIT Press).

Prelims

Function 1
SUSTENANCE: Soil as provider of food, biomass and all forms of nourishment

Function 2
REPOSITORY: Soil as source of energy, raw materials, pigments, and poetry

Function 3
INTERFACE: Soil as site of environmental interaction, filtration and transformation

Function 4
HOME: Soil as habitat, biological hotspot, and gene pool

Function 5
HERITAGE: Soil as embodiment of cultural memory, identity, and spirit

Function 6
STABILIZER: Soil as platform for structures, infrastructures, and socioeconomic systems

Function 1

SUSTENANCE: Soil as provider of food, biomass and all forms of nourishment

Sustenance

The first section of the book introduces the most widely recognized function of the soil—that as great provider. Approximately three-quarters of the Earth's surface is covered by water and only about half of the remaining quarter of land surface is available for human use. Of that land available for human use, three-quarters is either too rocky, too wet, too dry, too hot, or too cold for food production. An estimated ninety-nine percent of human nourishment comes from that last remaining three percent of available land. Despite this already slim margin of available land, the Food and Agriculture Organization of the United Nations (FAO) has warned that over a third of this land is already "moderately to highly degraded through erosion, salinization, compaction, acidification, chemical pollution and nutrient depletion." (See reference on Soil Solutions website cited December 1, 2017: https://soilsolution.org/10-soil-facts/.) This section of the book presents the production function of soil as a critical factor in human existence.

The section opens with one of the most iconic artworks of the twentieth century, Agnes Denes' *Wheatfield—A Confrontation*. The revered eco-art pioneer Agnes Denes exchanges views on soil, hunger, and world peace with the esteemed president of the International Union of Soil Sciences, Rattan Lal. Terroir is conceptually explored as a means of communicating soil's productive capacity in chapters by Laura Parker and Tom Willey, Lou Preston and Scott Burns, an interview with Brent Clothier and Matthew Moore by Alexandra R. Toland, and a recipe for a hearty root stew by star chef and sustainable soil activist Sarah Wiener. Meal culture is proposed as an alternative concept to soil and food security in the chapter by Parto Teherani-Krönner and Roxanne Swentzell, who argue for more attention to feminist and indigenous perspectives in soil protection debates. Problematic issues of agricultural industrialization, disrupted nutrient cycles, pollution, and the Anthropocene are brought up in chapters by Maria Michails and Ronald Amundson, and Bonita Ely and Richard MacEwen. As an antidote to these, experimental solutions in human waste recycling, urban "black gold" composting and artistic soil building are offered in chapters by Valentina Karga, Ayumi Matsuzaka and Stephen Nortcliff; Tattfoo Tan; and Sue Spaid.

Urban Farming

The New Green Revolution?

Agnes Denes and Rattan Lal in conversation with Alexandra R. Toland

Agnes Denes was born in Budapest, raised in Sweden, educated in the United States, and began her artistic career in the 1960s. One of the most prominent artists of our time, Denes is internationally known for works investigating science, philosophy, linguistics, psychology, poetry, history, and music, created in a wide range of mediums. Denes has had three museum retrospectives and participated in more than 600 exhibitions at prominent galleries and museums throughout the world including, among others, the Corcoran Gallery of Art, Washington, D.C. (1974); the Ludwig Museum, Budapest, Hungary (2008); the Biennale of Sydney (1976); documenta 6 and 14, Kassel, Germany (1977 and 2017); the Venice Biennale (1978); and Ends of the Earth: Land Art to 1974, Museum of Contemporary Art, Los Angeles (2012). She has completed public and private commissions in North and South America,

Title image: *Wheatfield—A Confrontation*, 12 acres of wheat, reenactment, Fondazione Nicola Trussardi, Milan, Italy, 2015.

Europe, Australia, and the Middle East, and has received numerous awards including four fellowships from the National Endowment for the Arts; the Eugene McDermott Award from MIT (1990); the Rome Prize from the American Academy in Rome (1998); the Ambassador's Award for Cultural Diplomacy (2008) from the American Embassy in Hungary; and she was a Fellow at the Massachusetts Institute of Technology (MIT) and Carnegie Mellon University and received a Guggenheim Fellowship in 2015. Denes holds honorary doctorates from Ripon College and Bucknell University, and lectures extensively at colleges and universities throughout the United States and abroad, and participates in global conferences on art, the environment, and the human condition. She is the author of six books and is featured in numerous other publications on a wide range of subjects.

Rattan Lal is a Distinguished University Professor of Soil Science and director of the Carbon Management and Sequestration Center, Ohio State University. With completion of education from Punjab Agricultural University, Ludhiana (BSc, 1963); Indian Agricultural Research Institute, New Delhi (MSc, 1965); and Ohio State University, Columbus (PhD, 1968), he served as a senior research fellow with the University of Sydney, Australia (1968–1969), Soil Physicist at IITA, Ibadan, Nigeria (1969–1987), and Professor of Soil Science at OSU (1987–present). His current research focus is on climate-resilient agriculture, soil carbon sequestration, sustainable intensification, enhancing use efficiency of agroecosystems, and sustainable management of soil resources of the tropics. Lal received the Hugh Hammond Bennett Award of the SWCS, 2005; Borlaug Award (2005) and Liebig Award (2006) of the International Union of Soil Sciences; as well as honorary degrees from Punjab Agricultural University (2001), the Norwegian University of Life Sciences, Aas (2005), Alecu Russo Balti State University, Moldova (2010), Technical University of Dresden Germany (2015), and University of Lleida, Spain (2017). He is currently president of the International Union of Soil Sciences. Among other posts, he was a member of the Federal Advisory Committee on National Assessment of Climate Change, NCADAC (2010–2014) and senior science advisor to the Global Soil Forum of IASS, Potsdam, Germany (2010–2012). Lal was a lead author of IPCC (1998–2000), and has authored/coauthored more than 868 refereed journal articles and 506 book chapters, and has written 20 and edited/coedited 68 books.

We invited Agnes Denes and Rattan Lal, two iconic figures in their own fields, to talk about their opinions on the future of urban agriculture. Bringing their voices together in one chapter is symbolic. Denes has been involved in several high-profile soil-related projects, including *Tree Mountain—A Living Time Capsule*, a monumental earthwork reclamation project in Ylöjärvi, Finland, and *Mathematical Forest* in Melbourne, Australia, consisting of 6000 endangered species. Denes' artistic practice is distinctive in terms of its engagement with aesthetics and highly visible socio-political impact. *Tree Mountain*, for example, was dedicated by the president of Finland upon its completion in 1996 and is legally protected for the next four hundred years.

As current president of the International Union of Soil Sciences and 2007 recipient of the Nobel Peace Prize Certificate by the IPCC, Professor Lal is a respected authority on the global mission to secure soil health and food sovereignty, especially as populations in urban centers continue to rise. According to Lal, healthy soil is not only integral to the stability and prosperity of nations around the globe, but is a key to world peace. Now more than ever, issues of social justice and environmental degradation associated with globalized industrial agriculture demand innovative answers from the sciences and arts. It is impossible to talk about feeding the world without imagining productive cities contributing to people's daily diet of grains, fruits, vegetables, and animal products.

The outgoing vision for the conversation begins with Denes' 1982 groundbreaking intervention, *Wheatfield—A Confrontation,* one of the first land artworks to address issues of food, energy, and world hunger. The site of *Wheatfield* was and is simultaneously a canvas, a commodity, a common good, a place of waste deposition adapted for food production, and a place of confrontation on many levels. Recently reenacted in Milan in collaboration with the *fondazione nicola trussardi* on occasion of the 2015 World Expo and International Year of Soils, *Wheatfield* provides a lens with which to imagine productive cities of the future and the soils that lie beneath.

Through a series of e-mail conversations, Lal and Denes generously shared their views on the future of soil health and the philosophical and political will needed to feed a hungry planet. In the preparation of the chapter, Denes invited her former agronomist helper and adviser for *Wheatfield,* John Ameroso, to share some of his on-the-ground insight about the project. His comments are included here, as well as Denes' original philosophical statement about *Wheatfield.*

Alexandra R. Toland: I want to start out with the concept of "soil health" and ask how it has changed over the last few decades, especially in cities where many rural people have been forced to relocate. Looking forward, as populations grow and cities grow with them, how do you both envision the future of soil health? Is it feasible to imagine wheat fields like Agnes Denes' artistic intervention appearing on the rooftops of schools and parking lots?

Rattan Lal: During the 1990s, soil health was defined as "capacity of soil to function as a critical living system to sustain biological productivity."[1] In this definition, however, some recent and emerging uses have been omitted. Important among these are climate change, urban agriculture, recycling of nutrients in urban waste, and importance of soil physical/hydrological properties.[2] The importance of modern urban agriculture to food security cannot be overemphasized, especially in the context of a close link between soil health and human health.[3] To be inclusive, the term soil health must encompass these emerging issues. This can be appropriately defined as the capacity of soil to function as a vital living system to sustain multiple ecosystem services of natural and agricultural (rural and urban) ecosystems to advance Sustainable Development Goals of the UN [United Nations] (Agenda 2030), such as improving soil carbon sequestration to mitigate the climate change and enhancing health of urban soils (Anthrosols) to increase crop yield.

John Ameroso: Looking back on *Wheatfield*, it is important to remember that the term "urban agriculture" did not exist as a common understanding in the USA until the 1990s when urban community gardening projects shifted their attention from "gardening" to production agriculture. In 1982, urban ag was not even thought about as a means to produce food in this country. Meanwhile,

in "third-world countries" most of the food consumed in cities has been produced on the perimeters of the city. I think today a two-acre wheat field would not draw too much attention. It has become common practice in cities in the US and the world to "cover crop" tracts of land where houses are razed as well as other unused spaces. What was unique about Agnes' *Wheatfield* was its location, which became even more symbolic after 9/11. Today, an impressive art/science project would be to grow a highly nutritive food crop on poor soils that currently do not support vegetation to its potential, while sequestering carbon at the same time. Through aerobic decomposition, the carbon in plants and soil organic matter is converted back to CO_2 and returned to the atmosphere. As Professor Lal mentions, the key is sequestering carbon in Anthrosols (urban soils lacking fertility). For me, the definition of an "urban soil" is anything that functions like a soil, that is, it provides nutrients, air, water, and can hold a plant's roots in place. That could be building rubble, or construction fill, as in the case of the *Wheatfield*. But how do you sequester carbon in these soils? How could these soils provide additional places for carbon storage?

Alexandra R. Toland: Agnes, it has been almost exactly 35 years after harvesting of *Wheatfield* in New York City. Looking back, what kind of impact do you think the project had? Do you think audiences would have different reactions to the sight of a wheat field in the city in 2015 as they did in 1982? Would you like to see *Wheatfield* enacted again (and again, and again …)? And, in light of how soil health was understood in the 1990s as Rattan mentions above, how do you think the conversation on food production—especially in cities—has changed over the last few decades?

Agnes Denes: Rattan didn't seem to answer your question about planting city rooftops, probably because they can't hold the necessary water and soil depth for growth of any consequence. I proposed that over 40 years ago but it would require major infrastructural reconstruction and rethinking the status quo. Little backyard gardens, yes, gardens in city parks and designated areas, and elevated constructions to grow vertically, yes. But you also need to do "crazy" things like *Wheatfield* to call attention to important issues. And they need to be done again and again.

People ask, why did you do it? And I say no other act can call attention to things that we know are harmful yet keep on doing unless it's powerful and big and can take your breath away. Then they understand that it had to be done and it needs to be done over and over again. I remember the people who stood around and cried during my harvest. Why cry? Why not celebrate? Did they cry for the earth being mishandled, for the hungry who don't have enough food due to mismanagement of resources, did they cry for the earth, the soil, the abuse, the greed? Did they cry for the children who go hungry or die from malnourishment? Or was it because it was the end a beautiful field in the middle of a metropolis? Or was it the silent tears of gratitude for the soil that yields its gift and treasure, even if not deserved?

To answer your question, my work speaks to everyone. Its philosophical underpinnings are carried in its visual beauty and easily understood by the young, even by children whose attention to these issues is most important—they are the future. I am presently planning to create a paradise for New York City, a forest of 100,000 trees and agricultural fields on a 127-acre landfill in the middle of the city, and schools to teach people how to plant in cities.

Alexandra R. Toland: Responding to what Agnes says, it seems like philosophical and political will is a driving factor of paradigm change. One of

the UN's Sustainable Development Goals for the year 2030 is to "end hunger, achieve food security, improve nutrition, and promote sustainable agriculture." I wonder about the philosophical underpinnings of the SDGs. Rattan, you have highlighted, for example, the importance of restoring soil organic carbon and adopting "nexus" thinking as two fundamental approaches to confronting world hunger. Why are these approaches so important?

Rattan Lal: The world population of 800 million at the time of Malthus in 1798 has increased 10 times to 7.6 billion in 2017 and is predicted to increase 14 times to 11.2 billion by 2100. The global grain production has already increased by 10 times over the last 200 years and may need to be doubled again by 2050. Such an increase in agricultural productivity must come from improvements in endopedonic processes rather than from horizontal expansion of land area under agroecosystems or from increase in exopedonic inputs of water, fertilizers, pesticides, and energy. Thus, restoration of soil organic carbon (SOC) concentration to above the threshold level (1.5%–2.0%) in the rootzone[4] and adopting the nexus approach of interconnectivity to advancing food security are essential to sustainable intensification of agroecosystems.[5] Enhancing SOC concentration and adopting the nexus approach are critical to "producing more from less" by reducing losses and enhancing the eco-efficiency. An example of the nexus approach is the use of grain legumes in rotation to improve soil health and human health and well-being.[6]

Alexandra R. Toland: Agnes, your work also speaks to the sustainable development goal of ending hunger and securing sustainable agriculture. The wheat you harvested in 1982 traveled to twenty-eight cities around the world in an exhibition called *The International Art Show for the End of*

World Hunger. What role do you think art can play in helping end world hunger?

Agnes Denes: To end world hunger you need science, but beyond that, understanding, intelligence, knowledge, compassion, and willingness to change things. It also takes good leadership.

Those art exhibitions reached people on all levels. The seeds from the harvested wheat was given to people from all walks of life who planted them in solidarity with the concept.

To show the importance of enhancing productivity I would have liked to plant sunflowers as a second field to enhance nutrients for another wheat field, but of course I was unable to do that. My contract was for one season only.

To end world hunger, you need powerful voices and very powerful art that visualizes concepts that can be quickly grasped. You need different kinds of understanding to fight greed, corruption, and mismanagement. You need different approaches to different problems. This art has tremendous impact. It is a powerful voice because it's honest and offers the best in the human psyche, while clearing the weeds that interfere with problem solving while it visualizes the solution.

When I started doing this new form of art in the late 1960s nobody was interested in ecological concepts. I am glad it is so widespread now. And even if the political landscape is busy elsewhere, its confused attitude will probably iron itself out when personal attitudes cease interfering with the essentials: the needs of a growing human population.

Alexandra R. Toland: I would like to address the imagery necessary to inspire change. Rattan, one of the images of the 21 WCSS prospectus shows a stock photo aerial view of a typical industrial agriculture scene: 32 combine harvesters glide in perfect alignment over a field of grains like a flock of geese parting the skies. The aesthetic framing is powerful and precise. In the middle of this image, highlighted in red block letters reads: "SOIL SCIENCE: Beyond food and fuel."

Between the march of the combine harvesters and the ubiquitous cupped hands holding a seedling, soil security has become visually coded into cultural understanding. What role do you think imagery plays in soil protection efforts? What influence do pictures have on public opinion, but also on scientific and political discourse? Would you like to see scientists work more closely with artists to develop a more comprehensive and effective visual language for protecting the soil?

Rattan Lal: My keynote presentation at the 20th WCSS was titled "The Soil-Peace Nexus."[7] The following diagram (see p. 9) published in the article shows the dependence of national and international security on soil health. Despite its vital role to human well-being and nature conservancy, there is an utter lack of awareness in the general public and among policymakers about the importance of soil and its sustainable management to national and international security.

In the June 1, 2017 note "From the Desk of Rattan Lal," I made a call for a prominent "Global Icon of Soil" just as a panda has served very well for advancing the mission of the WWF. Yes, indeed, imagery and pictures play an important role in enhancing soil protection efforts and in influencing public opinion and changing the mindset of policymakers.

It would be essential to see soil scientists work closely with artists in developing such a "soil icon" and a comprehensive visual vocabulary necessary to meet the ever-increasing challenge of protecting soil from anthropogenic perturbations.

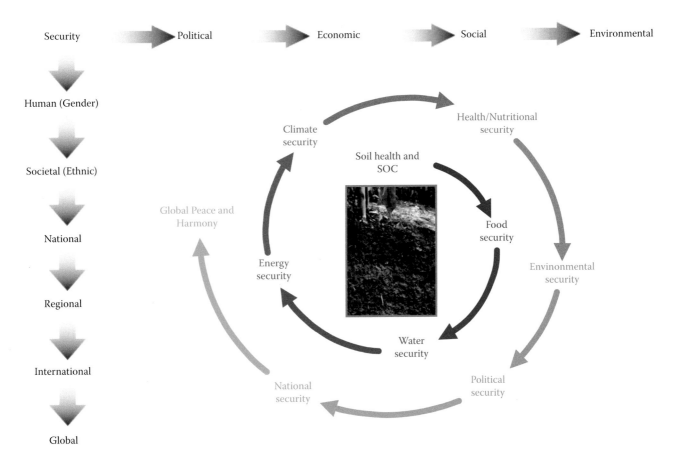

Security → Political → Economic → Social → Environmental

Security → Human (Gender) → Societal (Ethnic) → National → Regional → International → Global

Climate security
Health/Nutritional security
Soil health and SOC
Global Peace and Harmony
Food security
Energy security
Envinonmental security
National security
Water security
Political security

The dependence of national and international security on soil health.

From Lal, R. 2015. *Soil Science and Plant Nutrition* 61:566–578.

Alexandra R. Toland: Agnes, *Wheatfield* has become part of cultural memory, in the art world and beyond. Indeed, for some it is an "icon" of sustainable agriculture and soil health that has served as an example to follow, and to some extent signaled an impulse for a new "green revolution" in cities everywhere. Many people think about urban agriculture and picture you wading through a golden field of wheat with the backdrop of the New York City skyline looming large. Many artists have followed your lead and planted corn, potatoes, rice, and vegetable gardens on deserted lots and public plazas as a form of artistic practice and political activism. How do you think your work has shaped the image of agricultural landscapes, real and imagined? Can you imagine working with soil scientists in further developing such a "visual vocabulary" for soil health in cities?

Agnes Denes: Regarding imagery, it's interesting that you brought up the picture of me walking through the field, because just recently somebody had that image tattooed onto his arm. Or was it a her? I can't really tell from the picture they sent me. It was a great compliment, although I don't look for that. It showed solidarity with the ideas behind the work.

I am glad if Rattan and others feel the necessity of working together with artists.

I love to work with scientists and have collaborated on many projects. John Ameroso, with his background, was an invaluable contributor to *Wheatfield*. I've never approached a project without thorough research and knowledge of all its aspects. The expertise of a wide variety of fields is essential for an idea to be fully developed. Mutual respect emerges and we learn from each other.

My work is usually hard; I get dirty with hands-on work. Working on *Wheatfield*, I did the digging and the planting, managed an underfunded project and photographed the growing field lying in the mud, fighting daily interference with the construction guys who just couldn't stop making my life miserable. They stole my equipment and locked the gates so the volunteers or assistants could not enter. I remember digging a ditch under the gate so that the volunteers could roll under it and come out on the other side. Working a field is hard whether you are a city person or a farmer, a man or a woman. Working on a landfill or a forest, or a wheat field, your heart needs to be warm and nutritious like the soil, your endurance and belief as strong as the rays of the sun or the wind. Your "inner soil" has to be rich with nutrients. One calls it soil health, the other fertility. I call it life and fertile thinking.

Alexandra R. Toland: It is so important to hear about some of the on the ground challenges you faced, Agnes. I think sometimes the aesthetics of the imagery hides the hardship of the process. The waving fields of grain you managed to grow in New York and Milan give the impression of effortless soil fertility, which is often far from reality in cities. Truthfully, you painstakingly cleared these fields of debris and then carted in truckloads of topsoil to realize *Wheatfield*. Reflecting on some of what Rattan has said about soil health, it would be interesting to hear more about the challenges of planting grains downtown.

Agnes Denes: I designed self-contained environments in my Future Cities project in 1980 that stood up to weather conditions, housed thousands of people, grew their own food, and were totally self-supporting. But that's in the

future. For *Wheatfield* it was hard. After bringing in hundreds of truckloads of soil, we had to worry about high crosswinds off the New York harbor that we feared would blow the seeds out of the soil. So we dug deep furrows and planted the field by hand. I only had enough money for one inch of topsoil, so we had to enrich it, irrigate it, fertilize it, pick wheat fungus by hand, and a million other things that were different in the city from an open field in the countryside. We faced challenges every day and worked them out.

Scientists may describe the difficulties, but the solutions, if any, seem to remain in the footnotes. We have to bring any expertise out of the footnotes and into realizations. My *Wheatfield* was a calling to account and calling to arms. A strong statement is as powerful as a tsunami. Working on *Wheatfield* with all the intrusions was overwhelmingly powerful, but I had a mission. I had to succeed to show the world it can be done while calling attention to the problems we are all facing. We can fight the negatives as we can fight corruption and shortsightedness, and those who believe only in their own power, not the power of an idea. Believing in yourself is okay but should be for the good of others. Every work I have ever done was to help humanity in some form, to solve a problem and offer benign solutions if I could. Looking back at my works, as different as they are, they all served humanity's well-being in some form. I love humanity, and sometimes feel sorry for us. I certainly worked hard all my life to make people believe in themselves, keep their heads high in believing in what's good in us: the brilliant, the enduring, the resilient, like the soil that renews itself, and offers you its riches.

John Ameroso: Regarding the challenges of planting the *Wheatfield*, the soil brought in was not topsoil, but "gley" from wetlands placed on sandy fill, which allowed proper drainage. Ideally the two should have been properly mixed and a larger crop and yield would have been realized. But we did not have the equipment to do that, so we made sure we had plenty of irrigation as we started later than we should have (by about 2 weeks) to take advantage of spring rains. The soil was not fertile, so we used an inorganic fertilizer. If composts in quantity had been available back then, we would have incorporated that into the poor soil and possibly eliminated the use of synthetic fertilizer except for additional nitrogen. By the end of the project I am sure the fertility of the soil along with the amount of carbon increased due to our cultivation techniques.

As a participant in this project, I was thrilled to be able to plan and plant something that had not been done in NYC on such a large scale (2+ acres) since the nineteenth century, and then had only been done as potato fields. Also, the challenge of using a desolate piece of land to become productive, proves you can grow anything anywhere, as long as you provide stability, fertility, air and water for the crop.

From a New Yorker's perspective, there on a piece of bulkhead, where ships once anchored, a grain field arose. And called attention to itself. From a Biblical aspect, the "Wheatfield" symbolized that which sustained civilization—"Manna," the bread of life. The grains (and I don't mean "gluten" free) are what sustained human life throughout the ages and still keep civilizations thriving. Without it, there is no past or future.

Alexandra R. Toland: Rattan, Agnes mentions some of the on the ground challenges from her experience. What do you see as the biggest challenges—and opportunities—of farming in the city? Can city people sustainably grow grains, vegetables and meat AND improve soil health? Is it feasible to live, work, and farm side by side in the same space? Do you envision agriculturally productive, self-sufficient cities of the future?

Rattan Lal: Can city people grow food in urban lands? Yes, it has been done and must be done. The percent of the total population living in urban centers has increased from 2% in 1800 (when the population was only 800 million) to 50% in 2008 (when the population was 6.7 billion), and it will increase to 84% (when the projected population will be 11.2 billion in 2100).[8] The number of megacities (population >10 million) was 3 in 1971, 31 in 2016, and will be 83 in 2100. A city of 10 billion people requires 6000 tons of food per day.[9] Therefore, all the necessary plant nutrients (especially phosphorus) contained in gray water and other urban wastes must be recycled to produce food within urban and peri-urban ecosystems. The challenge lies in (i) eliminating biological hazards, (ii) minimizing chemical pollutants (heavy metals), (iii) developing efficient production systems, and (iv) improving the health of drastically disturbed urban soils.

There are other challenges too. It is possible to produce vegetables and fruits, and fish, etc. But, the space is not enough to produce grains and livestock. Thus, there is a scope of developing a close nexus between conventional agriculture and urban agriculture. But first, megacities must have a goal of producing 25%–50% of the food within their borders. This can be done by developing modern and innovative systems of urban agriculture.[10]

Alexandra R. Toland: Agnes, looking forward, what do you think? What advice would you give young urban farmers or artists, or social activists, or schoolteachers, or others wanting to farm the city?

Next pages:

Wheatfield—A Confrontation, summer 1982, 2 acres of wheat planted and harvested by Agnes Denes.

What is essential about soils in the city and why is it necessary to protect them?

Agnes Denes: We face many challenges and growing in the city is just one of them, and can be overcome.

In my work I plant the seeds both in soil and in the minds of people.

People can get infected with bad thoughts; just as urban soil can be contaminated, people can become polluted with fear and bad influences. Just as we need to eliminate biological hazards and chemical pollutants from the soil, we need to cleanse people's thinking from pollutants, bad influences, and not believing that healthy solutions are possible and necessary for human survival.

One must have the knowledge of a scientist and the soul of a poet to make a difference. One without the other may not be sufficient to change things.

Whether you plant on a farm or in a city or in a human mind, strong seeds and good soil are essential for a healthy harvest.

I see soil as a thin layer around the globe beyond the mantle where all life takes place. Less than a hundred feet where we humans live and all the miracles take place, and only a fraction of that is soil. So little, so much. The "earth" where things grow, giant sequoias, golden wheat, a thousand different fruits and trees, and sustenance for billions. The rest is rocks, mountains, and substrate. This precious substance, more precious than any gem, any ore, soft, good smelling brown earth, an aroma to inhale and know it will bring forth a harvest, small green shoots that nurture you, your children, and the future. The soil between the clouds and the mantle, a thin layer not covered by water. The soft soil, between air and rock, between mountains and the fire inside the earth, this thin layer of heaven is THE SOIL.

The Philosophy

My decision to plant a wheat field in Manhattan instead of designing just another public sculpture grew out of a long-standing concern and need to call attention to our misplaced priorities and deteriorating human values.

Manhattan is the richest, most professional, most congested, and, without a doubt, most fascinating island in the world. To attempt to plant, sustain, and harvest two acres of wheat here, wasting valuable real estate, obstructing the machinery by going against the system, was an effrontery that made it the powerful paradox I had sought for the calling to account.

It was insane. It was impossible. But it would call people's attention to having to rethink their priorities and realize that unless human values were reassessed, the precious quality of life, even life itself, was perhaps in danger. Placing it at the foot of the World Trade Center, a block from Wall Street, facing the Statue of Liberty, was to be a careful reminder of what this land had stood for and hopefully still does.

My work usually reaches beyond the boundaries of the art arena to deal with controversial global issues, questioning the status quo and the endless contradictions we seem to accept into our lives, namely, our ability to see so much and understand so little, to have achieved technological miracles while remaining emotionally unstable; our great advances, desirable, even necessary for survival, that have interfered with evolution and the world's ecosystem; or for that matter the individual human dilemma, struggle, and pride versus the whole human predicament.

Wheatfield was a symbol, a universal concept. It represented food, energy, commerce, world trade, economics. It referred to mismanagement, waste, world hunger, and ecological concerns. It was an intrusion into the Citadel, a confrontation of High Civilization. Then again, it was also Shangri-la, a small paradise, one's childhood, a hot summer afternoon in the country, peace, forgotten values, simple pleasures.

The idea of a wheat field is quite simple. One penetrates the soil, places one's seed of concept, and allows it to grow, expand, and bear fruit. That is what creation and life is all about. It's all so simple, yet we tend to forget basic processes. What was different about this wheat field was that the soil was not rich loam but dirty landfill filled with rusty metals, boulders, old tires, and overcoats. It was not farmland but an extension of the congested downtown of a metropolis where dangerous crosswinds blew, traffic snarled, and every inch was precious realty. The absurdity of it all, the risks we took, and the hardships we endured were all part of the basic concept. Digging deep is what art is all about.

Introduce a leisurely wheat field into an island of achievement-craze, culture, and decadence. Confront a highly efficient, rich complex where time is money and money rules. Pit the congestion of the city of competence, sophistication, and crime against open fields and unspoiled farmlands. The peaceful and content against the achiever. The everlasting against the forever changing. Culture versus grassroots.

Wheatfield affected many lives, and the ripples are extending. Some suggested that I put my wheat up on the wheat exchange and sell it to the highest bidder, others that I apply to the government for a farm subsidy. Reactions ranged from disbelief to astonishment to being moved to tears. A lot of people wrote to thank me for creating *Wheatfield* and asked that I keep it going.

After my harvest, the four-acre area facing New York harbor was returned to construction to make room for a billion-dollar luxury complex. Manhattan closed itself once again to become a fortress, corrupt yet vulnerable. But I think this magnificent metropolis will remember a majestic, amber field. Vulnerability and staying power, the power of the paradox.

The Act

Early in the morning on the first of May 1982 we began to plant a two-acre wheat field in lower Manhattan, two blocks from Wall Street and the World Trade Center, facing the Statue of Liberty.

The planting consisted of digging 285 furrows by hand, clearing off rocks and garbage, then placing the seed by hand and covering the furrows with soil. Each furrow took two to three hours.

Since March, over two hundred truckloads of dirty landfill had been dumped on the site, consisting of rubble, dirt, rusty pipes, automobile tires, old clothing, and other garbage. Tractors flattened the area and eighty more truckloads of dirt were dumped and spread to constitute one inch of topsoil needed for planting.

We maintained the field for four months, set up an irrigation system, weeded, cleared out wheat smut (a disease that had affected the entire field and wheat everywhere in the country). We put down fertilizers, cleared off rocks, boulders, and wires by hand, and sprayed against mildew fungus.

"We" refers to my two faithful assistants and a varying number of volunteers, ranging from one or two to six or seven on a good day.

We harvested the crop on August 16 on a hot, muggy Sunday. The air was stifling and the city stood still. All those Manhattanites who had been watching the field grow from green to golden amber, and gotten attached to it, the stockbrokers and the economists, office workers, tourists, and others attracted by the media coverage stood around in sad silence. Some cried. TV crews were everywhere, but they too spoke little and then in a hushed voice.

We harvested over 1000 pounds of healthy, golden wheat.

—Agnes Denes, *Wheatfield—A Confrontation, Battery Park Landfill*, downtown Manhattan, 2 acres of wheat planted and harvested, Summer 1982. © 1982 Agnes Denes

Endnotes

1. Doran, J.W., M. Sarrantonio, and M.A. Liebig. 1996. Soil health and sustainability. *Advances in Agronomy* 56:1–54.

2. Lal, R. 2016. Feeding 11 billion on 0.5 billion hectare of cropland. *Food and Energy Security Journal* 5(4):239–251.

3. For more on the role of urban agriculture in achieving food security, see Lal, R., and B.A. Stewart, 2017, *Urban Soils*, Taylor & Francis, Boca Raton, FL; Despommier, D.D., 2010, *The Vertical Farm: Feeding the World in the 21st Century and Beyond*, St. Martin's Press, New York, p. 400; Kemper, K., and R. Lal, 2017, Pay dirt! Human health depends on soil health, *Complementary Therapies in Medicine* 32:A1–A2; and Brevik, E.C., and T.J. Sauer, 2015, The past, present and future of soils and human health studies, *Soil* 1:35–46.

4. Lal, R. 2013. Food security in a changing climate. *Ecohydrology & Hydrobiology* 13(1):8–21.

5. Lal, R. 2016. Global food security and nexus thinking. *Journal of Soil Water Conservation* 71:85A–90A; Lal, R., D.

Kraybill, D.O. Hansen, B.R. Singh, T. Mosogoya, and L.O. Eik (Eds.), 2016, *Climate Change and Multi-Dimensional Sustainability in African Agriculture*, Springer, Cham, Switzerland.

6. Lal, R. 2017. Improving soil health and human protein nutrition by pulses-based cropping systems. *Advances in Agronomy* 145:167–204.

7. Lal, R. 2015. The soil-peace nexus: Our common future. *Soil Science and Plant Nutrition* 61:566–578.

8. United Nations. 2017. *World Population Prospects: Key Findings and Advance Tables, 2017 Revision*. Division of Economic and Social Affairs, United Nations, New York; Lal, R. 2017. Managing urban soils for food security and climate change. *SUITMA* 9, 22–27 May 2017, Moscow, Russia.

9. Lal, R., and B. Augustin (Eds.). 2012. *Carbon Sequestration in Urban Ecosystems*. Springer, Dordrecht, Netherlands.

10. Lal, R., and B.A. Stewart. 2017. *Urban Soils*. Taylor & Francis, Boca Raton, FL.

Taste of Place
Terroir as Experience
Laura Parker in conversation with Tom Willey, Lou Preston, and Scott Burns

Laura Parker is an American visual artist whose work tells stories and helps others make experiential connections to the land. Addressing themes of food, agriculture, and soil, she is best known for her interactive installations, drawings, paintings, and tapestries. She works to create relationships between her work, gallery visitors, and the land creating opportunities for discovery that are part of social practice work. Parker spent 40 years working as a graphic designer, and directing her own graphic design studio. She lives and works in San Francisco and rural Sonoma County, California. http://lauraparkerstudio.com

Tom Willey operated a 75-acre farm in the fertile Central San Joaquin Valley in Madera, California, with his wife, Denesse. The Willeys have been farming since 1980 and Certified Organic by California Certified Organic Farmers (CCOF) since 1987.

Title image: Laura Parker, Taste of Place, 2006–ongoing. Soil tasting in 78 varied locations.

Photos: David Matheson, 2009. Images reprinted with courtesy of the artist.

Lou Preston learned to grow grapes and make wine at UC Davis in the early 1970s. For 45 years, he has run a 125-acre farm in Dry Creek Valley, California with his wife, Susan. Using a biodynamic philosophy, they have expanded their vineyard to include olives, stone fruit, apples, vegetables, nuts, grains, pasture, hedgerows, sheep, pigs, and chickens.

Scott Burns is an environmental geologist and soil scientist at the Portland State University. His research interests include environmental geology, soils, landslides, engineering geology, hazard mapping, terroir of wines, and the Missoula Floods.

Terroir refers to the "set of all environmental factors that affect a crop's phenotype, including unique environmental contexts, farming practices, and a crop's specific growth habitat. Collectively, these contextual characteristics are said to have a character; terroir refers to this character."[1] Terroir is a marriage of environmental and experiential factors of a place that result in the particular qualities of taste, smell, and appearance said to be epigenetic with a plant sourced from that place. For this chapter, Laura Parker collected statements from three soil practitioners, a soil scientist she was introduced to for the chapter, and two long-term collaborators and organic farmers, to create a portrait of terroir in words to accompany images of her terroir-based artworks. Soil scientist and field geologist, Scott Burns, from Portland State University, speaks from the point of view of a field scientist. From the perspective of Tom Willey of T&D Willey Farms and Lou Preston of Preston Farm & Winery, the soil is first and foremost experiential through the senses. For people who have their hands in the soil every day, the authors approach the soil primarily through an embodied knowledge of place. The conversation centers around Parker's ongoing performance and installation work, *Taste of Place* (2006–ongoing), which offers participants soil tasting experiences as if they were tasting fine wines. After smelling, tasting, and touching the soil, participants are presented with a variety of vegetables, fruits, and cheeses grown from those soils.

> A major problem is that some of our assumptions about the world and the nature of things lie deep within us, often at the unconscious level. When we attempt to consider other possibilities, these hidden assumptions reassert themselves, usually in unexpected and often undetected ways.
>
> Wes Jackson[2]

I have been working specifically with the terroir of wine for over 50 years. The name came from French monks four hundred years ago in Burgundy who realized that wines from certain vineyards always tasted the same, no matter who made the wine. That is the "taste of the place." Today, we describe wine terroir—each bottle has a different taste. The factors that affect the taste are going to be the wine grape type (pinot vs. cabernet), the climate (water, sunshine, heat), the soils/geology (that give the 12 essential elements to the grapes), the aspect of the site (the elevation and slope orientation and angle), the soil biota (fungi, bacteria, etc.), and soil water holding capacity if there is no irrigation. These all together develop the taste of the place. Two other factors also affect flavor but are not considered terroir, and those are the winemaker (do you use oak, what type of yeast, etc.) and vineyard management (orientation of the rows, types of trellis, and cover crop or not). Terroir differences are best expressed in cool climate grapes like pinot noir, chardonnay, and riesling. It is fun to taste differences in terroir if you choose wines from the same winemaker in the same year so the only difference is the soil creating the different flavors.

Scott Burns
Environmental geologist and soil scientist

My journey with terroir began when I was quite young and unaware of the experience.

Terroir meant nothing to me. In our family, it was just dirt.

REFLECTION ONE: 1950s, Iowa. When I think back my first encounter with terroir was walking the fields with my grandfather. This must have been the 1950s. We would walk the rows from one side to the other, he would periodically pick up a hand full of soil, look at it, rub it in his fingers, put his nose close to the little mound in the palm of his hand and smell. He would open his fingers and watch how it fell to the ground, finally he would taste the residue leftover. This was the closest my grandfather got to a lab. This process told him everything he needed to know about his soil.

Next pages:
Laura Parker, *Grape Root, Grass Roots*, 2014, 4' × 4', acrylic on wood.

Text from "Mother Earth, Letters on Soil addressed to Sir George Stapledon by Gilbert Wooding Robinson," Published by Thomas Murby & Co, London 1937, reprinted 1947.

Photo: Trish Turney, 2014. Images reprinted with courtesy of the artist.

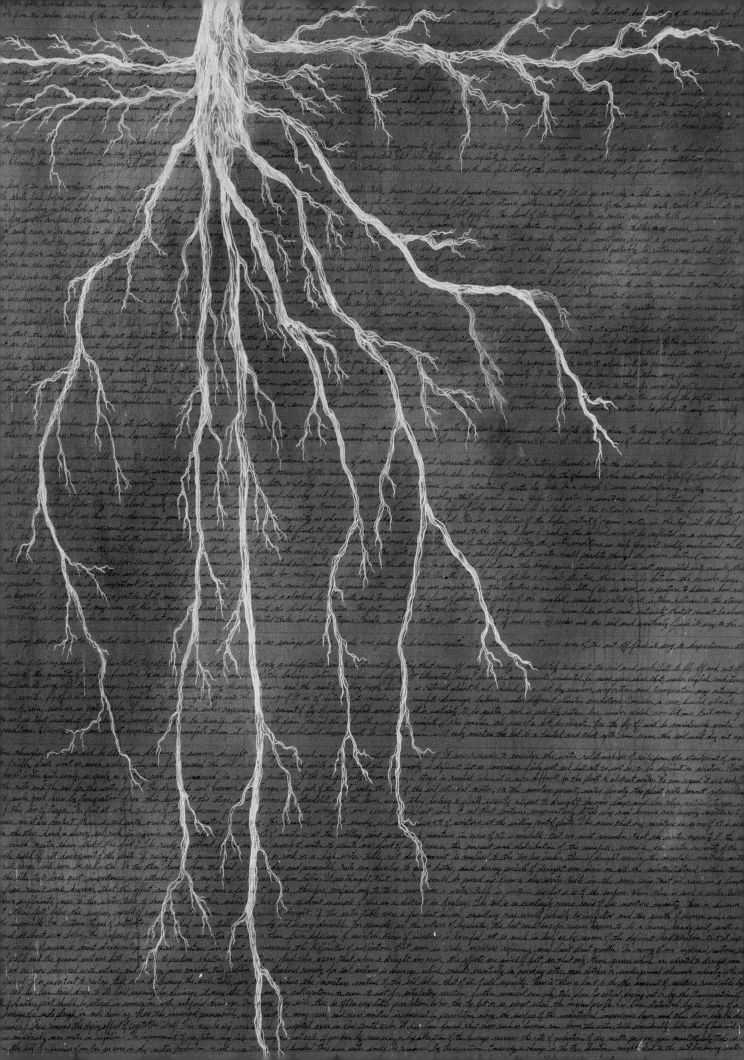

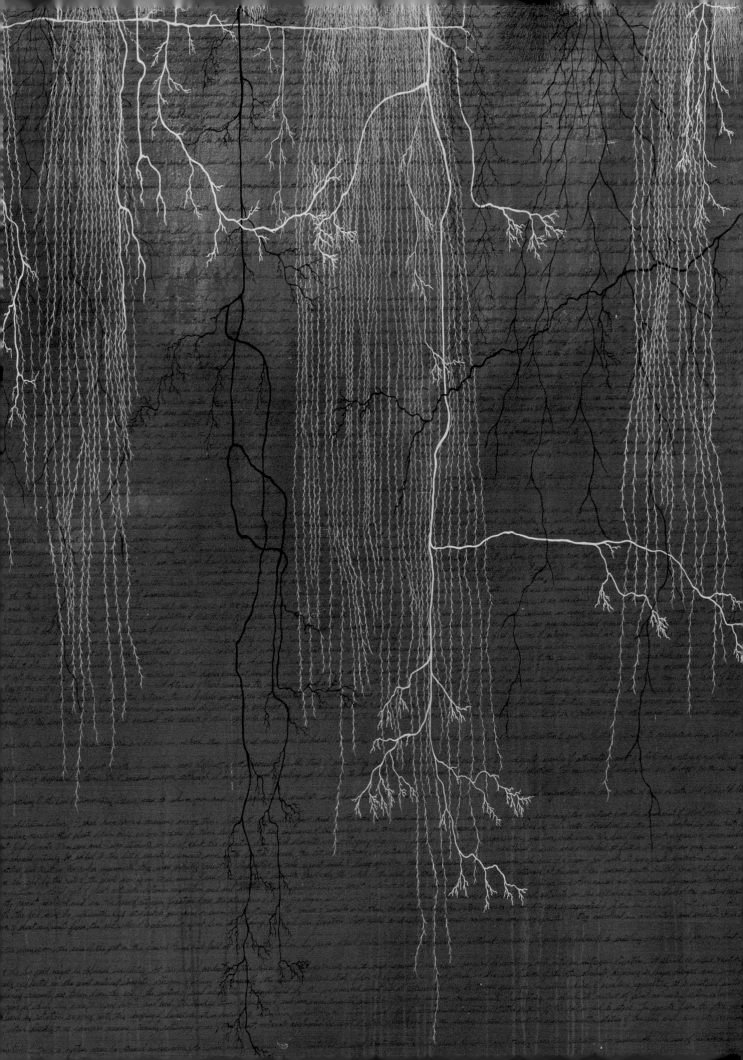

To a child, it seemed like magic. From this he would tell me what crop was next in his rotation, where the mineral content was and I think he also talked about salt. I wish I could ask him about that process now.

REFLECTION TWO: 1980s, France. I was in the Deux-Sèvres which is in the west, the middle of the west about 45 min from La Rochelle. I was traveling with one of my best friends, Robert Reynolds, an extraordinary chef from San Francisco. He was friends with Louis Marie, who had a herd of goats with his brother and made an exquisite cheese that was in high demand from the affineurs of France. The brothers divided their labor, one taking care of the pastureland, the other taking care of the goats and together making the cheese. What surprised me, city girl that I was at the time, was that Louis Marie combed every plant in his pastures to make sure that the combination of herbs and grasses was always stable and essential to the flavor of his cheeses. No "mauvaises herbs" allowed. This was the essence of their cheeses. One day I sat on the ground, drawing each of the pasture herbs within the reach of my arms while Louis Marie named them off. It had never occurred to me that flavor was so direct, controlled, and guarded. This was a very simple revelation but pivotal for my future work.

REFLECTION THREE: 2001, San Francisco. I was working on an exhibition called "Landscape, the Farmer as Artist," bringing the products of the field into the gallery just like the drawings I made from them. I was spending a lot of time with Tom and Denesse of T&D Willey Farm in the California Central Valley. One evening Tom leaned over the dining room table with his hand out drawing a circle in his palm. "If my hand was filled with soil, there would be more micro-organisms there than there are people on this planet." "WHAT?" I said, "we all know there are a lot of microorganisms, but really, equal to the number of people on the planet, in the palm of your hand?"

At that point, my work took another turn, and it was a turn to the soil.

> Farmers broker marriages between particular crop cultivars, soils, and microbiomes within a promising climate zone. Every farmer, then, is a matchmaker. The marriage might be short-lived, resulting in disastrous discord, or endure for a lifetime, delivering a memorable "taste of place." In rare cases, a marriage might outlive its matchmaker, defining a legendary terroir or even an appellation.
>
> **Tom Willey**
> *Organic farmer*

REFLECTION FOUR: *Taste of Place*, 2006–present. Am I expected to eat dirt? No, you will smell the soil and taste the food grown in it. When you marry the two experiences through the back of the palette you can taste the soil. *Taste of Place* is an interactive art installation focusing on the awareness of a particular place and the connections one makes with it. These references are stimulated through direct experience, conversation, memory, and context. What is a soil tasting? Soil is put into a wine glass with a little water and stirred. The participant smells, swirls, and sees the color, texture, and water absorption capacities of the soil. Then one eats produce raised in that exact soil, merging the two experiences to taste the place. The purpose of this installation is to ask two questions: How does soil touch our lives and affect our food? And why does it matter? *Taste of Place* is meant to stimulate public dialogue about food production and the role of soil in our lives. To date, 78 different soils have been included in the project.

If terroir is geography, geology, and climate as Scott says, how can I as an artist experience these elements through my senses: sight, hearing, smell, taste, and touch?

GEOGRAPHY/TOPOGRAPHY: What senses do I use here? How do I experience geography? I am in Northern California, in a place called the Bennett Valley. I can *hear* the birds, the frogs at night, the hummingbirds, bees, and yes, the flies too. I can hear the sheep bleating next door. I live just over the crest of the hill that begins our valley. The landscape is mixed vineyards and pastureland on rolling hills. Elevation ranges from 200 to 1200 feet.

GEOLOGY/SOIL TYPE: The soil here is clayey. When damp, I just have to push it together for a few seconds and it sticks into a little ball. This pasture has been fallow for a long time. I think someone may have had a few sheep and chickens here at least 15 years ago. According to the USDA Soil Survey, our land is mostly "Raynor clay, seeped, 2 to 15 percent slope, (RcD)" and behind me is an organic dairy farm that is "Goulding clay loam, 15 to 30 percent slopes, (GgE)."[3] I would say my front field is 5%–15% slope but the back pasture is more like 15%–30%. I make my acre and a half sound like a lot. To me coming from San Francisco, it is the first time my ideas about soil can be explored and implemented on my own land. Back to the Raynor clay. Given the California drought, I am very happy this soil is seeped clay because it retains water longer. "Depth to bedrock is 45 to 60 inches. Fertility is moderately high. The available water capacity is 6 to 9 inches." This kind of soil is mostly used for grazing. So, the Raynor series in general consists of well-drained clays underlaid, at a depth of 20–60 inches, by volcanic andesitic rocks. Vegetation is chiefly annual grasses (though I have been told many grasses on my land are nonnative invasive grasses), and a few scattered oaks. In a

Laura Parker, Taste of Place, 2006–ongoing. Soil tasting in 78 varied locations.

Photos: David Matheson, 2009. Images reprinted with courtesy of the artist.

typical profile, I *see* the surface layer is black and olive gray. "It is slightly acid to moderately alkaline clay about 47 inches thick. At a depth of about 47 inches is pale-olive, moderately alkaline very cobbled and stony clay. Basaltic cobblestones and stones are at depth of 56 inches."[4]

CLIMATE, AIR/WATER/SUN: I am in the macroclimate of California; the mesoclimate of Sonoma County; and the microclimate of Bennett Valley. It is a little warmer in the winter and a little cooler in the summer than in the valley. Sitting on my hill toward the end of January, it is late morning, cool, 60°, and foggy. I *smell* dampness, the grass, wet earth, and *feel* the sun just coming out. Annual rainfall is 22–35 inches. I have measured 15 and 3/4 inches of rain so far, this year. Annual temperature is 58°F–60°F, and the frost-free season is 260–290 days.

This approach of qualifying sensory perception of the land is a bit of a mechanistic deconstruction of something that is alive and not so amenable. Terroir might be a snapshot in a point in time using the 5 senses to fix the image, but it is a moving target and a dynamic system. The farmer also is not the only relevant audient of the system. It can be a child, an old man or woman, a prophet, musician, poet, genealogist, anthropologist, a dreamer. Is there a sixth sense that comes into play? I think so.

I think hope and intention are important pieces of the terroir matrix. Truth is in the eye/ear/nose/mouth/touch of the beholder, it's not absolute. Think of your own chronology, your changing perception of the land. You changed and your perception changed.

Fleeting or universal? The source? Silly now but before I did not really understand that without soil, in any of its permutations none of us or nothing could exist. Not you and me, the beautiful farms of Sonoma County or the grease on the streets of our largest cities.

Where does the conversation go after "terroir," which is very quantified at the moment?

I think we are too influenced by the wine world that tends to classify and categorize too much in its pursuit of commercial relevance. But I do remember an apt explanation of the concept of terroir in an early lecture at University of California Davis 45 years ago. They spoke of the triad of influences that you include in your piece: Geography, Geology, and Climate. But they added one more, the human element, the steward of that land, the interplay of culture with agriculture. In that there is room for hope, intention, and dreaming … desire?

Lou Preston
Organic farmer and vintner

The concept behind "Taste of Place" was really the beginning of this conversation. Ten years ago, it was merely about "awareness." I think a great deal of that has been accomplished. What now?

Endnotes

1. See definition in *Wikipedia*: https://en.wikipedia.org/wiki/Terroir

2. Wes Jackson, *Becoming Native to This Place*, Counterpoint, 1994, p. 37.

3. US Department of Agriculture Soil Conservation Service, Soil Survey of the Eastern Stanislaus Area, California, 1964. https://www.nrcs.usda.gov/Internet/FSE_MANUSCRIPTS/california/CA644/0/ca_East_Stan.pdf

4. Soil Survey, Sonoma County California; United States Department of Agriculture, Forest Service and Soil Conservation Service, in cooperation with University of California Agricultural Experiment Station, May 1972.

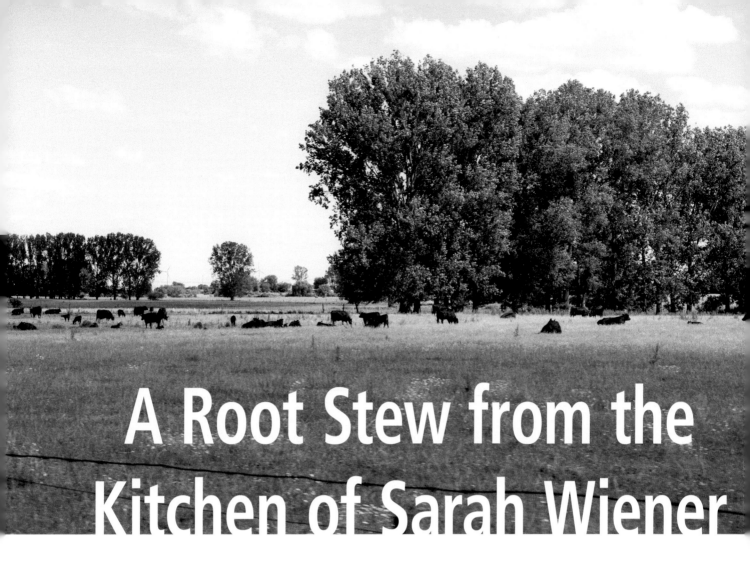

A Root Stew from the Kitchen of Sarah Wiener

Sarah Wiener

Sarah Wiener is an Austrian (TV) chef and sustainability icon. She advocates for preserving our natural resources, and an ethical and ecological awareness regarding nutrition.

Having spent her childhood and adolescence in Vienna, Austria, Sarah Wiener learned the trade of cooking in the late 1970s in artists' restaurant in Berlin, Germany owned by her father. Today, she runs a successful catering company, a restaurant with exclusively regional suppliers, and an organic wood stove bakery. As she truly believes in closing the circle from

Title image: Gut Kerkow.

Photo: © Sarah Wiener GmbH.

field to plate, she fulfilled a youthful dream of hers and bought, together with partners, an organic farm north of Berlin.

Alongside her entrepreneurial work, she campaigns for several causes. She is an advisor for Cradle 2 Cradle, ambassador for biological diversity, and a patron of the campaign People 4 Soil. With her foundation, she aims to inspire children to eat good and versatile foods. The Sarah Wiener Foundation recently launched Germany's biggest practical nutrition initiative *Ich kann kochen* ("I can cook") in 2016.

Having contributed to the volume of essays *Von Ganz Unten—Warum wir unsere Böden besser schützen müssen* (*From the Bottom Up—On Why We Need Better Protection for Our Soils*), edited by Gerd Wessolek on occasion of the 2015 International Year of Soils, we invited Sarah Wiener to take part in this volume as well. Her vision for regional field-to-plate food production and cuisine exemplifies the creative potential of the food and biomass production function of the soil. The recipe below is featured in the cookbook, *Sarah Wiener's Feierabendküche*, which promotes seasonal and regional cooking at home and donates part of its profits to increase the share of organic farmland in Germany. She kindly shares it with us here.

Cooking is an art: Produce grown in our soil represents a true color and taste palette in its own. Through cooking you can create a beautiful, but temporary work of art which appeals to all senses. And you even have a delicious meal.

Sarah

Farmer Josef Braun of Freising, near Munich, is a true agricultural hero. His fields are teeming with earthworms; he probably has more earthworms in his soil than any other farmer around. Why do these useful little creatures feel so at home in farmer Braun's organic farm? Because he does not plough the land, cares for it extremely gently, uses creative planting methods that loosen the soil, and never uses any poison whatsoever. Josef Braun treats his soil like a living being—and rightly so: there are more living creatures in a single handful of soil than there are people on earth. Two thirds of all the species that exist on our planet live beneath the earth's surface.

During the United Nations' International Year of Soils, 2015, I also became a farmer, joining with partners to buy an agricultural estate in the Uckermark district of Brandenburg, in the north of Berlin, which we run as an organic farm with the people on site. My dream of producing food from seed and knowing exactly where and under what circumstances it is produced is now coming true. We close the circle from healthy soil to the table, or "field to plate." Running our own farm was the most logical step in that direction. We are working toward running a sustainable system on our farm, conserving heritage breeds, and producing high-quality organic meat. The animals are fed on site with our own hay, grass, and grass silage, and are butchered on the premises. Altogether, we thus ensure that the process is transparent, and cut out unnecessary transport. In the case of pigs, we have introduced a method whereby the meat is processed while the meat is still warm. For this reason, we do not need to use phosphates (the conventional method) or citric acid (the organic alternative) during processing, as the meat binds together on its own. We also have a dream of growing our own vegetables soon.

Our aim is to show that it is possible to farm locally, regionally, and organic in a manner that is good for plants, animals, and people, and also in harmony with nature. This is where the future of food lies: It will make us, our food, and our soils more stress resistant, diversified, and independent of global shifts.

Riesen-Rüben Eintopf (hearty root stew).

Photo: © Veronika Lindlbauer.

From the recipe collection *Sarah Wiener's Feierabendküche* (*After Work Cuisine*), which originates from a cooperation with the social business "Feierabendglück" ("After Work Happiness")

(Serves 4)

2 beetroots
3 medium carrots
2 medium parsnips
4 large potatoes (waxier variety)
2 medium onions
Olive or rapeseed oil
Untreated salt
Black pepper (mill)
1/2 tsp whole caraway seeds
200 g cream
1 bunch of fresh coriander or parsley

Preparation:

Peel the beetroot. Wash carrots well, cut off stalk. Peel the parsnips. Cut all roots into 2 cm cubes.
Peel the potatoes, cut into 2.5 cm cubes.
Cut the onions into 1 cm pieces.
Fry onions briefly in the oil in a large saucepan until translucent, add all the vegetables, just marginally cover with water and mix in all spices. Cover and simmer for about 20 minutes.
In the meantime, wash coriander/parsley and chop into small pieces. Add the cream and herbs and cook until soft. The potatoes should start to fall apart to give the stew a slightly thicker consistency. The beetroot makes the stew a lovely pink shade.
Season to taste and serve.

Tip: If you want, you can also add 1 tsp of marjoram and decorate the stew with a dab of crème fraîche.

Artisanal Soil

Sue Spaid

Associate editor of *Aesthetic Investigations*, **Sue Spaid** recently published *Ecovention Europe: Art to Transform Ecologies, 1957–2017*, her fifth book regarding ecological art. In addition to "A Philosophical Approach for Distinguishing 'Green Design' from Environmental Art" in *Advancements in the Philosophy of Design*, her work has appeared in *Philosophica*, *Arte y Filosofía en Arthur Danto*, *The Journal of Somaesthetics*, *The Journal of Aesthetics and Art Criticism*, and *Rivista di Estetica*. Spaid defended her doctoral dissertation "Work and World: On the Philosophy of Curatorial Practice" in 2013.

To emphasize the connection between soil and food, some artists produce artisanal soil, which they typically employ in their artist-farms. Each of these artists has his/her own process and reasons for producing artisanal soil, characterized here as an "artist-initiated process for ameliorating top soil." To capture the distinctness of approaches devised by ten artists/teams working in the United States and Europe, I offer four artworks for each of three categories: (1) farm chain, (2) regenerative soil as public art, and (3) outdoor studios/experimental farms. Artists who opt to exhibit these practices as their art encourage art institutions and their publics to appreciate soil on par with museum treasures.

Society's Willful Disassociation of Food from Soil

While visiting Roma the first week of January 2015, I accidentally discovered the giant banner slung over the Food and Agriculture Organization's world headquarters announcing the United Nations' "2015 International Year of Soils." I figured that the UN had selected this particular topic to connect with "Feeding the Planet. Energy for Life," the theme of Expo Milano 2015, set to open four months later and welcome 20 million visitors. In my mind, these two events had been devised to pave the way for COP21, that year's pièce de résistance, where 196 parties were expected to meet in Paris to sign the accord committing the world's nations to limit global warming to less than 1.5 degrees Celsius and to achieve zero emissions sometime between 2030 and 2050. As I shall soon describe, none of these events transpired as I envisioned, despite the fact that fully half of the 158 Intended Nationally Determined Contributions (INDCs) drafted ahead of COP21 "ascribed importance to the agricultural sector. In particular, African and Asian countries are aiming for more sustainable uses of soil and land. In fact, soil remained mostly invisible in those contexts."[1]

Fully expecting healthy soil to be on the table in May at Expo Milano 2015, I was buffaloed by national pavilions bent on promoting exotic exports like crocodile burgers from Zimbabwe, and novel technologies such as hydroponics and aquaponics, which offer zero carbon sequestering opportunities. Stranger still, Italy barred the Belgian Pavilion from offering insect tastings to accompany its video installations promoting insect farming as a feasible, "future" protein source, even though "Green Bugs" were already on offer in its national supermarket chain.

Even the massive Pavilion Zero representing the United Nations neglected nutrient-rich soil, as if healthy human diets don't depend on soil (only a handful of pavilions bothered to mention water's crucial role). One possible explanation for this oversight is that in 2012 the United Nations adopted Ban Ki-Moon's "zero hunger challenge," which emphasizes zero loss and zero waste. Worldwide, one-third of all food produced is either wasted due to poor storage facilities (harvests decay before shipment), poor distribution (inadequate roads or unaffordable transportation), or poor management (tossing spoiled food), resulting in a food loss of 1.3 billion tons.[2] Unfortunately, the UN's zero-waste strategy, which rightly addresses postharvest malfeasance, totally ignores regenerative soil's greater role in stimulating plant growth and sequestering carbon. We need both!

Six months later, I fully expected regenerative soil and carbon sequestration to be central to COP21's agenda as a viable tool for stabilizing climate change. Once again, soil remained off the table, save for the launch of "4 Pour 1000," an initiative signed by about 60 nongovernmental organizations and nations committed to increasing soil's organic matter by 0.4% each year. I even naively fantasized President Obama springing "regenerative soil" on the nation at the very last minute, during his very last State of the Union address, as part of his administration's strategy for mitigating against climate change.

Apparently, the connection between regenerative soil and climate change is lost on those committed to fixing food storage, distribution, and management, yet agricultural practices contribute up to one-third of the greenhouse gases responsible for exacerbating climate change.[3] Moreover, Pavilion Zero's "fix" presumes food production capacities stay the same, but achieving this necessitates farmers' continued access to healthy topsoil, a precious resource that must be locally generated, widely available, and constantly rejuvenated. As long as current trends continue, whereby topsoils are routinely depleted due to overtilling, overgrazing, and overcultivation, and cause wind and water erosion, salinization, and/ or loss of organic matter; the planet will heat up and water supplies will dwindle worldwide, as reduced production levels leave fewer rooted plants capable of absorbing and transpiring water.[4]

Artists' Willful Reconnection of Food and Soil

In light of my worry that too few view soil as the "fix," whose greater attention would improve food quality and climate change alike, I now turn to artist-farmers for whom a healthy planet is the cornerstone of their artistic practice, enabling them to demonstrate soil's significance, if not inspire the public to prepare and employ homemade soil by composting table scraps and leaves. For my purposes here, *artisanal soil* is simply some artist-initiated process for ameliorating topsoil. To be clear, most artists discussed here use the "outdoors" as their studio, and then exhibit some aspect of their farm work indoors as art. Most consider farming to be their primary artistic practice.

Given the rich history of ameliorative art practices, artists who produce artisanal soils typically view their art as the process itself, not the resulting soil or produce. All artist-farmers depend on rich soil to produce their works, yet only a handful handmake artisanal soil, just as not all painters build their own stretcher bars. What artists call "soil" is likely considered soil amendments among soil scientists. Nonetheless, artists who handmake top soil draw on skills developed by artists over millennia to produce artisanal glazes and paints, and naturalists who cultivated those very flowers that have inspired painters for centuries. A many-layered process, soil production combines research, good judgment, and the successful implementation of set procedures that enable one to sustainably produce and nourish soil.

While researching *Green Acres: Artists Farming Fields, Greenhouses, and Abandoned Lots*, I noticed that artists-farmers make exceptional farmers precisely because of their art school/studio training, which helps them to develop acute observational skills and a knack for tedious activities, two skills that prove handy for farm work. Artist-farmers tend to approach farming the way fastidious painters approach painting, with great patience, dedication, and focus. Finally, art-farms benefit from artists' analytical skills such as pattern recognition, as well as their capacity to understand and maneuver systems. Even though each artist-farmer recommends a different artisanal recipe for producing homemade amendments, the outcomes are similar: more nutritious, living topsoil, with a greater capacity for carbon sequestration than ordinary soil that is repeatedly exposed to chemical fertilizers and pesticides, and routinely depleted of organic matter.

Because growing plants glean minerals and nutrients from whatever soil they inhabit, knowledgeable farmers must rejuvenate and replenish topsoil, otherwise the nutritional values of fruits and vegetables (grams of available minerals and vitamins) will continue falling far below those provided by produce just 50 years ago.[5] To remedy this, numerous artist-farmers have developed special strategies for producing artisanal soil "from scratch"; that is, nutritious soils prepared locally by hand with the help of local animals and microorganisms. This approach benefits not only human beings and other animals eating the plants, such as bees and butterflies pollinating them, but billions of microorganisms seeking habitat. Most important, these amendments keep topsoils alive, even in cities, improving their capacity to absorb rather than emit carbon.

Not surprisingly, artist-farmers for whom soil is the bee's knees consider society's willful disassociation of food from soil extremely dangerous. Artisanal soil production is primarily motivated by a preference for nutritious food grown in healthy soil, rather than crafting trends or a desire to optimize fertility. Moreover, these artists deem soil an invaluable *medium*, since it is the mechanical substrate in which plants take root, the supportive environment enabling seeds to absorb nutrients as they grow into plants, and the quintessential material for rainwater absorption and evaporation. Art lovers experiencing art-farms likely focus more on the visible plants than soil's invisible inhabitants, just as people tend to focus more on the overall painting than its less noticeable support mechanisms, but each artist knows deep down what propels plant growth.

Newton Harrison/Harrison Studio, *Making Earth* (raking, digging, eating), 1970.

To better grasp artisanal soil's role as an artistic medium, I divide artists' approaches into three categories. I begin with the most basic example, that of artistic systems that employ natural decay or composting processes to form a "farm chain," engendering enriched soil as a byproduct of food consumption and/or production. I next discuss artists whose public works broadcast the importance of regenerative soil and demonstrate easily implementable ways to avail it. Third, I describe how artists' outdoor studios double as experimental farms, some verging on agronomy labs.

The "Farm Chain"

The artistic effort to reconnect food to soil began with *Making Earth* (1970), an action/performance by artist Newton Harrison. He gathered different kinds of "manure, sewage, sludge, sawdust, vegetable matter, clay and sand, to create seven piles of earth that he watered and worked each day until they smelled so rich that he could put the soil in his mouth."[6] This action not only sounded alarm bells about society's need to focus on soil, but his handcrafted resource was used expressly to grow nourishing food for a herbivore, whose nontoxic poop he aimed to use to nourish more soil, becoming the first step of the "farm chain." Thus was born *Survival Piece #1* (1971/2012), a sustainable food source for a hog, whose poop was subsequently used to produce more artisanal soil. Harrison next worked with two algologists to grow brine shrimp in open-air tanks, thus providing food for farmed fish, whose fish scraps were served as fertilizer for growing fruits and vegetables. In collaboration with his wife Helen, the Harrison Studio visibly linked pastures to pig poop and shrimp to fish feed, using multiple animal layers to demonstrate the various steps needed to create the "farm chain." Their structures for growing fruits and vegetables eventually necessitated *Worm Farm*, a trough for composting table and garden scraps, which they first exhibited as part of *Full Farm (Condensed)* (1972), their installation of the chain's seven stages.

With *Soil Factory* (1998), N55 provide an easily implemented system for vermicomposting indoors, enabling households and office mates to sharply reduce organic waste. Approximately 1000 worms living in the second module from the top do all of the work as they move through holes from tray to tray. According to N55 cofounder Ion Sørvin, "After approximately 6 months, the material in the lowest tank is transformed into a black, soft substance mostly consisting of worm castings, a large part of which is humus."[7] N55 recommend using the worm *Eisenia fetida,* since it is among the soil-surface dwelling or compost-preferring species.

In contrast to *Soil Factory*, whose vermicomposting process is opaque, disguised as it is inside a repurposed black filing cabinet, J.J. McCracken proffers an extremely transparent indoor system, whereby people deposit their daily food scraps into gorgeous Plexiglas vermiculture boxes, thus enabling worms to fashion dynamic "functional paintings," befitting of dining rooms everywhere. Her elongated boxes have lids that slide off, enabling everyday eaters to easily insert scraps for the worms residing inside. As contents change and evolve, so do these fascinating "paintings."

Tattfoo Tan is so enthusiastic about artisanal nutrients that he regularly exhibits them as art, even without plants. Tan calls his brand of artisanal fertilizer *S.O.S. Black Gold* (2009), and he prices each 10-oz jar to track the market rate for an ounce of gold, which makes each jar worth $12,710 at today's gold rate (November 2017). First exhibited at the Bronx River Art Center, his fertilizer contains pure worm casting, the result of his having worked with worms that decomposed food and garden scraps in an undisclosed cellar in a secret location. When farming in his Staten Island garden, he uses *5pm Poop*, chicken manure that he composted with a mix of food scraps and fall leaves and then let sit a year. "It is amazing. So good, my garden now produces so many raspberry and various herbs."[8]

Regenerative Soil as Public Art

For his contribution to the Austrian Pavilion at the 53rd Venice Biennale, Lois Weinberger displayed garden scraps in a blue shed behind the Austrian Pavilion. Regularly gathered from the Giardini, the ever-mounting and shrinking pile was in an ongoing process of becoming soil-enriching compost. His artist's statement accompanying *Laubreise* (Leaves Travel) (2008–2009) captures soil's deep connection to life (and art):

> In the dissolved condition of the plant (such that insects are the essence of the blossom), the incomprehensible beauty of nature mutates into a rich ground of sensations. This collapse, an in-between space, produces time, whose arbitrariness is realized through repetition. The space of art reflects the space of the existential—that of "making good soil"—(as a cultural act). In fact something like an alchemistic process happens—leaves travel—into the complexity of the indeterminate. (The engagement with nature finishes at your own body.)[9]

To regenerate the soil of several abandoned Cincinnati lots, Permaganic Co, a nonprofit after-school program cofounded in 2010 by Luke Ebner and his wife Angela Stanbery-Ebner, first added one foot of fresh soil.

Next pages:

Lois Weinberger, *Laubreise*, 2009, Heap of rotting plants, 350 × 250 × 170 cm, Venice Biennial Austrian Pavilion, Venice, IT.

Photography credit: Herta Hurnaus.

Ebner and his team of teen gardeners next created "windrows" (linear piles of biodegradable material) from animal manure and wood chips. To further nourish the soil, they grew basil, "popcorn," tomatoes, and pumpkins. Twice a year, they flip the soil back on its bed.

Eager to present an artwork related to soil generation for "Green Acres" (2012), Ebner sought a way to capture the deep connection between soils and produce, resulting in *Soil Olympics*, two back-to-back contests between farmers, first in Cincinnati (2012) and then in Arlington, Virginia (2013). After convincing nearly twenty local farmers to participate in his soil competition, he requested them to supply their farm's prize produce alongside soil samples plus a written list of the soil's known ingredients, indicating each farmer's "soil recipe." He then typed each recipe onto a paper strip, and mounted it onto a wall map with a colored pin. Using the same color yarn, he linked each recipe to a same-colored pin designating that farm's location on the map. On a long table, he displayed both the soil and resulting produce in hand-blown glass display jars, whose glass lids, entwined in the same-color thread, could be removed for people to experience the relationship between vegetables, legumes, and herbs grown in different soils. Ebner selected soils ranging from local artist-farmers Homemeadow Song to dead soil found in an abandoned lot, cluttered with rusty nails and concrete chips. As far as I know, no one won a gold medal, but several samples looked quite scary!

An edible garden rooted in permaculture principles, George Mason University's SoA Green Studio, initiated in 2010 by Mark Cooley, doubles as public art and farm atelier amid a sprawling university campus. This garden hosts medicinal plants, as well as plants particularly beneficial for pollinators, which feed the school's bee population living in nearby hives. Noting that 24% of municipal waste consists of garden scraps, Cooley considers any notion of organic "waste" a misnomer, since it should be used to feed soil, rather than trucked to landfills, where it generates 18% of all methane emissions (yet another reason for COPs to include regenerative soil). SoA Green Studio provides George Mason's student body a viable model for ecological land management.

In addition to "chop and drop," a method comparable to windrows, whereby garden scraps are moved to beds, where they decay and are immediately reinvested, Cooley and his eco-art students mostly use sheet mulching to amend the campus's Virginia red clay. To generate their soil, they cover cardboard scraps provided by the university's cafeteria with leaves collected by maintenance workers. From Cooley's experience, as the soil gets richer, more and more earthworms help with the composting, eventually forcing out undesirable weeds. Despite this sluggish process, he notices how plants that "struggle[ed] for a couple of years now seem to be thriving."[10] Most important, everything is kept "onsite so there's no dumping of organic matter anywhere else."[11] Five years later, voles are back aerating the soil and rabbits are nesting in untrimmed perennial vegetation.

Sited since 2013 at Université Catholique du Louvain-La-Neuve in Belgium, Jean-François Paquay's *Portager*® du CREAT (centre de recherches et d'études pour l'action territoriale) grants passersby immediate access to biodiverse ecosystems. To produce the artisanal soil used in his biointensive portable farm, he mixes equal parts: (1) leaf mold that has already undergone three years of decay in artisanal leaf cages, (2) horse manure that has been composted with straw for at least two years, plus (3) garden and table scraps that have been composting a minimum of two years. To the last third, this ceramist/gardener first sieves the material and then adds a bit of potter's clay (1:24 ratio), because the local soil is very sandy. To maximize the soil's biodiversity, he composts all plant materials together, including weeds and invasive species, with the view that whatever survives composting adds value to the mix. Paquay notes that even though young compost (under three months) has higher levels of nitrogen, seasoned compost is preferable, since it releases nitrogen at much slower rates.

Outdoor Studios/Experimental Farms

Offering dozens of annual workshops, public gatherings, and full-day, nine-month school programs (8:30–3 pm) for home-schooled children, Homeadow Song offers far more than a prototype for farmers keen to adopt biodynamic methods. Homeadow Song cofounder Vicki Mansoor takes her tamale cart around Cincinnati, where over sales of blue and green tamales she discusses the importance of preserving ancient corn varietals, growing them locally using biodynamic methods, and soaking and cooking corn in an alkaline solution, which increases its nutritional value since it increases calcium and protein quality, reduces phytic acid, which blocks mineral absorption, and enhances niacin levels, thus preventing pellagra and mental illness. Situated amid Cincinnati's historic agriculture belt, Homeadow Song features a small pond stocked with fish, a flock of chickens, a small apiary of beehives, three wethers, and a rabbit, which all provide "fertilizer," food, fiber, pollination, and the opportunity to practice caring and appreciative relationships."[12] Mansoor remarks how "the animals form a constant exchange: sheep grazing, moving, and breathing low on the land; bees flying to every point where there

Spora Studios, 2017, Four stages of transforming a useless lawn into a medicinal herb garden for healing the land and body, Shenandoah Valley, Virginia.

Photo credit: Mark Cooley.

is a bloom; all the manure being moved all about."[13] Her partner Peter Huttinger ages low-odor sheep manure with barn bedding for six months to a year to produce a slow-release fertilizer rich "in phosphorous and potassium [which] is essential for optimal plant growth, helps with strong roots, defends against pests, and [encourages] vibrant plant growth."[14] Homeadow Song's specially designed border gardens and permaculture meadows attract beneficial insects and birds.

Although cows are considered essential for biodynamic farming, Homemeadow Song is too small to support cows, so a nearby farmer supplies them cow manure. As followers of Rudolf Steiner's 1924 farming practices, Homeadow Song participants have tested and regularly use Steiner's biodynamic preparations, such as the well-known "Preparation 500" (cowhorn packed with fresh cow dung that ferments when buried during cooler months), "Preparation 501" (cow horn filled with ground silica and rainwater buried in spring and dug up in autumn), "The Three Kings Preparation" (apply a mix of ground gold, frankincense [Boswellia], and myrrh with rainwater and glycerine on January 6 [Three Kings Day]), and "Pfeiffer compost starter" (inject microorganisms). Additionally, they employ classic permaculture procedures such as planting guilds (companion planting), making hugelkultur beds (branches buried in mounds that absorb water as they decay), and planting green manure (transfers nitrogen from air to soil and uptakes excess nitrates). Homeadow Song is particularly focused on creating an environment where critters find habitat, so as to discourage them from pilfering human food.

In light of numerous European cities' goals to grow 25% of produce locally, portable farms are sprouting up on balconies, backyards, and driveways. Eager to provide city dwellers access to nutritious soil, Jean-François Paquay proposed using a 50–50 mix of mole-hill soil (scavenged from urban parks, lawns, green spaces and the countryside) and commercially available organic leaf mold. After testing for the presence of organic material, minerals, and water, he discovered that this 50–50 mix approximates the volumetric proportions of organic and inorganic material found in his artisanal soil. This "epiphany," as he calls it, demonstrates that a mole-hill/soil mix could offer urban farmers a readily available alternative to time- and space-intensive artisanal soils.[15] In reappraising mole hills as invaluable time-saving gifts from nature, Paquay's research proves that the traditional notion of molehills as "eyesores" is shortsighted. Just as artworks routinely reveal our adaptive sense of beauty, Paquay's "epiphany" suggests that we might treasure many more naturally occurring irregularities were we not so bent on wasting time and resources trying to eliminate them.

Like Paquay, Mark Cooley oversees a public project at his university, as well as a backyard "farm-studio." Effectively an intervention on suburbia, Flawed Homestead (2005–2017) collaborators Cooley, Beth Hall, and their daughter Celia worked steadily over the years to gradually transform an initially lifeless, quarter-acre suburban lot into a permaculture homestead, replete with chickens, pond life, composting worms, rabbits, and a medicinal garden whose products are sold at market. Cooley remarks that "[t]he suburbs are a place where deer are hated for existing, and extreme defensive measures are taken to prevent squirrels from eating 'bird food' made from the very seeds of plants [ordinarily] banned by suburban homeowner associations, [yet] shipped in from thousands of miles away. … The crowning achievement of the suburban landscape is the lawn, a giant invasive garden where a handful of nearly useless exotic grasses reign, while edible, medicinal, and habitat forming 'weeds' are sprayed with toxic chemicals."[16] Cooley has built both rabbit and chicken tractors, enabling animals to supplement diets with fresh vegetation, while depositing droppings that fertilize future meals.

Flawed Homestead collaborators employ vermiculture to transform organic materials like yard debris, yet they rely on chicken poop to convert food scraps into nutrient-rich fertilizers for their artisanal soil. Hoping to influence neighboring lawn-owners, they spread their "suburban" grass clippings and leaves atop paper sheets, thus smothering undesired vegetation, while accelerating decomposition. They also experiment with both active and passive hugelkultur, but Cooley prefers the less laborious, passive version that merely requires moving fallen branches to beds, where their decay nourishes soil. In 2017, Coley and family moved to the Shenandoah Valley, where their half-acre experimental farm, Spora Studios, addresses permaculture.

In lieu of adding animal waste, Vera Thaens has been experimenting with magnets and energy-absorbent cables to increase soil's paramagnetism, thus strengthening soil's capacity to hold a magnetic field. Her experiments with electroculture (magnetic and electric forces) are aimed at "boost[ing] soil fertility and plant growth."[17] Rather than digging holes in the ground to plant seeds, she energizes the seeds by covering them with pyramidal soil mounds, shaped using a pyramidal mold. Not only do plants grow larger, faster, and more cheaply, but magnetic energy, as developed by Belgian agriculture engineer Yannick van Doorne protects plants from "disease, insects, and frost."[18]

Conclusion

Scores of artist-farmers are working across the globe, yet only a handful has been driven to produce artisanal soil. Some are motivated by soil

Homeadow Song, Fruit Trees
Painted with Biodynamic
Tree Paste Containing Cow
Manure Preparation 500, 2012,
Cincinnati, US.

Photography credit: Homeadow
Song.

degradation (some scientists predict that topsoil will all but disappear by 2070), while others aim to overcome the dearth of topsoil in their immediate environment. Doubling as soil activists, these artists have taken up the mantle to broadcast the importance of agricultural practices that ameliorate soil, thus significantly boosting water absorbency. The prospect of artisanal soils holds a vast opportunity for artists and the public alike to experiment with various materials, methods, processes, and time frames, so as to optimize nutritional food, produce availability, water absorbency, and even biodiversity. Of course, pros and cons must be weighed and balanced, so that end-users can decide which materials must be purchased as is (off the shelf), and which ones will require special attention.

What's radical about "artisanal soil" is hardly artists' recipes or approaches, many of which are adapted from permaculture, biodynamic farming, and even indigenous practices. What's radical is that by exhibiting these practices as their art, they demand both the public and art institutions to value soil on par with museum treasures. As several of these artists have lamented, had society not disassociated food from soil, they would never have felt impelled to make art that reconnects them.

Endnotes

1. Global Landscapes Forum, "Healthy Soils and Climate Protection at the Global Landscapes Forum," accessed November 4, 2017, http://www.landscapes.org/fr/healthy-soils-climate-protection-global-landscapes-forum/.

2. Food and Agricultural Organization of the United Nations, Food and Agricultural Organization of the United States, accessed November 4, 2017, http://www.fao.org/zhc/detail-events/en/c/246110/.

3. Natasha Gilbert, "One-Third of Our Greenhouses Gas Emissions Come from Agriculture," *Nature*, October 31, 2012, accessed November 6, 2017, https://www.nature.com/news/one-third-of-our-greenhouse-gas-emissions-come-from-agriculture-1.11708.

4. Sue Spaid, *Green Acres: Artists Farming Fields, Greenhouses and Abandoned Lots* (Cincinnati: Contemporary Arts Center, 2002), 80.

5. "Dirt Poor: Have Fruits and Vegetables Become Less Nutritious?," *Scientific American*, April 27, 2011, accessed June 15, 2015, https://www.scientificamerican.com/article/soil-depletion-and-nutrition-loss/.

6. Sue Spaid, *Ecovention: Art to Transform Ecologies* (Cincinnati: Contemporary Arts Center, 2002), p. 90.

7. N55, "Manual for *Soil Factory*," 1998, accessed November 5, 2017, http://www.n55.dk/manuals/SOIL_FACTORY/SOIL.html.

8. E-mail correspondence with Tattfoo Tan dated June 18, 2015.

9. Lois Weinberger, "Leaves Travel," *Laubreise* artist statement, Austrian Pavilion, 53rd Venice Biennale, 2009, translated by Felix Kindermann, June 24, 2015.

10. E-mail correspondence with Mark Cooley dated May 2, 2015.

11. Ibid.

12. Vicki Mansoor, "Artist's Statement," 2012.

13. E-mail correspondence with Vicki Mansoor dated June 25, 2015.

14. E-mail correspondence with Peter Huttinger dated June 18, 2015.

15. Jean-François Paquay, "A New Direction for Sustainable Art: Farming with Mole-Hill Soil in the Portager," in *Sustainable Art: Facing the Need for Regeneration, Responsibility, and Relations*, Anna Markowska (ed.), (Warsaw: Tako Publishing, 2015), 291–295.

16. Mark Cooley, "Making Dirt," unpublished document received April 30, 2015.

17. "Electroculture, Good Vibes for Agriculture," accessed November 4, 2017, http://www.electrocultureandmagnetoculture.com.

18. Ibid.

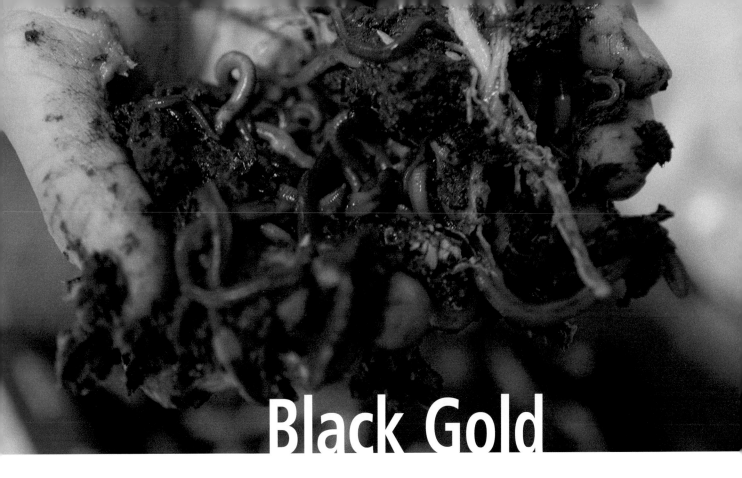

Black Gold

Tattfoo Tan

Tattfoo Tan was born in Malaysia and currently resides in Staten Island, New York. Tan's practice focuses on issues relating to ecology, sustainability, and healthy living. His signature work, Sustainable Organic Stewardship (S.O.S.), is project based, ephemeral and educational in nature. Tan has exhibited at venues including Ballroom Marfa, Creative Time, Aljira, Project Row Houses, The Laundromat Project, Philadelphia Mural Arts program, and the Contemporary Arts Center in Cincinnati. He is the recipient of grants from Robert Rauschenberg Foundation, Art Matters, Joan Mitchell Foundation, and Pulitzer Arts Foundation.

The S.O.S. brand *Black Gold* takes the concept of Piero Manzoni into the twenty-first century by canning worm castings instead of artist's shit (Merda d'artista, preserved produced, and tinned May 1961). By purchasing this artwork one is confronted with the dilemma of using the compost as a plant fertilizer or maintaining its status as a work of art. Piero Manzoni also priced his work based on the value of gold (around $1.12 a gram in 1960). Tattfoo Tan is similarly selling his Black Gold based on the current price of gold. Coincidentally, compost is called Black Gold by gardeners because of its value in improving garden soil. In an undisclosed location, a secret cellar is in the process of brewing the most creative compost ever. The curing process takes about a year, but the final results are pure gold. Black gold to be more precise. The special blended concoction is under the skillful hand and eyes of Master Composter Tattfoo. The first limited edition vintage Black Gold was available in early spring 2010. You can still reserve your bottle now! www.tattfoo.com

How to make an indoor compost bin (vermicompost)

Drill some holes into a used plastic tote. The worms need to breath.

Shred newspaper.

3. Add worms into it new home. They are red wigglers (Eisenia Fetida).

Add organic food scraps, but no meat and dairy products.

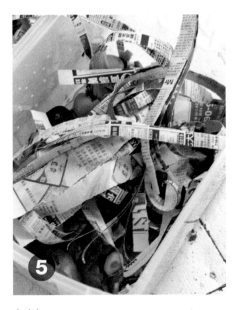

Add more newspaper or water to make the entire content moist but not flooded.

Store in a dark place. Worms love to work in dark environment.

S.O.S.   www.tattfoo.com

S.O.S. Black Gold VINTAGE 2009 PURE
WORM CASTING, BOTTLED IN THE ESTATE.
Composted by Master Composter Tattfoo Tan.

S.O.S. Black Gold "Live Animal" case exhibited at
Green Acres: Artists Farming Fields, Greenhouses
and Abandoned Lots, Contemporary Arts Center,
Cincinnati, Ohio, 2012.

S.O.S. Black Gold demonstration at Round 34: Matter of Food, Project Row Houses, Houston, Texas, 2011.

S.O.S. Black Gold demonstration at Bronx River Art Center, New York, 2009 and the Urban Wilderness Action Center at Eyebeam, New York, 2010.

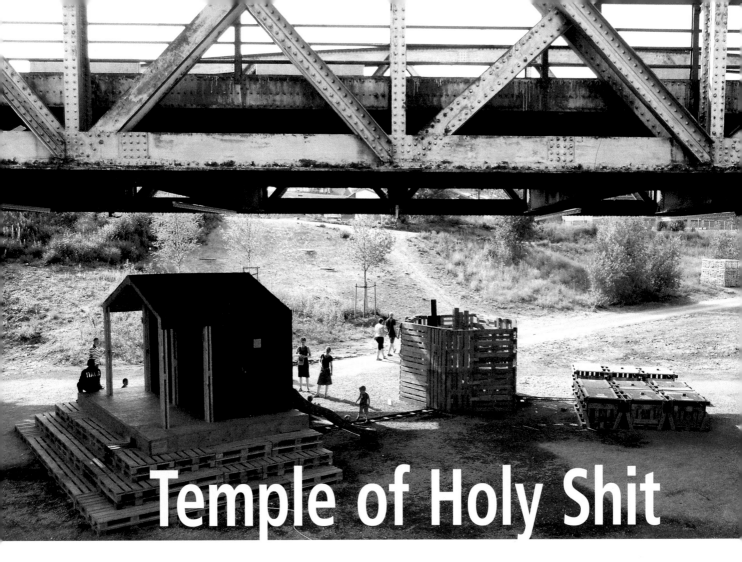

Temple of Holy Shit

On Human–Soil Nutrient Cycles and the Future of Sustainable Sanitation

Valentina Karga, Ayumi Matsuzaka, and Stephen Nortcliff
in conversation with Alexandra R. Toland

Stephen Nortcliff is Emeritus Professor of Soil Science at the University of Reading, United Kingdom. For the last three decades, he has worked across a wide range of topics focusing on sustainable soil management with a particular focus on the use of recycled organic materials as soil amendments and their benefits to soil nutrient status and soil physical properties. He is a scientific adviser to Sanitation First.

Title image: Temple of Holy Shit.

Image courtesy of Collective Disaster (Andrea Sollazzo, Valentina Karga, Pieterjan Grandry, and Louisa Vermoere).

Valentina Karga, born in Chalkidiki, Greece, is an artist and architect based in Berlin, Germany. Karga's projects encourage engagement and participation, facilitate practices of communing, and are concerned with sustainability. She is a founding member of Collective Disaster, an interdisciplinary group that works in the interstices of art, architecture, and the social realm. She has been a fellow at the graduate school, University of the Arts Berlin and she has been awarded the Vilém Flusser Residency for Artistic Research. Among others, her work has been shown at the National Museum of Contemporary Art in Athens, Greece, the transmediale festival, the Athens Biennial, and Kiasma museum.

Ayumi Matsuzaka is founder of DYCLE—Diaper Cycle based in Berlin, Germany. The project turns baby diapers into rich soil substrate, ready to plant fruit trees in local communities. It is one of the case studies of Zero Emissions Research and Initiatives. She originally trained as a conceptual artist and realized several nutrients-cycle art projects with Terra Preta soil scientists in Europe and in Asia. She joined Collective Disaster's public toilet project as a Terra Preta sanitation adviser in 2014. The "Temple of Holy Shit" was a project realized as part of Parckdesign 2014 in Brussels, Belgium, supported by the Ministry of Environment Brussels.

Soils sustain our existence. The soil is expected to provide humans, as well as all living terrestrial creatures, with food. This expectation is clearly defined in national and international policies worldwide. The production of biomass for food, fibers, and fuels is recognized as one of the major functions and services of the soil. But what sustains soil when human impacts render it no longer self-sustaining? Could sustainable sanitation be seen as a way of feeding the soil—a form of reverse sustenance to offset motions of degradation and desertification? In a series of e-mails and Skype interviews, the authors discuss case studies of small-scale sustainable sanitation and the challenges for planning and managing dry toilets in the city.

The title of the conversation references the "Temple of Holy Shit," a project by Collective Disaster initiated in Brussels in 2014 by Andrea Sollazzo, Louisa Vermoere, Pieterjan Grandry and Valentina Karga.

Alexandra R. Toland: One of the United Nation's Sustainable Development Goals (SDGs) for the year 2030 is to ensure the availability and sustainable management of clean water and sanitation for all humans. This includes the end of open defecation pits and the improvement of water and sanitation facilities from local to regional to national levels.[1] At this point in 2017, the UN reports that 4.9 billion people (over two-thirds of the world's population) have improved sanitation facilities, but these tend to be concentrated in urban and suburban areas. People without access to clean sanitation (about a third of the world's population) live overwhelmingly in rural areas, especially in Central and Southern Asia, Eastern and South-Eastern Asia, and sub-Saharan Africa.[2]

Stephen, the organization you have been involved with, Wherever the Need, has worked to provide drinking water services and construct clean sanitation facilities for communities in Africa and India. Some of this work includes the installation of urine diverting dry toilets (UDDTs) to provide compost for gardening and tree planting projects. What do you think are the main challenges of meeting the UN's goal for universal access to sanitation and clean water by the year 2030?

Stephen Nortcliff: First, let me provide a brief history of my experience with Whenever the Need (which has since changed its name to Sanitation First) for context. I became involved with the organization in 2008 following contact with David Crossweller, who presented a promotional video outlining the waste of a good resource through the failure to recycle human waste. He particularly pointed out that this failure was happening where access to other soil amendments was not possible. By separating human faeces and urine, new sources of plant nutrients and the addition of organic materials as soil conditioning amendments could be offered.

From this time, I acted as an adviser on aspects of soil management. In 2013 I was invited to Pondicherry in India where Whenever the Need had established a series of projects and also an experimental plot with different treatments (of urine and composted faeces additions). I also attended a conference organized by Whatever the Need, Pondicherry, and Annamalai University on "nonconventional organic inputs to soil" in which it was stressed that whilst there were government subsidies for fertilizers, there was no such subsidy for alternative nutrient inputs. Also, although there are subsidized fertilizers available in India, these are beyond the budgets of the small, essentially subsistence-type farmers.

The target recipients of the resulting compost (faeces plus wood ash) were poor farmers who farmed less than one hectare. The experimental trials seemed to show good yield improvements with compost and urine additions. I went to a small village where one family collected the urine and faeces and in addition to using them separately they also leached the compost with the urine to get a liquor that they told me was an excellent fertilizer, although I have no experimental evidence.

The availability of organic type nutrients and soil amendments as a result of the recycling of human excreta has the potential to make a significant contribution to improving the yields and sustainability of smallholder farming systems. This applied to most of the farmers I encountered in my visit to Pondicherry.

This visit to India illustrated the magnitude of the task to meet the UN SDGs on sanitation. UDDT's seem to offer the only likely alternative for improving sanitation in rural poor areas in countries like India. The toilets themselves are not cheap, but many orders of magnitude cheaper than alternatives. The strategy of using the urine and faeces as soil amendments offers an incentive, but the materials must be carefully handled if they are to be safe. For example, the composting process has to be managed to reach sustainable high temperatures to ensure the materials are sanitized. Whilst I made no measurements of the water, I observed numerous examples of toilets depositing human waste straight into water bodies, in addition to direct urination and defecation into the water. This water was often the source of drinking water to many people of the population. UDDTs can avoid this.

Woman of Pondicherry outside her UDDT.

Image courtesy of Stephen Nortcliff.

Alexandra R. Toland: Valentina, in the boldly named Temple of Holy Shit (TOHS), you and other members of the group Collective Disaster[3] erected a UDDT composting toilet as a pilot project in a public park in Brussels that was part public sculpture, part performance and event space, and part community education center. What kind of feedback did you get from people in the park? I mean, what would motivate city people to use a "dry" toilet if they already have a "wet" one? In your interactions with the public, did you get the feeling that people in cities are even aware of sanitation problems in other parts of the world, such as the villages in Pondicherry Stephen mentions?

Valentina Karga: Well, I don't know exactly what would motivate city people in general, but in our project, it was the only option in the public park, so basic needs encouraged visitors to try it out. But no, I don't think many people in cities, at least in Europe, know about the problem of sanitation worldwide because it's just so normal to push a button. If you haven't traveled to places without sanitation or you haven't seen a documentary or some kind of reporting on the issue, you just wouldn't know. So, Temple of Holy Shit was really intended to influence people's mentality about toilets and about, well, shit, and to show them that it doesn't have to be something dirty, that we could see it as a valuable resource instead. So that was the transformation we wanted to happen in people's minds. And I think many people got it! When we were doing maintenance there, people would come and hang out and chat with us. Some kids from the neighborhood were like, "Oh! What is this? Shit? Blech!" And then others, especially people with little kids and families, were very open to see the process and wanted to know how it's done. I was actually really surprised how many people were *not* disgusted. Some teachers wanted to put it in schools because they thought it would be a good thing for kids to learn about how soil can be regenerated.

Alexandra R. Toland: Ayumi, what about you? Do you think the UN's vision for universally accessible sanitation is commonly shared by people living in cities who might not think twice about functioning toilets? Do you think bringing sanitation into the consciousness of urban populations in richer countries could help achieve the UN's goals?

Ayumi Matsuzaka: I was thinking about three things regarding this question …

I go to the International Dry Toilet Conference every three years, and you often hear from people from developing countries who present their projects on compost toilets. But there's a big frustration because engineers from northern countries, or "developed" countries, come to discuss dry toilets while still using flush toilets in every daily life. The question is why should developing countries adopt composting toilets and not "developed" countries. They're like, "You guys propose compost toilets—but why can't I use a flush toilet too? It's so comfortable." So, there was an important discussion we had the last time that focused on developing a compost toilet for "developed" countries. Otherwise it doesn't seem fair. So, this is one reason that we should develop such small experimental projects (like TOHS) in our homes and gardens and city parks. It's a good idea for everybody!

The second thing to consider is global warming. Big disasters can come when we expect them and when we don't. Once such a disaster comes, the question is who will have the knowledge of making such toilets and how can you deal with this collective problem of managing excretion in a hygienic way. How can you encourage the use of composting to avoid potential destruction from natural disasters? I'm thinking about the floods right now in the Philippines and so on. All that water immerses you and you can no longer separate dirty, black water, and drinking water. So, people

need basic knowledge about hygiene and sanitation. Through these small experiences, like the project in Brussels, citizens can learn how to separate urine and faeces using simple UDDT systems.

A third reason is that there are shrinking populations in many "developed" countries. People in big cities in Japan, for example, have fewer children than they did maybe ten or twenty years ago. Falling birth rates mean fewer taxes collected over time. In, let's say, thirty to fifty years the city could have a big problem maintaining the whole sewage system. Because when you have fewer taxes, you need to prepare some alternative system because the city uses quite a big part of its budget for maintaining the sewage system. Once the city does not have enough money to keep up the system, it becomes a danger waiting to happen. One city in the Hokkaido area already had a very big problem because it could not keep enough water in the system any more. So, sanitation is not only a problem in "developing" countries, but also in "developed" countries that are facing climate change, population change, and disaster risks. So, it is really important to use small projects as a way to change people's minds, to show citizens that this is a city topic as well as rural one for people in all parts of the world.

Alexandra R. Toland: Wherever the Need appears to be working on a village scale, installing toilets for individual households and small neighborhoods. The TOHS worked on a similar scale, consisting of two Terra-Preta Toilets in a public park.

Stephen and Valentina, could you both please reflect on what you learned working at this scale and comment on the possibilities of upscaling. As urban areas grow and cities turn into megacities, it is interesting to think about how Terra-Preta toilets and eco-san technology could be adopted as larger-scale sanitation systems? While the UN sets its sights on providing access to sanitation for one-third of the world that still lacks sanitation, what are the challenges of converting the other two-thirds of the world to more sustainable systems? What scales do we need to think about to achieve not only *sanitation* worldwide but more *sustainable sanitation* worldwide?

Valentina Karga: In our experience, the people who were genuinely interested in the benefits of the UDDT system were mostly people with gardens. This is an important point in the context of cities because these kinds of projects need space. At one point I was really thinking about how I could transform my own apartment to include a compost toilet there, but I realized it's very difficult. For one thing, what do you do with all of the material if you don't have space for it? The practical problem is where do you put the resulting compost if you're not a farmer or gardener. If there's no company or something to collect it, you're on your own. So, the logistics have to be figured out first.

Stephen Nortcliff: As I mentioned earlier, UDDTs are a much cheaper alternative to a mains base sewerage system, but still very expensive. All those I saw in Pondicherry were paid for by the WTN charity (for example the group of c. 20 were paid for by a charity dinner—the two women who organized the dinner were to visit the site four weeks after my visit!). Another part of the problem is social attitudes. The women I met in Pondicherry told me that whilst their young boys used the toilets, once they were young teenagers the menfolk told them not to be "sissies" and do as they did in the street! Whilst purely anecdotal, I got the impression in Pondicherry that there was a reluctance by local officials to accept the lack of sanitation as a major problem, particularly as it was a problem of the poor and very poor. The strategy of WTN of bringing together the use of UDDTs with the provision of low-cost (or free) soil amendments provided an incentive for the adoption of this form of sanitation.

Alexandra R. Toland: Ayumi, you've explored scale in terms of our tiniest members of society—babies. The scale of small-scale waste is often overlooked, but you point out that each child produces around 4500 dirty diapers before she or he starts using toilets. The *DYCLE* (Diaper Cycle) project uses Terra Preta technology to address the waste streams of our smallest community members by transforming waste streams into Terra Preta black humus for fruit orchards. Could this project be upscaled to give parents worldwide an alternative to feeding landfills?

Ayumi Matsuzaka: My first idea about "upscaling" is to publish a DIY guideline on the diaper cycle. I will open source all of the engineering and designs, so you can really download recipes for making your own one hundred percent plastic-free, bio-based, diaper inlay by yourself in English and other languages. I will also include how to make a planter out of the baby diaper and show how you can use its nutrients for planting trees. I've received requests from people in more than twenty countries so far who would like to start the diaper cycle. Most of them are in developing countries with huge garbage problems. These people are also very good at making their own mechanisms for production. So, this is the second thing—producing a low-tech mechanism as our next milestone. But this actually means downscaling, not upscaling. Waste streams are local problems. We do not like to rely on high-tech mass production. We don't need six million new mass production assembly lines, but rather low-tech, simple mechanisms that can be reproduced in decentralized systems. Take, for example, ten communities in the pilot project in Berlin. That would be about one thousand babies. That would be the scale in which we could produce diapers

DYCLE System in action: (above) tree planting in Brandenburg, Germany; (below) Terra Preta bins with instructional diagram.

Images courtesy of Ayumi Matsuzaka.

together as a community. This can also create job opportunities for local people. And this model could be adapted in Tahiti, or South Africa, or the Philippines.

I actually brought one of our inlays with me today.

Alexandra R. Toland: Wow! I used one like this during your prototyping phase a couple years ago when my son was only a few months old. I was amazed at how the Terra Preta absorbed the smell.

Ayumi Matsuzaka: Yes. This is a really simple one, made from natural, local raw fibers from Brandenburg, next to Berlin. We've experimented with several non-chemical-treated fibers and powders and found the good combination for absorption. So, the idea is to stay small scale but to spread worldwide. I don't want to stay in Germany. It's better that everybody starts this type of cycle in their own community and in their own country using local materials, like maybe fibers from sugarcane or coconut skin.

Alexandra R. Toland: I can imagine local resources like the ones Ayumi mentions and local

environmental conditions play a huge role in our management of human waste streams.

Stephen, could you please talk about how environmental factors play into choices of sanitation design, construction, and management. How does climate, soil type, and land use play a role in sustainable sanitation?

Stephen Nortcliff: The systems I observed in India worked well because there are high atmospheric temperatures so the composting was rapid and if well aerated reached the temperatures sufficient to sanitize the materials. I have had some contact with other systems where the urine and faeces were fed into an aerobic digester to generate methane for fuel and the digestate was used as a soil amendment (providing a carbon source and nutrients); one cautionary warning on this if the digester is not effectively sealed the leakage of methane will have a major "global warming" impact. Anaerobic digestion is possibly an option in colder climates where the normal composting will not achieve temperatures to sufficiently sanitize the materials. Anaerobic digestion requires relatively large inputs and an efficient system of waste collection.

Temple of Holy Shit. Making Terra Preta out of bodily waste.

Image courtesy of Collective Disaster.

Alexandra R. Toland: Ayumi, Brussels is obviously cooler than India. You were working with an anaerobic system, correct? How did that work?

Ayumi Matsuzaka: Does this anaerobic digestion mean biogas production? If so, I might not be the right person to transfer knowledge of biogas production in cold countries. What I do is Terra Preta method, which is a combination of anaerobic fermentation and humification. The soil temperature will not go higher than 22–23 degrees. So, it is "cold" compost compared to normal hot compost. Anaerobic fermentation helps a lot because the humification process after fermentation is much smoother than aerobic hot composting systems. The benefit of adding the anaerobic fermentation process is that you do not lose important nutrients that can evaporate into the air. In Brussels, we fermented longer— for about three months. After that, we had more microorganisms, more fungi, as a result of the fermentation process. That increased the conditions for humification, which can be combined with composting worms and other insects and so on.

Valentina Karga: Andrea actually did an analysis of our compost to see how good it was. What we found out was that that it was good, but not the super fertile "black gold" that we thought it would be. It was not the rich humus we were expecting.

Ayumi Matsuzaka: Yeah, it was good but it wasn't holy.

Valentina Karga: I think the less-than-holiness was because we mixed it with the soil that we found there in the park, which was very sandy soil. Somehow a lot of the nutrients got lost. So, in the end we were just experimenting, but the important thing was that it was safe to use. There were no pathogens and it could be used in the neighboring gardens.

Alexandra R. Toland: So, beyond environmental factors, I'd like to hear your ideas about design and maintenance as fundamental factors in the development of sanitation systems. Valentina, what kind of research did you do in the planning of TOHS? How did you decide on the final design of your toilets, including two slides, not for diverting excrement but for children to play on? Beyond the elaborate design, talk about maintenance issues and the challenge of planning and managing waste in a public space.

Valentina Karga: So, the initial design team consisted of myself, Peter, Andrea, Luisa, and Ayumi, and then we collaborated with a lot of other people for different parts of the project. Regarding design research, there were some reference materials but we were mostly interested in the idea of integrating the composting toilets into the architecture of the park rather than designing toilets as objects in and of themselves. We were more focused on systems thinking than object design, if you know what I mean. We had been thinking about the park itself and what other uses the structure could have, beyond just being a toilet. It had slides, it had a stage, there were places to sit and eat and gather. Then we came up with the Temple idea to make it grand and exemplified, and "Temple of Holy Shit" just sounded fun. So that's how things came together. It was more of like a collage of different conceptual layers, practical and technical layers, than simply a toilet.

Regarding maintenance, that was very difficult because in the beginning the ministry said they would have people to take care of it, but this didn't happen. Andrea and Luisa live in Brussels, so they were there a lot to fix things and take care of things. We had also hired Marcelo, an illegal immigrant from Romania who was living in a small garden shed and was earning a little money from other projects in the park and taking care of

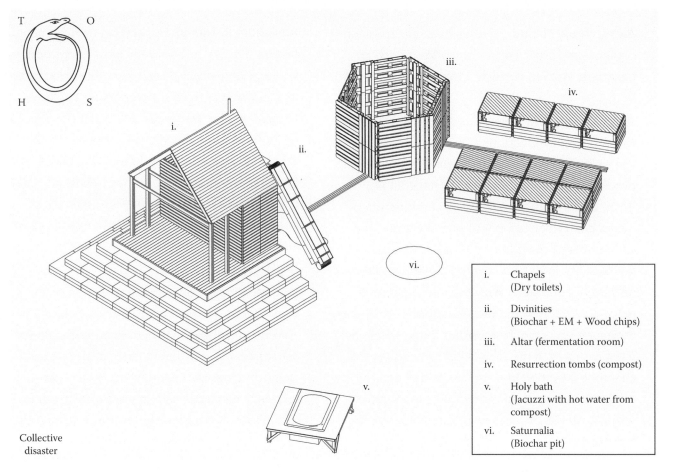

T O
H S

i.

ii.

iii.

iv.

vi.

v.

i.	Chapels (Dry toilets)
ii.	Divinities (Biochar + EM + Wood chips)
iii.	Altar (fermentation room)
iv.	Resurrection tombs (compost)
v.	Holy bath (Jacuzzi with hot water from compost)
vi.	Saturnalia (Biochar pit)

Collective
disaster

Diagram of functionality of the Temple of Holy Shit.
Image courtesy of Collective Disaster.

the gardens. He would take care of our toilets and do other things, but he was not paid or officially hired by the ministry. Actually, nothing was very clear in our communications with the ministry and I don't think the curators who decided to give us the project thought through all these processes.

Ayumi Matsuzaka: Maintenance is a very important point, especially in public places, because people don't know who has ownership. There's the question of who owns the toilet and who owns what comes out, and how does that get managed. If you're not connected to gardening or agriculture activities, you won't have the benefits out of the compost toilet.

Bringing the discussion back to the soil, the connection between sanitation and food production needs to be made. If we have the knowledge to design and use these systems, we can really increase soil biodiversity, which is necessary to produce good healthy food. Meanwhile, the price of fertilizers is getting higher and higher. If we all had nutrient cycling systems and food production in local communities, then much could be saved. I guess that's what many people want to have in the future. Of course, it takes time to make these connections, but there are good examples of working together. One farmer in Brandenburg really loved the DYCLE idea and tested it on her plants. So, we planted fifty fruit trees there. Farmers can easily appreciate the value of these projects, because they are interested in good soil too!

Alexandra R. Toland: Issues of design and maintenance are in part culturally determined.

Every culture has its own local traditions of growing and sharing food. As culture has become globalized, so has food production and consumption practices. In a sense this may be true for sanitation as well. I wonder if bringing back old sanitation technology could be akin to bringing back old heirloom species and recipes? How do you think sanitation practices and facilities are rooted in cultural tradition? What can we learn from sanitation models from different cultures and older times in our common human quest for sustainable sanitation? I was hoping you could each talk about the cultural dimensions of sanitation you've encountered.

Ayumi Matsuzaka: In some rural areas of Japan, human excrement is still an important resource. The grandparents' generation can still explain how they would use human compost for their vegetables and in the rice fields. This was important during the Second World War, of course. But what is also interesting to me, coming from the city, is the use of dry toilets as a way of adapting infrastructure to earthquakes and tsunamis in the region. You cannot use a flush toilet if your house is shaking. This issue has become a really serious topic since the disaster in 2011, when there were a lot of people without electricity, without drinking water, and without sanitation. That is one issue that the university has worked on—developing ways to build very cheap toilets, let's say 60–70 Euros so people can make one themselves.

Stephen Nortcliff: I saw an amazing project in a place a few kilometers out of Pondicherry. There was a community of about twenty families where Whenever the Need had installed about twenty UDDTs. I was shown every single one of the toilets by the very proud women of the families. Whilst they were interested in the use of the urine and faeces, the provision of the toilets had a much more significant impact on their lives. Through an interpreter, I was told that they no longer had to make trips outside the village when they wished to defecate, which made them feel safer. Previously, in the wet season, they would eat and drink less so they did not need to make the journey to the edge of the village in the rain! Interestingly, their children were using the toilets, but the menfolk still insisted it was "cleaner" to urinate and defecate in the street (they did not go somewhere "private" like the women!)

Another project I did not see but which was reported to me was where WTN established a sort of "municipal toilet" predominantly for men to avoid urinating and defecating in the street. The urine and faeces were collected and processed by one family and the compost was sold to farmers. I have asked for details on this.

I was also told of another project where UDDTs were built at a girls' school. A local social worker told me that when young girls begin to menstruate they are embarrassed and often leave school. The provision of the toilets provides some privacy and first indications were that the girls continued school longer.

Ayumi Matsuzaka: Gender issues are an important aspect of public toilets. I think it is important that there are men and women involved in designing and building these projects. Most of the men in a small community situation can convince the landowner where the toilet should be built. They have the knowledge of where you can get sawdust or how you can transport materials. So, men can make those kinds of connections in the decision making. Women, on the other hand, are very good at finding practical solutions and they always bring in the opinions of old and disabled people and children. Toilets are often too big for children. Men do not often think about children's size, height, and where you can hold yourself. Also, washing and hygiene is very important for women and children. And maintenance, yeah, oftentimes that's women work too. This is very natural—the blood comes every month so it's really easy to talk about cleanliness amongst women because you have to deal with it all the time. So, this is a recommendation in many guideline books—make sure you have men and women working together in such community projects.

Alexandra R. Toland: Valentina, what's your take on the social and cultural dimensions of sanitation?

Valentina Karga: So, in the park project we experienced a really big backlash from people in the neighborhood, which was very interesting. The project was in a mainly Muslim neighborhood in Brussels, and they explained that there is a big taboo in their religion regarding filth and cleanness. When they heard the words "temple" and "shit" and "holy" together in the same sentence, they were like, "You guys are idiots. You cannot come here to our neighborhood and say that shit is holy. Are you making fun of religion here? Because god is very important to us and you cannot joke about that. It's a no-go." So, they didn't get the humor and lightness about the subject we wanted to communicate, or the idea that transforming something into something else could be holy (we believed it was holy).

So, it was really difficult in the beginning. Some people were even threatening to come and burn it down. But then something changed. There was a little cafe in the park where the older Muslim guys would hang out and drink tea. One guy from the cafe was very sympathetic and he was also involved in the whole project of the park design. So, he tried to connect us with the community. One day Andrea and Luisa organized a public meeting with the residents to rename the project. "Temple of Holy Shit" is still the unofficial title that we use because we love it, but in the

end, we officially renamed the project for the neighborhood and in all the signage: *Usine du Trésor Noir*, which means "Factory of Black Gold."

Alexandra R. Toland: Hmm. That's kind of poetic. It's curious that Stephen also mentioned there was a name change from Wherever the Need to Sanitation First. It appears that one major challenge of establishing sustainable sanitation isn't only about finding technical solutions and social and political solutions, but also finding the right language. And UDDT is about as linguistically elegant as WC.

Valentina Karga: Language is so important. That's what I love about the arts—making concepts and then finding the right words to communicate things, because only then can you begin changing the way we make mental connections about things. We proposed a bunch of different terms and people could vote on what they liked and somehow this was the resulting combination of words that came out of that meeting. After that, everything was easier because the local community realized that we were willing to change. In the end, there was acceptance and we saw lots of people using it … women with their kids, younger park visitors, etc. At night when the park was officially closed it also became a kind of meeting place for some of the young Muslim guys to hang out. They were like dealing drugs and things like that, people were having sex in the toilets, yeah, it was a kind of temple …

Alexandra R. Toland: … a temple (or officially factory) of night soil—another allegorical term for the collection and recycling of human waste! I wonder if these visitors recognized the deeper meaning of the project, even if they didn't get the lighter one. As you explain in the promotional video of the project,[4] it's all about connecting the dots between biological flows, material flows, knowledge flows, and cash flows; it's about

Temple of Holy Shit inauguration.
Image courtesy of Collective Disaster.

educating and creating awareness through means of architectural intervention, performance, and public participation. How did you make these connections visible?

Valentina Karga: We did many events and workshops and guided tours explaining the project. Louisa also did some games for kids to explain what compost is and how to make it. There was a nice treasure hunt game. Of course, we also had a detailed printed diagram that was carved into Plexiglas so that it worked outdoors. It explained everything in Flemish and French. We had a great coding system that Peter designed, which related back to the bigger diagram. So, everything was covered in some kind of signage,

and people could figure it out even if there was nobody there to explain things. The compost, the buckets, had poo or pee drawn on top. We tried to design every part so that it spoke for itself, so that it was self-explanatory. And we put signs everywhere, even inside the toilet cabins.

Alexandra R. Toland: Stephen, "soil connectivity" is a term that has become popular within the soil science community in recent years. Can you talk about what that term means to you and discuss how Wherever the Need fosters connectivity? What about artistic approaches to connectivity? What do you think Temple can bring to the table that perhaps science and NGOs (nongovernmental organizations) cannot?

Stephen Nortcliff: I have always found it peculiar that we praise manure from animal as wholesome and a strong positive when discussing agricultural systems, but when we discuss human manure the reverse reaction is often true and often very strongly so. I had a discussion with an archaeologist friend last week who is working with some coprolite samples to determine diet. She said that there is no squeamishness when dealing with these samples, but if you mention contemporary faeces the reactions change dramatically. Even when I have spoken at meetings of organic farmers, green campaigners, and the like, and endeavored to explain the obvious benefits of recycled human waste, I have often finished with only a very few converts. I have explained that we apply animal manures with little or no treatment, but human manures are sanitized either through prolonged high temperature composting or high temperature anaerobic digestion, but the responses change very little.

As I have mentioned earlier, the linking of the adoption of UDDTs to the provision of soil amendments, at least in the context I was involved with, was a positive one. The projects I saw involved some local organization among the ten or so households. One person organized the collection of urine and faeces and the composting and in one case the leaching of the composts with urine to produce a nutrient-rich liquor (WTN provided the facilities). The liquor was distributed to farmers who sprayed it on their fields, the leached compost was distributed as a soil amendment on the farmers' soil, which were very low in soil organic matter. This is one of the few examples where I have seen a nearly closed circle: crop—food—excrement—fertilizer—land—crop.

When you look at these systems they are so obvious that you wonder why they are still relatively rare. Whilst NGOs have a role, they simply do not have the global outreach to achieve this. I am not exactly sure how to reply to the art approach … but anything that can change people's mindsets is important. This is a topic that many people avoid or wish to avoid, but it has to be become one which we address as a priority. We have to remove our first-world "squeamishness" about dealing with human waste; it is simply too valuable a resource to waste!

Endnotes

1. See the UN's targets and indicators sheet for SDG 6 Water and Sanitation: https://sustainabledevelopment.un.org/sdg6.

2. See the UN's 2017 progress report for SDG 6 Water and Sanitation: https://sustainabledevelopment.un.org/sdg6.

3. Project initiators are Andrea Sollazzo, Valentina Karga, Pieterjan Grandry, and Louisa Vermoere; with Terra Preta expertise from Haiko Pieplow and Ayumi Matsuzaka; and supportive collaboration from Ward Delbeke, Sonia Saurer, Tcharmela & Big Ben, Ane San Miguel, Sandra Guimaraes, Maria Ilia, Farmtruck & Rirbaucout, Jonathan Ortega, Caroline Claus, Mohamed's Teehaus, and LUCA school of architecture.

4. See promotional video on Vimeo: https://vimeo.com/106395914.

S.OIL

Maria Michails and Ronald Amundson in conversation with Alexandra R. Toland

Maria Michails is a Canadian interdisciplinary artist and new media writer. Her work bridges the arts, science, and technology to address environmental concerns. She creates human-powered installations that invite participation and inquiry into industrial and ecological processes. She recently shifted from interactive installation to community-engaged co-creative processes to tackle localized environmental problems. She has published and exhibited internationally, and her work has received grants and awards from institutions and government agencies in Canada and the United States. She is currently a PhD candidate and doctoral fellow at Rensselaer Polytechnic Institute, Troy, New York.

Ronald Amundson is a professor of pedology at the University of California, Berkeley. Amundson was raised on a farm in South Dakota, and attended South Dakota State University.

Title image: *The Handcar Projects: On The Grid*, Maria Michails, 2011. Detail view of houses with corn image.

Reprinted with permission from the artist.

He received his PhD at the University of California, Riverside. He has been at Berkeley since 1984. His research focuses on soil processes in deserts, and impact of life and water on soil processes and properties. He has used this work as a basis for using the soil chemistry of Mars as a tool to understand more about its climate history. He has long had an interest in human interactions with soil, and the role of soil in our climate system and as a societal resource.

The following conversation on agriculture in the Anthropocene is inspired by an artwork, or series of artworks, Maria Michails created from the years 2008 to 2015. *S.OIL* (2012) is a human-powered interactive installation that reflects on the history of industrialization and energy use in relation to farming, biofuels, water quality, and topsoil erosion. Mechanical and electronic systems are combined with living plants (experimental perennial food crops) and video. The installation entrusts visitors with operating a railway handcar to power an electronically controlled irrigation system to water the plants and the video monitors for viewing a contrasting video of corn growing and processing. Borrowing simultaneously from the Earth's living systems and industrial processes, the installation creates a space that is part laboratory, part factory, and part farm. The following dialogue was realized as a string of e-mail threads surrounding three central questions.

Agriculture and the Anthropocene

Alexandra R. Toland: The human-powered railway handcar is used as point of departure in Maria's interactive installation *S.OIL*: "The handcar references the train, an early mode of transportation, which was pivotal to the quick progression of the Industrial Revolution … The first use of the steam engine as a railcar makes its appearance in 1804. The combination of steam locomotion and industrial machines made production and distribution (including farming) easier and specialized, changing the mercantile system to a system of efficient mass production."[1]

The powerful metaphoric image of a handcar powering an industrial agriculture laboratory brings up at least two key questions pertinent to discourse on the Anthropocene: When did it start and what are its main drivers? Paul Crutzen, Jan Zalasiewicz, and other authors from the geological sciences have sought to pinpoint the beginning of the Anthropocene, from the intensification of agriculture over 2000 years ago,[2] to the worldwide distribution of the Boulton-Watt Steam Engine and subsequent industrialization processes in the late 1700s,[3] to the beginnings of nuclear technologies at the end of World War II.[4] I wonder if you could both weigh in on these questions. Why is it necessary to pinpoint the beginnings of the Anthropocene? What are the ethical implications of current agricultural production practices if agriculture is indeed one of the main drivers of the Anthropocene? And as Maria asks (paraphrased), how can the relationship between contemporary farming and manufacturing be honestly reassessed to ensure ecologically sustainable production for future generations, perhaps offsetting the negative trajectory of the Anthropocene?

Ron Amundson: Since the concept of the Anthropocene is the basis for a proposed new geological epoch, identifying the boundary between the end of the Holocene and the beginning of the new Anthropocene takes on academic and stratigraphic importance. Arguments are indeed made for the steam engine (manual versus machine labor) and the atomic age (pre- versus post-WWII). All these are important inflection points in human history and earth history. However, the most fundamental impact that humans have had on the planet is commonly overlooked in these attempts to focus on recent technological change. By far, the most profound and radical human impact on the planet is farming, which somehow was invented within the past 10,000 years. Despite the modern romantic view of the rural life and living off the land, farming is the most unnatural act imaginable to our planet. Pre-farm soils and ecosystems operated in a largely steady-state condition. Soil production from rock and sediment tended to match erosion. Carbon and nutrients added to soil roughly matched the losses. Farming—and one can make the case that the early artisanal attempts are the worst—completely removes the existing ecosystem, replaces it with one or a few plants, and physically disrupts (on a continuous basis) the upper layers of the land surface. The resulting erosion, CO_2 emissions, and biodiversity loss by farming are staggering.

I think the reason that agriculture as a human domestication of Earth is so often overlooked is due to what some call "environmental amnesia." We are simply accustomed to seeing a "natural" world with farms and crops. What is more pleasing to the eye in many locations? Yet, what we perceive as the "norm" due to our own exposure to this environmental baseline is far different than what existed through billions of years of Earth history. Of course, no living human being was around to witness the original innovation and land transformation that farming brought to the planet. However, there is a growing number of scholars and writers who now hypothesize that the Biblical telling of the creation and its aftermath in the Book of Genesis is actually the story of the tension between hunter-gatherers and pastoralists versus the upstart farmers—told by the pastoralists. Eden was, very clearly, a hunter and gatherer's paradise. Through misguided activity, God sentenced Adam and Eve to a life as agriculturalists. Sweat, labor, and pain in childbirth were their rewards—all the things associated with what we now know were part of the transition to farming. For example, frequency of childbirth increased among agriculturalists since woman did not need to carry infants around while traveling (thus increasing the "pain"). Cain and Abel are a straightforward example that God favored the fruits of the pastoralist, but not those of the farmer. Yet, ironically, since Cain killed the one pastoralist son of Adam and Eve, we are all "descendants" of the rogue agricultural upstart Cain.

Modern conservation biologists now argue that we have so radically transformed the planet, we should just begin to accept it and get over it by embracing forms of adaptation and accepting new "niche ecologies" in urban, industrial, and farmland areas. By "getting over it," they mean that instead of continuing the dichotomy of wilderness versus humanized landscapes, we should realize our fingerprints are everywhere, and we must develop a more integrated and comprehensive approach to managing and maintaining the Earth's systems. We live in an epoch, the Holocene, that began with the dawn of farming, and is ending with our final steps of domesticating all suitable parts of our planet for farms. Now the hard work begins: We cannot convert more of nature to farms, we have to effectively manage what we now have.

Maria Michails: When I first conceived of the series, *The Handcar Projects*, of which

S.OIL is one of the major works within it, I was not thinking in terms of the transition between geological epochs. I think of these terms (Anthropocene, Holocene, etc.) and their study as more of an academic exercise. What I wanted to focus on was making some parallels and working within the realm of metaphor and representation, what I consider connecting the dots. *S.OIL* and another project called *On the Grid* explicitly speaks to the parallels I see between monocrops, assembly-line manufacturing, and cookie-cutter housing, which results in the takeover of agricultural lands, and is synonymous with the loss of biodiversity. This phenomenon, whether intentionally planned or as a consequence of Western "progress," should be a starting point to discuss or reassess our current models of farming and manufacturing. Matters are even more complicated when we include the laboratory in this production cycle.

Ron's description puts a whole new light on farming, not just in terms of human impact in geologic terms but the Biblical interpretation by scholars wildly contradicts what some farmers believe. In my experience speaking with farmers during the research phase of *S.OIL* I learned that many feel that technology has helped farmers produce greater yields that can better feed the world. Of course, it also means that the farmer can improve his financial situation, although this has proven to be a double-edged sword for many. I found it troubling that corporate salesmen would play on the moral imperatives and obligations felt by devout (Christian) farmers in order to convince them to buy into this type of (factory) farm mentality.

To add to the "getting over it" conundrum, how much of this is scientifically based and how much of it sits within the realm of the aesthetic? Perhaps we need to "get over" both. We know that we need to better manage our soils, find another approach to agricultural production and with it the underlying economic and distribution systems. What role can science play in this that can better address, ethically as well as scientifically, concerns about soil sustainability? Changing or correcting public perception about what constitutes "nature" is a topic of great debate within humanities scholarship for some time now. I address the "wilderness" question more in a project I am currently working on, but the "idea" of nature was not one I had thought of much before because I didn't view nature as an "idea." I was caught in the division between nature and nonnature. Wilderness, for me, has a specific meaning and it is based in aesthetics as well as science, I suppose. As Aldo Leopold said, and I'm paraphrasing here, aesthetic value is not just the beauty of a place but the pleasure is also derived in knowing that the land is "fit." How can we know a land is fit? Wilderness, for me, isn't contained within a national park or a preserve. Perhaps it is naïve to think that wilderness—that romanticized wild nature untouched by humans—even exists. One would think that the land outside the park might be "wild" but in reality it is "nonwild" and okay for exploitation, including farming. Wilderness then is perhaps an

imaginary place. To some people, golf courses are natural environments, therefore, it's not hard to understand how many would consider making a "natural" association with farms.

Biomass Production and the Food-versus-Fuel Dilemma

Alexandra R. Toland: One of the main functions of the soil is the production of food and biomass. In ecological circles biomass is defined as the total amount of organic mass of all life forms in a given area. *Soil biomass* refers to the entire depth of the soil where organisms are present, and is subdivided into phytomass, zoomass, and microbial mass.[5] Biomass is also industry shorthand for fuel grown from a variety of plants, including eucalyptus, palm (palm oil), sugarcane, poplar, and sorghum. Although biomass production is classified as a renewable energy source, for example by the UN and EU, it can result in higher CO_2 emissions based on production methods and plant type. It also puts farmers in a difficult position as they are confronted with the choice of growing food crops or energy crops (and sometimes both). As former agricultural landscapes are rapidly being replaced with energy landscapes, what do you see as the main challenges and potential solutions to the food-versus-fuel dilemma of the ecosystem services debate? Beyond that, how do you envision sustainable allocation of land for both food production and energy production when farmers are at the same time under pressure to develop land for human settlement as cities expand and encroach on rural landscapes?

Ron Amundson: Broadly speaking, most major schemes to decarbonize the economy involve land and soil, for example, in the siting of wind turbines and solar installations. Sometimes both land owners and environmentalists oppose these efforts or programs for a variety of reasons. Farmers and landowners (despite significant

The Handcar Projects: S.OIL, Maria Michails, 2012. Installation view. Exhibited at the Art Gallery of Southwestern Manitoba, Brandon, Manitoba, October 4–November 24, 2012.

Reprinted with permission from the artist.

financial benefits in some cases) take a "not in my backyard approach." My home county in South Dakota, for example, will likely soon vote to have such restrictive offsets for turbines that it will be simply unfeasible to invest in that county. Meanwhile environmentalists in desert regions argue that equal areas of solar panels can be placed on rooftops in cities, preventing the use of desert land for panels. They ignore the reality that getting millions of private home and business owners to rapidly adopt this need is nearly impossible and due to current zoning regulations also unrealistic.

My view is this: We are in a race with time to drastically reduce our CO_2 emissions, and we are already starting too late. We (or more correctly, the folks alive in a few decades) are facing an environmental catastrophe of such magnitude that we simply have yet to grasp both the climatic and social upheavals it will produce, and in some places, is already a reality. We have to respond in kind to the magnitude of the size and rate of problem. We must build solar installations and install more wind-generating facilities. There has to be some give and take on this if we are to thrive at the end of this century. I don't think most people fully comprehend the scope of what we must do, and do very, very rapidly.

Maria Michails: "Energy landscapes" is an appropriate term in describing farmland today. In my recent visit to Saskatchewan it was not hard to notice the multiple outputs of energy sources extended by the soil. Pump jacks coexist, in rare places, with wind turbines and solar panels surrounded by and sitting in golden canola fields. Canola, a genetically engineered plant, is primarily processed for vegetable oil but with the European Union implementing biodiesel policies, similar to the use of corn in North America, much

farming land will be converted to growing this crop. There is still a back and forth among studies conducted as to whether or not corn ethanol is net energy positive or negative (in other words, whether it requires more energy to produce than it delivers). What isn't really discussed in biofuel debates is soil degradation and what the true cost that would amount to in the long run. Furthermore, I wonder how much good farmland has been put out of production because it's become more lucrative to use it for oil extraction, or wind turbines and solar installations for that matter. I wonder how much land has been converted to growing subsidized crops as raw material for ethanol or other fuels.

Ron Amundson: All farming is extractive because any product removed from the field contains some of the soil's store of nutrients and elements. Returning these resources back to the land has always been a challenge, and is even more so as the use of crops is more concentrated and centralized, making the return of crop residues or by-products costlier. It is not easy to project the future of biofuels. Much research is devoted to finding ways to use nonhuman food plant materials for biofuel stock. Yet, the challenge of nutrient recapture and return remains part of this evolving technology.

Maria Michails: There is a bit of confusion regarding the word "energy." Energy in the form of electricity is generated by many sources: solar, nuclear, wind, natural gas, coal, and hydro are the more common. All of these have a net energy value, as well as CO_2 count, because they all have materials that are extracted from the earth and require a multitude of energy types to be manufactured. Biomass, like petroleum, is not generally used to create electricity but is mainly used for transportation. What would happen if economies of scale flooded the market with electric vehicles? How would we charge them? Are we ready to install the necessary infrastructure and where will that electricity come from? I agree with Ron, we need to move very quickly. But can we? Do we shift from growing ethanol to building new manufacturing plants so we can increase solar panel, wind turbine, and electric vehicle production? It seems to me that it's a question of choice and prioritization. The technology is there and if we are to sacrifice land, which we are already doing at a rapid rate, it will likely take policy change to mandate these changes on the ground.

Production versus Extraction

Alexandra R. Toland: Maria suggests that there are inherent links between manufacturing and agriculture. Both are in the business of *production*. If we insist on defining soil as a natural resource (for human use), does that not also imply its potential to be mined, exploited, and depleted? In light of this discussion, do you think the ecological and ethical implications of industrial agriculture can be compared to the extraction of coal, oil, and gas? *Is agriculture an extractive industry?* Is extraction an adequate metaphor for large-scale, chemically based industrial agriculture? On the other hand, the claim that large-scale agriculture is necessary to feed a booming worldwide human population seems to underscore the more positive metaphor of production (and most radically the different iterations of the "green revolution"). What makes soil different to other natural resources is that extraction can be offset by inputs, by giving back what has been taken out. Talk about the power of these two different conceptions of the soil—production and extraction—and tell us your visions for a more sustainable agriculture of the future, including better metaphors for resource use and its repletion through good practice. Talk about examples of a *repletive*

agriculture that is large scale but also soil friendly and "climate ready."

Ron Amundson: The first farmer that ate and processed oats or some other grain was extracting P, Ca, N, K, and other nutrients form the soil. No matter what practices he or she followed, they were never able to return all these extracted elements back to the soil fully. Thus, "extraction" is not a modern phenomenon, it is a phenomenon of the history of agriculture. Today, of course, our collective pool of extracted nutrients (known to most of us as sewage) is highly concentrated and largely considered a disposal issue.

Alexandra R. Toland: Ron, what about no-till, agroforestry, and permaculture attempts to keep nutrients in the ground?

Ron Amundson: People who practice this also remove the food, and concentrate some of the crop residue around their farm sites and homes. Even when this manure and residue is manually returned to the field, it will have undergone biological decomposition and likely some leaching, making it less nutrient rich than the original food or fiber that was harvested. While some farming schemes, which include legumes as cover crops, can add nitrogen back to the soil, the soil-derived nutrients such as P, Ca, and K can become slowly removed over successive years of crop and food production and removal.

The challenge for us is to work more effectively to close the nutrient loop in farming by developing ways to convert "waste treatment" to "nutrient extraction." We are a long way away, but engineers are indeed developing novel ways to accomplish this. But it will never be a completely closed cycle: N is lost as gases, and it's impossible to extract and return every atom back to the soil (no one has ever succeeded in this, unless they acquired someone else's waste to add to their own). Integrated nutrient management is likely not a high-profile or sexy topic, but it is one that is critical to living in a world where we have pushed its limits.

Maria Michails: With *S.OIL* and all of my other human-powered works, I wanted to convey the importance of the feedback loop, so I avoided including battery storage but rather built the system to provide instant feedback on the participant's energy expenditure. When you operate the mechanism (the handcar) you will know immediately that you are powering the installation. You can modify how much force you put into the pumping action depending on how much power is actually needed so you don't waste your energy. *S.OIL* is also an example of a closed-loop system, represented by the water, which returns to the reservoir tank if the plants don't use what is being pumped. Natural systems integrate feedback loops and, therefore, self-correct to achieve homeostasis, much like our bodies. If manufacturing were to integrate similar systems, they would waste less and be more efficient in their production. Many today have indeed done this as it also has a huge impact on bottom line accounting. Waste management of course is one of our greatest challenges. Nature uses all its waste, nothing is wasted, and the very concept of waste is an illusion. But we have yet to figure this out.

The Handcar Projects: S.OIL, Maria Michails, 2012. Installation view with participants. Exhibited at the Art Gallery of Southwestern Manitoba, Brandon, Manitoba, October 4–November 24, 2012.

Reprinted with permission from the artist.

The Handcar Projects: S.OIL, Maria Michails, 2012. Detail of planters and electronically controlled, closed-loop irrigation system. Plants from left to right: perennial sunflower, perennial sorghum, and perennial wheat. Plants courtesy of The Land Institute, Salina, Kansas.

Reprinted with permission from the artist.

The Handcar Projects: S.OIL, Maria Michails, 2012. Detail of video with GMO corn footage.

Reprinted with permission from the artist.

Our current large-scale agricultural system is comprised of primarily five staple crops, all of which have varying amounts of subsidy (corn being the number one). For me, dense monoculture crops are akin to extraction, particularly on par with surface extraction such as mining. Mining doesn't return much back to the soil, even when reclamation efforts after the fact are well-meaning. There is, though, an agricultural system being developed by The Land Institute that has the potential to be large-scale, soil-building and "climate ready." The plants in *S.OIL* are hybrid perennial sunflower, wheat, and sorghum provided by The Land Institute. Through traditional breeding of wild perennial prairie plants with their domestic annual cousins, the researchers aim to preserve the benefits of the native characteristics, such as deep root systems and heartiness against drought, to produce a robust seed that will yield food grains that no longer rely on fossil fuels or even farming practices other than harvesting. Could this system give enough back to the soil over subsequent generations? Can this system produce large yields and avoid the pest or fungi problem? I think this system of agriculture has

a great deal of potential. I believe, if I'm not mistaken, some artificial inputs are used currently, perhaps to speed up the process. In the long run, I do think this system is sustainable and it certainly gives me a sense of hope.

Endnotes

1. See exhibition statement, Michails, M., 2012, http://treiastudios.net/Treia_Studios/Projects/Pages/S.OIL_files/SOIL_statement.pdf.

2. See, for example, Certini, G., and Scalenghe, R., 2011, Anthropogenic soils are the golden spikes for the Anthropocene, *The Holocene* 21(8), 1269–1274; Ruddiman, W.F., 2003, The anthropogenic greenhouse era began thousands of years ago. *Climatic Change* 61, 261–293; Ruddiman, W.F., 2005, How Did Humans First Alter Global Climate? *Scientific American* 292, 46–53.

3. See, for example, Paul Crutzen's original proposition of the Anthropocene from 2002 as well as Steffen, W., Grinevald, J., Crutzen, P., and McNeill, J., 2011, The Anthropocene: Conceptual and historical perspectives, *Philosophical Transactions of the Royal Society A* 369, 842–867. doi: 10.1098/rsta.2010.0327.

4. Zalasiewicz et al. 2015. When did the Anthropocene begin? A mid-twentieth century boundary level is stratigraphically optimal. *Quaternary International*, 383, 196–203.

5. Canarache, A., Vintila, I., and Munteanu, I. 2006. *Elsevier's Dictionary of Soil Science*. San Diego and London: Elsevier.

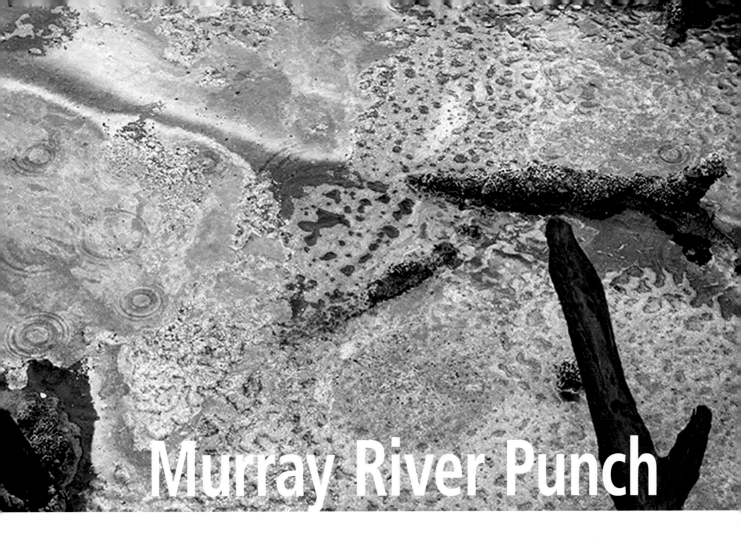

Murray River Punch

A Conversation on Changes along the River

Richard MacEwan and Bonita Ely

Australian artist, **Dr. Bonita Ely**, was raised on the Murray River in Robinvale, Victoria, which profoundly influenced her art practice. Interdisciplinary artworks typically address environmental and sociopolitical concerns, often using humor and forensic investigation to stimulate viewers' engagement. Her artworks are in national and international collections. She has represented Australia in significant survey exhibitions, most recently at the Documenta 14 in Athens and Kassel, Germany. An honorary associate professor in the Art and Design faculty, University of New South Wales, she is a member of the Environmental Research Initiative for Art (ERIA). She is represented by Milani Gallery, Brisbane.

After gaining an honors ecology degree from Edinburgh University, **Richard MacEwan** pursued spiritual studies while working for the Beshara School of Esoteric Education for

Title image: Iron floc and acid sulfate conditions in a billabong of the Murray River (2007).

Reprinted with permission from Bonita Ely.

fifteen years at Chisholme House near Hawick in Scotland. Addressing the expired "use by date" of his first degree, MacEwan gained postgraduate qualifications in soil science before emigrating with his family from Scotland to Australia where he subsequently worked as a university lecturer and as a research scientist in the Victorian government. Regarding himself as an "accidental soil scientist," MacEwan has maintained his interests in the broader questions of our relationship to nature and ways of communicating this through science and art.

Richard MacEwan and Bonita Ely met at Bonita's studio in Sydney in September 2014 to talk about their experiences with and definitions of the soil. They toured Ely's exhibition at the Gunnery Artspace and began to chart out common ground to develop a narrative that could be relevant for others. Bonita sent Richard many pictures and descriptions of her work and childhood growing up along the Murray River in the State of Victoria, Australia. She even sent a parcel containing soil from her family farm—a rich brown sandy loam from the cultivated paddock, and a red, dense sample from the adjacent roadside. MacEwan admits to repressing his initial compulsion to subject the samples to analysis. Instead he provided responses to Ely's story, interspersing and interrupting that narrative in a constructive way to add scientific detail to her vision of a unique and changing landscape. MacEwan's research of postwar archives reveals assessments of the fertility of soils where Bonita grew up, in a place called Robinvale, in the Mallee of the Murray River. Many World War One Soldier Settlement Schemes were disastrous failures, so after World War II when veterans, including Ely's father, were granted properties to grow grapes, rigor was applied to assess the soil fertility for agricultural schemes and the applicants' agricultural experience. Ely's immersion as a child in this iconic Australian landscape, its cultural significances and ecology, profoundly influenced her art practice.

Richard MacEwan: As a pedologist, soil is a big part of my life. I study soil as a natural body, as an element of the landscape. I study how it differs according to place and what this means in terms its productivity for human use and its susceptibility to degradation. My interest in soil began in the final year of my ecology degree in Edinburgh in 1972, so I am very focused on the ecological relationships in soil systems and our place as humans in that ecological nexus.

Soil Map of the Robinvale Irrigation Area (Skene 1951). Scan of colored map found in the Agriculture Victoria map archives in Bendigo.

The image is used with permission from the Victorian Government Department of Economic Development, Jobs, Transport and Resources.

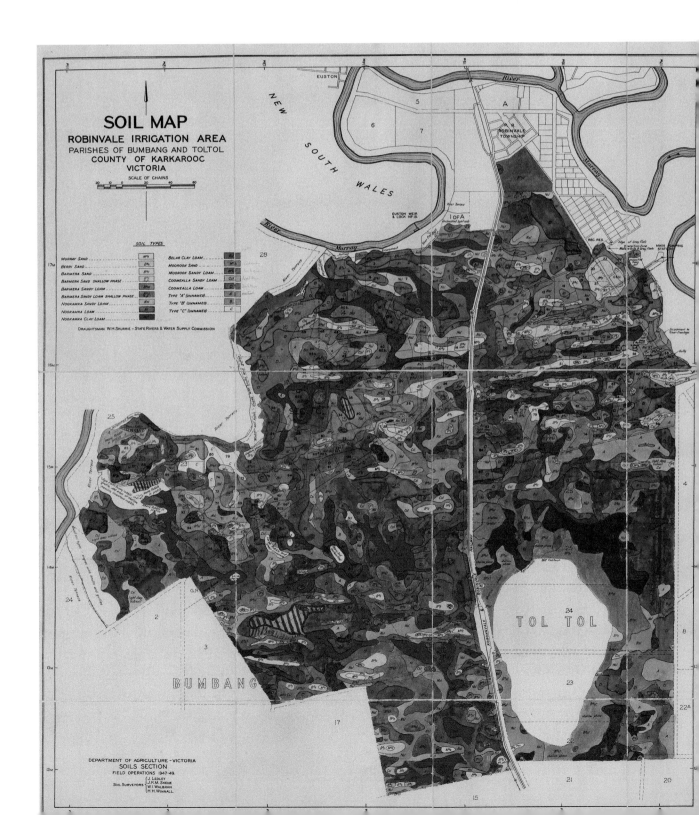

The Robinvale township and surrounding land in 2003 showing irrigated and dryland farming and remnant riparian forest alongside the Murray River, which marks the border between Victoria to the South and New South Wales to the North.

Image reproduced from Google Earth.

The above figure illustrates the extent of irrigated vineyards and citrus and almond orchards near Robinvale and Euston in New South Wales, as well as remaining natural riparian areas along the Murray River.

I am also fascinated about the use of soil in art, whether its occurrence is incidental or intended. I love the variety of color and form that appear naturally in soils, but my aesthetic curiosity is primarily scientific—by observing differences I seek explanations for those differences in form and color, and their implications for soil processes, land management, and agricultural production.

Bonita Ely: Earth, dirt, soil, and the permutations these words conjure have been a focus of my art making since the early days of my practice in the 1970s, when environmental issues were emerging in my consciousness. But perhaps my interest in soil developed much earlier. I was brought up in the Mallee[1] on a Soldier Settlement "fruit block" on the Murray River in the State of Victoria. Immediately after the Second World War, my parents applied to the Soldier Settlement Commission[2] for a property. Along with hundreds of other veterans' families, they were offered thirty-two acres of undeveloped land on which to grow grapes and oranges at a place called Robinvale.[3]

Richard MacEwan: In a soil survey of 14,000 acres in the Robinvale irrigation district carried out from 1947 to 1949 by the Victorian Department of Agriculture, approximately 5500 acres of soils were identified as suitable for growing vines and citrus fruit with irrigation from the Murray River. Twelve different soil types were mapped. The landscape there consists of east–west ridges of dunes and swales. The geology deposits are Pleistocene sands, frequently calcareous, and of lacustrine, fluviatile, and aeolian origin (Skene, 1951).[4] Soil sodicity and salinity are common in this part of Victoria and the survey identified areas with high salinization hazards, excluding them from irrigation development. Drainage was advocated but, in most cases, blocks were developed quickly without drainage that was expected to be installed at a later date. Most of the land had already been cleared of native vegetation and transformed for cropping and pasture.

Bonita Ely: Citrus had been grown successfully in the area before, and dependable water from the Murray River, the world's third largest river after the Amazon and Nile, had been made more "drought proof" in the 1930s with the installation of the Euston Robinvale Weir. The weir ensured the success of the settlement, but the real guarantee of success was the extraordinary effort invested by the "blockies," as the veterans came to be known, and their partners who "worked like men" to establish not just their livelihood, but a township—including a hospital and ambulance service, sporting facilities, garbage disposal, girl guides and boy scouts, the Country Women's Association (CWA), the Returned Soldiers League (RSL), and the ubiquitous Scottish Pipe Band.

Our block, 2B, was close to town. It was a trapezoid-shaped plot stretching broadly up a gentle hill where underground pipes linked back to a huge (in a child's eyes) pumping station on the river to feed the thirsty vines and citrus trees. At school, we were taught the average annual rainfall in the Mallee was 31 centimeters—a semidesert climate. The soil was a compact, red sandy loam.

Richard MacEwan: I found the original hand-colored map for the 1947–49 soil survey in our Victorian government soil map archive in Bendigo, and I can see five soil types on your block. A band of Nookamka Sandy Loam dominates east–west across the middle of the block, running into Nookamka Loam on the lower land (north) toward the road. Belar Clay Loam (referring to the vegetation of Belar trees), Barmera Sand and Sandy Loam and Berri Sand are minor components. Bonita, I wonder if you remember any problems with soils on the block?

Bonita Ely: Yes, there was a flat section of land left fallow adjacent to the road because, as I recall, the soil was of poor quality. Many of the blocks had fallow land like this. Long, white spear grass with clumps of hop bush and Boronia grew there, indicating what the countryside was like before it was irrigated. Now every square inch of land is cultivated so there must have been a solution found since then to fertilize poor soil.

Richard MacEwan: I'm guessing that fallow land would be some of the Nookamka Loam. There are two houses and lots of shedding in that part of the block now.

Bonita Ely: Yes, the block is different now, the owners maximizing production in ways that were not possible for veterans as there were restrictions applied by the Soldier Settlement Commission, plus agricultural methods have improved. The original 32-acre blocks have also been bought up and merged to improve financial viability. The large sheds pictured contain a cooling storage and working spaces for preparing fresh fruit for city markets, rather than producing dried fruit. My parents retired in the early 1970s and an Italian family bought the block. Amongst other nationalities, Italians and Greek migrants provided labor in the '50s and '60s and then stayed on—the Mediterranean climate and agriculture

Block 2B in 2003 showing vine plantings.

Image reproduced from Google Earth.

Dethridge wheels, now phased out of use. The photograph was taken near the Robinvale Pump Station.

Photo: Reprinted with permission from Bonita Ely, 2009.

The irrigation pipelines that deliver water to the blocks from the Robinvale Pump Station.

Photo: Reprinted with permission from Bonita Ely, 2009.

were very familiar. Vietnamese refugees, people from all over, have settled in Robinvale's multicultural community.

Bonita Ely: Originally each block had a water wheel to individually measure the water used. Water poured first into a central open chamber where we would soak up to our necks, fully clothed on stinking hot summer days. I wonder if our parents knew …

Richard MacEwan: The wheel was invented by John Dethridge in Australia in 1910. Dethridge was then commissioner of the Victorian State Rivers and Water Supply Commission. The wheel consists of a drum around an axle with four spokes originating from each end of the axle. Eight v-shaped vanes are fixed to the outside of the drum, which then spins around. The revolving wheel measures the flow of water from the irrigation supply channels into the farm channels. This provided the basis upon which farmers were charged for irrigation water but is progressively being scrapped under water reforms.

Bonita Ely: The true wheel turning the fruit blocks was the continuous engagement of the community

itself. In the settlement, the cooperative discipline of army training kicked in. Neighbors helped each other in work gangs to establish the labor-intensive infrastructure of each block. The first job was to plough and plant the grape vine cuttings. Then the irrigation system and straight, measured rows of posts and wire were put in as climbing support for the vines. Initially watering was done by hand from a water truck driven down the rows. Meanwhile everyone lived in tents with no shade, no stoves, and no amenities through the summer's forty-degree heat waves and dust storms, and then the icy frosts of winter. Later, a tin hut replaced tents until well-designed weatherboard houses were built for the families.

On summer evenings, the adults partied outside into the night beside huge bonfires with a beer barrel on tap and delicious home cooking. They organized community events such as dances, raffles, and picnics to raise money to improve the town's facilities. Children played outside all day long. In the springtime, we picked wild flowers for my mother in the scrub—Boronia, Hop Bush, paper daisies, wild orchids and violets, onion weed, and wattle. We fed our dolls mud pies and built cubby houses. We popped paddy melons, picked delicious wild mushrooms growing in the Mallee scrub's mulch. We engraved big drawings and hop scotch grids into the packed earth of the backyard. At the end of hot summer days, the family drove down to the river on the Fergie tractor[5] and trailer for dinner and a swim, returning in the dark, kids half asleep watching stars sparkle up in the black sky. My father swam me across the river on his back and taught me to swim at the sandbar we called St. Kilda Beach.

The track to the sandbar switched at the top of a steep incline from red dirt to the gray clay of the riparian environment. We passed the pumping station, followed the river, then cut across the flood plain through the soft tones of scrub, past a huge river red gum tree with a canoe scar inscribed the length of its trunk. At school, we learned the boat was made by the Latje Latje people—the people who walked along the railway line into town, worked on the blocks, trapped rabbits (as we all did), fished, hunted, and followed their traditional ways. "Under the wire," they lived among the black box and lignum on the river flats in humpies made from corrugated iron and hessian bags salvaged from the rubbish heap, away from the eyes of the Aboriginal Protection Board. Later, huts were built for them away from the river flats, out of town, out of sight, behind the railway line. Later still they were given permission to live in town and this substandard, segregated housing was abandoned.

Richard MacEwan: I think, Bonita, that you had a much closer relationship to the natural landscape than I did growing up. I spent my childhood days growing up in an urban environment in the English cities of Coventry, Leeds, and Loughborough, playing in the road, on old bombed-out land and, later, in Beech and Sycamore woods near our house in Yorkshire. The woodlands were formative for me and I spent a lot of time up trees thinking about life and the adventures I would have. My love of

nature began there and I variously went through fossil collecting, entomology, and botany to end up in Edinburgh studying ecology. At the end of my ecology degree I didn't know very much about anything except that it seemed like there was a big mess out there, which was as much a spiritual crisis as an ecological one. I was feeling very righteous about it too—ready to defend the planet against pollution. Books like Ian McHarg's *Design with Nature* and the writings of Buckminster Fuller and Loren Eiseley inspired me to look deeper into the science I was exposed to in my ecology degree program. Edward Goldsmith began publishing *The Ecologist* in 1970 and pollution was the big issue, not soils, though "acid rain" became a conceptual link. Ironically, reduction in air pollution had a negative effect on some soils, which subsequently became sulfur deficient resulting in lower agricultural production (Prince and Ross, 1972).[6]

Around the time that I graduated in 1973, I visited an exhibition of Mark Boyle's work *Journey to the Surface of the Earth*[7] at a gallery in Glasgow and that made a huge impression on me. After looking at his pieces, every detail of every surface outside the gallery became an aesthetic object for me. I actually considered the possibility of going to art school to teach ecology! I ended up in the Borders of Scotland, helping renovate and manage an estate, and studying "mystical" literature consisting largely of Sufi texts. Thirteen years later I returned to the University of Reading to study pedology, where I came across the "art" of soil peels—thin layers of soil glued to fabric. These are not easy to make but are beautiful pieces in their own right. In 1986, the Boyle family[8] had an exhibition at the Hayward Gallery in London where I asked Mark Boyle how the family made such wonderful representations of the earth's surface, but he would not reveal much of their secrets only saying, "I have always admired the pedologists' peels." In 1988 I came to Australia, which seemed to need soil scientists at that time, so I became one, letting my artistic interests, mystical pursuits, and ecological aspirations take on a more hidden role in my life.

I wonder, Bonita, what was your trajectory as an artist like? Did your work always have a focus on the environment, developed from your experiences growing up on the Murray?

Bonita Ely: I compulsively drew and painted all through my childhood and adolescence, then studied art in Melbourne. In 1969 in London, the class system was firmly in place, pollution hovered, rubbish accumulated, rock 'n' roll reigned, rain drizzled, and white snow was soiled brown. One of my earliest sculptures, *The World is My Ashtray*, says it all. In New York, in 1974, I again encountered a soiled Earth—dirty purple-brown horizon, blurred red sun, and more pollution. I created an installation there called *C. 20th Mythological Beasts: At Home with the Locust People*.

On my return to Australia, I became aware of environmental matters closer to home with reports in the 1970s of increasing levels of salinity in the Murray River watershed caused by irrigation and deforestation. Both factors raised the level of subterranean groundwater, drawing up sea salt from the ancient seabed that underlies the Mallee's land mass. The situation affected me on a personal level that greatly informed my work as an artist.

Richard, were you aware about the situation of salinization on the Mallee?

Richard MacEwan: Salinity in the Murray-Darling Basin was the major issue when I arrived in 1988 and I have seen many additions to knowledge of the system, its interaction with land management, and improved advice to managers since that time. My initial research in Victoria was focused on drainage for agriculture, so salt and nutrients in drainage water was also a concern of mine.[9]

According to hydrogeologists, much of the salt from the ancient sea has long since left the Murray River basin, and the salt in the system now has been accumulating over millennia from small amounts added in rainfall.[10] Because the underlying geology of the basin is a relatively closed system, groundwater cannot escape except via base flow into the Murray River. Groundwater rise in the basin is a natural consequence of increased recharge, due both to tree removal and excess leakage from irrigation. As the river progresses through Victoria, New South Wales and South Australia, water is extracted for irrigation on the neighboring land such as the developments at Robinvale, so the river becomes increasingly saline downstream due to these irrigation withdrawals and intruding salty groundwater. This has led to severe salinity problems for the agricultural land and the river ecology.

We have some monumentally important groundwater systems and saline discharge areas in Victoria, with several times the salt concentration of seawater. Lake Tyrell in the Mallee region is world famous in this respect. Phillip Macumber published a comprehensive report on Tyrell and related groundwater systems and salinity in Northern Victoria, which is well worth reading.[11] Water reforms, particularly the ability to transfer water rights, and the creation of piped rather than open-channel infrastructure are doing much to alleviate the problems. There is also a much better understanding of water requirements for crops and pastures, and pricing and scarcity have encouraged more efficient irrigation practices.

Bonita Ely: In 1977, I began a series of artworks to address the river's plight, often using soil as subject and medium. First, I took a field trip to photograph the river's changing terrain from the mountains to its estuary. The Murray begins in Mount Kosciuszko National Park. Fed by

The Murray River (1979), handmade paper, soil, etchings. Approx. 4.5 m W × 1.5 D. Collection, Queensland Art Gallery.

Reprinted with permission from Bonita Ely.

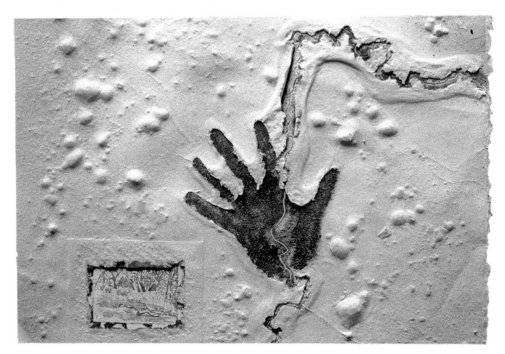

Detail: The Murray River (1979), handmade paper, soil, etchings. Approx. 4.5 m W × 1.5 D. Collection, Queensland Art Gallery.

Reprinted with permission from Bonita Ely.

springs and creeks, it gathers momentum then slowly winds its way 2520 km westward across flat plains,[12] fed by tributaries, forming lakes, anabranches, and billabongs. In Morgan, South Australia, it turns south through spectacular cliffs of calcareous rock full of fossils, formed over a period of 130 million years, then flows on into Lake Alexandrina, Lake Albert, the Coorong, and then out to sea.[13]

The floor piece, *River* (1979) tracks the Murray from mountain to sea, connecting it to the passage of our lives. At the places documented on the field trip, imprints of hands across the river's embossed line begin with a tiny baby's, aging through time and distance to a skeleton at the estuary. The hands also quote hands often stenciled in Aboriginal rock art, referring to the Aboriginal heritage along the river. Beside each handprint is an etching the size of a postcard showing the landscape at each place.

This artwork developed from my cartographic examination of the changing geography of the river that I witnessed on the 1977 field trip. Beginning in the foothills near Corryong, where the river is a fast-flowing stream over polished granite rocks,[14] I made a grid of four sections along the river's edge. The rectangles correspond with the dimension of a camera's viewfinder. I photographed the sections as if for mapmaking, mimicking the process of aerial photography. This process was repeated in the Barmah Forest, where the river's edge is a fine, silky gray clay. Next, at the junction of the Murray and Murrumbidgee Rivers, the river sweeps around in a wide arc, depositing a beach of fine, white sand over millennia. Near Swan Reach, the river's edge

is a lacework of pock-marked, eroded rocks at the foot of huge cliffs. The sandy edge of Lake Alexandrina was the last place documented.

Many artworks evolved from this material over the years. I repeated the cartographic process as a comparison at the same locations thirty years later in 2007, documenting the effects of an extended drought. I also documented the tiny soak in Mount Kosciuszko National Park at the official headwaters of the Murray, a dried-up Lake Boga near Swan Hill, and a billabong called Bottle Bend near Mildura, where the water had turned to acid sulfate. I contextualized the grids with extensive photographs of these and other key locations, such as blooms of blue-green algae in Mildura.

Richard MacEwan: I love this piece of work, and your practice of performed, or lived, cartography. I think more people could try this approach to become aware of changes in the landscape. It reminds me of a time in 1960–61 when I followed a brook in Leicestershire that flowed into the

Boundary Bend: Grid,
2007, 44 cm H × 29.5 cm W.

Reprinted with permission from Bonita Ely.

River Soar. I followed it to its source, observing birds, trees, water creatures, and the fields that it passed through. Most, if not all of that land, would be under housing now and the brook in a culvert. The character of the landscape everywhere is changing rapidly under anthropogenic influences, but also because of changes in climate, which are also humanly driven.

The Murray River, it is well known, is an overused water resource with preference of supply for human use. The lack of rain in our "decade of drought" ending (pausing?) in 2010 has brought extra pressures on the river and its riparian ecology. The occurrence of acid sulfate conditions is a result of the river bed drying, causing oxidation of iron sulfides and the formation of sulfuric acid.[15] With so much irrigated land and fertilization, the eutrophication of the waters in the Murray is accelerated and results in algal blooms.

Bonita Ely: In 1980, these concerns were dramatized in a performance called *Murray River Punch*, a cooking demonstration using the pollutants from the river as the recipe's ingredients to make a drink. The performance was reprised in 2010, renamed *Murray River Punch: The C20th*. This time the recipe was a sticky paste as there was so little water in the river after the Ten-Year Drought. In 2014, it was performed again as *Murray River Punch: The Soup*, a parody of cooking shows once again, mocking our disregard for nature. Plastic bags dripping clay from the riverbed mixed with cigarette butts, bong water, a sliced-up Coca Cola can—every ingredient was found littering a "picnic area" in Mildura.

Richard MacEwan: I do like your punch analogy. Humor is so important in getting serious messages across, and lampooning TV cooking shows can be a lot of fun. My colleague Mark Imhof and I created a parody of *Iron Chef* for our research group at a social event. It was called "Iron Pedologist," in which we commentated as two of our scientist colleagues were challenged to make an interesting main course and dessert from soil materials using a mix of laboratory instruments

Murray River Punch: The Soup, performance with Emma Price (2014), Brunswick Street Gallery, Melbourne.

Reprinted with permission from Margaret Bell-Winford.

and cookware. The hydrogeologist challenger to our "Iron Pedologist" contest was the winner. The commentary included discussions on soil texture, color, aggregate stability and dispersion testing, pH, hydro-pedology, and soil layering. It had a loose educational component.

Bonita Ely: It's good to know that scientists also sometimes use humor in their work.

In his publication, *Sapiens: A Brief History of Humankind*, Y.N. Harari points out we are animals like any other. He says, "Like it or not, we are members of a particularly large and noisy family called the great apes."[16]

We are great apes whose intellect, inventiveness, greed and ambitions have outweighed our insights and sensibilities when weighing up priorities associated with nurturing, regenerating, and preserving the planet's ecologies. My art practice aims to communicate that we do not and cannot rule the Earth. We are, like all other life forms, Earthlings dependent on its bounty.

Richard MacEwan: Collectively and politically it is true that "intellect, inventiveness, greed and ambitions" have compromised ecology, but individually we all make our own choices. Acknowledging our dependence is a key to making good choices. In the end, I believe the processes of nature will dominate, however much we may try to dominate nature. Understanding these processes is fundamental for the translation of science into practice for the management of soil and land. Recognition of sacred and spiritual values conferred on the landscape through different cultures adds another dimension that we need to consider. There are so many examples where we have done the opposite and created nothing but mess. It is time to change all that—we have the knowledge, but economic pressures and politics have largely conspired to keep society on a rather

destructive track. Fortunately, there are exceptions and raising awareness through a combination of science and art is a vital counterpoint.

Richard MacEwan (postscript): This dialogue with Bonita as well as reflection on works by Elvira Wersche[17] presented as a poster in Jeju Korea in 2014 at the WCSS, inspired me to think about engaging the public in art and soil through painting and mapping. In preparation for World Soil Day in 2015, the International Year of Soils, I encouraged my colleagues to help with collecting soil samples from Victoria to place on a 1:250,000 scale soil map. I saw this map installation as an analogue of "people and place" visualized though "soil and landscape" and as a potential way to engage people in thinking about these differences in constructive ways. The result was to be a colorful and textural mosaic, representing the soils and land of Victoria but using the actual soil, carrying the spirit of the land and whatever meaning it has for people culturally, economically, and spiritually. The samples, previously separated by tens or hundreds of kilometers, were brought close together, crowded for space at a vastly reduced scale, compared to the extensive country where they formed and where they may have rested for thousands of years. Where we had no sample for a grid square, neighboring samples were brushed into the voids rather like kriging soil data to provide complete cover from separate data points. The subtle or gross differences in color, texture, and structure of the samples are unified in a final installation, and it is an analogue of our own diverse communities that can engender reflection on our cultural differences in unique ecological spaces. Contributing soil to an installation like this unites all participants in an exercise of adjacency, a community where distance is compressed into a space in which knowledge and ideas can be more readily exchanged.

Soil samples placed on a 10-by-10 km grid of a soil map of Victoria at 1:250,000 scale. Federation Square, Melbourne, World Soil Day 5 December 2015.

Photo: Richard MacEwan 2015.

Endnotes

1. The Mallee district is in North West Victoria. http://www.murrayriver.com.au/explore-the-mallee/.

2. http://researchdata.ands.org.au/soldier-settlement-commission/148021.

3. http://www.abc.net.au/landline/content/2010/s2882249.htm.

4. http://vro.depi.vic.gov.au/dpi/vro/malregn.nsf/pages/soil_robinvale_irrigation.

5. Ferguson tractor.

6. Prince, R. and Ross, F.F. 1972. Sulphur in air and soil. *Water, Air and Soil Pollution* 1(3), 286–302.

7. Boyle, M. 1970. *Journey to the Surface of the Earth: Mark Boyle's Atlas and Manual*. Hansjörg Mayer, Cologne.

8. http://www.boylefamily.co.uk/boyle/about/index.html.

9. MacEwan, R. J., Gardner, W. K., et al. 1992. Tile and mole drainage for control of waterlogging in duplex soils of south-eastern Australia. *Australian Journal of Experimental Agriculture* 32(7), 865–878.

10. House of Representatives Standing Committee on Science and Innovation Science. 2004. *Overcoming Salinity: Coordinating and Extending the Science to Address the Nation's Salinity Problem*. Commonwealth of Australia, Canberra.

11. Macumber, P.G. 1991. *Interactions between Groundwater and Surface Systems in Northern Victoria*. Department of Conservation and Environment, Victoria, Australia.

12. The Euston/Robinvale weir is only 47.6 m above sea level, less than the length of an Olympic swimming pool tipped on its end, yet the distance across country to its mouth at Goolwa into the Southern Sea is about 400 km, but approximately 1050 km by river.

13. http://www.murrayriver.com.au/about-the-murray/how-the-murray-river-was-formed/.

14. http://dbforms.ga.gov.au/pls/www/geodx.strat_units.sch_full?wher=stratno=4755.

15. http://www.mdba.gov.au/what-we-do/research-investigations/acid-sulfate-soils-risk-assessment.

16. Harari, Y. N. 2014. *Sapiens: A Brief History of Humankind*. Harvel Secker, G.B. P 5.

17. www.elvirawersche.com; see Chapter 14 in this volume by Elvira Wersche.

Yield

Matthew Moore and Brent Clothier in conversation with Alexandra R. Toland

Matthew Moore is a multimedia artist and fourth-generation farmer based in Phoenix, Arizona. His video and installation artworks are exhibited internationally. He lectures on art and agriculture across the country from San Francisco to New York, and most recently appeared at the 2014 TEDx Manhattan. His work has been exhibited at the Phoenix Art Museum, the Sundance Film Festival, the Walker Art Center, MassMoCA, the World Congress of Soil Science in Korea, and Nuit Blanche in Canada, among many other venues. He has been featured in publications including Art Forum, Art in America, Art Lies, Metropolis, Dwell, and Architecture. Moore's art practice explores the broad issue of place making, often by creating large-scale installations and environments with the goal of achieving a state of wonderment, contemplation, and invitation to change. For Moore, "farming is an endeavor

Title image: Te Whau Vineyard on Waiheke Island in New Zealand.

Photographer: Briar Shaw. Reprinted with permission from the New Zealand Institute of Plant and Food Research Ltd.

of questioning and exploration, the same negotiations inherent in the practice of making art. For me, making art is creating questioning observations of how we live, and illuminating those choices to incite new discourse on what it is to be human." (See artist statement: http://matthewmoore.com/about-my-work/.)

Brent Clothier is a principal scientist with Plant & Food Research in Palmerston North, New Zealand. He is a world-renown soil scientist, who was awarded the Don & Betty Kirkham Soil Physics Award of the Soil Science Society of America in 2000, the J.A Prescott Medal by Soil Science Australia in 2001, and the L.I. Grange Medal by the New Zealand Society of Soil Science in 2014. Brent is editor-in-chief of the international journal *Agricultural Water Management* and has published over 300 scientific papers on the movement and fate of water, carbon, and chemicals in the root zones of primary production systems, irrigation allocation and water management, plus sustainable vineyard and orchard practices, including adaptation strategies in the face of climate change. Clothier has been, or is still involved in various water-related aid and development projects in the Pacific and Indian Oceans, as well as in the Middle East and Africa.

The following dialogue chapter was developed from an interview with Matthew Moore from October 2013 by Alexandra R. Toland. For the purpose of this book, new questions were added and addressed to Brent Clothier, who, with the insight and experience of a soil scientist, reflects on many of the topics of scale and social pressures of farming brought up by Moore. Since the time of the interview, the digital farm collective Moore speaks about has evolved into a farm-to-table hospitality business that partners with city parks. The group has created one- to two-acre urban farming education centers in and around Phoenix, Arizona, with restaurants that pay for these activities and deliver meals at an affordable price for local communities. In both Moore's and Clothier's work, the question of scale is key—under the pressures of climate change and increased urbanization, innovative solutions are needed to "scale-up" food production and create systems of value that honor the services of the soil rather than its speculative yield.

Alexandra R. Toland: Matthew, you've talked about YIELD as a fundamental value of the American dream, referring to profits gained from farming but also to suburban encroachment on agricultural land for even higher profits. The very concept of yield seems to be inherently at odds with soil conservation. Can you talk a little bit about the culture of *soil conservation* from your perspective as a farmer but also from your perspective as an artist?

Matthew Moore: So, first I want to clarify what soil conservation means for me, as it means two different things: one as a farmer and one as an artist. As a farmer, soil conservation is usually related to policies and practices that we have here in the United States, which are defined and regulated by the NRCS (U.S. Department of Agriculture's National Resources Conservation Service). There's been a lot of knowledge developed in, for example, saving topsoil so farmers can produce sustainably, and through best practices like rotations, and all sorts of technical stuff that we do in growing to make sure that we're still able to grow in another five years, let alone a hundred years.

And to be clear, my farming practice has always been about vegetables and staple crops, like cotton or corn. "Futures crop" is what we call them out here. So, my relationship to soil and soil conservation—as a farmer—is different from a dairyman or whatnot. I've grown conventionally and I've grown organically. I've done CSAs [Community Supported Agriculture], direct market farming, all in an effort, really, to know what it means to be a farmer in the contemporary society that we live in today. And it was because of my *artistic practice*, trying to discover the meaning of soil itself and my relationship to it, that I realized I don't really know what it means to be a farmer anymore, that there is a disconnect between practice and that knee-jerk reaction of what it means to lose soil, to lose your own land.

You see, farming is a risky business. It becomes more and more complicated the longer you do it. And then you have kids and then they're connected. And then all of a sudden, you're 89 years old and a guy comes up to you and offers you a price for your land that is more than you've ever made out of every year of yield. *What* do you do? You have grandkids, you have great-grandkids, sons and daughters. And you have *one* grandson who came back to run the farm. There's this inevitable problem with the way we value land. And the thing with contemporary farming is, it's always a measure of investment. So, we go out and we tame a landscape. Then we plow it down and make it flat. We make it produce. And then at a certain point, just like a bay or a port, people are drawn to that economic activity and a sort of evolution of growth ends up taking over farmland. It's a perpetual process. It's just part of the Western economic structure that we live in.

As an artist, on the other hand, I've discovered the idea of conservation is a little bit more complex. For me, it means access to knowledge of what soil is and what soil means for humanity in terms of our connectivity and sense of place. It's about the raw knowledge of how we grow our food. So, as an artist I think about soil conservation on a conceptual level. I think it's about confronting that loss of knowledge but also a loss of identity, a loss of history, a loss of connection to the soil, of cultural history that is threatened not only by suburban encroachment but also by global climate change. Losing soil is losing an essence. It's like a cut that you can't seal up, and you're losing your lifeblood. So, it's important to talk about best practices and what not, but I think soil conservation also goes to a gut level much deeper than all the policies that we see in farm bills and NRCS practices, and as an artist I can speak on that level.

Alexandra R. Toland: What do you think about worldwide sustainability movements or the

Landcare initiatives in Australia and New Zealand, or the kind of grass-farming operations that have been popularized by award-winning author Michael Pollan and the locavore hero Joel Salatin? Do you think those kinds of holistic, locally based practices are becoming more accepted in society? Do you see your own role as a farmer, which is made public by your work as an artist, as a way of endorsing these kinds of practices? Do you even see yourself as a kind of visual activist for a new kind of farming?

Matthew Moore: Well, I'm a pragmatist. I appreciate Michal Pollan's work, but I feel like there's a false sense of security because of all the attention that we see given to certain movements in film and television documentaries. It's as if we have the farming culture licked, you know, and we're on the way to recovery. I mean, I've given lectures in Brooklyn, far removed from the flyover parts of America where a lot of those seemingly accepted ideas about nutrition and sustainable agriculture are not easily accepted, no matter how logical they are. And it feels like CSA-based locavorism is sometimes like that really cool leather jacket that you wore in the eighties, that you can kind of think back on, like, "I remember the time I did farming, you know, I was a farmer for a little while."

I'm way more of a Wendell Berry kind of guy. One of his most powerful statements, that sort of changed my life, is the idea that you have

to ask the land what it wants to do. And we've been telling the land, and I've been telling the land, for the longest time what it should do. So, we can argue all we want about how organic is better than conventional, but the argument has become so polarized at this point that it defeats the discussion of what we really need, and what the land really needs. And as a conventional farmer looking at population growth, I know I can't feed the world with a CSA. The demands of society push you into unsustainable growth and into business places that you're not capable of dealing with on your own. But that's where I believe art can be a practice of talking to and meeting people and building communities. I think we can be a source, hopefully, for change. In many ways, artists are problem solvers. And that's where science and art can come together … to better understand how we can function as a society, and to help move ourselves forward.

In terms of some of the art projects that I do, I think those activist discussions are better won by other people. And there's good work to be done on that, but from my perspective, it's much more about trying to *do* something with the knowledge I've gained over the years from looking at different problems, and talking to people, and traveling around. For me, the scariest thing is this sort of multigenerational knowledge that disappears, and that's what I'm trying to address in this nonprofit project called the "Digital Farm Collective". As a farmer, you grow a certain set of crops, and you can grow them your whole life. So, you get to know your crops, and you get to know your land. You get to know where it's weak. It has its own personality. You understand what to look for, and that takes many years to develop and then you pass that on. The imperatives of the Digital Farm Collective are very basic. It doesn't get into any of those controversial discussions, but simply documents the phenomenon of plants growing. It puts them on a time scale that relates to human beings—using digital technology to visualize that our food grows in the ground and that the sun moves across the sky. I mean, that's how basic it is. We're at ground zero.

Alexandra R. Toland: Can you talk about the kind of knowledge transfer that you engage in as an artist? Beyond the wisdom passed on to you by your parents and grandparents, you have acquired a hands-on knowledge from working both in large-scale commercial agriculture as well as smaller scale CSAs. You describe yourself as a pragmatist, as a problem solver, so I wonder how this kind of practical knowledge as a farmer has influenced your work—your methodologies—as an artist, and maybe vice versa, how has your creativity and intuition as an artist influenced your farming practice?

Matthew Moore: I guess the most important thing to me is that I'm constantly put in a place where I'm forced to have to adjust and to learn and to meet and destroy my own assumptions, because I have an infinite amount of them. And that's what you do in farming every day. You wake up and you have no idea what you're gonna be presented with and you have to just figure it out. It might be a flat tire. You might not have the money to buy a new tire, so you have to learn how to pack a tire. Or you might have some massive fungus that's attacking a field that you've never been seen before. And so, you go to your community to figure out what it is and apply that knowledge however you can. And that's a lot like being an artist. We have to go out and figure out how we're gonna communicate the best way we can to be successful, to be able to wake up the next day and have another idea and another project. So, they're really intertwined for me.

On a personal level, I'm using my art to discover what it is like to think about loss … I have a two-year-old son [at the time of original interview he was 21 months; now I have two

children, aged four and six], and when he grows up he's not gonna be able to go out on that land. And that tears me up. There's an importance of connectivity there that I try and use my artwork to figure out, as a tool to inform a greater public on why the decisions we are making as a society are shortsighted. There are water shortages and growing populations, and we need to reconcile so much to move forward rather than waiting for disaster and then recovery. So, my art gives me a sort of one-step backward … a lens to look through reality. And I use it completely as a way to reveal the realities of social and economic processes. In this way, it's not merely about agriculture. It's about land use and ecology on a greater scale, and about a different way of relating to soil, land ownership, and all the assumptions we have around those things, especially as Americans.

Another thing I've learned is that art is a business. A lot of the counterculture or punk rock stuff I identified about being an artist a long time ago has fallen away. Art is more of a way of thinking, or a life mission, and in order to make it a sustainable endeavor; it's a lot like farming. For example, having five acres and doing a CSA is a small world … it's this really beautiful connective community experience, but it needs to exist in a thirty-mile radius to be sustainable. It has a manageable scale. But once you get into about 100 or 150 shares, your scale needs to make a business switch. The same way that artists are really unsuccessful sometimes in business endeavors, small farmers often can't make that jump either. So, there's this growth barrier that happens where you hit the scale of having to expand. I mean this is true for any business, but I find art and farming to be very similar. At some point, you have to start asking for help, and then you have to start paying for help, and then you have to negotiate getting bigger to be able to subsidize this sort of *scaling up*. But at any scale, you're providing a service and you're doing a good for the community, is how I look at it.

Matthew Moore, Digital Farm Collective. Planting squash seed; lettuce growing; time lapse units in farm; Lifecycles screening at Nuit Blanche, 2012.

And farmers, we're a bunch of gamblers. We'll have seed every year and we look at our land and look at our seed and think "I'm gonna *kill* it this year." And then everything is just a reduction from yield. The first time you plant, you may have eighty-percent germination, and then you have a storm, and then you just start subtracting, and it's just like a terrible gamble in a lot of senses. And for me, farming is complicated because of the land scale and the manner in which I have to manage the farm with my grandfather still around marketing the land for the greater family of forty people. So that sense of levity—I don't really know if it exists on that sort of scale anywhere else in the world—that scale of time and space in contemporary agriculture. It's just a rough game to play.

Alexandra R. Toland: We've talked about yield and loss, scale and social pressures. Finally, I was hoping you could give me your personal definition of soil. What is that stuff we grow our food and hopes and dreams in? What does soil mean to you at a very personal level?

Matthew Moore: My experiences and realizations as a farmer greatly inform my ideas about what soil is. There's a saying in farming that once you get the dirt under your fingernails you can't get it out. That means that, you know, there is always this sort of return that happens. There is this cycle of hopes and dreams, the birth and death of every plant, every season, every year. And then there is the work and a life of grit that becomes your identity through all of the trials and tribulations. Man … If I think about it, it's honestly everything. It's my entire life, from the smell of it to the sight of it … And I want my son to be able to know something like that, to participate in that cycle. Sorry, I'm not giving you the best definition … For me, soil represents everything about not only our conceptual relationships but our true physical relationships

to the earth. It's a vehicle. It's identity. It's home. It's prosperity. It's hope. It's why we're here. It's why we can exist here, *and* it's why we're gonna leave.

Alexandra R. Toland: Brent, you have quantified things like terroir, in an attempt to evaluate the particular qualities of place and practice that give foodstuffs their unique character and create identity for the people that grow and consume them. Could you talk about the value of soil conservation "on a gut level," as Matthew suggests? Is it enough to quantify the "cultural services" of the soil, such as terroir, or are qualitative, narrative, and even visual methods needed to reach people at their place of identity and personal essence? When we conceptualize yield not only in terms of kilograms of grain per hectare but as real estate units for sale, how can environmental measurement and monitoring ease the sense of loss felt by so many farmers today in order to actually inspire change in policy?

Brent Clothier: Terroir has a specific meaning applied to grapes and viticulture. Simply, it can be quantified by the price differential for a bottle of wine.

So, it does have a component of "yield" as mentioned by Matthew. You need a certain kg/ha of grapes. But there's also other elements mentioned by Matthew—soil, and the need to conserve the soil over centuries (as in European vineyards), the artistic appeal of plants growing as "hedges" with grassed rows in between, the cultural link of grapes to wine and to history. That's part-and-parcel of terroir.

Also, terroir is the price difference between *vin de maison* and a *grand cru*. Not just yield as Matthew notes. And there are many shades in between. It's not so much about how many kilograms of grapes are harvested per hectare, that is "yield," but rather what is the "value" of those kilograms. We can do a straightforward economic analysis of the provisioning service provided by terroir based on price per bottle. But there's more. There is a less tangible cultural service provided by terroir and that is of aesthetics and recreation. Te Whau Vineyard on Waiheke Island in New Zealand, pictured next, plays to its strengths of visual appeal. [I would also add that the vineyard provides a valuable regulating service of delivering clean water to the streams that flow into the harbor in the background. Those aesthetic and regulating ecosystem services are, as yet, difficult to value economically.] Another economically importantly cultural service provided by terroir is the nonviticultural use of vineyards, as for example in New Zealand. The presence of fine restaurants and the staging of music concerts. There is a cultural ecosystem–service link between terroir, cuisine, and the recreational enjoyment of music.

So, simply the notion of the yield of grapes per hectare is not a good indicator of the ecosystem services, from provisioning, through regulating to cultural provided by a vineyard. This is captured in the notion of terroir—the *je ne sais quoi* value of vineyards, soils, climate, and the mastery of the vintner.

In a general sense, the notion of terroir can be applied beyond viticulture. I am working in the deserts of Abu Dhabi, where since time immemorial, Arabs have been growing dates. Dates were very important in Arab nutrition in yesteryears. They're still a very important part of the diet of modern Arabs, and they are culturally valued very highly. Again, it's not so much about the yield of kilograms of dates per hectare, it's about the variety of dates and how they are treated postharvest before being eaten, and then how many are eaten. An odd number should be eaten at any time. It is considered by some that there is a hadith (report) that the Prophet Muhammad used to break his fast by eating an odd number of dates.

The scientific challenge is nonetheless to understand how much water is needed for their production, in order to maintain the yield of primary production that Matthew talked about. Groundwater in the Emirates is in short supply, and becoming increasingly saline. So, we're carrying out detailed scientific investigations to assess how much water is needed. This work with modern technology does nonetheless provide a cultural service to the scientists, and other observers, as it possesses aesthetic appeal, as shown by my PhD student Ahmed Al-Muaini downloading data. His results and analyses are changing the water policy of the Abu Dhabi Government so that date production can be sustainable … and beautiful. There's the beauty of the sand, and the date palms. Their centuries-long cultural history as Arabic food and sustenance, and then there's Ahmed in his traditional dish-dasha (robe) and keffiyeh (headscarf) attending to twenty-first century electronic equipment. The sun's rays frame this scene beautifully. Indeed, it is this sunlight that drives the yield of dates, and demands the need for irrigation water. Ahmed's research will lead to changes in centuries-old practices to lead to sustainable date farming in the twenty-first century.

Alexandra R. Toland: Brent, what are soil scientists doing to guarantee a handing-down of knowledge to future generations in order to avoid the loss Matthew speaks about? What have you learned from the Landcare movement in Australia and New Zealand or the recent push toward "soil connectivity" from scientific organizations around the world regarding knowledge transfer and food security, or as some like to call it, food sovereignty? And where do you see soil science, and indeed your own practice, as part of that wider conversation on social change?

Dr. Ahmed Al Muaini from Environment Agency – Abu Dhabi downloading information for data loggers in a date-palm experiment at the International Centre for Biosaline Agriculture (ICBA) near Dubai in the United Arab Emirates.

Photographer: Brent Clothier.

Brent Clothier: I was interested that Matthew described himself as a Wendell Berry kind of guy. I'm similar, but I would liken myself more to an Aldo Leopold kind of guy. Leopold wrote *A Sand County Almanac* in which he developed the notion of a land ethic. He noted that "the land ethic simply enlarges the boundaries of the community to include soils, waters. Plants and animals, or collectively: the land." He added that "land, then is not merely soil; it is a fountain of energy flowing through a circuit of soils, plant and animals."

We need to pass on our knowledge of how this "fountain of energy" works to the next generation so that they can maintain the land ethic and develop new ways to sustain this fountain from which flows our livelihoods.

As a senior scientist, I take seriously the role of mentoring emerging scientists and supervising students. As well as Ahmed's photo earlier on, I am involved in supervising another PhD student in Abu Dhabi. Wafa Al Yamani is doing her PhD on how to manage treated sewage effluent for irrigation of amenity forests in the Emirates. These arid forests provide a huge range of valuable ecosystem services: provisioning of wood, regulating of sand blowing, creation of habitat for gazelles and birds, provisioning of aesthetic appeal for all.

Transferring knowledge of the biophysical functioning of the critical zone of our soils and vegetation is important: for conservation, production, regulating services and aesthetic appeal.

I am also working in Africa working with a local company to improve avocado production by the small-holder farmers in the Central Highlands of Kenya to alleviate poverty. We are working to plant new seedlings of the correct variety, improve the management of water and nutrients on farms, and to ensure that the fruit are picked in prime condition to maximize their value. The knowledge we are generating and the measurements we are making are being passed on to help farmers and extension workers from the company.

Knowledge generation is vital to ensure we conserve our soil and water resources for future generations. Passing this knowledge on is vital.

Alexandra R. Toland: Brent, could you talk about scale from an ecosystem services perspective? Has the scale of agriculture—and the productive services of the soil—become too big compared to the scale of regulative, supporting, and cultural services of the same landscapes? What lessons can be learned from the risky business of large-scale agriculture for the "business" of large-scale conservation efforts? And in terms of on-the-ground management but also global sustainable development goals such as food security and climate protection, what are the advantages or disadvantages of approaching soil ecosystem services on a localized, smaller scale?

Brent Clothier: Matthew mentions that farmers are gamblers. As the stakes of climate change and soil degradation rise, there are chances that "farmers as gamblers," and we as consumers, will both lose. So, we need to develop ways to take the risk out of farming, so that rather than gamble—which implies "losers"—we need to develop systems of food production through which we're all winners.

Here's a winner … The systematic framework of ecosystem services enable us to think widely about our environment and its natural capital stocks that we depend on to deliver us valuable provisioning, regulating, supporting, and cultural services. Even if we cannot economically value these and put a price on all of them, we do recognize their value, and it enables us to assess the impacts and trade-offs about making decisions. There is no optimum use of our soils. An optimum is in the eye of the beholder, and there are many beholders. So, we can use an ecosystem service framework to have discussion with the community and stakeholders

Wafa Al Yamani of Environment Agency—Abu Dhabi measuring the stomatal conductance of a Sidr tree in the western desert of Abu Dhabi near Madinat Zayed.

Photographer: Brent Clothier.

about future options and decision making. These discussions might be fraught and challenging, but they will enable us to develop policy and practices to achieve what the community wants. We need to know the upsides of decisions and balance that off with the downsides. And if we see an unacceptable degradation of any one of the services, we can use this to change policy and practice.

The challenges ahead of us are huge. Our water and soil resources are finite and coming under increased pressure from human activities. Meanwhile, climate change will exacerbate these pressures. We need science and knowledge to ensure we make the right decisions about using our lands and waters. The ecosystem system services framework helps us with that.

Alexandra R. Toland: With these challenges in mind, could you give us your definition of soil? What does soil mean to you on a personal and professional level?

Brent Clothier: Like Matthew, I've got dirt under my fingernails that I can't get out. But unlike Matthew I'm not a farmer doing it for profit. Like Matthew, I recognize clearly that soils provide valuable provisioning services—for farmers first and for us as consumers next. We know that, as economists can easily value yield

from our soils. But they also provide valuable and essential supporting and regulating services—often for free.

On a prosaic note, soils have provided me with a career—I'm a soil scientist. I got into it by accident and have no regrets. I love what I do. I love the people that I work with, and those who I have come into contact with. As a soil scientist, I've seen large tracts of land in many countries of our stunning world.

And finally—soils are beautiful. The soil below is a rich mixture of colors and shapes, and it's got layered horizons to add to its visual appeal. It contains a history of how it got there too—layer by layer! The value of the soil's aesthetic ecosystem services is unable to be calculated—it's unknowingly big, as it's massively invaluable to us.

The profile of the Kawhatau soil of the river terraces in the Manawatu province of New Zealand. There are greywacke boulders and gravels at depth, with a friable, brown, silt loam surface horizon.

Photographer: Quentin Christie Reprinted with permission from the New Zealand Society of Soil Science.

On Corn Mothers and Meal Cultures

Ecofeminist Alternatives to Food and Soil Security

Roxanne Swentzell and Parto Teherani-Krönner in conversation
with Alexandra R. Toland

Roxanne Swentzell is an artist, permaculturist, builder, seed collector, author, community organizer, mother, and grandmother based in the Santa Clara Pueblo region of New Mexico. Stemming from a long line of renowned potters and sculptors, Swentzell has devoted her life to making art that reflects the complete spectrum of the human spirit. Swentzell focuses a lot on interpretative female portraits attempting to bring back the balance of power between the male and female, inherently recognized in her own culture and hopes that her expressive characters will help people get back in touch with their surroundings and feelings. Swentzell is well known for her prolific practice as a ceramic and bronze sculptor, having exhibited pieces at the Heard Museum, the Denver Art Museum, the Smithsonian Museum of the American

Title image: Tamales pictured in *The Pueblo Food Experience Cookbook: Whole Food of Our Ancestors*. Roxanne Swentzell and Patricia M. Perea, Museum of New Mexico Press, 2016.

Indian, and many other museums and galleries across the United States. She is the cofounder and president of the nonprofit Flowering Tree Permaculture Institute and teaches courses at the Institute of American Indian Arts in Santa Fe. In an effort to keep local art at the Pueblo and encourage the creativity of young artists she founded the Tower Gallery and Studio and helped establish the Poeh Cultural Center & Museum in Pojoaque, New Mexico.

Parto Teherani-Krönner has worked as a lecturer at the Faculty of Agriculture and Horticulture in the division of Gender & Globalization and is now a guest scientist at the Faculty of Life Sciences at Humboldt University of Berlin. She holds an MA in development and rural sociology and a PhD in environmental sociology. Since the early 1990s she has established women and gender studies in rural areas as a field of study in Germany. Her areas of research are the sociocultural dimensions of sustainable development, the engendering of agricultural policy, and the cultural ecology of meals and food security. She has conducted a number of field studies in Iran, Sudan, and Germany, and has organized within the last decade international summer school programs in Omdurman, Sudan, and Berlin, Germany, with participants from countries in Africa, Asia, and Europe. She is a member of the scientific board of the German Society for Human Ecology and is currently working as a consultant to the research project Diversifying Food Systems: Horticultural Innovations and Learning for Improved Nutrition and Livelihood in East Africa (HORTINLEA).

Discourse on the United Nations' Sustainable Development Goal to end hunger focuses on issues of food security, which has led some scientists to argue for soil security as a fundamental prerequisite. Soil security is defined as "the maintenance and improvement of the world's soil resources to produce food, fibre and freshwater, contribute to energy and climate sustainability, and maintain the biodiversity and the overall protection of the ecosystem."[1] While the maintenance of soil ecosystems is at first an obvious goal, the language of both terms—food security and soil security—is inherently defensive, conjuring up militaristic measures to prevent some kind of assault. What is often missing from the soil and food security debate is reflection on the cultural contexts in which food is grown as well as the agency of those who farm, distribute, and finally cook or otherwise prepare the fruits of the earth. As Parto Teherani-Krönner points out, it is largely women who farm and prepare food in many rural communities around the world. Local and geopolitical decision making regarding food production and soil conservation policies need to therefore recognize the relevance of meals and meal culture as well as food and soil. Roxanne Swentzell's work demonstrates how artistic tradition plays a central role in meal culture, from the way food is planted and sown to how and where it is cooked to the vessels that hold it and the rituals surrounding its consumption and enjoyment over generations. I interviewed Roxanne and Parto via a series of e-mails and telephone calls in a kind of virtual round table discussion about the role of meals as they relate to the food and biomass production function of the soil.

Alexandra R. Toland: To begin the conversation, I'd like to hear a little bit about the different practices that influence your work. Roxanne, maybe you could go first. You are well known for your artistic practice with ceramics and bronze. I was fascinated to read about your practice as a permaculturist, seed collector, cookbook author, adobe builder, and community organizer, not to mention mother and grandmother. How do all these different practices relate to one another? For example, are there certain things that you have learned as a sculptor that helped you in your pantes oven-building project, or have there been moments in your garden where you have developed new ideas for sculptures?

Roxanne Swentzell: I am a person who has always seen the way everything in life seems connected. To call myself an artist, or builder, or farmer, only seems like saying the mountain is a rock. What about the trees and the birds and the grass and the dew and the spiders and the deer and the sounds and smells and fog? They are all the mountain. I seem to have many hats but it's all one head I put them on. I think this is what attracted me to permaculture. I understood the part about patterns and how everything is connected. In permaculture, you don't just see a chicken; you see the whole system that happens to have a chicken part of it. So, when I build an oven or house, sculpture or pie, I am still coiling up walls with dirt/clay/dough to shape a container … what the container is holding affects the shape of the walls whether it be a person, bread, or a figure to tell a story. As I'm weeding the field of corn, I watch corn leaves blow in the wind, their roots dig down in the wet soil. I love this dance of corn, air, water, sunlight, desert heat, grasshoppers, birds, and me standing in the dirt, sweat pouring off my face, thoughts drifting into food thoughts, prayers for rain to quench all our thirsts. My whole life is studying how it all goes together and how to be part of as much of it as I possibly can. What I

know for sure is that the more I connect to all the parts of my life, the deeper the mystery becomes and the more it matters. Everything starts to be part of a bigger picture instead of objects disconnected to place and time.

Alexandra R. Toland: Parto, you have a diverse research background with interests in the fields of feminist studies, cultural ecology, agricultural science, food systems science, environmental justice policy, and sustainable development. Could you tell us how these different theoretical schools of thought have influenced your concept of "meal culture"? In your book, you propose that eating is more than the ingestion of food, but a cultural and social phenomenon.[3]

Parto Teherani-Krönner: It is quite obvious that people usually do not eat raw agricultural products like rice, corn, and barley, but prefer to eat prepared meals. "All living species need food for survival, but human beings are governed by cultural norms and taboos regulating this process of incorporation of natural products. Even under difficult circumstances, human beings will not accept and eat anything just to supply their need for calories, vitamins, proteins and minerals, even if they are hungry. … This process of human accommodation is tied to the normative system of a culture, no matter how economically wealthy or poor its people are" (Teherani-Krönner 1999).[4]

So, as a rural sociologist and human ecologist, I try to acknowledge human activities and thus put people first in developing my thoughts about meal culture. This means that in order to understand agricultural production and rural development and the whole environment for that matter we should not only concentrate on crops, animals, and technology but also look at those who fulfill that work. From an ecofeminist and gender studies perspective, it is important to recognize that much of that work is carried out by women, especially

Dishes pictured in *The Pueblo Food Experience Cookbook: Whole Food of Our Ancestors*. Roxanne Swentzell and Patricia M. Perea, Museum of New Mexico Press, 2016.

Said one among them—"Surely not in vain
My substance of the common Earth was ta'en
And to this Figure molded, to be broke,
Or trampled back to shapeless Earth again."

After a momentary silence spake
Some Vessel of a more ungainly Make;
"They sneer at me for leaning all awry:
What! did the hand then of the Potter shake?"

Whereat some one of the loquacious Lot—
I think a Súfi pipkin – waxing hot—
"All this of Pot and Potter—Tell me then,
Who is the Potter, pray, and who the Pot?"

As under cover of departing Day
Slunk hunger-stricken Ramazán away
Once more within the Potter's house alone
I stood, surrounded by the Shapes of Clay.

From the Rubaiyat of Omar Khayyam (1048–1131)[2]
Translated by Edward FitzGerald (1859)

Next pages:

The Corn Mothers are Crying. Original clay sculpture by Roxanne Swentzell, 17″ H, 16″ W, 16″D. This piece was created in response to being asked, "Using your art to speak for you, what would you say to our tribal leaders?" First, all information is seeds, so the symbol of a seed pot seemed appropriate. I come from a pueblo tradition so corn is essential to who we are. My seed pot holds corn. It is not only a seed pot but a prayer for the healing of our communities. The Corn Mothers face their directions as they watch over the world. The pot is symbolic of the Earth. The Corn Mothers lean outward trying to reach the turned backs of "the children."

One "child" is suffering from bad nutrition and health. Living in a world of fast foods overprocessed ingredients, products grown using pesticides and herbicides, and now GMO foods, we are truly in trouble for our health. Diabetes, cancer, birth defects, are just a few of the symptoms of this tragic choice, And the Corn Mothers are crying. ...

The second figure has turned his anger toward the world. Trying to pretend that he doesn't care, he destroys and vandalizes. The need to hurt those around him is a reaction to how he feels hurt. And the Corn Mothers are crying. ...

The third figure has turned the destruction on himself. Using drugs and alcohol, he tries to escape a world and a self that he is not happy with. And the Corn Mothers are crying. ...

A mother hides from the screams of her child left to fend for itself. How is it possible for unwanted, abandoned, and abused children, to grow up and raise children of their own? This has become a generational crippling of our society. And the Corn Mothers are crying. ...

I asked tribal leaders of our communities to help bridge the gap that exists between our cultural ways and our lost children. The sense of belonging and spiritual fulfillment that are essential to our cultures can help heal the wounds of our people.

in countries of the Global South. Women are the ones who nurture the world. In many agrarian societies in African and Asian and Latin American rural areas women still shoulder most of the fieldwork. In her book *Kitchen Politics*, Silvia Federici[5] mentions that 80% of what is consumed and eaten in Africa is produced by women. And let us not forget that nearly all over the world it is usually women who take responsibility for everyday preparation of meals at the household level. Unfortunately, modern agriculture and the implementation of new technologies in food production have not taken these vital social dimensions into much consideration. Women's contribution to agricultural production and indeed the whole care economy is mostly ignored or even overlooked. The gender order can be seen as a key to understanding our food system, or as I prefer to call it our meal culture.

From an anthropological point of view, cooking is a primary innovation of human kind that helped make it possible to save our energy for digestion and allow us to spend our time on other inventions.[6] Our cultural development is in other words based on the art of meal preparation. Cooking is only one aspect of this development. Ingredients are important elements, but from a sociocultural point of view the processes and rituals around the meal are equally important components of eating and sharing meals. Human interaction is deeply seated in the social construction of meals. Human relations start with drinking and eating together.[7] If we were to lose this everyday ritual we would lose an important cultural space for socialization and building relationships. The question will remain as to where that loss might be compensated. Thus, meal culture is not about what we eat only, but about the greater cultural context, the social setting, and the structure of a meal as a communication system. Meals are the materialized symbols of social networks[8] and women have been and

somehow still are the masters of this social interaction system. Adam was born when Eve started to cook.

Alexandra R. Toland: Roxanne, you bring a lot of these ideas to life in your artwork. In 2010 you created a symbolic sculpture titled *The Corn Mothers Are Crying*. One of the figurines in the sculpture suffers from obesity and diabetes, food-related diseases that disproportionately affect Native Americans. Last year, you published a cookbook and series of essays with Patricia Perea about the food preparation practices of Pueblo peoples: *The Pueblo Food Experience Cookbook: Whole Food of Our Ancestors*.[9] The cookbook was based on an experiment you and other members of your tribe made in which you adopted a precontact diet for three months to see if native foods could improve health issues such as diabetes, heart disease, and even depression. It worked! The experiment illuminated deep connections between place and individual and community health as well as the need to protect and support the intangible cultural heritage of food preparation.

Given the genetically diverse "melting pot" of the Americas, do you think contemporary hybrid cultures and peoples could benefit from the *Pueblo Food Experience Cookbook*, or in Parto's terms the Pueblo meal culture? What wisdom can the Corn Mothers offer a country hungry for nourishment at many levels?

Roxanne Swentzell: *The Corn Mothers Are Crying* is a piece about heartbreak concerning our Pueblo people. It shows how broken and lost we have become. My attempt was to show the pain so we can start the healing process. We have to look at it and cry to figure out what happened. I believe it is because of the disconnect from our culture, due to colonization, genocide, disease, and Western culture's views on objectification and worth. When we forget who we are and where we came from, we are like uprooted plants. We slowly die. The Corn Mothers are our original mothers who took care of us for so long. They are shown leaning toward their children but the children have forgotten them and no longer hear their songs. I created this piece as a seed pot, not just seeds of corn but seeds of knowledge and remembering. It's about finding our way back home. So, the mythology of the Corn Mothers is absolutely affecting my journey to finding my way back to our traditional diet. I prayed and they came.

The interest in the cookbook and the message behind it has been very well received. I keep hearing about other organizations trying similar things, so I believe it is right time for it. We are starving in more ways than food. What I tell people at my talks on the Pueblo Food Experience is not to copy our ways but to find their own. Everyone is indigenous to this Earth somewhere. I ask them to find out where that was. Where was the last place your genes were some place for more than 20 generations? What did they eat? What did they do? Feel your ancestors in your blood and how they were connected to place. The Pueblo Food Experience is our example of what that was for us. It's a map for Native peoples of the Southwest and Northern Mexico. The deeper we go "home" the more nourished we will feel.

Alexandra R. Toland: Nourishment looks a lot different when it is framed, as Roxanne says, in terms of "finding our way back home." Parto, could you talk about where nourishment fits into the overall idea of meal culture and its accessibility to those who have forgotten their proverbial Corn Mothers? I'm thinking about elderly people in poor rural communities with no access to organic supermarkets, or to low-income youth in cities who acquire most of their calories from fast food chains, or urban singles who consume "Community Supported Agriculture" products alone in their kitchens, posting plate-selfies to

share with friends on social media. Would it be fair to say there are "sustainable" meal cultures and "unsustainable" ones, or is there something to be learned in any context in which people gather, physically or virtually, to sit and consume food?

Parto Teherani-Krönner: Coming and joining a meal together is the key to a healthy diet and social well-being. Eating is one of the first forms of sharing and an important step in human socialization. This has been underlined more than a hundred years ago by the sociologist Georg Simmel (1910) who wrote an article about the sociology of the meal (Soziologie der Mahlzeit). So, I would say meal culture is everywhere—even in fast food chains, but there are meal cultures that are closer to sustainable practices than others. For example, there is general agreement that too much meat consumption is neither suitable to human health, nor is it compatible to environmental protection (Worldwatch Institute 2017).[10]

Some generations ago most communities were dependent on their own direct environment for nourishment. There were moral barriers to killing animals in most societies. It was often combined with rituals to legitimize the slaughtering of animals; not seldom combined with symbolic sacrifice and meat distribution (Rappaport 1967).[11] But globalization has changed the way we eat dramatically. Now societies can transport commodities and waste from and to other regions without borders. The direct connection to environmental resources at "home" gets lost and becomes no longer immediately perceptible. It takes some time before people get confronted with the consequences. But then it might be too late.

In European countries as well as in the Global South meat consumption is increasing with processes of industrialization and modernization of agriculture. Meat consumption was once a matter of prestige, dedicated to special occasions, whereas now it has become an everyday expectation in many parts of the world. This development is accompanied by overstressing environmental resources and consuming huge amounts of energy. Producing a unit of meat needs eight to nine times more plant-based energy. There is also the social context to consider. Nutritional recommendations focus mostly on telling us what to eat and what to avoid. But this is a too narrow view on meal culture. From my cultural background, I would say: "Eating alone—having no one to share food with—is a sign of poverty." This is reflected in the structuring of meals as well as their nutritional ingredients. Hamburgers are not only nutritionally deficient, they are also just designed for one person, whereas a soup or a stew is much more amenable to sharing. This means that the composition and the structure of a meal give us the scope of action for our social interaction and communicational opportunities around our meals.

So, the challenge is organizing and designing new ceremonies around meal cultures with delicious dishes without or with only little meat; not only in vegetarian or vegan restaurants but in schools, in canteens, and home for the elderly. We need a more holistic approach when it comes to understanding our nutritional practices, customs, and taboos. Our nutrition and meal preferences are deeply rooted in our cultural traditions and normative systems, and our rituals for eating together are ideally accompanied by some sort of social control that can help avoid things like swallowing fast, obesity, and waste—the cultural disconnections Roxanne's Corn Mothers were lamenting. We need to rediscover or create new sustainable consumer habits in meal culture. We need to find new spaces to cook and enjoy our eating together and communicate and socialize around meals.

Alexandra R. Toland: This challenge is grounded in aesthetics as much as it is grounded in social

practice and knowledge transfer. Roxanne, I wonder what your response to this might be as an artist. You have not only compiled recipes in your cookbook but planted gardens, saved seeds, and erected a cooking house with traditional millstones and an adobe oven. What role does aesthetics play for you in the practice and protection of meal culture? Why is it important to consider beauty in the planting, tending, harvesting, storing, preparing, processing, and consumption of foods? And how do specific practices, places, and people provide an aesthetic framework for meal culture, in Pueblo cultures and beyond?

Roxanne Swentzell: I believe everything matters, especially our intentions behind our actions. They create energy, good or bad, life producing or life killing. When we do things with care and mindfulness, respect and gratitude, we can't help but nurture life. Doing things "artistically" is something that is a great gift we have as human beings. We can create things of great beauty (or not), but we have that choice. In the Native world, when you walk with a good heart, and be respectful of all Nature's creations, you will be taken care of. It is very reciprocal. The more you give, the more you get. To take the time to make something carefully is to love it. That love becomes embedded in that clay, or dough, or wall or dirt, or each other. So, it is very important to practice creating beauty.

Alexandra R. Toland: Parto, I had the honor of eating a shared meal with and by you at the Kreuzberger Salon last year.[12] This was an incredibly aesthetic experience. Over rice pilaf with barberries you talked about the cultural differences in meal preparation and consumption in Germany, Iran, Sudan, and Vietnam. Could you talk about some of the specific practices that are passed on and embedded in cultures over generations, despite trends toward globalization and loss of agricultural knowledge? Could you perhaps offer an anecdote from your own background in Persian meal culture? Are there certain foods or meals that represent an ancient terroir for Iranians around the world, perhaps like the corn recipes of the Pueblos?

Parto Teherani-Krönner: There is a special art of cooking and preparing food in Persian culture. It is not only the quality of rice, but the way it is prepared which is somehow "Persian." Those who have had the chance to eat a Persian meal might be familiar with the very special crunchy part of the rice. This is known as "tahdigh" (the end of the pot) and is in a way the pride of Persian cuisine (Zubaida and Tapper 1994).[13] In fact, we name all the different ways rice is prepared with specific terminologies. The white and plain one is called "Chelo—Kateh—Dami" according to the method of preparation. "Polow" is a rice dish mixed with ingredients like barberries or green beans. All these dishes are prepared with raw rice, but specified according to the way it is cooked and served.

In Iran, we also differentiate everything that you can eat with the categories of "cold" and "warm." This classification belongs to an old philosophy and is a basic concept of traditional Persian medicine still alive in everyday knowledge. Everything we eat is either "warm" or "cold" and this dichotomy can be applied to our illnesses and diseases as well. Our well-being and recommendations for recovery will correspond to the rehabilitation of the warm and cold equilibrium in our body. A good meal is supposed to keep these two components in mind. A very typical Persian meal like "Choreshte Fesenjun" is a stew prepared with walnuts that are warm in combination with grenadine syrup, which is cold. This classification of local knowledge is still omnipresent in everyday life. Restaurant servers, for example, are usually not willing to serve fish

together with yogurt or a yogurt drink as both are classified as cold. Recently a colleague and I were on our way to the Caspian Sea in the northern part of Iran and were confronted with such a recommendation: it is better not to eat fish and yogurt together, else we get a stomachache the next day.

This dual system of cold and warm is not only well known in Iran; it is similar but not identical to yin-yang principles in China. This philosophical principle can be followed throughout the Silk Road and found in far away destinations. Obviously, the cultural connections and exchange of knowledge about foods existed way back in historical periods. In Nepal, for example, we can learn from the ethnomedical wisdom that the first question of the medical doctor is not where do you have pain but what did you eat yesterday (Heller 1977)[14]

Alexandra R. Toland: I'm going to make a jump here from meal culture to agriculture, from food to soil.[15] I was hoping you could both weigh in on the issue of soil security, Parto coming from a perspective of feminist rural sociology and Roxanne from your perspective of community work and permaculture practice in indigenous communities in present-day New Mexico. Is "soil security" an important concept for either of you and how would you define it in other terms? For example, some have balked at the term "security" and suggested "sovereignty" or "sustainability" or simply "care" instead of invoking a term nuanced with militaristic self-defense. What does this mean to you?

Parto Teherani-Krönner: I believe that our perceptions of eating and nourishment will change tremendously by having "meals" in mind instead of "food." Changing the food security debate, highlighted by the SDGs, into a meal security debate will have consequences with regard to discourse on hunger and malnutrition as well. So, we need to think about meal politics and meal culture, as well as food systems and agriculture. But these terms are rarely used within the scientific community, even though their practice is culturally embedded in our everyday lives. I have introduced the concept of meal culture in my research because I felt a strong need for a new view—a new paradigm that reflects the social, cultural, and environmental embeddedness of our gender relations in the nutritional and agricultural sciences (Teherani-Krönner and Hamburger 2014).[16]

On the one hand, I have followed the food security debate within the agricultural sciences on the international level. I realized that the discussion was mostly about how much wheat, rice, or corn is produced. The raw products were figured in yields per hectare or by calculating in kilocalories. For the most part, the agricultural sciences concentrate on increasing yields by using pesticides and fertilizers, and legitimizing themselves by referring to the pressures of population growth. But that what people eat, and how they eat, is more than a matter of quantity of raw products.

On the other hand, the nutritional sciences look at the substance of content, analyzing the ingredients and the vitamins, minerals, and chemical composition of foodstuffs, but this again is too narrow. The meal culture concept harbors a human and cultural ecological perspective in which the whole process of producing and preparing, serving, and eating a meal is included. The ingredients, the raw agricultural products, are but only one component of the social and cultural construction of a meal.

The food sovereignty approach has taken care of environmental as well as regional cultural aspects of agricultural production systems and seeks to support local communities. The meal culture approach approves these ideas, though I think it pays additional attention to the unattended practical requirements of everyday life. It encompasses the whole dynamic cycle from field to plate, including all the necessary rituals and side effects (like leftovers). Looking at the gender dimensions, from the production to the preparation and serving of meals, is a fundamental part of our meal culture concept as a social construction that encompasses labor power relations and social stratification.

So, in light of the food security debate, we also have to recognize that meal preparation requires clean water that is not always available everywhere. It also requires utensils to wash, cut, chop and grind ingredients, as well as the fuel source for actually cooking. Therefore, a whole set of technologies is necessary in addition to a person who spends time with the required knowledge, recipes, and formulas for using the right ingredients and spices. And last but not least there are always cultural criteria for eatables and enjoyables. Our normative system is part of the social context and cultural space that matters. Such a broader concept of meal culture security will definitely help us to understand our perceptions of eating habits as well as our connection to food and soil.

Roxanne Swentzell: For me, "soil security" sounds so cold and militant. It says a lot about the mind-set of the people who speak about their mother in those terms. For indigenous peoples, the soil, the dirt, the ground is our Mother. … I know Western minds think of it as dirt that can be moved around and done whatever they like to do with it next, but truly if they stop and feel Her, she is not a resource to devour and spit out like everything else they touch. Nothing has been more disrespected in Western culture than the female. This includes, women, the Earth, containers of all sorts (look at our disposable packaging sickness) and communities. As long as the individual is placed above all else and the consumption of whatever the individual wants without thought about where and how it came to be, we will be on a suicide journey.

I know I cannot convince many that the Earth is a living being and She will get rid of us if She finally gets tired of how we treat Her. Maybe creating laws to try and protect the biodiversity and health of the soil is the next best thing. But I still believe if people can remember how to feel the world around them again, they naturally, without rules and laws, cannot hurt something they are empathic with. It hurts to hurt our world, so you just don't do it. This makes me wonder why we are so afraid to feel. This fear has disconnected us so dangerously with our beautiful world that we may not survive ourselves. I do believe, if we even have a chance to continue, that women will have to lead the way to healing. No computer, no scientific calculations will have the answers. We are organic creatures living on an organic living planet and we need to feel our way through this darkness. We have to find our own roots again and start growing ourselves in soil that is able to nourish us. I suppose when the main culture takes care of its women better, we might have a chance to speak our knowledge and our wisdom as containers of the Mother energy, the intuitive, the soil, Home.

Parto Teherani-Krönner: In her book *Death of Nature*, Carolyn Merchant[17] introduces an interesting ecofeminist position toward science and modern technology that has destroyed the holiness of our soil and earth. The exploitation of nature for economic interest is seen as a fundamental reason for this. Mining and digging in the earth with heavy machineries made it necessary to give up the former worldview of the sacredness of soil and worship of Mother Earth.

But the ethical and aesthetic care for our natural elements is fundamental to human life. This is evident in all cultures, from the Greek mythology of Gaia the Mother Earth to what Roxanne has described of the Corn Mothers in the Pueblo tradition. In former times, according to the old religion in Persia soil was holy and one of the four elements that needed to be kept carefully neat and clean. Cleanliness of the elements of nature—soil, water, air, and fire—are the religious and ethical principles of Zoroastrians, the ancient religion of the Middle East and Persia long before Islam. Believers still live in some parts of Iran like in the town of Yazd and in some communities in India, where they are called "Parsi." They do not even bury their dead in the soil to avoid polluting it.

In the Islamic tradition, the Koran states that humans were even formed of clay: "And indeed, we created the human form from dried (sounding) clay of altered black mud [min hama'in masnoon]" (al-Hijr 15:26). Clay pottery also has a high symbolic value in Persian literature and in poetry besides being one of the oldest arts known to human kind. Coincidentally, this was also likely a domain where women played an important role. Pottery vessels for serving and storing food have been found from ten thousand years ago in the Middle East and Iran with artistic designs and patterns on them. In terms of food culture and soil culture, Roxanne is really at the heart of an everlasting human activity and artistic tradition!

Alexandra R. Toland: Indeed. I think if we all stored our food in clay vessels and ate from handmade plates we would immediately reconnect to the meal cultures of our past, wherever that might be.

Thank you both so much for sharing your insight in this chapter.

Endnotes

1. Koch A et al. 2013. Soil security: solving the global soil crisis. *Global Policy* 4(4):434–441.

2. Omar Khayyam was a Persian mathematician, astronomer, and poet, most notable for his work on cubic equations and calendar reform.

3. Teherani-Krönne, Parto, and Brigitte Hamburger. 2014. (Hrsg.) *Mahlzeitenpolitik Zur Kulturökologie von Ernährung und Gender.* Münich, Oekom Verlag.

4. Teherani-Krönner, Parto. 1999. Women in Rural Production, Household and Food Security. An Iranian Perspective. In: Manfred Kracht and Manfred Schulz, (eds.), *Food Security and Nutrition.* The Global Challenge. Lit, Münster, p. 189–218.

5. Federici, Silvia. 2012. *Aufstand in der Küche*, edition assemblage. Münster.

6. Pollan, Michael. 2013. *Cooked: A Natural History of Transformation.* Penguin Press, New York.

7. Douglas, Mary. 1972. Deciphering a meal. *Daedalus* 101(1):61–81.

8. Teherani-Krönner, Parto. 1999. Women in rural production, household and food security. An Iranian Perspective. In: Manfred Kracht and Manfred Schulz (eds.), *Food Security and Nutrition.* The Global Challenge. Lit, Münster, p. 189–218.

9. Roxanne Swentzell and Patricia M. Perea, Museum of New Mexico Press, 2016.

10. Worldwatch Institute. 2017. Is meat sustainable? From http://www.worldwatch.org/node/549 (accessed September 24, 2017).

11. Rappaport, Roy. 1967. *Pigs for the Ancestors: Rituals in the Ecology of a New Guinea People.* New Haven, London.

12. The Kreuzberger Salon is an informal regular meeting platform organized by Miriam Wiesel and Axel Schmidt to discuss the political and cultural dependencies between rural and urban communities, especially regarding agricultural practices. Parto Teherani-Krönner presented some of her research on meal culture at Kreuzberger Salon in December 2016.

13. Zubaida, Sami, and Richard Tapper (eds.). 1994. *Culinary Cultures of the Middle East.* New York, London.

14. Heller, Gerhard. 1977. Die kulturspezifische Organisation körperlicher Störungen bei den Tamang von Cautara, Nepal. Eine empirische Untersuchung über die Hintergründe kulturbedingter Barrieren zwischen Patient und Arzt. In: Rudnitzki, G., et al. (Hg.), *Ethnomedizin – Beiträge zu einem Dialog zwischen Heilkunst und Völkerkunde.* Detlev Kurth, Barmstedt, S. 37–52.

15. The number two UN Sustainable Development Goal (SDG2) for the year 2030 is to end hunger, achieve food security and improved nutrition, and promote sustainable agriculture.

16. Teherani-Krönner, Parto, and Brigitte Hamburger (eds.). 2014. *Mahlzeitenpolitik. Zur Kulturökologie von Ernährung und Gender.* oekom Verlag, München.

17. Carolyn Merchant. 1980. *The Death of Nature: Women, Ecology and the Scientific Revolution.* New York, HarperCollins.

Function 2

REPOSITORY: Soil as source of energy, raw materials, pigments, and poetry

Repository

Soils have an unprecedented capacity to store an abundance of materials, including nutrients weathered from rocks and the water that carries them; gaseous chemical compounds such as methane and nitrous oxide; construction materials like gravel, sand, and clay; as well as mineral pigments for artistic and industrial applications. The second section of the book explores the idea of a repository as a fundamental function of the soil. Human concepts of savings banks and extractive machinery emerge, together with images of sacred vessels, secret underground vaults, dusty archives, and buried treasure. These stored materials are the building blocks of pedogenesis, the process of soil formation over time. The authors in this section reflect on various meanings of the soil as source and sink, storage, and archive.

To begin the section, the narrative, poetic, and ecological aspects of pedogenesis are explored in chapters by Veronique Maria and Ólafur Arnalds, and Ulrike Arnold and Thomas Scholten. Elvira Wersche and Alex McBratney discuss pedometrics in terms of pictures and poetry, and soil science in terms of *nescience* or that which cannot be quantitatively known. The transcendent properties of soil minerals are explored in three very different practices: by the glass artist Sarah Hirneisen, the printmaker Ekkeland Götze, and the pioneering ecological artist herman de vries in their respective dialogues with Donald Sparks and Jason Stuckley, Winfried Blum, and Nico van Breemen. The painter and educator Peter Ward offers a guide for collecting and working with earth pigments for teachers and artists. Aesthetic and ethical aspects of carbon storage are brought to light in chapters by Laura Harrington and Jeff Warburton, and Terike Haapoja and Taru Sandén. The repository function of city soils is celebrated as a potential resource in the chapter by Margaret Boozer and Richard Shaw, while the technocultural "archiving" of nuclear waste in clay substrates around the world is posed as a repository for potential conflict in the chapter by Dave Griffiths, Sam Illingworth, and Matt Girling.

Soil Genesis
A Dialogue for Creation

Veronique Maria and Ólafur Arnalds

Veronique Maria is a UK-based artist, psychologist, creativity coach and artists' mentor who has worked extensively with volcanologists and deep ecologists exploring relationships between earth and body. Maria collects soil and volcanic ash to inform her art and integrates these into live art, video, and large-scale paintings as part of her art and science project called Orogeny (genesis—birth of a mountain). She considers poetry to be a key element to her research and she delights in collaborations. www.veroniquemaria.co.uk

Ólafur Arnalds is a professor of soils and a former dean at the Agricultural University of Iceland. He is an expert on soils of volcanic areas. Arnalds lead a project to evaluate soil

Title image: *Untitled* by Veronique Maria, 2017, 150 × 100 cm, mixed media (including fired earth and raw pigments) on canvas. Reprinted with permission from the artist Veronique Maria.

erosion in Iceland, for which he received the Nordic Council "Nature and Environmental Award." His research activities also involve land use impacts, land condition, and ecological restoration. He is the author of *The Soils of Iceland* (Springer) and the main author of soil maps of Iceland and the Nytjaland GIS land cover database of Iceland. Recent research includes characterizing aeolian processes in Iceland. Public outreach programs include the publication of *How to Read and Heal the Land*. He has edited several international books on his subjects. His website is www.moldin.net.

For their chapter contribution, Maria and Arnalds were given the task of discussing their thoughts about soil genesis. The work started one Friday afternoon on Skype, developed into an exchange of e-mails, keeping in mind that first conversation, and resulted in a series of short poems. The choice of poetry as medium for "expressing the inexpressible" has a long history in the geosciences, from Goethe's metered musings about the earth and skies to the proletariat "bard songs" of the Leningrad Mining Institute's (LGI) geologist-poets. While Arnalds has generally been more involved with music and the short story format rather than poems, he began the work by expressing his soil ideas in poems, based on their conversations and inspired by Maria's art, with Maria adding more ideas and modifications, and gradually the poems were completed by cooperation between the two.

Veronique Maria: I think it would be good to explore the creation of something new together right here in our conversation so that the dialogue itself echoes the idea of "genesis." What do you think about exploring how our process of verbal and written exchange might mirror the process of genesis in some way? Soil genesis and the genesis of all life. How does that sound to you?

Ólafur Arnalds: I like that idea, as I see that this project offers an opportunity to get off the beaten path and explore new views on soil formation processes. I am interested in different forms of writing, for scientific purposes but also creative purposes. I don't know how this will take form but I am very willing to explore.

Veronique Maria: I am delighted to have this opportunity to exchange ideas with you on the subject, as I have no formal scientific background. I have been trying to express these processes in my artwork (my paintings, performances, and creative writing) for several years now, where I explored the relationship between earth, body, and psyche. As an artist with a professional training in psychosynthesis psychotherapy, I tend to work from a psychospiritual perspective. I work from the feeling of the thing (rather than the look of it) so I consider myself a kinesthetic artist. I try to absorb the sense of the thing, be it an idea, an emotion, a dynamic, or an environment, for example, and then express this in my artwork.

I am interested in what it is like to live on edges, in ever-transforming landscapes, such as Iceland, and how one might regard the rumblings from the center of the earth as a mirror of the human

You Set Me On Fire, 150 × 100 cm, mixed media (including fired clay and raw earth pigments) on canvas, 2017.
Reprinted with permission from the artist Veronique Maria.

DWELL by Veronique Maria, 120 × 120 cm, ceramic on canvas.

Reprinted with permission from the artist Veronique Maria.

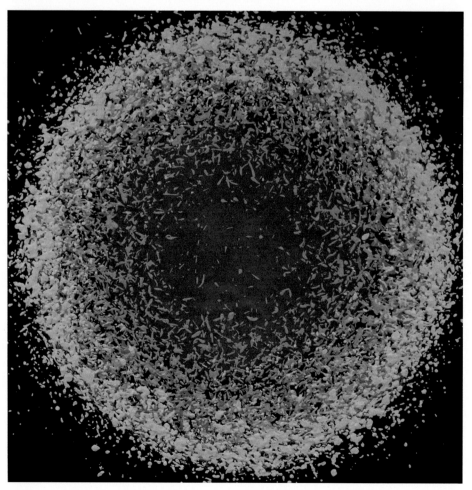

Home Coming by Veronique Maria, 120 × 120 cm, mixed media on canvas.

Reprinted with permission from the artist Veronique Maria.

psyche. For example, one might consider the grinding of tectonic plates, and violent eruptions of magma forcing through meters of ice or from within the ocean bed, and the formation of new mountains on the landscape, as mirroring creative processes some artists and scientists encounter in themselves in order to do their work.

Ólafur Arnalds: I guess my professional work is more down to earth—but within the subject of my scientific field there are questions of creation, formation, and the cycles that maintain life on the surface of this planet. A fresh approach to this very essence of existence is truly worth the try.

Conventional scientific text is not the only way to approach, teach, and communicate.

Veronique Maria: Then we'll dive into the unknown from the outset and try to express what is essentially the inexpressible. I mean, how does one express the "genesis" of "earth"?!

Ólafur Arnalds: *Sol*, the root of the old Latin word for soil means earth. That is where everything begins—and ends in due time. So really, we are creatively exploring, in a sense, the beginnings of earth.

Genesis

The very beginning …
first whispers of a conversation,
a breath—inwards, outwards, a pulse
loose words,
ideas forming (or hoping to form),

fragments in an ocean of what seems to be "nothingness"

elements of knowledge within the sense of unknown. …

Primordial

Primordial
energy—atmosphere—water

rocks—solids—solution—life
chemistry—precipitation

Body—a being
from earth, of earth, to earth
a living death
a knowing and a not knowing
a longing, hoping, wanting
a deep yearning
and an existential ache

deliberation
procrastination
thrusting, pulling, heaving

Creation

Softening, grinding, opening,
connection,
friction,
abrasion,

volcanic eruption,

glacier leaving the land,
a rush of a flood,
land rising from ocean,
reception, conception,
transformation.

Birth
birth of a land
birth of an infant—soil

No development
only unequal dealings of the passes of future
soil's fate
the pain of birth
the environment awaits.

Myriad numbers of soil individuals
earth takes its breath
developing
changing
growing, shrinking
absorbing

Environment of birth
play for stakes
for greatness
fertility
riches.

A struggle
play for stakes
home in hard rock
frigid temperatures
dryness
poverty.

Cycles

Soil becomes of age
character
expressed by the properties it assumes along the way.

Maturing
becoming a media of trade
receiving deliveries of energy, water and nutrients
processing, storing and delivering
workers on the floor
soil biota

Soil may prosper
become rich with stocks and bonds,
trading energy, water and nutrients
with vegetation
thriving and turning around the goods
surely and rapidly.

Or

a soil of hardship
supplies in short
limited resources
water and energy
adjusting to bad tidings
still vital

Taken over by alien forces
farmer with plough,
seed and fertilizers
grazing animals
overwhelming
disrupting
collapse
the cycles are broken.

Development

Development by energy and life
time
Parent material
making of a soil

Dissolution of solids
ions floating in water
leaching
elements making bonds
precipitation of creation
tetrahedra houses of silica
octahedra buildings of aluminium
the endless vistas of the phyllosilicate
clay—the produce of soil
Argillic horizon—the world of clay

A quest for energy
treaty with the Kingdom of Plants
provision of green fuel
for its untold numbers of workers
soil biota
in exchange for water and nutrients

Prosperity is founded on cooperation and trade
everyone benefits
equality
society—culture.
Home.

Individuals

The multicolour face of a Boreal forest soil
sun-baked soil of the tropics
pale looking soil of the desert
deprived of water
salty

Dark-brown soil of the grassland
strong in character
fertile and energetic
wealthy from harvesting the sun
resilient to abuse of the plough
Mollisol

Youthful offspring of volcanoes
diverse and affluent in character
fertile
mystic, dark in complexion,
Andosol

Old Age

Ancient civilisation
with eons of experience

red and rich in clay
lacking youthfulness leached and consumed
tired

A Kind of Soil Genesis on Canvas

Ulrike Arnold in dialogue with Thomas Scholten
Facilitation and text: Bettina Dornberg

Ulrike Arnold has been an international recognized artist for almost 37 years. She was a student of the master class of professor Klaus Rinke at the Düsseldorf Art Academy. Ulrike Arnold paints within nature and with the help of nature, usually far from the tracks of human civilization. She has painted on all continents: in sand and salt deserts, on top of volcanoes, in front of ancient caves, and in the midst of rock formations. Arnold wants to capture the essence of place. Natural landscapes form into works of art, works of art that in turn become landscapes. Since 2003 she amplified her earthen palette of pigments with meteorite dust. Ulrike Arnold won the Eduard von der Heydt Award of Wuppertal, Germany, in 1988 and

Title image: Ulrike's soil collection (Earth-based installation: 500 linen bags filled with soil samples from five continents—covering the entire spectrum of colors—dug up by the artist on her travels.)

Photo credit: Ulrike Arnold.

the Viola Award of Flagstaff, Arizona, in 2015. So far, she has had 36 solo exhibitions and has participated in 78 group exhibitions.

Thomas Scholten is a professor of soil science and geomorphology at Eberhard Karls University Tübingen. He regularly conducts fieldwork in Europe, Africa, and Asia, and has received grants from the European Union (EU), the German Research Foundation (DFG), and the German Federal Ministry of Education and Research (BMBF). As a widely published author with 7 books and 158 refereed papers, he currently serves on the editorial boards of several environmental and geoscientific journals and has acted as guest editor for *Geoderma* and *Catena*. Thomas Scholten was president of the German Soil Science Society (DBG) from 2012 to 2015 and has been a council member of the European Society for Soil Conservation (ESSC) since 2004.

Bettina Dornberg is an author, journalist, and ghostwriter. She has a master in communication and theatre sciences of the Freie Universität, Berlin. Her beloved genre in journalistic writing is the portrait. She won the Karl-Theodor-Vogel Award 2012 for investigative journalism. Bettina Dornberg also works as a specialist in public relations and for corporate media as well as a consultant for strategic communications. She met Ulrike Arnold at one of her exhibitions almost 13 years ago. Since then they have worked together on several projects.

The idea was to gently moderate a joined session between the artist Arnold and the scientist Scholten in order to generate a real creative dialogue—with art and words. Bettina Dornberg shortened and finalized the resulting transcript without changing the character of the dialogue, just to pronounce the process in the sense that everyone can feel the magic.

In their childhood, they liked to play with the earth, and both made it their profession: Ulrike Arnold became an artist and Thomas Scholten became a soil scientist. Would that be the key to their dialogue? Perhaps they could build a sandcastle on the Rhine River or paint a picture in Arnold's studio together?

The latter turned out to be easier. They met in Arnold's Düsseldorf studio on December 8, 2014. Some five hours lay before them, time to spend eating and painting together: two strangers, at home in two different professional disciplines, brought together by a shared and deep interest in soil.

In the middle of Arnold's studio there is an earth-based installation: 500 linen bags filled with soil samples from five continents—covering the entire spectrum of colors—dug up by the artist on her travels. Scholten is instantly enthusiastic: this is the first time that he has embarked on the unknown terrain of art. The joint activity is also unknown territory to Arnold. She has worked with other artists, but has so far never worked with a scientist that she has only just met. What connects them is their unanimous passion for the earth.

First, they eat together and talk about their specific experiences and ideas about soil. Scholten explains: "In soil science, we use the term 'soil genesis' when earth forms from solid rock and the living world, that is, from plants and animals. Earlier, in the 1960s, it was assumed that soil forms in one place. Today we know that 80% of soils have not been formed only in the place where they are found. Instead, the soils have a background from another place, transported by the wind and water. That means they are made up of materials that come from many places."

The first thing that comes into Arnold's mind when she hears the word "genesis" is the Bible, the Old Testament, which starts out with the world being created. She describes her artistic approach to soil genesis: "That makes me think of the origins of the earth, its primordial state. I understand my work as a tribute to that primordial state, when it was still virginal, unspoilt and paradisiacal. Those are the places I seek out; places that are still untouched and exert a magical attraction. I would like to portray that primordial state with the material that I find on the spot, and transfer that idea from my mind to the picture."

Scholten describes his personal approach to soil genesis as an urge to better understand the world: "I always wanted to know why there were mountains and valleys. I was always looking for the beginning. At some point in my studies, I came to the conclusion that there was never any primordial state. When I understood that there was no simple starting point for it all, I started to observe and take in what there actually was. I won't ever understand everything, but I have managed to tackle and understand the existing situation."

After eating together, they turn to the act of painting. Arnold has already got everything ready: an unprimed canvas (100 × 120 cm in size) is placed on a table; next to it is a chair with bowls for the paints and a mug with brushes of different thicknesses. On the floor, there is a bucket containing transparent binders mixed with water, another bucket of water so that they can wash their hands from time to time, and towels to dry their hands. There is also a smock for Scholten. Arnold invites Scholten to use any color from her collection of soils from five continents kept in 500 linen bags. The dialogue begins ...

Opposite page:

Typical painting setting of Ulrike Arnold, here in "White Pocket," Utah.

Photo credit: Victor Van Keuren.

Earth paintings in a show, Museum of Arts, Solingen Germany (2016).

Photo credit: Victor Van Keuren.

Ulrike Arnold: When I am outdoors I always perform a little ceremony: Before I start working, I try to imagine the primordial state. At that moment, I try to release myself from everything; from problems and thoughts, which are not connected to the place where I am, and to focus my awareness on nature. That is, to be free and not think of anything else … in art there are no laws.

Thomas Scholten: Yes. (*And chooses a red earth.*)

Ulrike Arnold: That red is from Karkoo, Australia. That red is the most magical color there is; such an intense element. The Aborigines use that red for healing purposes, to paint their bodies. I had to get permission to dig it up.

I will of course react to that red—and now choose this gray.

Thomas Scholten: Good.

Ulrike Arnold: Isn't that texture great? It's almost like …

Thomas Scholten: … flour.

Ulrike Arnold: Yes, flour, but as if it had fat in it. It comes from Brazil, one of the big iron mines. It's so incredible to me; you want to just spread it onto your face …

Thomas Scholten: When soil scientists are out in the field, when they're out walking around, like we're walking around this picture now, they do a "finger test." That's what we call it. In other words, I take the material between my fingers and rub it. Really carefully. And then after years of experience I can determine, relatively quickly, what the particle size is, whether it's sand or silt or clay. If I rub it then I can feel the individual grains. Or if I hold it up to my ear—that's the best thing of all …

(*Thomas Scholten rubs the earth between his fingers and holds it up to Ulrike Arnold's ear.*)

Ulrike Arnold: Oh wow, I've never done that!

Scholten holding soil up to Arnold's ear and Arnold listening.

Photo credit: Christoph Berdi.

Thomas Scholten: That's fine sand, it has a particular grain size.

Ulrike Arnold: And that?

Thomas Scholten: That's incredibly fine. I can't even hear that grate. You can't actually hear anything, unless you press so hard you can hear the ridges of your fingers rubbing together. It's a bit greasy. It's made up of very fine flakes. They're not round. That's why it feels greasy, because the flakes fall into parallel positions when I press them together.

Ulrike Arnold: There's mica in that, isn't there?

Thomas Scholten: Yes. It's a thousand times finer than other mineral soil materials. If we put it under the microscope, it would probably be in the range of 2 micrometers in size, and the other between 0.5 and 1 millimeter.

Ulrike Arnold: I also do the finger test. I dig the earth with my fingers, but I do it intuitively. I don't know much about particle size or other properties of the earth, but I think that sometimes it's good if you go about things with a certain amount of naivety, isn't it?

Thomas Scholten: Yes. In principle that's what scientists do too. They say "I don't know what's happening here and I'm going to see what I can find out."

(*Ulrike Arnold describes the different ways in which earth can be used when painting—whether the pigment is first applied to the canvas and then fixed with the binder, or whether the binder is already spread on the canvas, then the pigment is scattered on top, or whether the colors are applied with a brush, or the hands, or if they are splattered or tapped on.*

The red soil that Thomas Scholten is painting with reminds Ulrike Arnold of blood. Thomas explains the various iron modifications in soil that are responsible for the color. Terms like "hematite" and "goethite" come up.)

Ulrike Arnold: Does goethite have anything to do with Goethe?

Thomas Scholten: Yes, he was the first to describe it.

(*Ulrike Arnold splatters her gray pigment on the picture.*)

Scholten and Arnold examining the gray soil with their fingers.

Photo credit: Christoph Berdi.

Ulrike Arnold: I'm getting in touch with the red, that's important to me. Now that red from Australia is mixing with the gray from Brazil.

Thomas Scholten: Color is something that can often be deceptive. It's the thing our eyes recognize, as a reflection. That makes it dangerous. Sometimes the color reflects the soil's content, but sometimes it doesn't. For example, the gray could be a mineral, but it isn't necessarily.

(*Ulrike Arnold comments on the painting technique and Thomas Scholten chooses a red for the second time.*)

Ulrike Arnold: That red is from Brazil, and it is almost pink. I'm going to take a green sample now. This is from Armenia. It's a very strong green. And I'll take another brush. This is a rough, earthy material now. Sometimes I break it down with the hammer or mortar and pestle, and sometimes I don't. I find the differences fascinating. The contrasts of coarse and fine, of light and dark, of cold and warm colors. And the interesting thing is how the color changes when it is dry.

(*Thomas Scholten and Ulrike Arnold now work as one creative force on the painting, using brushes and their hands, flinging and scattering paint. For a while they say nothing.*)

Ulrike Arnold: Do you touch the earth with your hands when your work?

Thomas Scholten: I do everything with my hands. Only recently, at the age of 54, I realized that my hands can contaminate the soil, for example, if I want to determine its age. There are flakes of skin on my hands, which could mix in with parts of the sample. Younger scientists are increasingly starting to wear gloves when taking samples, even in cases when there is no need for that, in terms of preserving the chemistry.

Ulrike Arnold: I do everything with my hands. I'd go mad if I had to use gloves.

Thomas Scholten: I do everything with my hands, too. I tell my students if they are going to investigate something new in a new terrain, when research is just starting out, they should ideally

Scholten and Arnold working on the painting from a distance.
Photo credit: Christoph Berdi.

take a couple of days off to just walk around and do nothing at all and touch the soil.

Ulrike Arnold: Are they able to do that?

Thomas Scholten: They have to. I make them do it. They have to learn how to observe and let things sink in. The worst thing they can do is just reproduce what they already know. When you're out in the field, there are no experiments. If I am walking in the countryside, the only question is should I take a sample or a measurement. After all, I can't pack the entire country into a bag and take it home with me. Now there are various options. I either have a certain design based on mathematics or statistics, for example, I take a sample every ten meters. Or I take earth from places where the countryside or the surroundings strike me as

relevant. That's called "expert-based" sampling. So, it's often the case that I don't measure the soil down to the finest detail, but just let the landscape sink in.

Ulrike Arnold: I never would have thought that. I go on walks and always start out by searching and looking and watching. It's a process.

(*Again, there is silence for a while as the two continue painting.*)

Ulrike Arnold: I sometimes pour something on the canvas here and let it flow. Some of the composition is determined by chance …

I've been to Asia a lot. They have these elaborate spice markets. At first I collected and kept my soil samples in plastic bags. But then I remembered the spices and recreated that aesthetic. I had some old sheets at home, and my mother made them into little sacks. Five hundred sacks, and my mother sewed them all. Jars were too prosaic and stiff for me. My linen sacks are so unspoilt, and with me it's all about being unspoilt, getting back to the primordial.

Thomas Scholten: It's interesting, scientific soil samples used to be kept in sacks like this. Later on, only little plastic bags were used. Once a fellow Polish scientist brought along some linen sacks to use in the field. And what do I do with them today? They are so special, today my wife and I use the little bags for our Advent calendar at Christmas time.

(They both laugh and keep on painting.)

Thomas Scholten: … If you have strong colors in the soils, then those soils have undergone a lot of weathering, and have been eroded very intensively and/or for a very long time. That is, the color in the soil, if it is very strong, is in some ways a description of the amount of energy that has gone into it.

Ulrike Arnold: What about the absence of color? In cave paintings, there is no green. I've always asked myself why. Did the green rot away, or was it perhaps not used for religious reasons?

Thomas Scholten: I'd make that question even more complicated: What is the possibility that something was there that is no longer present? Green is predominantly something that characterizes plants. If the plant is no longer alive and stops photosynthesizing, then the green goes away. You can see it every autumn in the leaves on trees. And I could imagine that that's something you might have to consider here. In other words, green is something that belongs to life and to living organisms, and is not as lasting as a piece of stone.

Ulrike Arnold: But green earth pigments do exist, because of the copper content.

Thomas Scholten: Yes, but unlike copper, iron is ubiquitous. That is, it is found everywhere. According to the chemical composition of the earth's crust, iron is the fourth most common element after oxygen, silicon, and aluminum. That also explains why there are all sorts of red and reddish-brown shades in soils all over the planet, much more than green shades from copper, or yellow shades from sulfuric compounds.

Ulrike Arnold: There are cave paintings in many places in the world, and most are painted red. Maybe people just like painting with red.

Thomas Scholten: We human beings do a lot of things intuitively, but I think that there are often very pragmatic and simple reasons behind what we do. Iron is bound in the rock. It is released through weathering and combines with oxygen, resulting in these wonderful iron oxides. Many rocks that are this color are nothing less than soils which might have formed millions of years ago, were then eroded, then solidified again to form rock. The main elements that we have in the earth's crust are silicon, oxygen, iron, aluminum, and then a bit of calcium, magnesium, potassium, and sodium. The remaining elements of the periodic table account for less than one percent of the total volume. It's like when you used to play with Lego pieces and tended to have an abundance of primary colors: rocks are made up of these main elements, which are abundant across the planet.

Ulrike Arnold: … I need some contrast. What's missing alongside these earth pigments is some meteorite dust, something unearthly. What is it with meteorites? They fall to earth, after all. Shall we add that as a compositional element now too? Elements from other planets and cosmic bodies?

These are just iron too? Or nickel? The cosmic place of our planet is important, isn't it? The earth and its neighbors are connected though similar origins.

Thomas Scholten: The meteorite hits the earth and is no longer alone, no longer a separate entity.

Ulrike Arnold: Impact minerals.

Thomas Scholten: Yes. There are impact structures, and there are also impact minerals, such as coesite, a polymorph of quartz formed under high pressure.

Ulrike Arnold: (*Handling the pigment.*) … This material is extraterrestrial and beyond our world of experience.

Thomas Scholten: Bringing things back down to earth and speaking of the material behavior of existence here, if we were to put this painting out in the open air, in 10,000 years these swaths of black meteorite dust will also turn red.

(*The two talk about the power of nature, tell one another stories of their personal experiences of elements and landscapes, such as earthquakes and volcanoes, talk about climate change and about both the natural and anthropogenic greenhouse effect; and about the possible danger of meteorites striking the Earth again.*)

Thomas Scholten: What is the working process like as a painter? I'm having a lot of fun, but there's another question that's bothering me … It must be hard to find the right moment when a picture is finished, mustn't it?

Ulrike Arnold: Exactly. Enjoying the working process is the most important thing of all because it is so unpredictable. I love it when a picture

runs. I mean, when I'm painting outdoors, as I always do—today is an exception, of course—the elements often change the structures of the painting overnight. I always say very consciously that I work in cooperation with nature. I consciously allow natural influences such as rain or animal tracks.

… I have the feeling that we are nearly finished. But my feelings also tell me that we need to add a bit more green.

Thomas Scholten: Green?! … I thought we needed a bit more red.

(*He laughs. Both stop speaking. Ulrike Arnold throws her color onto the picture.*)

Ulrike Arnold: It's important to go around the picture and observe from every side. Basically, you can hang the picture any way up, as it has been painted from every side. And we have a view from the top, in contrast with other artists, who paint vertically. Here, you really have a view from above, onto the ground below. To me, that's what's so lovely about art: it's extremely intuitive and flexible.

Thomas Scholten: I think I'm finished now.

Ulrike Arnold: Me too. Just a little more green and, ah, another little sprinkle of meteorite dust. Really great.

Thomas Scholten: This was a really good alternative to building a sandcastle. I would have enjoyed that too though.

Ulrike Arnold: Yes, a sandcastle, we can still do that some other time. With painting, it's less

Photo of finished work.

Photo credit: Christoph Berdi.

predictable. You never know how it will turn out in the end.

(The experiment comes to an end. The soil–art dialogue in material composition and verbal exchange is complete. Ulrike Arnold and Thomas Scholten sign the painting. Both are satisfied with the result. They high five one another and leave the painting on the table to dry. At the annual meeting of the German Soil Science Society (DBG) 2015 in Munich, the work was exhibited along with a brief description of the discursive activity as creative experiment and cross-disciplinary knowledge transfer.)

Painting with Earth

Earth Pigments in North Devon;
A Guide for Teachers and Artists

Peter Ward

Peter Ward began his career in graphic design and illustration, engagement with subject being central to his practice. He has consequently worked with arts and environmental education bodies, including the Field Studies Council, Centre for Contemporary Art and the Natural World (CCANW), and YATOO (Korean Nature Artists' Association), to enrich and celebrate human experience in the natural world. Peter completed a Masters in Art & Environment at Falmouth University in 2012, receiving the Sandra Blow Award for

Title image: eARTh gown, painting together group work with Francesca Owen, Clare Thomas, and Sue Bamford, White Moose Gallery. Copyright Peter Ward 2015.

Outstanding Achievement. He shares a studio (eARTh) with his partner in Cornwall, UK exploring local pigments and materials through painting, installation, and workshops.

Since human beings have made marks and symbols to express our relationship with the world, we have been using pigments from the earth. Whether painting on cave walls or our bodies, we have found ways of utilizing colours from the materials found in our local environment. From minerals and plant dyes, animal parts, and more complicated processes we have explored and discovered ways to colour our world.

In 2008, at the invitation of the Appledore Arts Festival and through a personal interest in exploring the natural world, Peter Ward began researching the geology, history, and uses of earth pigments in North Devon in the United Kingdom. Earth pigments are simply coloured rocks and soils that may be used in the production of paints and colourings. This exploration has led to a rich interdisciplinary experience and understanding of the local area and beyond, involving geologists, soil scientists, historians, chemists, art restorers, ecologists, ceramicists, and other artists. It has inspired further projects with local museums and arts organizations and involvement with international arts projects to highlight the global importance of soil through the arts.[1]

With a little effort and some local knowledge, it is still possible to make our own paints from pigments gathered in our local environment, gaining insights into the history of painting and the industrial and geological history of an area, while deepening our connection with our environment and appreciating the use of earth pigments by different cultures around the world.

The following guide presents a good example of how such work may be directed at different audiences and learning levels. Through it we may learn to not only see the world differently but to touch the earth, to sense its age and complexity, and to express our joy as part of it.

The use of locally discovered pigments has brought my artwork intimately in touch with the idea of "artistic process"; from gathering the materials and mixing and creating my own paints, to producing images alive with the deep resonance of the natural world. In response to the materials, the imagery reflects the world on a molecular, energetic level, speaking in a language of its own. The work explores the spiritual and visual relationships between quantum theory and indigenous culture, reminding ourselves of the material beauty of this eARTh, and our own indigenous roots within local, contemporary contexts.

Peter Ward, 2009

Bideford Black

Bideford Black seam, Greencliff, North Devon, UK.

Copyright Peter Ward 2010.

Historic mining activity in North Devon.

Photo credit: local archives.

Raw Bideford Black, Greencliff, North Devon.

Copyright Peter Ward 2011.

Bideford Black Face, earth pigments.

Copyright Peter Ward 2013.

North Devon, in the southwest of the United Kingdom, has rich and varied geological formations dating from the Devonian era 450 million years ago, through Carboniferous, Permian, and subsequent glacial and interglacial epochs to the present day. Consequently the region has been endowed with a good range of accessible natural mineral colours and resources to use. While better known as an agricultural region, North Devon supported diverse heavy industry until the nineteenth century, alongside a globally significant wool trade.

Until 1969 a coal-based clay, known as Bideford Black, was mined and processed as a pigment to be exported around the world. Umber, mined in Berrynarbor in the north of the area until the 1790s, was highly sought after by Reeves of London, an artists' paint company. White "Ball" clay, significant in the region's famous potteries and used for a number of other ceramic applications, is still quarried and exported from pits at Peters Marland and Meeth. Red "Fremington" clay was mined until 2009, leading to the historic Brannam and Fishleigh Potteries. Copper and iron were extensively mined across Exmoor and exported to South Wales where it was processed, in exchange for coal and lime for agriculture and building. Railway lines, now mostly defunct, crisscrossed the area linking mining sites with ports.

Formed in lens-shaped pockets of clay alongside high-grade graphite coal between Abbotsham and Tawstock, Bideford Black is a unique pigment in the British Isles. The clay was formed from the lignin of tree ferns cloaking 8 km high mountains some 350 million years ago. It was subsequently buried 8 km underground as the landmass, Gwondanaland, originally situated south of the equator, moved steadily northward. The material occurs today as a sticky, intensely black clay, a harder coal-like substance and a fine black powder. It was mined using simple tools along a large network of tunnels following steeply sloping beds, then dried, ground, and packaged locally. It was used as a natural antifoul for timber boats, for artist and industrial paints, for printing inks, for colouring paper, stove polishes, cement products, bricks, floor tiles and linoleum, and rubber tires, as camouflage paint during WWII, and by Max Factor for mascara.

Like many of the other mining activities in North Devon there is little physical evidence left, but we may still meet people who remember the mines along with references to the industry in street names and buildings. Although the pigment was predominantly mined in Bideford, there is evidence of other mining activity along the length of the seam. In 2013, Peter led a project with the Burton Art Gallery and Museum with the aid of Heritage Lottery Funding, researching and documenting this locally significant industry, through artifacts, the knowledge of local experts, and aural accounts from ex-miners.[2] The project was followed by an Arts Council–funded residency program to explore Bideford Black as an artist's medium.

Other Pigments

Geological anomaly, Fremington Quay, North Devon.

Copyright Peter Ward 2010.

Permian Red Sandstone, Peppercombe Cliffs, North Devon.

Copyright Peter Ward 2009.

North Devon take-away (rocks, soils and clays).

Copyright Peter Ward 2014.

BURNT UMBER
Quaternary glacial deposits, 40,000 years old
Manganese dioxide
Fremington Quay; Grid Ref: SS 511331

BURNT SIENNA
Permian, 280 million years old
Iron oxide, formed as sandstone and conglomerate when land mass
was over the equator, retained along fault line
Peppercombe Cliffs; Grid Ref: SS 385246

YELLOW OCHRE
Quaternary glacial deposits, 40,000 years old
Manganese dioxide
Fremington Quay; Grid Ref: SS 511331

WHITE CLAY
Eocene secondary sedimentary glacial deposits, 50 million years old
Kaolin/granite/shale/sandstone/chalk run-off
Mined today as 'Ball Clay': used for bricks and potter's slip
Peters Marland, Grid Ref: SS 523098

FREMINGTON GREY
Carboniferous, 350 million years old
Manganese and carbon based mudstone
With traces of Iron Sulphide (FeS_2)
Fremington Quay, Grid Ref: SS 511331

'BIDEFORD BLACK'
Carboniferous, 350 million years old
>80% carbon, semi graphite coal measure
Mined in Bideford as a pigment until 1969
Greencliff, Abbotsham, Grid Ref: SS 406273

Six colours from North
Devon, pigment chart.

Copyright Peter Ward
2016.

On a small cliff on the River Taw estuary near Fremington Quay we can find at least four notable distinct colours: "Yellow Ochre" and "Burnt Umber" (manganese oxide deposits laid down in glacial lakes and streams 40,000 years ago), "White Ball Clay" (kaolin rich clay deposited in glacial lakes 500,000 years ago) and "Poor Man's Grey" (a carboniferous mudstone rich in organic matter, closely related to Bideford Black). At Peppercombe cliffs we find an iron-rich red Permian sandstone (quite like Burnt Sienna) captured in a fault line. The same sandstone is found across southeast Devon and West Somerset. This sedimentary rock was laid down in inland lakes and riverbeds 280 million years ago when the area, predominantly desert, was situated on the equator. The River Umber running through Combe Martin was named after the highly prized Umber pigment (iron oxide) that was mined there until the 1790s. Also in North Devon we may find a large variety of grays, browns and reds, and even a green from malachite (copper carbonate) at the old iron and copper mines at Heasley Mill near North Molton.

The majority of pigments may be accessed easily and gathered by hand where mineral seams and deposits are visible near the surface such as on river banks and sea cliffs, although industrially people would have dug a lot deeper to find them, following seams and deposits underground.

Wherever we are, the earth and its soils offer an incredible array of colour, indicative of a cumulative expression of a slowly shifting underlying geology and ongoing succession of flora, fauna, and climatic conditions. Throughout history, a region's cultural identity has been directly informed and shaped by the colours and materials found within it.

Making Paint

Until the 1830s almost all artist colours were made using natural pigments from mineral and plant sources. Mineral and plant colours were used in prehistoric cave paintings and have been gathered and processed by indigenous tribes for ritual and ceremonial purposes, for body painting, and decoration. Paint used in early classical art and what we might recognize as that of the "masters" would have been carefully selected for its specific qualities and bought from traders around the world. More precious and hard-to-find colours, such as ultramarine blue from lapis lazuli and imperial purple derived from rare mollusks, were used in imagery and textiles to signify power and beauty. Chemical experimentation and processes were developed to secure more varied, intense, and safe colours. Wars were fought for hundreds of years over colour—red pigment from the "cochineal" beetle was a closely guarded Spanish secret for 200 years until stolen by a Frenchman. Trade routes developed across the world to transport pigments. Colours, and their sources, are something that we very much take for granted today, but has been serious business throughout history and has influenced, informed, and enriched our civilization to this day. However, if we were to ask today where does paint come from the majority of us would not know.

By simply drying and crushing the raw materials, we have a starting point for making paint. The powders produced may be mixed with water, linseed oil, PVA (polyvinyl acetate), egg yolk, tree saps, animal glues and milk proteins, or other commercially available mediums to produce paint. Some of the more clay-like materials may be molded into pastels, others drawn with directly. Through experimentation we may find out which raw materials work best to produce a desired paint or effect. We can spend time examining their consistency, their luster and graininess, and their behavior on different surfaces. We may find how each rock and soil has not only a different colour, through the way its constituent minerals and structure reflect and refract light, but also a different way of behaving with each medium. As artists we can chose how fine we grind the pigments, where our materials come from and express ourselves through this knowledge. The phenomenal age of the pigments may be proof of their permanence and light fastness, although some mineral pigments are liable to oxidation over time.

Art is a space to enrich, celebrate, and transform our perception of our being in the world. The practical exploration of earth pigments through workshops, in the studio or in the field, may provide such a space. Through the input of geologists and local historians alongside hands-on experience of gathering, processing, and using earth pigments as paint, workshops offer a new and informed multisensory and site-specific interpretation of a place or region. Working with earth pigments in relation to geology and history inspires new perceptions of time and place. It may alter our understanding of where we live, raising questions about history and our relationship to place and colour. Working with earth pigments also opens up practices of vernacular art, from graffiti to cave art, and to the ephemeral natures of these diverse materials, their value, erodibility, and ecological history.

Earth Pigment Workshops

Painting together/ Bideford Black field trip, Greencliff, North Devon. It is important to remember when gathering pigments to think about the amount of raw material that is available to use and the visible and structural damage that may occur due to your "mining" activities. Some rocks, sediments, and minerals are also highly toxic, particularly in areas mined for heavy metals such as tin and lead. We advise extreme caution, respect, and local knowledge in our workshops.

Copyright Peter Ward 2015.

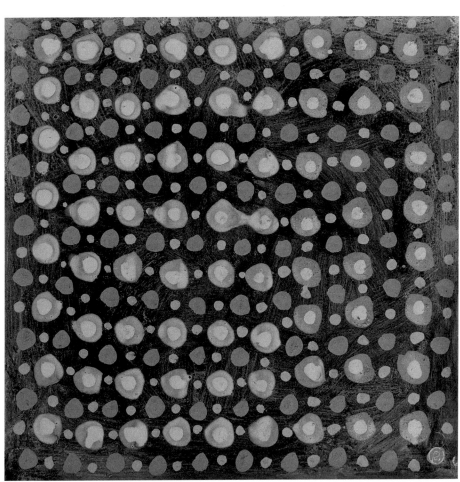

Potential II (earth pigments on paper).

Copyright Peter Ward 2009.

Selected Bibliography

1. *Colour: Travels through the Paintbox*, Victoria Finlay (London, Hodder & Stoughton, 2002).

2. *Bright Earth: The invention of Colour*, Philip Ball (London, VINTAGE, 2008).

3. *The Painter's Handbook*, Mark David Gottsegen (New York, Watson-Guptill Publications, 2006).

4. *Exmoor's Industrial Archaeology*, edited by Michael Atkinson (Tiverton, UK, Exmoor Press, 1997).

5. *Bideford Black: The History of a Unique Local Industry*, Sound Archives North Devon (Bideford, UK, SAND, 1994).

6. *Devon's Non-Metal Mines: Discovering Devon's Slate, Culm, Whetstone, Beer Stone, Ball Clay and Lignite Mines*, Richard A. Edwards (London, Halsgrove, 2011).

Acknowledgments

I would like to thank all those who have contributed and supported my research, especially Dr. Chris Cornford for his wonderfully creative geological knowledge; Sandy Brown and the Appledore Arts Festival for initiating the project and continued support; The Burton Art Gallery & Museum in Bideford and the ex-Bideford Black miners and local people for their stories; the White Moose Gallery in Barnstaple; and to everyone who has attended my workshops and enjoyed my artwork for keeping the project alive.

Endnotes

1. http://ccanw.org.uk/soil-culture/

2. http://bidefordblack.blogspot.co.uk.

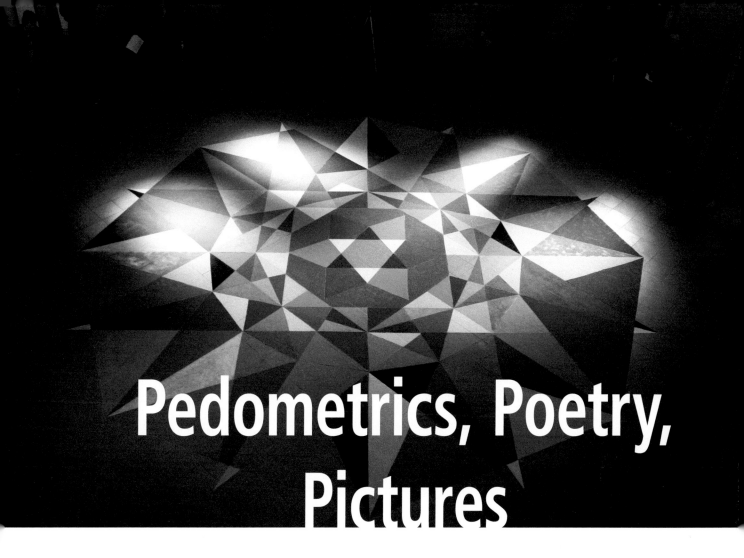

Pedometrics, Poetry, Pictures

Alex McBratney and Elvira Wersche

Elvira Wersche studied at the Hochschule für Bildende Künste Braunschweig at the Gesamthochschule Kassel, Germany, where she finished her study of fine arts and higher art education cum laude in 1972. Since then she has lived and worked as an artist in the Netherlands, having taken part in many international conferences and exhibitions, for example, at the Pergamon Museum Berlin, Germany; the National Museum of Antiquities, Leiden, Netherlands; SMIC Sharjah Museum of Islamic Civilization, United Arab Emirates; the Museum of Arts and Design, New York; and the Design Center of Istanbul, Turkey. Wersche often collaborates with musicians and dancers, and explores interdisciplinary relationships between art and science in her work. She has studied with experts in the fields of mathematics and Islamic design in Iran, Uzbekistan, Tajikistan, Morocco, and Turkey. She

Title image: Sammlung Weltensand, Installation and performance views of *Kazimierz*, 2005. Synagoga Kupa, Krakow, Poland. Audio Art Festival, 2005. Size: 7.5 × 7.5 m. Material: sand gathered from 500 locations around the world. Music: Horst Rickels and students of the Art Science Interfaculty of the Royal Conservatoire The Hague, The Netherlands.

Photographer: Elzbieta Kasperska; reprinted with permission from Elvira Wersche.

sees her main body of work, Sammlung Weltensand, as a work of art with intersections to geology, mathematics, and design, and she hopes it can serve as a catalyst for intercultural and interfaith harmony wherever it unfolds.

Alex McBratney holds BSc, PhD, and DSc degrees in soil science from the University of Aberdeen in Scotland, and the DScAgr degree from the University of Sydney in Australia for research in precision agriculture. He has made major contributions to soil science through the development of the concepts of pedometrics, digital soil mapping, and precision agriculture. After completing his PhD work at Rothamsted Experimental Station in the United Kingdom, McBratney spent seven years with CSIRO Division of Soils in Brisbane. McBratney joined the University of Sydney in 1989. He has just completed terms as Dean of the Faculty of Agriculture and Environment and chief editor of the global soil science journal, *Geoderma*. He is heavily involved with the activities of the International Union of Soil Sciences, having served as Deputy Secretary General, and the global digital soil map project, GlobalSoilMap. In 2014, he was awarded the VV Dokuchaev medal by the International Union of Soil Sciences. Currently he is helping to develop and promote the concept of global soil security.

For the ongoing project, Sammlung Weltensand (Collection of World Sand, 2001–present), Elvira Wersche has collected thousands of soil samples from all over the world, which she uses to create large geometric floor designs in meditative performances and installations in museums, churches, and other public buildings. After the design is complete, all colors are merged in a simple final performance as sand from all over the world is swept together and then given to members of the audience. With a passion for artworks and poetry, Alex McBratney looked at Wersche's art and was inspired to write some poems. In response Wersche wrote down some ideas and the dialogue process iterated from there. The two discussed soil diversity, the field of pedometrics, and the idea of soil nescience, or that which cannot be quantitatively known about the soil.

Sammlung Weltensand, *Taqsim—Division: quest for the other*. 2009. Landesmuseum für Natur und Mensch, Oldenburg, Germany. Floor installation on occasion of the exhibition *The Art of the Early Christians in Syria*. Size: 6.5 × 6.5 m. Material: sand gathered from 600 locations around the world.

W. Kehmeier; reprinted with permission from Elvira Wersche.

Dirt

Dirt
A four-letter word
For soil
For life
For ever

David van der Linden

Viento, agua, piedra (excerpt)

…

El viento esculpe la piedra,
la piedra es copa del agua,
el agua escapa y es viento.
Piedra, viento, agua. …

Octavio Paz

Oda a la sal (excerpt)

…
canta la sal, la piel
de los salares,
canta
con una boca ahogada
por la tierra.
Me estremecí en aquellas
soledades
cuando escuché
la voz
de la sal
en el desierto.
Cerca de Antofagasta
toda
la pampa salitrosa
suena:
es una
voz
quebrada, …

Pablo Neruda

Pedome

Pedome pedome pedome
Earth generator
Biodiversity guardian
Humanity stanchion
Nature amanuensis
Pedome pedome pedome

David van der Linden

Digging (excerpt)

Nicking and slicing neatly, heaving sods
Over his shoulder, going down and down
For the good turf. Digging.

The cold smell of potato mould, the squelch and slap
Of soggy peat, the curt cuts of an edge
Through living roots awaken in my head.
But I've no spade to follow men like them.

Seamus Heaney

The unity of I

Pedometrics is a search for a quantitative understanding of soil properties, distributions, and processes. It tries to move us from the qualitative to the quantitative but is not foolish enough to think that that is all—by doing this we hope eventually for a deeper knowledge of how the soil evolved in any place. It attempts to answer where and when is soil, but also what and why is soil. Quantification allows us to turn our ideas into formal mathematical models and moreover to test, reject, and improve those ideas. A corollary of this understanding is hopefully the ability to monitor and manage the soil as it reacts and modifies to the onset of human manipulation. In the pedometric search for understanding, the diverse and multiscaled spatial and temporal patterns of soil properties and soil materials as well as mental constructs such as systematic soil classification become tools of inquiry in our armory of applied science.

One of the objectives of pedometrics is to deal with the continuity and variability of soil. Soil scientists know that soil is notoriously diverse. Knowing one soil profile calls for a simple account, but the diversity of millions requires manifold discussion and explanations. Some scientists argue about the continuous nature of soil—there is no real individual, but rather a unifying continuum—one soil that covers the whole world. The poem "The unity of 1" encapsulates these ideas, which are important to pedometrics. While it could be adapted to any local soil in any language, this version from England also tells of the torment of a scientist whose job it is to conduct the soil classification for a country, except that this particular scientist abhors making decisions or drawing firm conclusions. There is sometimes an agonizing humanity of science at work.

The visual artist brings another realm of perception and understanding to the field of pedometrics. The soil material itself has a deep meaning and fascination to the visual artist in its natural state and its coincidental poverty and wealth. In addition to its geopedological character, the cultural history is important, because the soil carries layer by layer not only the long-term history of our planet but also our human history. The "mystical" dimension of soil touches and inspires humanity in religion and art. It is the storehouse, the collective memory of the earth. The ideas behind "The unity of 1" are reflected in the artwork of Elvira Wersche, who has worked with mathematicians to determine and in a sense "quantify" her aesthetic designs with soil color. The following image is from 2011, made for the Rijksmuseum van Oudheden, Leiden (Netherlands). The pattern is created from the division of the circle. The transformation takes place from the undivided unity with its invisible

Sammlung Weltensand, *Shamseh*, 2011. Rijksmuseum van Oudheden Leiden. Event held at the exhibition *Merchandise and Souvenirs: Islamic Art from the Rijksmuseum, Amsterdam*. Size: 7 × 7 m. Material: sand gathered from 600 locations around the world.

Photographer: Elvira Wersche; reprinted with permission from the artist.

The Unity of I

Once
There was only one soil
In all our worlds
For you
A Martock brown earth
In a cider orchard
On the Isle of Avalon
No need for maps
No doubts
No arguments

You were the one
Who often asked
Why did we have to
Find another
And another?

David van der Linden

center to diversity, a path of development that we also see in nature. This state of paradise—the undivided unity—without the argument of maps, is a place of longing for romantics. Everyone should have such a place of stillness and loneliness.

Sand Story

The intrinsic diversity of soil is evidenced by Wersche's large collection of sands from more than a thousand locations from all over the world, referred to as her "sand atlas." The collection is spontaneous and randomly assembled. It represents many countries from all continents. It holds samples from numerous landscapes, but it has no systematization. Although apparently random, the pedometrician would seek to find the relationships between them spatially, and more important, to order them according to the hidden patterns of natural pedogenic processes that are probably different from the patterns that please or vex the eye.

When Wersche works out her floor patterns, it is the aesthetic aspect of the color of the sand that plays an important role in composition. But this aesthetic ordering is not all. By showing the public the beauty and richness of the earth, the composition highlights ethical and political issues related to soil security—issues that soon transcend the domain of pedometrics and move from the mathematical, chemical, and biophysical to the philosophical, social, political, and metaphysical. The experience of this richness and diversity creates awareness and respect, because most people are very unconscious about soil. This awakening, attentiveness, and a sense for beauty can trigger critical and humane attitudes toward our environment.

When people bring sand and soil to Wersche from their travels, they ask her if she has ochre. Many people don't realize that there are countless hues of ochre, based on subtleties of mineralogy, and that soils of different locations are never identical. Pedometrics would still ask to which degree this is deterministic or random.

To the artist, soil diversity raises the question of the impact of the color, consistency, and composition of the soil on the character and identity of its inhabitants. Many soil scientists would agree with this notion of environmental determinism. One can imagine the heavy clay soil of the parts of the Low Countries affects people differently than the light sand of the Sahara. But does this identity inevitably change with alienation from the soil, as we all migrate to cities and then to virtual realities? Or is a connection to soil intrinsic to human nature? Wersche believes that art starts from a longing for another world.

Sand Story

Soil miscellany
Melodious atlas
Flurried and gusted
Collected and clustered
From citizen expeditions
To every pedome
All but every dab
In the bibliochrome

Extemporaneous
Driftlessly assembled
Classless haphazard
Angular tessellation
Or heavy hidden
Pedogenic patterns
Irking thought and vista?

Specious spangled sand
Constitution and harmony
Electrifying exquisitude
Transcends the algorithmic
To the metaphysical and ethereal
Bounty and multiformity soon
Attends awareness and approbation
Focuses noble quandaries
And political puzzles
Justifies soil security
This vivication
A proper sensibility
For the envisioned milieu

She has ochre and daub
In haunting hues
Subtleties of goethite and haematite
Humus and everyother-ite
Soil imprints
Character and identity
In its denizens
Modernity devolving autochthons
To soil aliens
Transpedological people
Severed from their earth
Pursuing art and explanation
Longing for their lost kinship

David van der Linden

Sammlung Weltensand, Installation and performance views of *Kazimierz*, 2005. Synagoga Kupa, Krakow, Poland. Audio Art Festival, 2005. Size: 7.5 × 7.5 m. Material: sand gathered from 500 locations around the world. Music: Horst Rickels and students of the Art Science Interfaculty of the Royal Conservatoire The Hague, The Netherlands.

Photographer: Elzbieta Kasperska; reprinted with permission from Elvira Wersche.

The transformation, performance Gallery Bunkier Sztuki, Krakow, Poland 2001, 5 × 5 m Material: pulverized coal, laser light, crystals

Photographer: H. Rickels; reprinted with permission from Elvira Wersche.

Soil Nescience

Soil nescience literally means "what we don't know about soil"—the antonym of soil science—and some would argue a much larger subject. This poem relates some of the things we don't know, like where does soil come from? Why is soil really here? It treats soil as a dark mysterious box; soil as a question mark; soil as a hidden dimension; soil variation itself as a resilient self-protection mechanism. These views underlie many of the questions that pedometrics—and art—pursue. This poem is illustrated by a mysterious imprint on its surface: What caused it? Why is it only this deep? How long will it last? What effect does this pattern have on what is happening underneath?

This reveals an apparent paradox of any science, including the science of the soil—as we learn more we ask more and more questions—so relatively speaking, do we know less and less?

Soil Nescience

Soil variability –
A multiform jewel
Or a capricious pig
In a heterogeneous poke?
God tossing the dice
Across the earth's felt
Games and throws of craps
Since time was created
The crazy paving of
An unknowable artisan
Or a nonlinear god
Toying with
Pedogenic parameters
Out on the edge?
Chaos realised
Over and again.
Or is it simply
Clorpting away
According to a plain
Old-fashioned determinism
With unknown factors
Yet to be revealed
By further understanding
And meticulous measurement?
And in this nescient state
Can we husband it
To a centimetre?
Or is the soil's diversity there
To protect us
—From ourselves?

David van der Linden

Catena

Catena, a topographic sequence of a soil and landscape, was described by Geoffrey Milne in the 1930s based on his field observations in East Africa. Catena is seen as a bejeweled necklace—a thing of great beauty arising from natural processes of transformation and translocation. This poem asks whether this sparkling view is the truth and whether we are fooled by the simple beauty and there are deeper relationships to be unearthed.

The "transformational" aspect of landscape processes evoked in the poem is picked up in the image from the performance/installation *Transformation* in the Bunkier Sztuki Gallery Krakow, Poland from 1997. The materials are pulverized coal from a coal mine in Dinslaken, crystals, and laser light. The theme of this work is the transformation of light to carbon as it happens in the soil. The photosynthesis in the leaf, the absorption of light (white) and the transformation finally to carbon (black). This is as an artistic visualization of the long evolutionary process of the Earth.

Catena

String of soils
Hanging in the landscape
Necklace of profiles
Adorning the neck of evolution
Jewels of transformation
Twinkling in the half-light
Of pedogenesis
Or schlock for deception
Glistering pretence
Of the real earth

David van der Linden

From Earth

herman de vries in conversation with Nico van Breemen

The visual artist **herman de vries** (1931) considers himself invisible in his work: "i have nothing to say, it is already here." This was the case for his first collages trouvés and white monochrome works in the late 1950s, and for his white wood reliefs, drawings, and collages, which he called *random objectivations* in the '60s. And it remained so when he began to collect and present objects from nature in the '70s. Few artists have created a body of work that is so completely and consistently based on a few simple principles while yielding such tremendous visual variety of art works.

Title image: View of the installation by herman de vries titled *musée des terres* ("earth museum") in Musée Gassendi, Digne-les-Bains, France.

Photo: susanne de vries, reproduced with permission.

herman de vries represented the Netherlands during the 2015 Biennale of Venice. His impressive exhibition in the spacious, light-saturated Rietveld pavilion featured, among other things: soil colors, sickles, and boulders from all over the world; tokens of natural phenomena and human activity collected throughout the Venetian Laguna; a large circle of fragrant rose buds on the floor; and a book with the names of all the plants that the artist remembered having consumed in the course of his life.

Soil features very prominently in the work of herman de vries. Over the past 40 years he has collected over 8000 soil samples. Since 2009 these are kept in the *musée des terres* within Musée Gassendi in Digne les Bains (France). In his studio in Eschenau, in the district of Hassberge in Bavaria, Germany, he maintains an archive of this collection: over 450 sheets of paper (35 × 25 cm), each with 20 small earth rubbings, and a separate list with collection sites and dates of all samples. A reproduction of the archive has been published as an artist book (herman de vries, 2016, *the earth museum catalogue 1978–2015*, volumes I (index) and II (473 p.), edition of 200 copies, signed and numbered by the artist, published by Peter Foolen, Eindhoven).

Selections from the collection are shown regularly in exhibitions as small heaps of earth in vitrines, as rectangular-shaped thin layers of soil put directly on the floor, or as earth rubbings on paper. These works are connected to the environment of the museum where he has a show. The earth rubbings are usually titled *from earth: …* with the sample location and the date of preparation.

For me (van Breemen) as an author, and as someone who worked for 35 years at Wageningen University as a soil scientist before starting an art gallery, the approach of de vries to soil (which he invariably calls "earth") is fascinating. Scientific and practical questions that automatically pop up in the mind of a soil scientist when observing a soil (for example, to what soil series/taxonomic unit does it belong, how was it formed, its ecological role, what crops will grow well on it) are taken over by meditative musings that make such pondering, at least temporarily, irrelevant. In early June 2016, I had a telephone interview with herman de vries about his soil-related work.

Nico van Breemen: herman, around 1975 you shifted gears in your work. Your *random objectivation* works of the ZERO years made way for aspects of nature: plants, stones, shells, and, of course, soil. What was behind this shift?

herman de vries: i worked with chance, using rigorous statistical programs that were based on observations in the real world. (Being opposed to any kind of hierarchy, herman de vries exclusively applies lower case in all his texts, including his own name.) for example, the distribution of scots pines in the national park "hoge veluwe," where the kröller müller museum is situated. together with my wife susanne, i made increasingly complex line and dot drawings. they didn't completely satisfy me; they were two-dimensional, while we felt we actually needed a third, and even a fourth dimension. then susanne and i looked at each other. we started to laugh and said, "is there actually a better model of reality than reality itself?" from then on i concerned myself with, for example, leaves and soil. not a single leaf on a tree is the same as another, each one is different, there is no repetition. this variation fascinates me. in a botanical garden in england i worked with botanists to collect leafs of dwarf birches. when i told them that i wanted to also collect earth they laughed and said it wouldn't make sense, the soil is just brown here, everywhere. they were

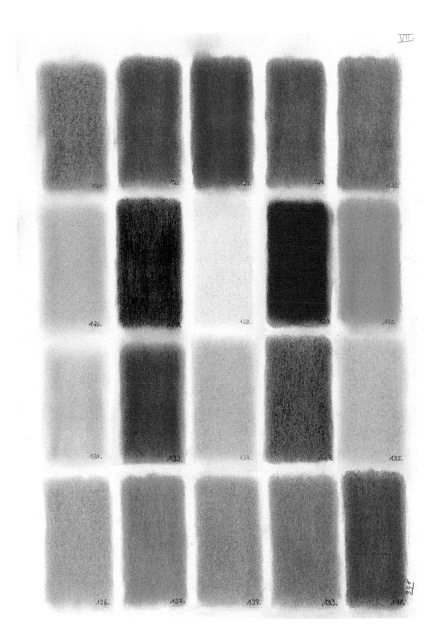

Reproduction of page 398 of herman de vries'
earth museum catalogue 1978–2015, Vol II
(samples taken in Goa, India), 313 × 228 mm.

Reprinted with permission by the artist.

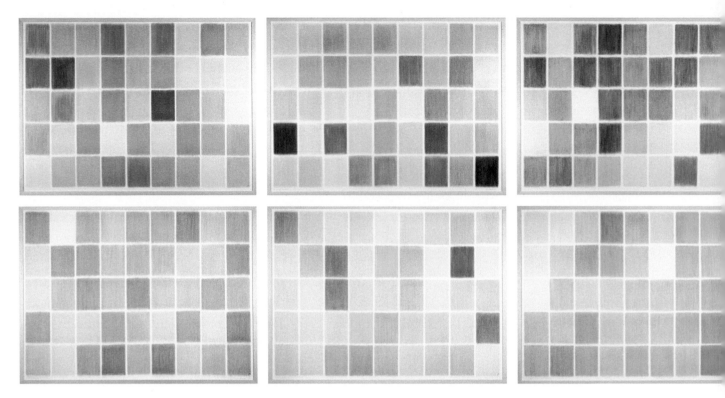

herman de vries, 1995–1996, *comparative landscape studies*, 12 panels of 73 × 102 cm, each with 45 earth rubbings.

Photo: studio herman de vries/Joana Schwender.

Reprinted with permission from the artist. Origin, from left to right and from the top down, from earth: eschenau & northern steigerwald, 1995; from earth: western ross, scotland, 1995; from earth: sicily, 1994; from earth: el hierro, 1995; from earth: senegal, 1995; from earth: oberwallis, 1995; from earth: from around erfurt, 1996; from earth: massif du tanneron, 1995; from earth: nepal, 1995; from earth: stiftland; from earth: leros, 1996; from earth: north groningen, 1995.

astonished to see the great variation in the colours of the samples that i had collected. and those were biologists, people who worked outside and closely observed the world surrounding them! in lausanne i had a similar experience. so, I came to the concept of "chance & change," interrelated ideas: no chance without change, a dynamic model of reality.

Nico van Breemen: In your *random objectivations*, the dice (or, in fact, statistical tables of random numbers) determined, for example, how many small blocks of wood you placed exactly how and where on a given wood panel. Some of those wood reliefs are visually closely related to your leaf fall works, which show tree leaves distributed on sheets of paper exactly as they swirled down during a day during the fall. Does chance also play an important role in your work with elements from nature?

herman de vries: the concept of "chance" has kept me busy for a long time. i've given it much thought, and considered many definitions. one i like is contained in the plain german word for it, *zufall*, that what befalls you, without knowing why. "chance" is a word for not knowing the causes behind events and processes. so, yes, it plays a role in the leaf fall works. we cannot come to a complete fundamental understanding what is ultimately behind natural phenomena. that is true even for science, which seeks to understand cause and effect chains. i consider art and science as two related approximations of reality. two adjacent approaches that sometimes fertilize each other.

Nico van Breemen: In your *random objectivations* you strived for "objective art," you wanted to

eliminate the personal. How "objectively" do you choose the colors in your *from earth* works? Does aesthetics play a role there?

herman de vries: the word "aesthetic" stems from the greek *aisthētikě* (*tekhne*) or the "art of observation." "to observe" translates into german as *wahrnehmen*, or "to take what is true," but, yes, i like earth colours and that plays a role in choosing them. i cannot and do not want to exclude my sense of beauty. if a soil that is mainly red has lilac or purple tints here and there, i find that fascinating, and i probably select those colours too. so, i'm not strictly objective. art, just like science, is never completely objective.

Nico van Breemen: In your works with patches of earth, with pebbles, with leaves, etc., the elements are arranged in a regular two-dimensional grid. Why? Do you arrange those at random or by aesthetic criteria?

herman de vries: by placing similar elements in a regular grid you can compare them directly. and it's the differences between them that i find interesting and important. for the same reason, they are arranged at random: i pick them blind so as to avoid any bias caused by color preferences or aesthetic aspects.

Nico van Breemen: The elements from nature that you consider are usually unobtrusive, they are very common and everyone can see them. You don't show impressive landscapes and species of rare "charismatic" fauna that feature in so many TV films about nature. Is that a deliberate choice or are there also practical reasons involved?

herman de vries: i consciously choose a "down to earth" approach: i pick the simplest things, the earth on which i walk, things that come my way, those things that i see and i can put together and can compare. it is the differences and changes that i see nearby, in front of me, that fascinate me. the colours of the earth under my feet differ between here and there, i find that captivating. in la gomera, a fully volcanic island, i collected 500 soil samples. if you look closely, you see they're all different.

Nico van Breemen: Your works *from earth* are made, after the crushing the soil in a mortar, by rubbing it in vertical strips with your fingers on paper. Why vertical?

herman de vries: in my early earth rubbings i tried out different ways and directions of application. rubbing the earth with my fingers in a vertical direction was most convenient and worked best.

Nico van Breemen: Does soil texture present limitations? Coarse sand does not seem very useful to earth rubbings.

herman de vries: yes, i try to avoid coarse sand; it provides hardly any color and it gives me bloody fingers when rubbing it!

Nico van Breemen: Soil scientists focus on the soil profile with its information on the vertical distribution of soil properties 1–2 meters below the surface of the earth. They use that information to characterize and classify the soil or to analyze the processes responsible tor its development. Do you also take deeper soil layers into consideration?

herman de vries: in forests i take the top layer away, the layer i call "substrate," to come to the earth below. substrata are another, separate, aspect in my work. normally i don't dig deeper. if there is an excavation at hand, i often do collect material from deeper layers, but i rarely made profile images. changes are important in my work, but i know little about the soil processes behind them. i just want to show their visual effect. they determine how the soil looks, and in particular its color, which is something that others usually do not see. i'm just using a very simple approach, but its results are that time and again they are met with great surprise; apparently almost everyone passes by these phenomena unnoticed.

Nico van Breemen: Preparing and making earth rubbings seems to be easy. What about the beautiful circular windows of polished rock that you made in rocks of different geological periods in the Réserve Naturelle Géologique in the Haute-Provence, France? They offer a surprising look through the drab, weathered surface layer, into the fresh unweathered rock and reveal beautiful details of minerals and fossils! Not so very simply made, those windows!

herman de vries: yes, for those i had help from a geologist for the geological information and from a stonemason of tombstones to mill the windows and polish them. thanks to the curiosity of an artist, geology students are now led past them on excursions so as to be able to a look directly at the distant past!

Nico van Breemen: Does your interest in earth also stem from its role for terrestrial life: as a recycler and supplier of water and nutrients and as a seedbed for future generations of plants?

herman de vries: yes, the earth is an important primary factor, for example, because it provides our food. that is an aspect of my forthcoming exhibition in the museum für konkrete kunst in ingolstadt, *stone, earth, wood*. earth and wood are fundamental concepts in the *i ching*, the ancient chinese book of changes. stone changes to earth, and earth allows the growth of plants which feeds all terrestrial animal life, and produces wood. in all these processes earth plays a central role. earth is the base.

Nico van Breemen: herman, a final point: Can you imagine that your soil museum will ever be complete in that it contains all samples of earth that you consider worthwhile for the collection?

herman de vries: no, it can always be added to. the greek island gavdos is the southernmost tip of europe, but geologically it belongs to the african plate. i recently visited it and collected 43 soil samples, many of which had a fascinating greenish yellow color. the gavdos samples will be used for earth rubbings in works dedicated to gavdos, and the rest of the material goes to digne to enter the *musée des terres* up there. that could go on and on forever.

Correlation Drawing/ Drawing Correlations

Margaret Boozer and Richard K. Shaw in conversation with Claire Huschle

Margaret Boozer lives and works in the Washington, DC, metro area. She received a BFA in sculpture from Auburn University and an MFA in ceramics from New York State College of Ceramics at Alfred University. Boozer developed an interest in digging native clays that has led to collaborations with soil scientists and work that explores intersections of art and science. Her work is included in the collection of the Smithsonian American Art Museum, the U.S. Department of State, the Wilson Building Public Art collection, and in many private collections. Boozer taught for ten years at the Corcoran College of Art and Design before founding Red Dirt Studio in Mount Rainier, Maryland, where she directs a business-of-art seminar and artist incubator. Recent projects include

Title image: *Correlation Drawing/Drawing Correlations: A Five Borough Reconnaissance Soil Survey,* by Margaret Boozer, installation view, as exhibited in *Swept Away: Dust, Ashes and Dirt in Contemporary Art and Design,* Museum of Art and Design, New York, NY 2012. 109″ × 103″ × 9″.

Photo credit: Suzanne DeChillo/The New York Times/Redux.

a large-scale mapping relief for MGM National Harbor in Maryland using clay and artifacts dug from the site.

Richard K. Shaw currently serves as a state soil scientist for the U.S. Department of Agriculture Natural Resources Conservation Service (USDA-NRCS) in New Jersey. His responsibilities include providing technical soils assistance to internal (NRCS) and external customers, serving as liaison to National Cooperative Soil Survey partners, and overseeing the management and distribution of local soils information. Shaw received a BS in natural resource management from the University of Maine, and MS and PhD degrees from the Department of Soils and Crops at Rutgers University. Shaw began his career in soil science as a laboratory technician at Rutgers University in 1979, providing technical support for soils-related research. He joined the USDA-NRCS as project leader in 1996, serving six years in northern New Jersey, followed by ten years in New York City.

Claire Huschle (dialogue moderator and editor) is a curator, writer, and educator living in the Washington, DC, area. She received her BA degree in art history from the University of Michigan and her MA in art history from the University of Texas at Austin. Huschle served as director of the Arts Management program at George Mason University in Virginia from 2013 to 2017, where she has been on faculty since 2007. She has held executive committee positions in the Washington, DC, chapter of ArtTable, the national professional organization for female executives in the visual arts. She has held leadership positions at the Arlington Arts Center; the Torpedo Factory Art Center in Alexandria, Virginia; and Duncan & Miller Gallery, in Washington, DC. She first worked with Margaret Boozer through Duncan & Miller Gallery, and the two have cocurated several exhibitions. Huschle is currently in residence at Red Dirt Studio, from where she runs Scaffold, a project management support firm for artists.

The city gives the illusion that the earth does not exist.

Robert Smithson

In 2011, the Museum of Arts and Design (MAD) in New York City invited Margaret Boozer to participate in an exhibition titled *Swept Away: Ashes, Dirt, and Dust in Contemporary Art and Design*, and specifically requested that she construct the work using New York City soil. As an artist, Boozer had long been interested in soil and its ability to capture the historical record, both geologically and anthropologically, so the request itself was wholly within her methodology. The challenge she faced was simply that she knew nothing about the city's soil. In order to create the most powerful artwork possible, she wanted to understand more about it. Through her network of soil scientist colleagues in Washington, DC, she met Dr. Richard Shaw, the lead scientist on the then recently completed New York City Soil Survey. He agreed to help collaborate.

At their second meeting, they passed through Shaw's former office on Staten Island, which his team had recently vacated, to pick up some equipment to use in that day's planned dig. Almost as an afterthought before leaving the office, he opened a storage room that held the correlation boxes from the survey, each containing soil samples. Boozer knew in that instant that she'd found the core material that would become the focus of her installation at MAD.

Correlation Drawing/Drawing Correlations contains the soil samples from all five boroughs, which were extracted over the span of more than 15 years. When fully installed, the soil samples are presented in crisp Plexiglas translations of their original cardboard boxes. Each box is divided into eight sections, reflecting the content and strata of soils present in a four-foot deep boring, and correlating to a given location. Correlation boxes are simple and functional tools for collecting and transporting samples. Translated into Plexiglas, they become tools to expose the beauty, variety, and complexity of the earth they contain.

The installation reveals a huge variety in the soils: green, rocky, serpentine soil unique to Staten Island; gray sparkly schist bedrock from Central Park; and a sandy soil series in Brooklyn and Staten Island, named Bigapple formed as a result of dredging activities in the coastal waterways and rivers. (The U.S. National Cooperative Soil Survey lists the Bigapple series as consisting of "very deep, well-drained soils formed in a thick mantle of sandy dredge spoils from dredging activities in coastal waters and rivers. These soils occur on modified landscapes in and near major urbanized areas of the Northeast. Saturated hydraulic conductivity is high or very high in the surface and very high in the subsoil and substratum." The taxonomic class is Dredgic, mixed, mesic Typic Udipsamments, and the typical pedon is Bigapple fine sand on a large smoothed manmade island of soil material on a 0 to 3 percent slope.) There are dark coal ash soils formed from the deposition of coal burning furnace refuse over 100 years ago. Other soils display bits of glass, brick shards, and twisted metal.

At MAD, the correlation boxes were installed in a roughly map-like fashion—all samples from the Bronx sat north of Manhattan, which sat to west of Queens and Brooklyn. Oriented in this way, viewers could find *their* borough and *their* neighborhood—and quite literally see the earth that existed beneath the movements of their everyday lives. Mounted on a wall next to a window facing onto neighboring Central Park was the park's correlation box. From within the museum, viewers could simultaneously comprehend that soil ran below both the natural environment they observed outside and the built one in which they stood: in one moment breaking the very illusion within Smithson's quote, "The city gives the illusion that the earth does not exist."

Detail, *Correlation Drawing*, installation view, as part of the exhibition *Swept Away: Dust, Ashes and Dirt in Contemporary Art and Design*, Museum of Art and Design, New York, NY 2012.

Reprinted with permission from Margaret Boozer.

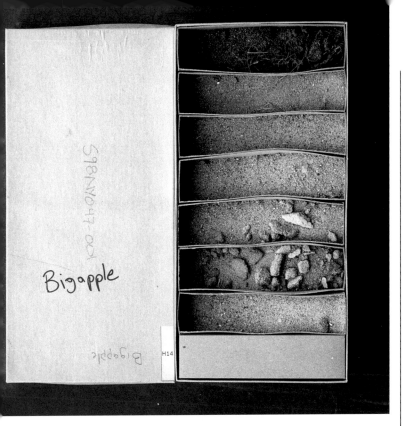

Original correlation box and contents, Bigapple. H14 sticker refers to new cataloging as part of *Correlation Drawing*.

Reprinted with permission from Margaret Boozer.

Finally, next to the Central Park correlation box was a QR code so that visitors could access more information about the Soil Survey, the individual samples, and more details about the science and the soils. The exhibition ran from February 7 to August 5, 2012.

Claire Huschle: The New York City Soil Survey is the inspiration for the art installation *Correlation Drawing/Drawing Correlations*. An extensive project covering all five of the city's boroughs, the Soil Survey assesses multiple functions of soil, serves as an important record of New York City history, and has become a model for future surveys and studies.

Richard, how did you get involved with the survey, and what excited you about being part of it? Margaret, what intrigued you about the results of Richard's work? When did you see the connection to your own work?

Richard K. Shaw: I was Soil Survey project leader for USDA-NRCS in Northern New Jersey when the project leader position for the New York City survey opened up in 2002. I was interested, but unsure whether I could handle the headaches of working in the city. I knew I might not see this kind of opportunity again, and decided I was not too old for some adventure.

In addition, I had visited the NYC Soil Survey crew on several occasions and always admired what they were doing. I thought then—and

Profile of Bigapple soil, formed in sandy dredged materials sourced from coastal waterways, bays, or rivers. Classification is: Dredgic, mixed, mesic Typic Udipsamments.

Reprinted with permission by Luis Hernandez.

still do—that urban soils were one of the final frontiers for our agency and our science, and that this work was particularly important since most of our population lives in urban and suburban areas. I also admired the network that the survey crew had established in the city: municipal, state, and federal agencies; colleges and universities; nonprofits and community groups. The program included soil sampling and research projects, site investigations, soils training and lectures, and provided volunteer and internship opportunities, in addition to soil mapping. It was about raising awareness of soil science, demonstrating that we could provide useful information, and establishing our presence in the urban environment. New York City Soil Survey information is now available online from Web Soil Survey, our agency's official source of soil survey information. This was the first survey of a major metropolitan area where so many human-altered and human-transported (fill) soils were mapped to the soil series level of classification—we identified 29 of them.

Margaret Boozer: I was doing some research on Richard's work, and I really connected with some of the language in the foreword of the Reconnaissance Soil Survey. The authors talked about their belief that the information provided would help lead to "a better environment and a better life." As an artist, I have these goals for my work, too. Certainly, it's on a very small scale, one person at a time. But I hope to encourage people to pay more and closer attention to their environment, to encourage awareness of our interdependence with it. Working with dug clays over the years, I'm interested in directing attention to beauty that is commonly overlooked … the ground under people's feet. I like the idea of making the ordinary beautiful, appreciating the commonplace, and creating a metaphor for how we can be satisfied with the ordinary imperfection in our lives.

To me, the information in the soil survey was really interesting, but its presentation to the world was a stack of paper or PDF documents. I wondered if I could use my skills in the visual arena as an invitation to consume the information. That day in Rich's office, I was instantly mesmerized by the stacked boxes, cataloged with handwritten letters and numbers; the random Ziploc bags of soils; rolled up maps and other documents. Each box was a distillation of some serious time and focused attention, not to mention labor. I imagined Rich and his crew down in a trench, with their tools, carefully collecting the samples and organizing them in this little portable package. It was like discovering buried treasure.

Richard K. Shaw showing Margaret Boozer the soil samples, unearthed from the storage closet. They had been kept, some for over 15 years, for their value as reference materials.

Reprinted with permission from Margaret Boozer.

Margaret Boozer unpacking the correlation boxes in Richard K. Shaw Staten Island office and taking a first look at the soil samples.

Reprinted with permission from Margaret Boozer.

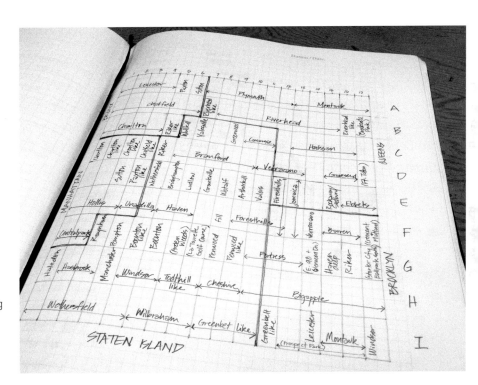

Margaret Boozer's sketchbook showing proposed arrangement by soil sample and borough.

Reprinted with permission from Margaret Boozer.

I knew that I had to find a way to make a work of art that preserved all that cataloging and exposed the beauty of the soils. My task became how to shift this science over into the realm of art, and how to do it with the least amount of disturbance to what was already there.

Claire Huschle: Obviously, both the Soil Survey and *Correlation Drawing/Drawing Correlations*

have classification as their organizing principle. What is most important about classification as it relates to your work, respectively?

Richard K. Shaw: I have spent much of my career in soil survey. A good classification system is a requirement for mapping soils, to establish relationships between them. By necessity, we adapted our system to handle human-altered and human-transported soils, in order to include them in our surveys of urban areas. But because the system is somewhat cumbersome, there is a real need for more widespread familiarity with soil classification, and our agency (USDA-NRCS) has begun some efforts in this direction. Margaret's project helped with that, in a way, by translating the science into something more visually accessible to a broader audience.

Margaret Boozer: I find the nature of classification compelling. I'm not really sure what is so fascinating about it to me. The illusion of order and control? Surely, something about the distillation and condensed nature of the activity, the tight little packets of information. Once the information is condensed, one can apprehend large amounts of information at a glance. Putting it all in the same format allows instant perception of differences. Those correlation boxes are compact versions of a four-foot deep soil sample, or monolith. They are simple and functional tools—but they also seem like little poems. Poetry has the same quality of being dense and portable, small but deep. You remember the poem, and so you can carry it with you. With time and further investigation, you can discover more complex connections and meaning.

There's a certain satisfaction in that. You get rewarded for close inspection.

Claire Huschle: With a project that so deeply depends on classification and compartmentalizing, talk about how you pushed yourselves to think using less familiar methods. Richard, how did you find the experience of working with a visual artist who had a different vocabulary, and in most ways comes to soil from a vastly different background? Margaret, talk about how you bridged the differences between science and art. How did you translate geological ideas to aesthetic ones?

Richard K. Shaw: As our first collaboration with an artist, this was a very positive experience: fun and rewarding. I enjoyed working with Margaret. She was an established artist who was already quite familiar with our science and had collaborated with some well-respected soil scientists before. In looking over her website, I thought she had used soil and geologic materials in some creative and attractive ways. The fact that she valued our soil samples from an artistic perspective meant something, and bringing them to a new audience at a museum like the MAD was exciting. For us, the project turned out to be pretty much ideal as it highlighted the diversity of the city's soils and provided a new kind of publicity for our survey.

Margaret Boozer: Frankly, I was pretty intimidated to try to incorporate the science. I didn't want to look like a dilettante, trying to add a cool sheen of science to my art, and I didn't want to try to get so deep into the science that I lost sight of what I had to bring to the project as an artist. But I did want to draw a connection between what I understand to be parallel systems of artistic process and scientific inquiry: being curious, asking questions, doing research, paying close attention, and then presenting something to viewers that is the result of all that activity.

Mostly, I just tried to pay attention to what I found compelling, and, as a nonscientist, to connect with Richard's work, and use that as a way to invite in an audience outside of those who saw the survey as a scientific tool, whom I take to be soil scientists, engineers, developers, etc. I think the art audience, the people who would usually look at my work, would normally not have much overlap with Rich's audience, and vice versa.

Claire Huschle: The MAD installation invited New York City audience interaction and self-discovery, from finding one's own neighborhood and comparing its soils to others, to using a QR code prompt to the survey's website to find more details about the soil under our feet. Talk about the importance of audience participation and understanding in your work.

Richard K. Shaw: From my perspective, I'd like the audience to take away a sense of the diversity of soils in the city, first of all, and then make the connection between the soil and the environment: that it matters, even in the city—*especially* in the city. The viewer doesn't have to be a soil scientist to appreciate the variations in color, texture, and consistence of the samples, all of which makes them visually attractive. The fact that they are locally derived should make them more meaningful or relevant. Associating a geographic area, a landscape or land cover with a soil sample brings them even closer to home.

Margaret Boozer: For me, it's really about stewardship … making something personal, getting people to take ownership of their soil, their little plot of existence, and then protect and care for it because it belongs to them. Though this work is not environmentally didactic, I do hope *Correlation Drawing/Drawing Correlations* might spark an interest in soil as a material, as a fundamental element of our life, which deserves much closer attention. And, I hope the work invites an awareness of the interdependence of things we often see as separate, like earth and city, art and science.

Claire Huschle: What do you both wish for the future of the project, in terms of audience exposure and education? Why is this important to you?

Richard K. Shaw: The samples will likely find a good home at the Department of Earth and

Environmental Sciences at Brooklyn College. I'd like to see them make some additional appearances in the city, especially to the grade-school audience. They might bring about an appreciation for the variety and beauty of the natural world or even provide some inspiration toward a career in the natural sciences.

Margaret Boozer: It's important to me that this project stay in New York City, for exactly the stewardship reasons discussed earlier. First, people in New York may not know that there is any soil under all that concrete, and then they

Luis Hernandez digging the Central Park sample. Hernandez, the first project leader for the NYC Soil Survey Program, began collecting the correlation boxes.

Reprinted with permission by Luis Hernandez.

are astounded by the beauty once they make that connection. The next step is that they search for how they fit into it, what soils belong to them and to their neighborhood.

I love the idea of the work being placed in an educational setting. There probably won't be room to re-create the map-like configuration that was on display at MAD, but now that the soil samples are in Plexiglas boxes, they can be on display even as they are in "storage." They can be tested for trace metals or accessed for other purposes. The work can move effortlessly back and forth between being art and being a science tool. Arranging a different configuration that's more appropriate to the new setting is pretty interesting to me too, from the standpoint of the work's evolution and adaptive reuse. And I'd get an opportunity to work with Richard again and to meet his colleagues at Brooklyn College. Who knows where that could lead?

QR CODE from MAD installation, link to mapping and soil sample information:

For further information on *Correlation Drawing/ Drawing Correlations*, please visit:

- More information on the art can be found at http://www.margaretboozer.com/ exhibitions2012.html.

- More information on the New York City Reconnaissance Soil Survey and other related New York City soil survey projects can be found at http://www.soilandwater.nyc/soil.html and http://www.soilandwater.nyc/urban-soils.html.

- New York City Soil Survey information can be found on USDA-NRCS Web Soil Survey at http://websoilsurvey.sc.egov.usda.gov/App/ HomePage.htm.

Mineral Traces
The Aesthetic and Environmental Transcendence of Soil Mineral Properties

Sarah Hirneisen, Jason Stuckey, and Don Sparks

Sarah Hirneisen is an artist based out of Austin, Texas. She began using soil in her artwork to document people and places in her life shortly after making a cross-country move to California. Hirneisen received her BFA from Rhode Island School of Design and her MFA from Mills College. Hirneisen has exhibited extensively throughout the United States and abroad, and has been the recipient of a Phelan, Murphy & Cadogan fellowship, an artist grant through the City of Oakland's Cultural Art's Program and was awarded a residency through the Hungarian Multicultural Center. Hirneisen continues to be inspired by the soil underfoot

Title image: Close-up of the bubbling reaction when firing soil in the glass.

Images courtesy of the artist.

in the creation of her sculptural and installation based work. When not in the studio, she is teaching sculpture or spending time with her two daughters.

Jason Stuckey is an assistant professor of environmental science at Multnomah University in Portland, Oregon. Stuckey's enthusiasm for soil originally stemmed from a broad interest in the Earth itself. He figured the best place to begin study of the Earth was at the surface, and that has kept him busy ever since. Stuckey received a BS in soil science at Cal Poly San Luis Obispo before pursuing graduate work that involved impacts of minerals on the fate of metals in the environment. Stuckey earned an MS in soil chemistry at Penn State and a PhD in soil and environmental biogeochemistry at Stanford University before performing postdoctoral work on soil mineral–organic carbon interactions at the University of Delaware.

Don Sparks is the S. Hallock du Pont Chair in Soil and Environmental Chemistry and director of the Delaware Environmental Institute at the University of Delaware in Newark, Delaware. He was born and raised in an agricultural community in Central Kentucky. His family had a farm, and from an early age, he developed a keen interest in the land and natural resources. In high school, he took a course in earth science and became fascinated with soils. He majored in agronomy as an undergraduate at the University of Kentucky (UK) and then received an MS degree in soil science at UK and a PhD in soil physical chemistry at Virginia Tech. He has spent his entire career at the University of Delaware, conducting research in soil and environmental chemistry, and preparing students and postdoctoral researchers for meaningful careers in academe, government, and the private sector.

Soils contain minerals with tremendous physical and chemical properties allowing them to serve as the primary medium from which food is grown and organic carbon is stored in the terrestrial environment. Soils may also play a prominent part in shaping one's impression or recollection of a particular place. Here two soil scientists (Sparks and Stuckey) and an artist (Hirneisen) explore the aesthetic linkage between soil properties and larger scale attributes of a geographic area evoked by firing soils from particular places with glass in a kiln. Soil properties, such as particle size and color, had a dramatic impact on the resulting artwork. The following essay stems from the dialogue between Sparks, Stuckey, and Hirneisen that occurred in the fall of 2014 through early 2015 by way of video conferencing, phone, and e-mail. Sparks and Stuckey also sent soil samples from their research site in Delaware to Hirneisen for firing with glass, and some of the resulting artwork is shown here.

Agiterra glass panels display soil the artist collected during a one-year period; gallery visitors were invited to take the panels and send their own soil sample to be added to the collection installation view at Monterey Peninsula College, 2005–2008; fabric, glass, soil, paper, 45″h × 53″w.

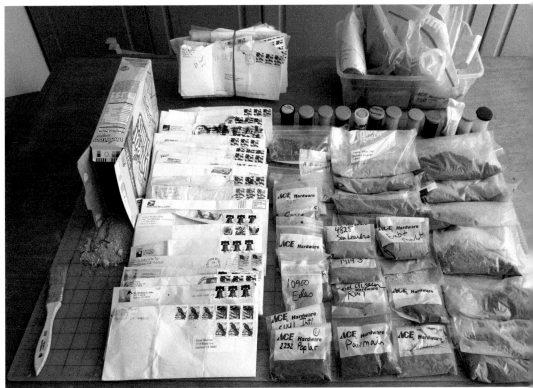

Image of Sarah Hirneisen's studio table with her collection of soil samples.

Images courtesy of the artist.

Hirneisen's work exploring the reactivity of soils in terms of how they react with glass in the kiln initiated a series of conversations concerning the notion of scale and the aesthetic symbolism of this interaction, including the soil's role in shaping our sense of home and identity. Physical, chemical, and biological phenomena involving soil minerals begin at the nanometer scale. However, the impacts of soil minerals can have a regional to global significance. Here we explore how fine-scale properties of soil minerals impact our local experience of home as well as the global challenges we face in the twenty-first century, including food security and climate change.

Soils are broken up into three different fractions decreasing in particle size from sand to silt and finally to clay. Sand particles are no more than 2 mm in diameter, whereas clay particles can be a million times smaller than that. This range in scale spanning orders of magnitude has huge implications for how different soils react and behave. Take for instance the mineral ferrihydrite, an iron oxide similar to rust. Ferrihydrite can have a specific surface area of 800 m^2/g. This means that if you add up all the surfaces in less than a handful of ferrihydrite, the total surface area would easily exceed that of an American football field! This massive surface area (and that of other soil minerals) is not inert, but contains charges—both positive and negative depending on the soil pH. The development of charge within soil minerals allows soils to retain nutrients, which are often charged themselves. Thus, besides serving as a physical medium for supporting plant growth, soils and their charged mineral surfaces trap nutrients and allow plants to access them. Soil mineralogy is therefore directly related to the fertility of the soil, and hence, food security.

In Hirneisen's earliest works dealing with soil, she was interested in using soil as an artistic medium to reference place and identity. She used it as a record-keeping device by collecting soil samples from different locations and thus mapping her movements through life. To give permanence to ephemeral actions, much like taking a photograph or buying a souvenir on a trip, she began fusing (or heating) the soil samples between layers of glass. The outcomes were surprising. One side effect of heating the glass and soil together was the bubbling of the glass. When Hirneisen began the correspondence with Sparks and Stuckey, she discovered that the extent of this reaction was most likely a function of the soil particle size and mineralogy. As the soil was heated, the particles expanded causing the glass to bubble. The soil samples that came from areas heavy with clay tended to bubble less while the samples with a higher sand content bubbled extensively. In almost all cases the soil samples caused the glass to bubble to some extent. The effect of the expanding glass during heating becomes a fitting conceptual link to the idea that these tiny soil particles combine to make up a much bigger place. The soil underfoot is home—the foundation of life. Our food grows in it, our houses are built on it, and we travel over it. We live, sleep, play, and learn on our home turf, and it has a strong influence on the people we become. Using soil as an artistic medium, Hirneisen seeks to illustrate an idea that is often overlooked—that soil is inextricably intertwined with our identity and place on this Earth.

Agiterra, detail of glass and soil panels.

Quartz is typically the dominant mineral in the sand size fraction of soils and has a high thermal conductivity relative to the average soil mineral (Van Wijk, W.R. 1963. *Physics of Plant Environment.* Amsterdam: North Holland Publishing Company), meaning it has a strong ability to conduct heat to the glass. Further, sandy soils are packed more tightly than clayey soils, and therefore have a higher bulk density (and a lower porosity), making heat transfer between the sand particles more efficient. Therefore, the mineralogy and structure of sandy soils may promote greater heat transfer, causing increased glass bubbling than in the glass–clayey soil reaction.

The soil's ability to retain plant essential nutrients extends to organic matter, which is comprised chiefly of carbon, nitrogen, oxygen, hydrogen, phosphorus, and sulfur. The growing consensus is that the predominant means by which soils retain organic matter is mineral protection. That is, minerals hinder microorganisms (and their enzymes) from degrading organic matter either by acting as a physical barrier or by binding to the organic matter in such a way that withstands or slows microbial enzymatic attack. As microorganisms break down organic matter they can release greenhouse gases, such as carbon dioxide, into the atmosphere. Therefore, soil minerals are fundamental regulators of carbon exchange between the land and atmosphere, and play a critical role in directing the Earth's climate system.

Projected changes in the Earth's climate will have feedbacks that impact the soil, including sea level rise and drought. First, sea level rise is projected to occur as the global mean annual temperature rises and polar ice caps melt, causing increased seawater intrusion into coastal groundwater. The increased salinity and periods of saturation will likely cause mineral transformations that will impact the cycling of nutrients and metals within soils and sediments. For instance, the mineral ferrihydrite mentioned earlier will transform into other minerals or dissolve completely once saturated soils turn anaerobic, and may release any associated nutrients or metals into the groundwater. Mineral transformation reactions under a sea level rise scenario may be especially important in industrial areas with legacy metal contamination. The changing chemical conditions in soils brought about by seawater intrusion will initiate complex reactions between soil minerals and metal contaminants, such as arsenic, that are part of Sparks and Stuckey's current research.

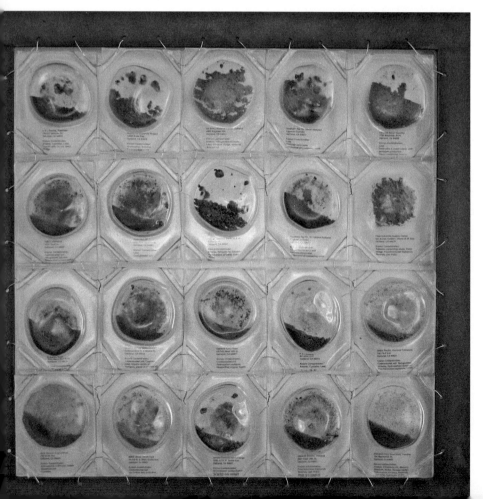

Contaminated soil collected from "Calsites" in Oakland, California, are contained in glass, installation view at Pro Arts Gallery, Oakland, California (2005), soil, glass, metal; 20"h × 20"w.

Image courtesy of the artist.

As Hirneisen continued to experiment with incorporating soil into her artwork, she noticed another side effect of heating soil in the kiln: changes in color. Assuming this was based on the chemical makeup of the soil samples, she decided to collect samples specifically from contaminated sites. California has a database of locations throughout the state with known soil contamination, called "Calsites." Many of the locations are now defunct manufacturing sites and government-owned properties. Lists of known contaminants were available from each site and ranged from arsenic and lead to hydrocarbon solvents, acids, and metals. After collecting soil from 20 Calsites, all in Oakland, she fired the samples into glass. The aesthetic outcome was rich with varying shades of reds, oranges, browns, and blacks. There was a beautiful irony to knowing such lovely colors came from tainted soil. Although contaminants were listed on the Calsites website, Hirneisen had no way of knowing what percentages of the contaminants were in each of the fired samples. Sparks and Stuckey volunteered to send along some samples of arsenic-contaminated soils from a site in Wilmington, Delaware, that may eventually be affected by sea level rise. Extreme bubbling and a shift from brown to bright reddish orange occurred when firing.

A second impact of climate change on soils is drought. An archetypal image of drought is that of cracks disseminated across the soil surface, forming a sea of parched polygons. This meter-scale phenomenon ultimately stems from the nanoscale behavior of a particular subset of minerals known at smectites. California is currently experiencing one of the worst droughts in its history, and contains smectites throughout much of the state. Smectite clays are part of the broader class of sheet-like

Studio image of glass panels fired with soils and sediments from the top 3.5 m of an industrially contaminated site in northern Delaware.

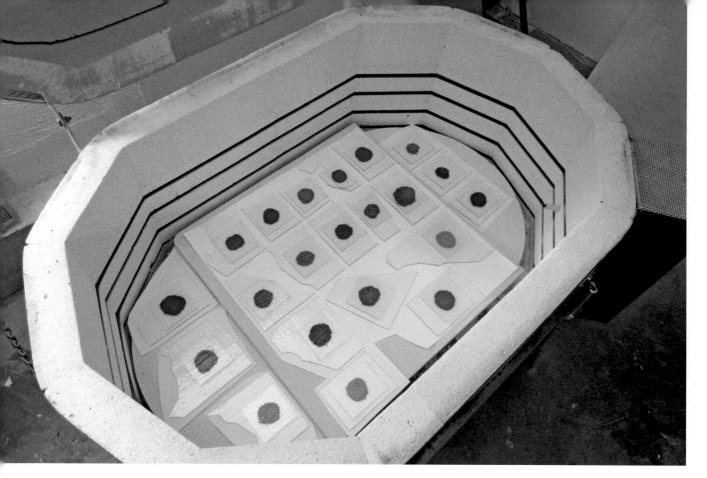

Drought Conditions image showing the glass and soil pieces laid out in the kiln before being fired, in artist's Oakland studio 2015.

Images courtesy of the artist.

silicate minerals called phyllosilicates, which are classified by the amount of fixed or permanent charge (known as "layer charge") inherent to the mineral structure. The layer charge in phyllosilicate sheets is negative, and the negatively charged sheets form layers that are held together by positively charged ions (cations). Smectites have a low layer charge, and therefore exert less "pull" on the interlayer cations. This relatively weak attraction between smectite layers and interlayer cations allows water to enter the interlayer, causing the mineral structure to expand. A saturation event in smectite-rich soils causes the solid phase to expand into the pores or void spaces, which are the main conduits for water movement. Therefore, a smectite-rich soil will retain water well, but will restrict water movement under saturated conditions. The opposite phenomenon occurs

Opposite page:

Drought Conditions. Soil was collected from regions throughout California affected by the drought and fused into glass, exhibited at Blue Lines Art Gallery, Roseville, California in 2016, glass, wood, soil, 24″ × 38″ × 3″.

Image courtesy of the artist.

when a smectite-rich soil dries up as in a California drought. The smectite interlayers contract, ultimately resulting in a larger contraction of the soil solid phase, creating cracks upward of 5 cm in width at the soil surface.

When Hirneisen, Sparks, and Stuckey's discussions on the above topics began during early fall of 2014 California was well into a severe drought that has continued to become one of the driest periods in California's history. In the San Francisco Bay Area (where Hirneisen was based at the time), January 2015 had no rainfall at all. This hasn't happened since the city of San Francisco began keeping records of rainfall in 1850.

Inspired to create an artwork that specifically responds to California's drought, Hirneisen traveled around the state collecting soil samples from some of its driest regions. The samples were again heated between glass layers and then arranged to form a map of California. The process of heating and firing the soil samples further evaporates any moisture left in the soil, therefore exaggerating the process that has naturally occurred due to the lack of rainfall. Hirneisen has also been experimenting with casting the voids left in the earth when the smectite inner layers contract. These chasms of dry cracked soil are reminiscent of the fault lines that run throughout the state and give a visual presence to the issue.

Hirneisen's hope is that by creating visual illustrations of drought through her artwork, more people might be reminded of the magnitude of the problem and the role they can each play in its mitigation.

A Snapshot in Time

The Dynamic and Ephemeral Structure
of Peatland Soils

Laura Harrington and Jeff Warburton

Laura Harrington is an artist who works across different media often in multidisciplinary research and collaborative environments. With a specific interest in geomorphology and the field of physical geography within northern climates/landscapes, her work is concerned with earthly matter—testing out the human and non-human positions at play within our experience of nature. In recent years Harrington has been developing dialogues with physical scientists as a means to provide a way into thinking among specific landscapes and these discussions are becoming the bedrock of her process-based inquiry. Recent work developed

Title image: Peatland landscape of Moss Flats, North Pennines, UK. The bare peat flat surrounded by areas of vegetated moorland.

Photo: Laura Harrington.

through a Leverhulme Trust Artist Residency with Jeff Warburton in the Department of Geography at Durham University, Durham, UK, explored eroded peat as a subject to consider our relationship to mundane and viscous matter. Recent exhibitions and residencies include HIAP, Helsinki, 2016; Nothing and Nowhere, 2016, (Hangmen Projects, Stockholm); Liveliest of Elements, 2015 (Durham University, Invisible Dust, Woodhorn Museum); A Lively Sense of Nature, 2014, (Figure Two at BALTIC 39); and Layerscape (peat bogs), 2012 (AV Festival). Her current AHRC funded PhD with Northumbria University "Upstream Consciousness—Re-Presenting and Re-Thinking Our Relationship with 'Distant Landscapes'" draws on geomorphology to consider the interconnectedness between humans and a "lively earth."

Jeff Warburton is a professor in geomorphology at Durham University. His research interests are primarily concerned with understanding fluvial, hydrological, and hillslope processes operating in upland environments. He has worked extensively on peat erosion (carbon balance) and peat mass movements. This work is underpinned by intensive field monitoring programs and sediment budget analysis. He has pioneered work on wind erosion of upland peat; transport and dispersal of eroded peat blocks in fluvial systems; peat mass movement mechanisms and processes; and sediment budget analysis of eroding peat catchments. In 2007, together with his coauthor, Martin Evans, he published *Geomorphology of Upland Peat: Erosion, Form and Landscape Change* (Blackwell, Oxford), which is widely regarded as a seminal work on this topic. He is an international expert in sediment transfer in upland fluvial systems and has served on committees of the European Science Foundation and the International Association of Geomorphologists.

This chapter was compiled as a result of conversations between Laura Harrington and Jeff Warburton, which took place as part of a Leverhulme Trust Artist Residency, February 2014–January 2015. Together they explore the morphology of peat soils through their distinct disciplinary perspectives, considering the diversity of views and similarities they share in their attentive study of peat. The focus here is on UK peat bogs, which have been badly damaged by a history of drainage, burning, overgrazing, and industrial pollution. The dialogue focuses on the Moss Flats in North Pennines, North East England, and is based around three questions: Why are peatlands important to me? What do I see when I look at the microscale morphology of peat soils? And what do I see when I look at peatland landscapes?

Peat soil is a heterogeneous mixture of decomposed plant material that has accumulated in a water-saturated environment in the absence of oxygen. The structure of peat varies enormously from partially decomposed plant remains to a fine amorphous colloidal soil.[1] Peatlands are landscapes with or without vegetation with a naturally accumulated peat layer at the surface. At first sight, many peatlands may be seen as expansive landscapes, which blanket the terrain with monotonous regularity. However, peatland ecosystems are dynamic ecological entities, constantly changing, growing spreading, and eroding.[2] Peat soils are a historic resource for fuel and provide space for the landscape to function: to regulate hydrology, climate, and water chemistry. Peat soils are a major carbon store, but can also release carbon in a degraded state, which adds to atmospheric carbon dioxide and significantly contributes to climate change.[3]

Based on a textual exchange using three questions, we can summarize our disciplinary views about peatland soils in a "word cloud" that visualizes the key words of our dialogue: peat, peatland, landscape, surface, and different. The first four are no surprise in an article about peat soil morphology, but the fifth highlights a common theme we both identify in peatland landscapes: difference. Difference refers to the heterogeneity of peat soil landscapes manifest at microtopographic and landscape scales. Difference also reflects the focus in artistic and scientific perspectives. In the scientist's word cloud, there are two groups of words that hang together. First, water, wind, and soil are grouped together as descriptors of the key erosion processes acting on the peat. Second, there is a group of terms that include topography, high-low, and climate, which summarize the driving mechanisms behind erosion. These word groups contrast with the artist's, in which key terms tend to be more descriptive, emphasizing change, connection, and place.

Geomorphology, understanding how landscape evolves and changes over time, is like art—a somewhat aesthetic and visual mode of inquiry. Both disciplines rely on an important key skill: observation. A keenness of eye and an attention to detail are essential attributes for both disciplines. Looking at the same features in slightly different ways provides an important synergy between artist and scientist, enabling a productive exchange of ideas. Both disciplines in this instance have made peatlands their area of study and in doing so share and strive to articulate a vision and understanding of the complex and intrinsic dynamism of these unique landscapes. As an art–science collaboration it was vital that we both took a performative approach to the production of knowledge and rejected from the outset the notion of art as an instrument or illustrative device to interpret science. While the artist is driven by the otherworldly nature

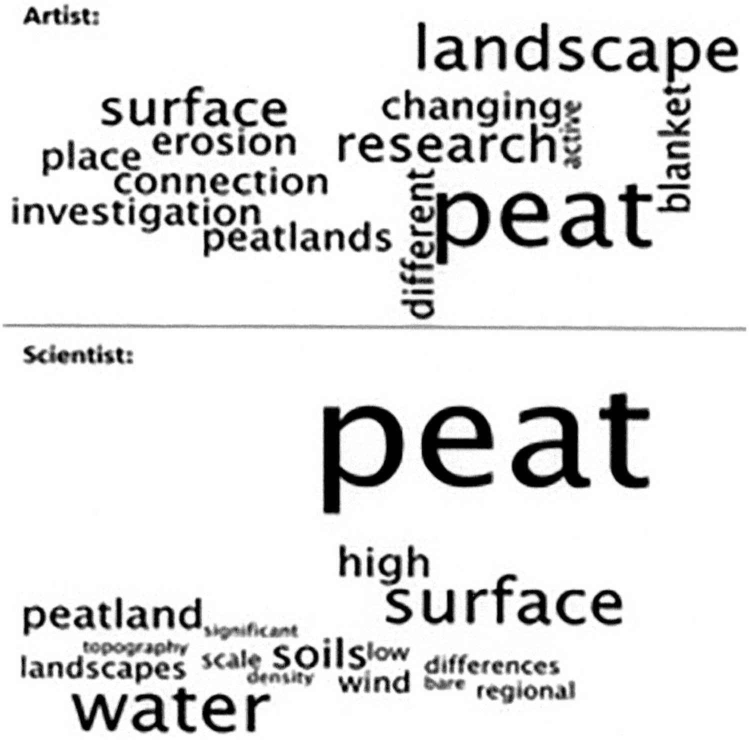

Artist:
landscape
surface
changing
active
research
blanket
place
erosion
connection
investigation
peatlands
different
peat
Scientist:
peat
high
surface
peatland
significant
topography
landscapes
scale
soils
low
density
wind
differences
bare
regional
water

Word cloud of artistic and scientific views on peat soil morphology.

of the landscape, its processes, and the emotions they can conjure up, the scientist is more driven by a desire to explain and interpret the evolution of these fascinating environments. At the microscale, the artistic and scientific views are far more similar as both marvel at the morphological intricacies of peat soilscapes. At the larger landscape scale, both recognize the importance of wind and water in moulding the surface topography. Where the scientist looks for an explanation of the main mechanisms driving environmental phenomena, the artist is more inclined to concentrate on the experience of being with such a landscape and how to creatively interpret the feelings that such phenomena produce. In the end, it is all about what we can actually see and how what we see creates meaning and appreciation for a unique type of landscape and the soils at its foundation.

Why are peatlands important to you?

Laura Harrington: Peatlands outside of scientific research could easily be described as both distant and elusive. However, since my first encounter with a peat bog, I've been compelled by their complexity, processes, and otherworldliness. A vital element in the earth's ecosystem due to their impressive ability to capture and store carbon makes them comparable in environmental importance to glamorous glaciers and rainforests but somehow these landscapes sit very differently within our psyche. A palimpsest of recent geological history (hundreds of thousands rather than millions of years), these ancient, breathing sponges with vast and continuously changing habitats can be appreciated, felt, and understood in many layers. It is within these subtle and intricate layers that I try to locate myself *with* the peat—observing and exploring their finer details, different processes and movements—from slow temporality to fast weather events. Through the making and presenting of different artworks, peat has become a lens and focus point used to attempt a recalibration of our human position on seemingly inanimate matter. This quiet and sensuous observation and creative research investigation of local peatlands enables an exploration into wider thinking with our lively earth.

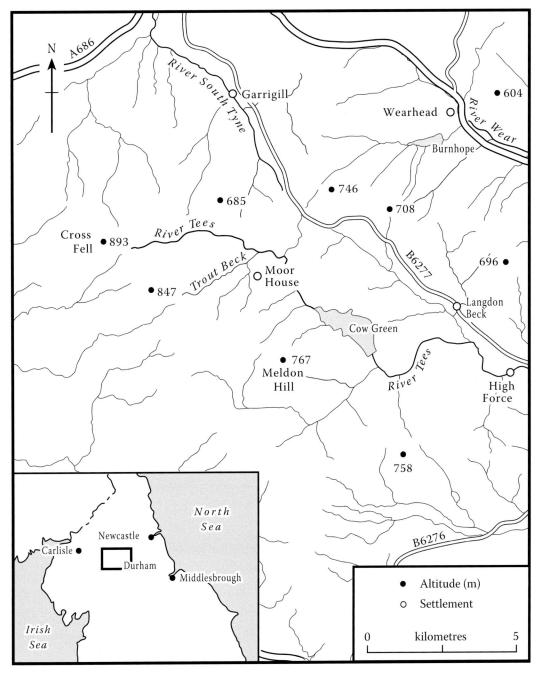

Map of the North Pennines and upper Teesdale.

Moor House is a National Nature Reserve located in the North Pennines and upper Teesdale with a large percentage of this reserve being blanket peatlands. A blanket peatland is typically a rain-fed upland landscape where peat covers most of the surface except for the steepest ground. Altitude ranges from 290 to 850 m and the sources of two of England's great rivers are there, the South Tyne and the Tees. Weather recording and research into the effects of climate change on the uplands has been going on at Moor House since 1932. It is recorded as one of the most well-understood uplands in the world. If we zoom in a little further into Moor House, upstream of Rough Sike we can find Moss Flats, approximately 3-ha of relatively flat, sparsely vegetated peat and the basis of my collaboration with Jeff during a Leverhulme Trust Artist Residency. The rawness, the presence of loss at Moss Flats instantly drew me to it. At first, this site might be thought of as akin to a lunar landscape—barren and stripped of life—but as one finds a way in, it quickly becomes a rich medley of energy and activity. It is this paradoxical feeling of loss—a sense of emptiness that comes from its openness—and uncertainty alongside strength and acceptance that excites me and drives my curiosity. It became clear, quite early on, that this site offered an opportunity to engage deeply with a specific eroding landscape that exists on both a micro-, meso-, and macroscale. As Jeff and I explore this landscape, both together and separately from the perspectives of artist and scientist we are learning about its intricacies and secrets. Through our different approaches, it will not only be data that will be gathered but the essence and spirit of the place too.

Jeff Warburton: Peat soils and their associated landscapes are fascinating from a geomorphological perspective. Their intrinsic interests are based on a combination of three factors that set peat landscapes aside from other environments. First, peat soils are a unique material with a number of distinguishing characteristics. Peat has a high but variable natural water content (c. 500%–1500%) meaning it is supersaturated and full of water. Because it is a material composed of decaying vegetation, it has a very high organic content (typically 90% and can be highly fibrous). Peat is full of small voids, which means it can hold and transmit water. Due to the combination of these factors, it is a material which has a very low density, resulting in a soil which has high compressibility and low strength. Although peat deposits are highly variable, the degree of humification, or the extent of biochemical decomposition of plant remains, is a key factor determining the overall behavior of peat. Peat soils therefore vary from highly fibrous deposits with barely recognizable decomposition to amorphous peat with no discernible plant remains. Secondly, water flow through peat soils is a very complex process. The old view that peat acted like a sponge, by virtue of the very high natural water content is a fallacy that has now been disproven. Peat soils do store large amounts of water, except for when extreme drought water tables generally remain high. This means that many peatlands will shed water quickly during rainfall because the incoming water cannot infiltrate the soil. The special water relations found in peat also determine the response of upland environments to hydrological extremes, whether they are floods or drought. This makes peat landscapes susceptible to the action of water runoff, often leading to erosion. Thirdly, peat is a low-density geological material and as such can be transported by wind and water in novel ways. A desiccated bare peat surface can be readily stripped of dry surface peat, which often forms small paper-thin sheets, which are transported on the wind almost like kites. A combination of wind and rain drives rain droplets into the soft bare surface, remoulding the peat into intricate surface forms and driving the eroded peat relentlessly in the direction of the prevailing wind. This results in wind-moulded geomorphological features called haggs and small terraces of peat, which creep across the landscape. Eroded peat may also enter streams and rivers, but due to its low density it will not sink and will be carried along "floating" in the water column for great distances. By examining the morphology of peat soils at the microscale and landscape scale it is possible to decipher the history of a particular peatland and determine its present status as a stable or unstable landscape.

A Moment in Time, 2014, Laura Harrington, plaster cast of a mould taken from the surface topography at Moss Flats (9 cm × 29 cm × 20 cm). Image courtesy of the artist.

What do you see when you look at an area soil close up?

Laura Harrington: Bare peat is a fascinating substance. Over the last year, on every visit to Moss Flats, the bare peat surface never once looked the same. Movement and stillness can be seen, heard, and felt everywhere. The bareness of Moss Flats gives a tangible understanding of the materiality of the peat. Behind the apparent surface stillness, the active decomposition and erosion processes taking place on and through it gives way to an incredible dynamism. The different formations present on the surface are like slow time-lapse recordings of wind movements, a snapshot of waves breaking and bubbling over a vista—a moment in time captured from within a whole landscape which is constantly moving, raw, rugged and geomorphologically active. The delicate subtleties of microinteractions present on the bare peat surface contain intricate mosaics of peat erosion patterns and forms, morphing, changing, and disappearing. The living energy and decay characterized by the dynamic forms of the tussocks and haggs hold a sense of the unknown and the mystical. *Hagg #1 and Hagg #2* is from a series of intricate line drawings—made with a 0.25 mm Rotring isograph pen (kindly donated by Jeff whose use for them for technical drawing was no longer required)—examining eroding forms within peatland landscapes through the process of drawing and transformation.

Jeff Warburton: The surface of bare peat provides a forensic blueprint of the geomorphological processes that have formed the intricate surface topography. The form of these small-scale features will vary depending on the nature of the peat soil and the magnitude and frequency of the dominant geomorphic processes sculpting the surface. Microforms can show enormous variability over the scale of a few meters, for example, the exposed top to the base of a peat hagg or the opposing aspects of an incised gully. These features are significant for three main reasons. First, they increase surface roughness, which has important impact on near surface sediment transfer mechanisms. Second, they indicate the relative magnitude and direction of sediment transfer. Third, they provide a mechanism by which surface peat is detached from the main peat mass and promote local soil loss. These small-scale surface forms battle with surface vegetation in trying to keep the soil surface intact.

What do you see when you look at peatland landscapes?

Laura Harrington: For me, the blanket peatlands of the higher reaches is an environment that encourages a deep inward connection *to* and *with* the landscape. Maybe it's the sense of my own smallness against their vast expanse or even uplands more widely having the ability to elevate one's thinking into and onto a wider vista. Perhaps their solitude and constantly

Hagg #1, 2014, ink on paper, 30 × 42 cm, Laura Harrington.

Image courtesy of the artist.

Laura Harrington, *Finely Wrought #15*, 2014, one of 54 images installed in six mini projectors.

Image courtesy of the artist.

Hagg #2, 2014, ink on paper, 30 × 42, Laura Harrington.

Image courtesy of the artist.

changing atmosphere gives space for a deeper engagement with the *moment* as it occurs; after all, solitude has long been connected with the practice of meditation or spiritual searching.

Uplands, where arguably some of the most interesting and rugged peatlands exist, can provide space for ourselves to reconnect with an inward gaze, to allow a connection back to what is sustaining us and how we should maintain it. Uplands have the quality of being both very far and very near, and are useful landscapes affecting us all. Our physical and relational connection to place interests me—connecting back to what is sustaining us; a feeling of being both close and distant. This sense of connection to the land expanding points of consciousness is something I am pursuing as an idea of "upstream consciousness." Peatlands are a perfect subject for exploring this.

At first glance, peatland landscapes seem visually bleak and almost monochromatic. Sonically they don't seem rich either. However, there are different forms and rhythms present, which all call for a different type of listening, looking, and patient practice—quite suddenly one can be overwhelmed by the richness of the layers, the colors, the subtlety and detail of sounds, and the diversity of such lands. The subtle changes experienced, from revisiting these different landscapes, appear like slow time lapses in my mind. One spends hours in a landscape with no one to be heard or seen. Trans-Atlantic flights and helicopters slice through the sky above, suddenly and momentarily reminding me of my geographical position in the world before my concentration returns to the soft, fragile, and slightly elevated feeling of being on and with peat.

The winds that buffet these uplands produce beautiful shifting patterns as its energy is transformed. As an example, the nature of how Moss Flats has formed over time by the predominance of wind with rain from the southwest is visually striking. This can be clearly seen within its complex topography—from the asymmetrical forms of eroding haggs to the formations of toothed surfaces and tiered terraces, reminiscent of an amphitheater where the tiered seating surrounds a central performance area. Haggs of varying scale are aplenty.

Jeff Warburton: As a geomorphologist, when I look at peatland landscapes my thoughts initially focus on the regional or catchment scale because the regional topography exerts a strong influence on blanket peat thickness and drainage patterns. The variability of slope angle across a particular landscape to some extent controls the accumulation depths of the peat soil, which is closely related to the ecohydrology of the peatland, which is inseparable from the prevailing regional climate (and water regime). Regional differences in peatlands arise due to local differences in both climate and topography. It is therefore important to recognize the differences that exist between different upland peatland systems at the regional scale.

Such differences have recently taken on greater significance as peatland scientists strive to improve estimates of the carbon storage of peat soils. Such endeavors require more precise estimates of peat soil depths across the landscape.

Endnotes

1. Moore, P.D., and Bellamy, D.J. 1974. *Peatlands.* Elek Science, London.

2. Joosten, H., and Clarke, D. 2002. *Wise Use of Mires and Peatlands.* International Mire Conservation Group and International Peat Society. NHBS Ltd, Devon, UK.

3. Evans, M.G., and Warburton, J. 2007. *Geomorphology of Upland Peat: Erosion, Form and Landscape Change.* Blackwell Publishing, Oxford, UK.

Carbon

Taru Sandén in dialogue with Alexandra R. Toland, postscript
and images by Terike Haapoja

Taru Sandén is a soil enthusiast who enjoys working with various stakeholders and leads the TeaTime4Schools project. She holds an MSc in environmental science from the University of Gothenburg, Sweden, and an MSc in environment and natural resources from the University of Iceland. Her PhD degree in geography, also from the University of Iceland, focused on soil aggregates and soil organic matter in agricultural soils. Currently, Sandén works as a senior expert at the Department for Soil Health and Plant Nutrition at the Austrian Agency for Health and Food Safety (AGES). Her works consist of Tea Bag Index studies in mainly agricultural settings, as well as several H2020 projects including LANDMARK, FATIMA, and AgriDemo-F2F.

Terike Haapoja is a Finnish visual artist based in New York. Haapoja's large-scale installation work, writing, and political projects investigate the mechanics of othering with a specific focus on issues arising from the anthropocentric worldview of Western modernism. Haapoja represented Finland in the 55 Venice Biennale with a solo show in the Nordic Pavilion. She is a recipient of ANTI prize for Live Art (2016), Dukaatti-prize (2008) and Ars Fennica prize nomination. *History of Others*, Haapoja's collaboration with writer Laura Gustafsson, was awarded a Finnish State Media art award (2016) and Kiila-prize (2013). Haapoja contributes to journals and publications internationally, and is the coeditor of the publications *Altern Ecologies: Emergent Perspectives on the Ecological Threshold at the 55th Venice Biennale*, *History According to Cattle*, and *Field Notes: From Landscape to Laboratory*. Haapoja is adjunct professor at Parsons Fine Arts, New York.

Introduction

The conversion of woodlands, prairies, and other landscapes into agricultural land has led to an increased loss of soil carbon worldwide. Land-use conversion and soil cultivation have been a significant source of greenhouse gases (GHGs) to the atmosphere, responsible for about one-third of GHG emissions. But what is carbon in essence? Is it possible to measure something but never really know it? There is a major potential for increasing soil carbon through restoration of degraded soils but also by forming new meanings around the concepts of carbon sequestration. The soil can be a repository for decomposed organic matter but also memories of forms once living. This chapter looks at the carbon cycle from the points of view of the Finnish soil scientist Taru Sandén and Finnish artist Terike Haapoja. After several e-mails the three of us managed to arrange a Skype meeting in April 2016. After speaking about the cultural and chemical complexities of decomposition for about an hour we parted ways. Ironically, the audio file had digitally decomposed and our recorded musings were lost. To best reflect our exchange, the chapter consists of an interview with Taru Sandén about her public awareness campaign and citizen science project, the Tea Bag Index (TBI), and a photo essay by Terike Haapoja featuring her work for the 55th Venice Biennale, *Closed Circuit—Open Duration*.

Alexandra R. Toland, 2017

It is not enough to show the workings of carbon in the ecosystem. We need to try to see what CO_2 means to us, how it works its way in our own inner reality, the reality of love, and bodily being, and death.

Terike Haapoja

Alexandra R. Toland: Taru, could you please explain the difference between organic carbon and inorganic carbon in the soil?

Taru Sandén: *Organic carbon* is the carbon stored in soil organic matter. Organic matter consists of many different compounds with varying structure, content, and recalcitrance. It originates primarily from dead plant remains, litter, and the microbial biomass on this litter. Organic carbon enters the soil through the decomposition of plant and animal residues, root exudates, living and dead microorganisms, and soil biota. *Inorganic carbon* in soil consists of elemental carbon and carbonate minerals such as calcite, dolomite, and gypsum.

Alexandra R. Toland: In Finland, forests cover 76% of the total land mass, which is a lot compared to other European countries. Is carbon sequestration a culturally determined phenomenon? What anecdotes, experiences, or data can be found on the floor of the Finnish forest in this age of accelerated climate change?

Taru Sandén: A country full of forests has a lot of forest owners who have many stories to tell about their environment. Mauno Forsström has collected a lot of anecdotes about Finnish forests in a book in the 1970s.[1] The forest owners in the south of Finland are mainly farmers, whereas in northern Finland the Finnish state is the biggest forest owner. In general, the forests in the north are more intensively managed than the forests in the south.

There are also many forest monitoring sites across Finland, where soil carbon among other soil parameters, has been measured. The European-wide monitoring program is called BioSoil.[2] It is clear from the results that the trees carry the largest biomass, but both in South and North Finland the soils have the biggest pool of carbon. The forest biomass is an important carbon storage, which had increased by 6.8 Tg CO_2 y^{-1} between 2000 and 2005. The forest soils in Finland are mainly a sink of carbon, but in the north where management is the most intensive, they are more or less neutral (no sink, no source).[3]

Alexandra R. Toland: You have been working on an ongoing citizen science project that seeks to understand the soil carbon cycle in new ways. Could you describe the Tea Bag Index?

Taru Sandén: The Tea Bag Index is an easy to do and standardized method to gather data globally on decomposition rates and carbon stabilization.[4] It was developed by researchers now working in the Netherlands, Austria, and Sweden. Participating Citizen Scientists use commercially available tea bags (green tea and rooibos) as standardized test-kits for carrying out simplified litter bag experiments. Tea bags are weighed, buried in the soil for three months, dug up, dried, and weighed again. In order to make it easy for schools to take part, a lesson plan has been designed for teachers. TBI requires only little means and knowledge, making data collection by crowdsourcing possible. Engaging Citizen Scientists as coresearchers increases the amount of data collected, increases awareness of soils and provides essential development in including soils more frequently into natural sciences and environmental curricula at schools.

Decay of organic material, decomposition, is a critical process for life on earth. Through decomposition, food becomes available for plants and soil organisms. When plant material decomposes, it loses weight and releases the greenhouse gas carbon dioxide (CO_2) into the atmosphere. Commercially available teabags containing plant material can provide vital information on the global carbon cycle, if we study their decomposition in soils. Terrestrial soils contain three times more carbon than the atmosphere and therefore changes in the balance of soil carbon storage and release can significantly amplify or attenuate global warming.

Teabags from the Tea Bag Index.
Photos courtesy of Taru Sandén.

The numerous Tea Bag Index data points collected will allow for a great leap forward in mapping decomposition, as well as understanding and modeling the global carbon cycle. The so-acquired Tea Bag Index provides process-driven information on soil functions at local, regional, and global scales essential for future climate modeling; and it is sensitive enough to compare data between different ecosystems and soil types.

Alexandra R. Toland: Your project goes beyond measurement to emphasize the visualization of something normally invisible—carbon. What interested you in visualizing the carbon cycle?

Taru Sandén: On a practical level, many members of the Tea Bag Index group had been measuring decomposition in their research with regular litter bags, which takes a lot of time in the field and a lot of effort in making the litter bags that you bury in the soil. Tea bags of polypropylene material, which does not break down in the soil itself, are wonderful tools for a researcher. It saves time and costs, and makes it possible to do many more measurements. The Tea Bag Index shows what happens in the soil during three months. Everyone can see with their own eyes that the tea bag weighs less after it has been buried in the soil for these three months. With having two different types of tea, easily decomposable green tea and more stable rooibos, will show that there are different kinds of organic matter in the soil. Some of it turns around very fast; whereas the more stable forms of organic matter take much more time to decompose.

On a level of human interaction, the Tea Bag Index gave us the opportunity to communicate with citizen scientists in an easily understandable way. We wanted to develop a method that is so simple to do that everyone can do it. Keeping things simple will also minimize errors that come when we receive data from different researchers and citizen scientists. By explaining what happens to the organic matter is the Tea Bag Index way to open the black box of soil. The keywords are curiosity, enthusiasm, joy, and asking questions. I hope that people will get more curious when they learn about our method, and feel that they want to be part of it. Curiosity and enthusiasm are some of the most important characteristics of a researcher and I hope Tea Bag Index can bring joy to many of our Citizen Scientists using the message "science is fun and everyone can do it!" If teachers want to carry out the experiment with their class, we provided guidance. The lesson plan is freely available online and includes discussion questions as well as tips for modifying the method. The method is so simple to do, that it gives the teacher a lot of freedom in how to include it in teaching. Thus, we will also learn from the citizen scientists, when they send us questions or when they describe how they have carried out their experiments.

Alexandra R. Toland: The Tea Bag Index project relies on the participation of hundreds (maybe thousands?) of tea drinkers worldwide. Who has participated? Where do they come from? What do you think motivates them to participate? Your participants provide an interesting source of qualitative data as well as quantitative data on decomposition rates. Have you or your team also gathered data on the opinions, perceptions, fears, or hopes of your participants regarding climate change and the carbon cycle?

Taru Sandén: So far, we have data points from approximately 2000 locations and more and more participants are getting interested. Our first Citizen Scientists were a scout group from

Antonia in agricultural fields, August 2015.

Photo credit: Taru Sandén.

Gars am Kamp, Lower Austria, back in 2011. In Sweden, my colleague Judith Sarneel worked with 250 schools in 2015. In 2015 and 2016 in the UK, a PhD student Sarah Duddigan encouraged roughly 300 people to participate though the Royal Horticultural Society. In Austria, roughly 40 school classes, 4 groups of farmers and 18 individual farmers participated in a Citizen Science Award competition in 2016, which created approximately 180 new data points. Between 2017 and 2019, roughly 150 Austrian school classes will be taking part in TeaTime4Schools project, and contribute to enlarge the database. One of the schools will also be involved in figuring out "who is who in decomposition" by extracting DNA from their soil and tea bag samples and following the next-generation sequencing of the microbial

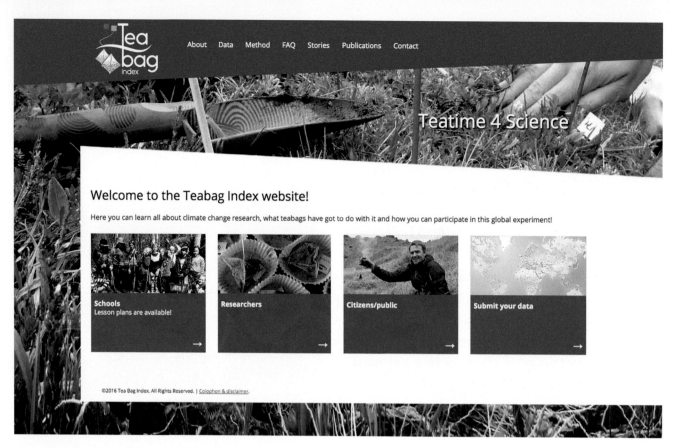

Screenshot of Teabag Index website http://www.teatime4science.org/.

communities of their own samples. Researchers are important, active users of our method, and we know of several initiatives worldwide that are helping us to collect worldwide data or that are carrying out their own experiments. Researchers and Citizen Scientists from Arctic countries in the north and all the way to South Africa and Japan have participated so far. The Tea Bag Index Team organizes an annual workshop for researchers who work with the method, in order to exchange ideas and keep track on the data being generated by various groups.

Since the tea bags are not available all around the globe, we either send the tea bags to the participants or guide them to where to purchase the tea. It is important that everybody uses the same kind of tea bags and that the tea bags are buried as they are, meaning they are not used as tea before. My colleague Judith Sarneel actually sent a couple of extra tea bags to the schools in 2015, for drinking the tea too. Our communication with people from all around

the world has shown us where the tea bags are available, and where are the regions where people need extra help finding them.

Judith Sarneel is currently working on her data from the Swedish schools' classes, and Sarah Duddigan with her data from the members of the Royal Horticultural Society. The feedback from the participants will be analyzed in order to make our future crowdsourcing efforts even better. The participants of Citizen Science Award 2016 in Austria filled in a questionnaire regarding their motivation to participate. The project TeaTime4Schools will also send a questionnaire to the participants in order to learn more about their motivation to participate. We wish to gain more insight into the reasons why Citizen Scientists want to participate in research and how we could motivate and help them even more. It will also be a way to get feedback on what the Citizen Scientists would like to learn from the project and on what they think about the soils around them.

Alexandra R. Toland: The dominant discourse around carbon sinks has been an economic one. We can imagine the soil as a bank for storing carbon stores and calculate the costs of nonaction or further business-as-usual land use. The term "black gold" has been used to describe biochar soil amendments as a strategy for sequestering soil carbon. Meanwhile, carbon markets have emerged, including agricultural soils in offset and trading schemes that often put small subsistence farmers at risk. What other metaphors exist for carbon beyond economic terms? How can soil carbon be a larger metaphor for the functioning of soil as a giant repository—for energy, for inspiration, for knowledge, for love, for future life-stores on the planet? How can humans come to see the soil repository as a resource for all life and not just human profit?

Taru Sandén: One of the most important functions of soil is to sequester carbon.[5] All soils have the capacity to sequester carbon, but some soils are better in performing this function than others. This depends on land use as well as the chemical, physical, and biological properties of the soils. The Tea Bag Index Team wants to visualize how fast the decomposition rates are and what are the stabilization factors around the globe, by making a map of the results that we collect. This can show us the hot spots of carbon turnover and the places where carbon is being stabilized.

I believe that maps can be a way to show the importance of soil carbon and the importance of soils. There are many stories of civilizations disappearing when they developed problems with their soils, and maps can help us see where we could have problems in the future and where we need to act in the present. Interactive maps can combine many disciplines and give concrete examples from soils around the world about "why carbon matters." By combining many layers, such as soil carbon, soil biology, and human stories, it would be possible to see how different disciplines interact and what kind of hot spots we could find.

The economic aspects are important for making changes in policymaking, but bringing in the human aspects would give soil carbon a voice from different parts of the world and would encourage people to think about soils more. Farmers are important guardians and managers of the soils, and they can show how regulations can be made reality in different agricultural regions of the world. They can show an example of how we could sequester more carbon in our soils. For example, the "4 per mille" initiative is trying to get different stakeholders, including farmers and policymakers, to commit to a voluntary action plan in order to sequester more carbon in agricultural soils.[6]

Soils are also a huge gene pool, and a handful of soil provides home to more organisms than there are humans on the whole globe. The microorganisms in soils are so small and difficult or expensive to measure that they don't always get the exposure and attention that would be needed. Being a habitat for soil biodiversity is one of the most important functions our soils have.

Alexandra R. Toland: Is carbon beautiful?

Taru Sandén: Of course! Carbon gives a beautiful color to the soil, and most soil-interested people have held a Munsell color chart in their hands to estimate carbon content based on the exact color of soil they hold in their hands. The darker the color, usually the higher carbon content the soil has. Carbon is beautiful from the beginning to the end: from green leaves of living plants that are producing glucose from carbon dioxide to the almost black color in the soil that has a high content of soil organic matter. I quite like the Carbon Tree project Terike was involved with, because you can see the flow of carbon and understand what a dynamic process the carbon cycle is. It fascinates and easily keeps you looking at it for a longer time.

Terike Haapoja: Inhale—Exhale

Durational sculptures, 2008/2013

plywood, glass, soil, CO2 sensors, sound

Programming: Aleksi Pihkanen, Gregoire Rousseau

Scientific advisors: Eija Juurola, Toivo Pohja

Closed Circuit—Open Duration, Installation exhibition, 2008/2013, Exhibition view: Nordic Pavilion, 55th Venice Biennale

Photo previous page: Terike Haapoja

Photo above: Sandra Kantanen

Photo below: Terike Haapoja

Terike Haapoja: Dialogue

Interactive installation, 2008

live trees, electronics, sound, light, CO2 sensors, breathing

Programming: Aleksi Pihkanen, Gregoire Rousseau

Scientific consulting: Eija Juurola, Toivo Pohja

Closed Circuit—Open Duration, Installation exhibition, 2008/2013, Exhibition view: Nordic Pavilion, 55th Venice Biennale

Photo top: Photo Terike Haapoja.

Photo middle: Photo Terike Haapoja.

Photo bottom: Photo Ugo Carmeni.

Inhale—Exhale

Carbon is the fourth most abundant element in the universe by mass, and it is present in all life-forms. Carbon cycle, the biochemical cycle in which carbon is exchanged among the different spheres of earth, is one of the most important cycles of the earth, allowing for carbon to be recycled and reused throughout the biosphere and all of its organisms. In the human body, carbon is the second most abundant element after oxygen.

As an organism dies, this element, essential for the ecosystem, is released and circulated back to the atmosphere. Our life is thus dependent on the vitality of the microbes that take care of decomposing. Soil can be perceived rather as an active process than a dead material; a process, in which matter is transformed into gazes—body into spirit. In biology this process is called "soil respiration," as the soil seems to "breath out" carbon dioxide. The process is necessary for the circulation of carbon in the global ecosystem and also plays an important role in climate change. As the earth warms, and falls become longer, warmer, and wetter, and the period of soil breathing becomes longer, adding to the accumulating process of global warming.

In the installation *Inhale—Exhale 3* coffin-size glass cases are filled with soil and dead leaves. Carbon dioxide that is produced by decomposing is measured with CO_2 sensors and the level is sonificated. Ventilation fans on both sides of the coffin are automatic, opening and closing in 20 s intervals. The ventilation fans function as gills, regulating the CO_2 level inside the coffin. As a result, the coffin seems to slowly inhale and exhale as the CO_2 level goes up and down.

Terike Haapoja

Dialogue

In the process of photosynthesis carbon dioxide is converted into sugars and other organic compounds. Nearly all life depends on photosynthesis either directly or indirectly as the ultimate source of energy in food. But photosynthesis is also essential for maintaining the normal level of oxygen in the atmosphere. As we breath out or release carbon dioxide into the atmosphere, photosynthetic organisms can fix it, releasing oxygen as a by-product. The interaction between species is thus physical, as we are in reality parts or the same metabolism.

The installation *Dialogue* enables an audible dialogue between breathing and the plants' photosynthesis process. When the visitor whistles to the trees they response by whistling back. The installation consists of a platform, through which live trees grow, and a bench for the visitors. The visitor is invited to whistle or breathe to the CO_2 sensor placed in front of the bench. The CO_2 breathed out by the visitor activates the light system and small measuring chambers attached to the branches of the trees. The decreasing of CO_2 level inside the measuring chambers, caused by photosynthesis, is audible as a whistling sound. Lights, sound, CO_2, and digital and analogue technology form a circuit of information that enables the viewer to perceive the interaction with the nonhuman environment not only as a physical process but also as flow of information.

I'm not actually interested *visualizing* the carbon cycle. I'm more interested in internalizing it and allowing others to do so. I'm interested in different forms of existing. So, I wanted to find ways to internalize what the carbon cycle is on a more human, personal level. It was more about finding out what it means to be human than directly about the soil, or more about finding an emotional connection to the phenomena in the soil that somehow remain beyond our direct experience.

Terike Haapoja, 2016

Endnotes

1. Forsström, M. 1970. Metsäherrat—kaskuja ja anekdootteja. WSOY, Finland. 142 pages. http://www.antikvaari.fi/naytatuote.asp?id=894222.

2. Tamminen, P., and Ilvesniemi, H. 2013. Extensive forest soil monitoring. In: Merilä, P., and Jortikka, S. (eds.), *Forest Condition Monitoring in Finland—National Report*. The Finnish Forest Research Institute. Available at http://urn.fi/URN:NBN:fi:metla-201305087584.

3. Tasanen, T. 2004. The history of Silviculture in Finland from the Mediaeval to the Breaktrough of Forest Industry in 1870s. metsäntutkimuslaitoksen tiedonantoja 920, Metsäntutkimuslaitos, Finland.

4. Keuskamp, J. A., Dingemans, B. J. J., **Lehtinen, T.,** Sarneel, J. M., Hefting, M. M. 2013. Tea Bag Index: A novel approach to collecting uniform decomposition data across ecosystems. *Methods in Ecology and Evolution*, 4(11): 1070–1075; **Lehtinen, T.,** Dingemans, B. J. J., Keuskamp, J. A., Hefting, M. M., Sarneel, J. M. 2014. Tea4Science. (http://www.soils4teachers.org/files/s4t/lessons/lesson-plan--tea4science.pdf).

5. Schulte, R. P. O. et al. 2014. Functional land management: A framework for managing soil-based ecosystem services for the sustainable intensification of agriculture. *Environmental Science & Policy* 38: 45–58.

6. Minasny, B. et al. 2017. Soil carbon 4 per mille. *Geoderma* 292: 59–86.

Deep-Time Moles

An Interdisciplinary Approach to Geological Archiving

Dave Griffiths, Sam Illingworth, and Matt Girling

Dr. Sam Illingworth is a senior lecturer in science communication at Manchester Metropolitan University, where his research involves developing dialogue between scientists and nonscientists. He does this primarily through the use of poetry and games (analog and digital). You can find out more about Illingworth and his research via his website: www. samillingworth.com.

Dave Griffiths is based at Manchester School of Art, United Kingdom, where he teaches interdisciplinary art, science, and design practice. He researches the aesthetic potential of

Title image: Dave Griffiths, *Deep Field (Unclear Zine)* 2016, installation view.

archival microfilm media to create speculative narratives about scientific and technological infrastructures and events. Griffiths' recent practice has developed through commissions with the Arts Catalyst and Rothschild Foundation, including the exhibitions *Perpetual Uncertainty—Contemporary Art in the Nuclear Anthropocene* (Sweden and Belgium, 2016) and *Finding Treblinka* (Poland and United Kingdom).

Matt Girling is an artist, prop maker, and comic creator living and working in Manchester. His work attempts to understand the role of the human race within nature as a whole. For the *Deep Field (Unclear Zine)* project he found aesthetic inspiration in the long-running sci-fi comic anthology *2000 AD* and *Heavy Metal* magazine. He is a firm believer that collaboration between science and art is vital to make positive change in society. Working with Dave Griffiths and Sam Illingworth opened a new avenue of inquiry within his practice that he continues to explore in zines and comic materials related to the pertinent subject matter of nuclear waste disposal in the Earth's substrates.

This chapter presents *Deep Field (Unclear Zine)*, a 2016 art-science work conducted at Mol and Dessel, two neighboring rural villages coexisting with sites researching geological nuclear-waste disposal in northern Belgium. Dave Griffiths produced a microfiche zine that probes and narrates the scientific testing and politics of decision-making surrounding controversial ONDRAF-NIRAS (Belgian National Agency for Radioactive Waste and Enriched Fissile Materials) projects: at cAt, a tumulus for encasing low-level waste, and HADES, a lab investigating the feasibility and safety case for geo-burial of high-level waste in Boom clay strata. Griffiths' fieldwork used qualitative and experiential methods such as ethnographic interviews with state scientists and independent monitoring groups, and photographic derive, to sense a wider Anthropogenic narrative of energy production, mineral extraction, and terrorist threat. Griffiths' findings were remixed through narrative responses by scientist-poet Sam Illingworth and DIY-comix artist Matt Girling, and archived as miniaturized microfilm. The zine attempts to translate the past, present, and future history of the repositories as folkloric sites of conflict, complexity, and unknowing, for the benefit of a far-future readership. The chapter discusses epistemological uncertainty around the survival and reception of crucial nuclear-security information in the face of inevitable material, linguistic and political ruination. We suggest that place markers, as monumental semiotic warnings to the future, along with digital archives, might also be augmented by decentralized analog fragments that promote ongoing memorialization of nuclear-heritage sites through intergenerational storytelling and rearchiving. The gesture of microfiche proposes an indeterminate, spectral archive for future citizens, that could be retranslated and reproduced many times through deep-time subject to a decision: to remember, or to delete?

Deep in the Congo alongside malicious soil,
Mined for men in Manhattan with ambitious soil.
A democratic route of being instructed,
"These sites will be placed on your inauspicious soil."
They actively remember to forget their past,
Buried deep amongst this blurry, fictitious soil.
Blood has been spilt by hands and by the writer's pen,
Tarnishing the compost of this now vicious soil.
The lone voices stand above, screaming silently,
Their objections now lost in the suspicious soil.
The heavy metal festival has been replaced,
With heavy metals seeping through nutritious soil.
The white sands flitter in and out of our vision,
Feeding the eagle with our surreptitious soil.
Amongst dirty, poisoned clay Mol moves unguided,
Searching for answers in this repetitious soil.[1]

The ghazal is an ancient form of Arabic verse dealing with loss and romantic love, and which was later adopted by Persian poets, gaining prominence in the thirteenth and fourteenth centuries thanks to luminaries such as Hafiz who used the form to convey both erotic longing and religious mysticism. Consisting of syntactically and grammatically complete couplets, the ghazal also has an elaborate rhyme scheme, with each couplet ending with the same word (or phrase), preceded by an additional internal rhyme. Traditionally, the last couplet also includes the name of the poet. This ghazal was written as part of an interdisciplinary archiving project from the point of view of an archivist-gatekeeper, the identity of who will be discussed shortly.

How might we communicate knowledge to far-future beings, about ancient hazards buried deep below the soils that they inherit? *Deep Field (Unclear Zine)* imagines a fictional future observer, from a society that may be culturally and technologically advanced or regressed, who encounters and attempts to understand surviving analog remains which record events and materiality in a nuclear landscape. The title of this artwork contains a deliberate mistranslation of the words "nuclear zone," as it has interpreted such a zone surrounding Mol and Dessel, two quiet communities in the Antwerpen province of northern Belgium. The place name "Mol" in Flemish translates into English as "mole," the burrowing animal, a name etymologically linked to the Old Norse "mold," Proto-Germanic "muldō" and Latin "modulus." The soil as dusty muck for planting, giving shape to a problem through thought, time, and technology.

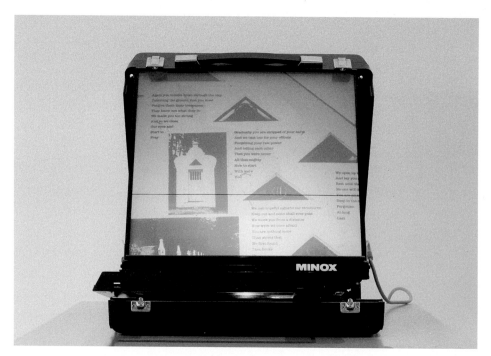

The people of Mol and Dessel participate and coexist with SCK-CEN, the Belgian nuclear research center that is surrounded by the local forest. Within this zone is cAt, a surface-burial project currently in planning stage, which aims to construct huge elongated tumuli to house 70,500 m^3 of monoliths for immobilizing low-level radioactive waste over a 300-year period.[2] Also inside the Atom Village is HADES (High Activity Disposal Experimental Site), a research lab investigating the feasibility and safety case for deep-time geo-burial of high-level nuclear waste with a half-life of over 100,000 years, where tests have been conducted since 1980 in a tunnel complex dug 225 meters below Earth in the boom clay strata.[3] HADES is not an operational repository, since the facility's research remit is to test geo-burial feasibility by simulating thermal conditions rather than using actual spent nuclear fuel ("radwaste").

Dave Griffiths produced *Deep Field (Unclear Zine)* in 2016 using new aesthetic approaches to microfiche, an old library media for storing photographs, data, and newspaper archives. This was exhibited in the touring exhibition *Perpetual Uncertainty—Art and the Nuclear Anthropocene*, at Bildmuseet in Sweden, curated by Ele Carpenter.[4] Griffiths also took part in a roundtable with ethicists and scientists from the Swedish and Belgian nuclear industries, alongside artists, curators and archaeologists. The show presented contemporary nuclear aesthetic responses prompted by twentieth and twenty-first century issues of civil energy production, disaster, and nuclear militarization. Related events discussed the role of contemporary artists in partnering scientists and engineers, whereby they contribute as cultural activists to raise questions about the role of the

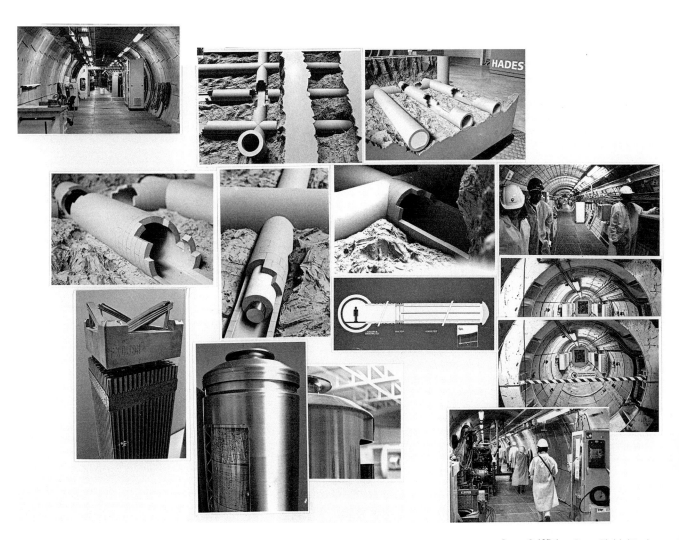

Dave Griffiths, *Deep Field (Unclear Zine)* 2016, microfiche detail showing photographs taken during Arts Catalyst/Z33 field trip to HADES.

Copyright Dave Griffiths.

human in coexistence with the nonhuman. In responding to radiological inheritance (the problem of long-term management of radwaste and its potential future biospheric impact), artists propose ways to make sense of the continuing present; for instance, in deconstructing scientific and social assumptions, language, and governance structures, and feeding into urgent contemporary debate about how to make, and speak to, the future.

In editing *Deep Field (Unclear Zine)*, Dave Griffiths' multimodal methodology combined transversal learning from the Belgian nuclear community through recorded conversations with experts; direct observation of research sites and the surrounding geographic area through photographic derive; and conducting archival research by collecting contemporary Twitter discourse on perceived nuclear threat in the region. This data-gathering stage generated a knowledge-brief for science communicator and poet Sam Illingworth, and the illustrator Matt Girling, in their subsequent mythological remix of the geo-burial narrative troubling this locality and northern Belgium.

To begin editing the microfiche-zine Griffiths first interviewed a safety expert from ONDRAF/NIRAS (Belgian Nuclear Waste Authority, Brussels) and two civil counterparts in STORA (Dessel), a citizen's group monitoring the government's projects. These conversations gave insight into the material behavior of radwaste; technological proposals for low- and high-level waste disposal; and the postwar origins of Belgian civil energy in relation to uranium trade with the Congo. Crucially these interviews provided understanding of both the international macropolitical emergency of radwaste proliferation, and local, micropolitical problem solving in which community representatives, SCK-CEN, and ONDRAF/NIRAS aim to consensually arrive at feasible geo-burial proposals and potential operational solutions.[5]

Boom clay materiality, tunneling processes, and concrete-steel radwaste supercontainers were explained and observed in the HADES underground facility during a July 2016 artists' field study organized by ONDRAF/NIRAS, Arts Catalyst (London), and Z33 House for Contemporary Art (Hasselt, Belgium). After this brief experience as a mole in the bunker, Griffiths staged a golf activity on a nearby driving range for the art-science delegation to unwind and reenact the leisure pursuits of expat nuclear-energy workers from the United States, who aided this industry's establishment in postwar Belgium. Griffiths crisscrossed the towns by bicycle to sense the wider site, taking photos in the forests, leisure spots, and industrial zones. He encountered a wider Anthropogenic narrative involving the nuclear zone, a local construction boom evidenced by mounds of soil and gravel at many street corners, and the community's other key heavy industry: exporting fine white sand to cement and fracking companies around the world. During Dave's visits in May and July 2016 the nuclear zone was roadblocked by armed patrols; earlier news articles had reported an Isis plot against an SCK executive as an element of wider terrorist operations in Belgium during 2015–2016.[6] Outside Dessel, near fields where the cAt tumuli are planned, Graspop Metal Meeting, an annual rock festival, was being rigged for over 100,000 music punters.

Through fieldwork Griffiths hoped to sense what was going on at the surface layer and what is embedded underneath. Together Griffiths, Matt Girling, and Sam Illingworth addressed the idea, discussed by experts in nuclear records, knowledge, and memory management, that geo-burial documentation should engage folklore to create active remembrance, so that radioactive sites may be safeguarded from future inadvertent or deliberate intrusion.[7] Girling, a zine illustrator from Manchester, drew cartoons in response to the sense of conflict and unknowing emerging from the conversations and photos that Griffiths gathered. They created the character of "The Mol", a terrifying, mutant underworld guardian of

// deep field [club golf nuclea Mol] //

the site who wards off transgressing humans. Girling also drew a future archaeologist-treasure hunter who navigates this territory through deep time, and a Neolithic band of nu-metal musicians staging future rock festivals dedicated to the merciless Mol. A far-future repository control center was visualized, in which a technocratic priesthood operates The Mol to monitor radionuclide migration and deter intruders.

Dave Griffiths, *Deep Field (Unclear Zine)* 2016, microfiche detail, showing Deep Field (Club Golf Nuclea Mol) performance.

Copyright Dave Griffiths and Matt Girling.

Similarly, Sam Illingworth created a poetry anthology, where traditional rhyming forms are used as a method for carrying scientific mythology reliably through time:

> *While making deep excavations we found some quaint bronze jewellery,*
> *Then discovered something old hidden in this pocketful of geography;*
> *We were panicky, losing all our favourite letters at a hideous rate.*
> *They leaked to the surface faster than the wet, pounding night;*
> *We couldn't keep up as they calligraphically disappeared.*
> *And then we gulped, as all we knew to write was:*
> *Do not go into that area, then kept to this:*
> *Danger—keep out!*
> *Danger![8]*

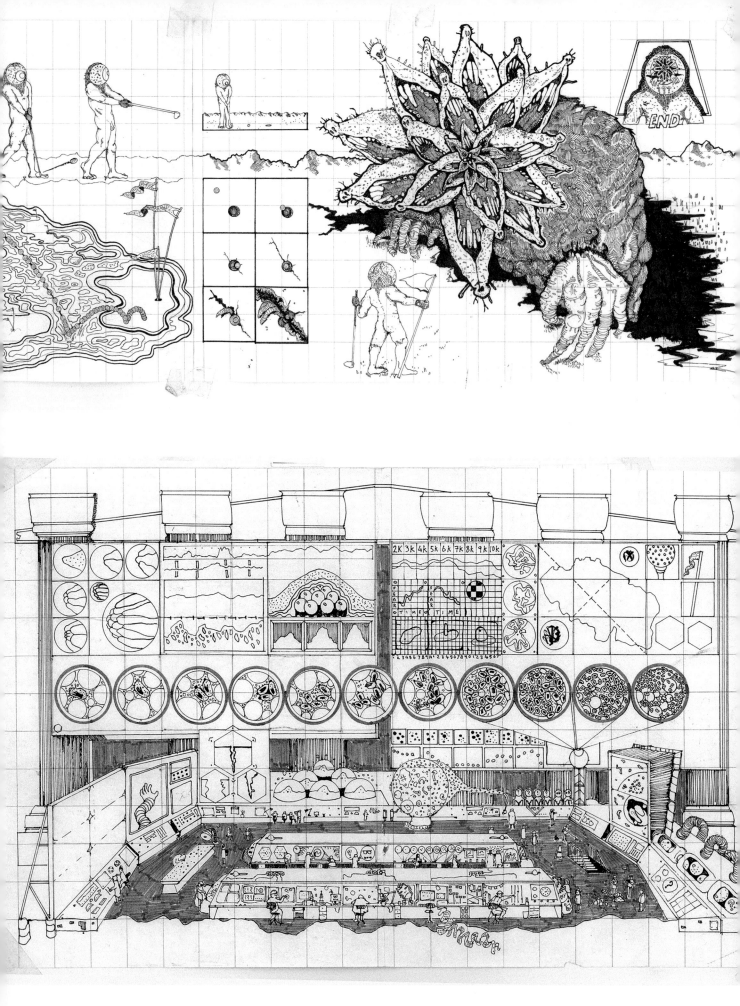

234 | Function 2   Repository: Soil as source of energy, raw materials, pigments, and poetry

This poem begins with a pangram using all letters of the alphabet. The line lengths then gradually reduce as letters leak out of the writing. This is referring to the possibility that human intrusion into thermal-stage radwaste supercontainers could release toxic isotopes that would migrate, through deep time, to eventually reach the surface biosphere. There's a reference to the discourse of place-design strategies proposed to mark dangerous nuclear sites.[9] Empty stone walls would allow future inscription of translated knowledge as language evolves over 100,000 years of repository storage, but vital knowledge might inevitably leak out.

Like artistic moles, we were sensing in the dark, trying to depict something not easily knowable. In order to record these findings for posterity we used documentary and speculative narrative modes to compress stories about material transfer through deep time. Using microfilm, the artwork was constructed as a zine, a throwaway subcultural format offering a contemporary thumbnail sketch, that may in time become collected and migrated into formal archive institutions. The work proposes a potentially useful historiographical resource for future observers. As an archaic celluloid media used to store and transmit information since modernity, microfilm is now nearly redundant due to the hegemony of digital technicity; however, it is also a durable medium that will survive floods, last 500 years, and can be copied as a near-exact analog. Microfilm may or may not outlive digital documents containing the same information, but to decompress and read microfilm's compressed contents requires only a simple lens and light.

There's a democratic impulse to the microzine—as voices of artists, state scientists, citizen monitors, and Twitter pundits are equalized in their description and speculation on future material becoming alongside dangerous nuclear waste. The textual and visual material was reduced to microfilm strips, then cut-and-pasted into the A6-sized fiche format, a collage complete with fingerprints, masking tape, hairs, and illegible sections. The making is fragile and uncertain, perhaps reflecting what seems to outsiders like unclear policymaking in knowledge management surrounding nuclear geo-burial.

We wanted to playfully rethink the potential of microfilm as a knowledge-transmission medium, and emphasize the complex problem of showing future humans what went on—and what lives on—in such international geo-burial efforts. The project asks if there can be a useful relationship

Opposite page top:

Dave Griffiths, *Deep Field (Unclear Zine)* 2016, microfiche detail, The Mol drawing by Matt Girling.

Copyright Matt Girling.

Opposite page bottom:

Dave Griffiths, *Deep Field (Unclear Zine)* 2016, microfiche detail, drawing by Matt Girling.

Copyright Matt Girling.

Dave Griffiths, *Deep Field (Unclear Zine)* 2016, installation view showing *Mounds of Mol* photo collage.

Copyright Dave Griffiths. Photo: Dave Griffiths.

between the archive and a collective care toward intergenerational equity in managing the future of waste produced by our energy consumption. Wide distribution of tiny microfilm fragments would contrast with monumental memorials in nuclear landscapes, and suggest a reconsideration of the spatiotemporal communication of vital records about decision-making and place-marker warnings we may want to send to future people. But the absurdity of microfilm is that successive like-for-like reprints, over the 100,000-year lifetime of the radioactive material, would erase legibility, via a myriad of mistranslations, losses, and deletions, to end as blank sheets of celluloid. Ultimately, it is a medium that is subject to ruination, just like digital records.

Manipulating the microfiche reader in the gallery is a searching gesture, where viewers might peer into the dimly lit glass screen to make sense of the post-colonial past, troubled present, and hazardous future entangled in the site. The audience is glimpsing fragments in the dark as they traverse the archival remnants above and below the soil. Hidden treasures, buried secrets, and forgotten memories all lie within this fractured space, waiting to be discovered by current and future archivists—if only they know where to look.

Endnotes

1. Sam Illingworth, "From Our Own Correspondent," in Dave Griffiths, ed., *Deep Field (Unclear Zine)*, microfiche, Manchester 2016. Copyright Sam Illingworth, reproduced with permission.

2. *The cAt Project in Dessel: A Long-Term Solution for Belgian Category A Waste*, masterplan, ONDRAF/NIRAS 2010.

3. Herman Damveld and Robert Jan van den Berg, *Discussions on Nuclear Waste: A Survey on Public Participation, Decision-Making and Discussions in Eight Countries*, report, CORA (Dutch Commission for the Disposal of Radioactive Waste) 2000.

4. *Deep Field (Unclear Zine)* was commissioned by Arts Catalyst, London, and funded by Manchester School of Art. Fieldwork was supported in Belgium by Z33 House for Contemporary Art, Hasselt; ONDRAF/NIRAS, the Belgian National Agency for Radioactive Waste and Enriched Fissile Materials, Brussels; and STORA, Dessel, a group representing local business, community, and politics in study and consultation around planned low-level radwaste storage in the municipality, and wider nuclear activities in the region. STORA is funded by ONDRAF/NIRAS to independently monitor the government's projects.

5. Christoph Depaus, recorded conversation with Dave Griffiths, ONDRAF/NIRAS, Brussels May 25, 2016; Geert Lauwen (STORA), recorded conversation with Dave Griffiths at ONDRAF/NIRAS cAt demonstration site, Dessel May 27, 2016; Katleen Dervaux, recorded conversation with Dave Griffiths at STORA, Dessel May 27, 2016.

6. cf. Samuel Osborne, "Isis Suspects Secretly Monitored Belgian Nuclear Scientist, Raising Dirty Bomb Fears," *The Independent*, February 19, 2016, http://www.independent.co.uk/ news/world/europe/isis-dirty-bomb-nuclear-scientists-paris-attacks-a6884146.html, accessed January 16, 2017; Cynthia Kroet, "Paris Attack Suspects Filmed Nuclear Official's Home,"

Politico, February 17, 2016, http://www.politico.eu/article/paris-attack-suspects-filmed-nuclear-officials-home-isis-terrorism-counterterrorism-isil-belgium-mol-plant, accessed January 16, 2017.

7. The importance of engaging intergenerational education and cultural heritage in actively producing folkloric mythology to promote security, place-marking, and remembrance of radiological burial sites is discussed in Thomas Sebeok, *Communication Measures to Bridge Ten Millennia*, report, Office of Nuclear Waste Isolation, 1984; Kathleen Trauth, Stephen Hora, and Robert Guzowski, *Expert Judgment on Markers to Deter Inadvertent Human Intrusion into the Waste Isolation Pilot Plant*, report, Sandia National Labs, Albuquerque, 1993; Jantine Schröeder, Radu Botez, and Marine Formentini, *Radioactive Waste Management and Constructing Memory for Future Generations: Proceedings of the International Conference and Debate, September 15–17, 2014, Verdun, France*, report, Organisation for Economic Co-operation and Development, 2015.

8. Sam Illingworth, "When Deep Geology Leaks Out," in Dave Griffiths, ed., *Deep Field (Unclear Zine)*, microfiche, Manchester 2016. (Copyright Sam Illingworth, reproduced with permission.)

9. Kathleen Trauth, Stephen Hora, and Robert Guzowski, *Expert Judgment on Markers to Deter Inadvertent Human Intrusion into the Waste Isolation Pilot Plant*.

Function 3
INTERFACE: Soil as site of environmental interaction, filtration and transformation

Interface

Between the soggy ceiling of the groundwater aquifer and the uppermost interface of earth and air lies the unsaturated space of soil particles and pores known as the *vadose zone*. Authors in this section explore the fascinating fluid world of the vadose zone and its interface function as a site of environmental interaction, filtration, and transformation. Questions of groundwater recharge, wetland management, and the flux of nutrients and toxic inputs are discussed in terms of long-term sculptural interventions that render artistic intention part of the landscape. Problems of river bank erosion, climate change–related flooding, and urban and industrial runoff are seen as design challenges to be addressed equally with aesthetics and engineering. Plants are used as biosculptural materials to draw out toxins from the soil, protect the soil from the erosive impacts of rain and wind, and prevent slopes from slipping into waterways. The landscape is seen as malleable canvas that can be restored by humans at the same time as it is degraded. The soil is understood as an interface of environmental processes and cultural responses.

Gerd Wessolek opens the section by honoring selected soil hydrologists and their vadose zone research in dedicated paintings. Lillian Ball and Ed Landa exchange opinions about renaturalizing degraded locations along the Bronx River. Aviva Rahmani and Ray Weil discuss the possibility of locating "trigger points" and meridians along the landscape body for restoring watersheds at regional scales. Stacy Levy looks for "backyard solutions" to acid mine drainage in a dialogue with Patrick Drohan, while Mel Chin discusses the social costs and some creative solutions to heavy metal leaching in an interview with Patricia Watts and Amy Lipton. Daniel McCormick and Mary O'Brien talk about their on-the-ground restoration works as "Watershed Artists" with Bruce James, and Lindsey Rustad, Xavier Cortada, and Marty Quinn reflect on their collaborative sci-art data visualization work, *WaterViz*, at Hubbard Brook Experimental Forest. Finally, the ephemeral nature of our own existence in Earth's nutrient cycles is brought to light in an essay by Farrah Fatemi and Laura Fatemi on their exhibition, *Rooted in Soil*, and in a touching dialogue between Maxine Levin, a soil scientist with the U.S. Department of Agriculture, and eco-art pioneer Jackie Brookner, who passed away in 2015, becoming part of the ephemeral soil cycle she spent a lifetime trying to understand and communicate.

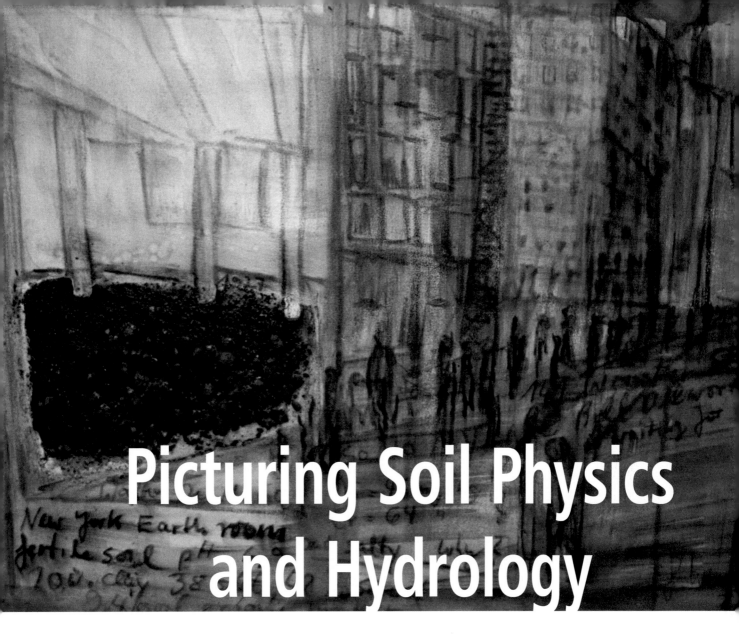

Picturing Soil Physics and Hydrology

Gerd Wessolek

As contributing author turned co-editor of this book, soil physicist, and hydrologist, **Gerd Wessolek** presents some of his own artwork as reflective introduction to the section on soil as interface. His paintings combine scientific results from his own and others' research with soil materials, symbols, and vibrant color. The paintings have been featured in soil textbooks such as one by Klaus Bohne (2005),6 in several calendars (2004-present), and scientific papers such as Feller et al. (2015).7 Gerd writes, "It is not common practice to express soil scientific data on canvas, but friends and colleagues have encouraged me to continue this kind of illustration. I believe this kind of artistic practice represents a unique way of training observation as a scientist and bridging disciplines."

Title image: Homage to Walther de Maria's *The New York Earth Room*.

From Field to Canvas

Soil scientists and artists do not commonly work together. Their fields are independent and separate from one another, from how they are taught in elementary schools and university courses to how they are practiced and made public later on. For as long as I can remember, I have somehow tried to move between these worlds and disciplines. In my scientific research projects, I try to integrate artistic approaches, specifically fine art painting techniques, to reflect on soil hydraulic theorems and applications in my paintings. By doing so, I try to uncover a deeper motivation for my scientific research and at the same time motivate others. This hybrid aspect of my work gives me a chance to honor my colleagues' work on a much more personal level than a footnote or citation in a journal paper would allow. For me, their extraordinary scientific impulses and genial ideas are alive. The vibrancy of their ideas are to be celebrated through a visual language that can be shared as art with the scientific community.

Over the course of the last ten years I have discovered that other soil scientists, some represented in volume, have a similar experience. Beyond out scientific work we soil scientists participate in exhibitions and artist residencies, take part in films, write poetry, and practice in groups.[1]

My own professional background is soil science, but I also have been painting since I was 12 years old. Painting has always been an important part of my life. Besides my studies in soil science, I also attended art classes in drawing, painting, print techniques, and art history at the universities in Braunschweig, Göttingen, and Kassel. Participating in a seminar of Josef Beuys on the occasion of the Documenta in Kassel was an absolute highlight of my time as a university student.

In the '90s, I started trying to combine both disciplines in my research and lectures as a soil science professor at the Technische Universität Berlin, Germany. This was welcomed because my department is embedded in the faculty of planning, architecture, and environment, in contrast to most soil science programs, which are embedded in faculties of agronomy. Within our faculty, disciplines of spatial design come together with disciplines of ecology and urban planning theory, offering space for more experimental ways of approaching the challenges of a changing world.

The following examples show how I use artistic approaches to frame my reflections as a soil scientist. These are organized into three distinct areas: (1) soil hydrological processes in the landscape, (2) soil taxonomy, and (3) urban soil properties.

Illustrating Soil Hydrological Processes

In my first example from 1993, I pick up a research topic of my colleagues Dr. Manfred Renger, Dr. Otto Strebel, Dr. Jürgen Böttcher, and Dr. Wim Duijnisveld. We worked together at the National Geological Survey, Hannover. At that time, water consumption in Germany increased dramatically and new wells of the water works were constructed in the northern parts of Hannover in order to supply the growing water demand. The wells were mostly installed in landscapes with gleyic soils and peat soils, which traditionally were used as grasslands or meadows because of the shallow groundwater table. As a reaction to the water pumping, the water table dropped. The former grassland sites became much drier and were consequently ploughed and then used as cropland. Thus, the landscape character changed completely: sites became larger, the land use more uniform, and grazing animals disappeared. This transformation process also resulted in an increasing percolation rate with nitrogen mineralization losses passing into the groundwater. Part of this scientific story is summarized in below figure showing four typical land use situations. The arrows show the direction of water and solute transport. This painting was used as the covering page of a scientific report on the effects of land use change (grassland into cropland) for the German Research Foundation (DFG), which financed this project. The scientific results were originally published in Duijnisveld et al. (1983), Böttcher et al. (1985), Strebel et al. (1984), and Wessolek et al. (1985).[2]

In my second example, I focus on the idea of preferential flow paths in soils. This phenomenon emerged as a mainstream research topic in soil physics and hydrology in the late '80s. Our working group was specifically

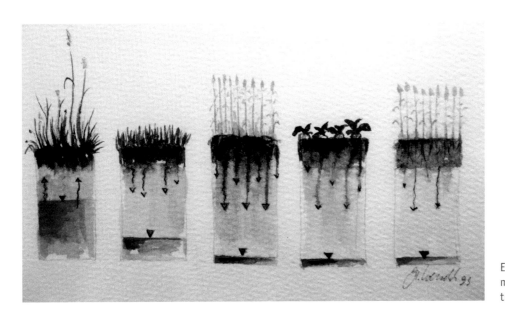

Effects of groundwater lowering on mineralization, water, and solute transport.

Water sorptivity (left) and preferential flow paths (right).

engaged in describing water transport under pine tree stands on sandy soil (Greiffenhagen et al. 2006).[3] Preferential flow paths quite often occur after heavy rainfall events in mostly dried-out soils with macropores such as clayey soils or sandy soils with a water repellent behavior. Hydrophobicity leads to a much faster transport of water and solutes into deeper soil layers, as we as soil physicists expected by describing the transport behavior using Richards' equation.[4] In my paintings I picked up aspects of soil water sorptivity and infiltration of water repellent soils in order to honor the scientific work of my colleagues Brent Clothier and others. In the second picture, I used acrylic color to express this preferential flow path in a well-structured clayey soil. The color makes the pattern of water transport visible. Under natural field conditions this could only be done using tracer. One of the most common tracers in practice is Brilliant Blue, a vibrant biodegradable dye used to characterize flow pathways and measure soil hydrologic functions in the field. Visually this tracer medium seeping though preferential flow pathways competes with International Klein Blue (IKB), a pigment patented by the artist Yves Klein in the '60s.

The next picture shows a soil profile with soil mapping aspects on the right side. I combined these with some fundamental equations and driving forces used in soil physics on the left side of the picture. The scientific part is also dedicated to Donald Nielsen and Rien van Genuchten for their outstanding scientific work. Explanations for the picture: both sides of the picture should demonstrate that mapping is basically needed for transferring soil reality into soil physical and hydrological units. The picture should also show that mapping and soil physics are traditional instruments of describing soils, but still are far away from describing the reality. Therefore, I put the soil profile itself into the center of the picture. Neither science nor

Next page:

Soil mapping meets soil physics.

Time–depth curves of soil water and solutes through a sandy soil—a homage to Dr. Wim Duijnisveld, who developed this concept.

art separately can explain natural phenomena. So the remaining question of the picture is, How is it possible to bridge these divergent perspectives?

In the '80s it became relevant to predict solute transport in soils for groundwater protection purposes. One of the upcoming topics was how nitrogen from arable soils is leaving the rooted zone and is leached into the groundwater. A suitable concept for making water flow visible and predictable is the so-called time depth curve procedure of my colleague and friend Dr. Wim Duijnisveld. He developed a numerical code to predict the time that soil water and solutes (like nitrate) needs to percolate to the groundwater table. This approach is very impressive, innovative, and successful because water and solute transport in soil became visible. The homage shows contour lines in brown color expressing the time depth curve starting at the soil surface and going continuously into deeper soil layers. The red-brown spots are soils from the so-called Fuhrberger Feld case (also see first example). The big yellow spot on the left is a nitrogen impulse that is already leached in deeper layers and is not any more visible on the right picture.

Soil Taxonomy

The next examples pick up the targets of soil survey: mapping soils, describing their properties and environmental conditions. For this kind of work, the soil science discipline has developed various soil mapping approaches such as the U.S. Soil Taxonomy and more international World Reference Base (WRB) taxonomic system. Soil scientists all over the world use these codes to "translate" soil properties into technical units

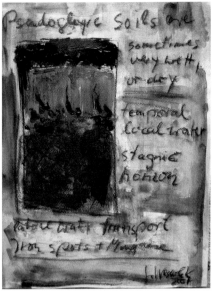

(left) Podsolic soil,
(middle) Pseudogleyic soil,
and (right) Gleyic soil.

and masses that can assist in recommendations for land use and help
formulate protection measures. Even though properties such as soil texture
and soil color are registered, such observations are of a technical rather
than aesthetic nature. This is a detriment from my point of view, because
aesthetic observation helps train a kind of field literacy that not only
helps scientists read their way through a landscape for research purposes
but garner appreciation for unique soil sequences. Moreover, aesthetic
interpretation of soil types could help taxonomists frame already existing
criteria in a view of uniqueness and rareness.

Depicts a "beautiful" but "poor" podsol, aesthetically pleasing to the. This
picture includes some specific comments, such as people do like podsols
but soil scientist always classify them as a poor soil for crop production.
Two more examples are presented for a pseudogleyic soil and gleyic soil
using soil scientific and aesthetic observations.

Urban Soils

The expression "urban soils" is often used to designate soils of anthropogenic
influence in cities. However, this definition is too restrictive to take into
account all areas that are transformed by human activities. For that reason,
the international SUITMA research group (Soils of Urban, Industrial,
Traffic, Mining and Military Areas) was initiated by Burghardt and Morel
in 1998 to encompass all soils under strong human influence. It could be said
that this group mainly deals with questions of soils in the Anthropocene.

Urban soils have become a conceptual focus of a handful of artists
visualizing human impacts on the environment, some of which are

Urban soil.

represented in this book. During several field excursions with students I have used painting as a means to discuss soil properties and encourage different methods of observation. On many of these sites, former industrial usage is still visible and detectable by the many artifacts embedded in the soil. Waste residues like cigarettes and chewing gum, buried tile, and ash become components of urban soil paintings.

The last example (see title picture) honors the work of the renowned land artist Walter de Maria, who was one of the first artists to shed light on soils in the city. The *Earth Room* in New York City was installed in a loft at 141 Wooster Street in 1977 and still remains intact and alive today. This soil sculpture is at least the third *Earth Room* installed by the artist, the first having been presented in Munich, Germany, in 1968. The second one was installed at the Hessisches Landesmuseum in Darmstadt, Germany, in 1974. *The New York Earth Room* is now the only one that still exists.

The sculpture shows a room that is nearly completely filled with a thick layer of black soil, which is carefully watered and tended to by employees of the DIA Art Foundation. The interior earth sculpture has the following dimensions: 197 m³, 335 m², 56 cm thickness, and a total weight of 127,300 kilos.[5] Coming in from the streets of New York, the room confronts viewers with a visceral smell and sight. There is nothing else but pure soil in a big light hall. At this moment one can feel that our whole human existence depends on the existence of soil. I visited the room and was allowed to take a sample for analyzing in my laboratory. I included the analysis in my homage to Walter de Marias' Earth work, a tribute to the potential of soil in the city.

Some Final Comments

My paintings combine scientific results from my own and others' research with soil materials, symbols, and vibrant color. The paintings have been featured in soil textbooks such as one by Klaus Bohne (2005),[6] in calendars (the first in 2004), and scientific papers such as Feller et al. (2015).[7] It is not common practice to express soil scientific data on canvas, but friends and colleagues have encouraged me to continue this kind of illustration. I believe this kind of work represents a way of training observation and bridging disciplines.

Endnotes

1. See, for example, works by Ken van Rees and Jay Stratton Noller.

2. Böttcher, J., Strebel, O., Duynisveld, W.H.M., 1985, Vertikale Stoffkonzentrationsprofile im Grundwasser eines Lockergesteins-Aquifers und deren Interpretation (Beispiel Fuhrberger Feld). Z. dt. Geog. Ges. 136, 543–552, Hannover. Duynisveld, W.H.M., Strebel, O.,1983, Zeit-Tiefen-Kurven der vertikalen Wasserbewegung und Verbleibzeit des Wassers im ungesättigten Bodenbereich. Mitt. Dtsch. Bodenkundl. Ges. 38. 77–82.
   Strebel, O., Böttcher, J., Duynisveld, W.H.M., 1984, Einfluss von Standortbedingungen und Bodennutzung auf Nitratauswaschung und Nitratkonzentration des Grundwassers. Landw. Forschung. Kongressband. J. D. Sauerländer's Verlag Frankfurt, 34–44.
   Wessolek, G., Renger, M., Strebel, O., Sponagel, H., 1985, Einfluss von Boden und Grundwasserflurabstand auf die jährliche Grundwasserneubildung unter Acker, Grünland und Nadelwald, Z. F. Kulturtechnik und Flurbereinigung 26, 130–137.

3. Greiffenhagen, A., Wessolek, G., Facklam, M., Renger, M., Stoffregen, H. 2006. Hydraulic functions and water repellency of forest floor horizons on sandy soils. *Geoderma*, 132(1–2):182–195.

4. Richards' equation is a nonlinear partial differential equation used to represent the movement of water in unsaturated soils. It was formulated by Lorenzo A. Richards in 1931 and is still used today.

5. See description from Dia Art Foundation, 2016.

6. Bohne, K. 2005. *An Introduction into Applied Soil Hydrology* (Lecture Notes in GeoEcology). Catena.

7. Feller, C., Landa, R.A., Toland, A., Wessolek, G. 2015. Case studies of soil in art. *Soil* 1: 543–559.

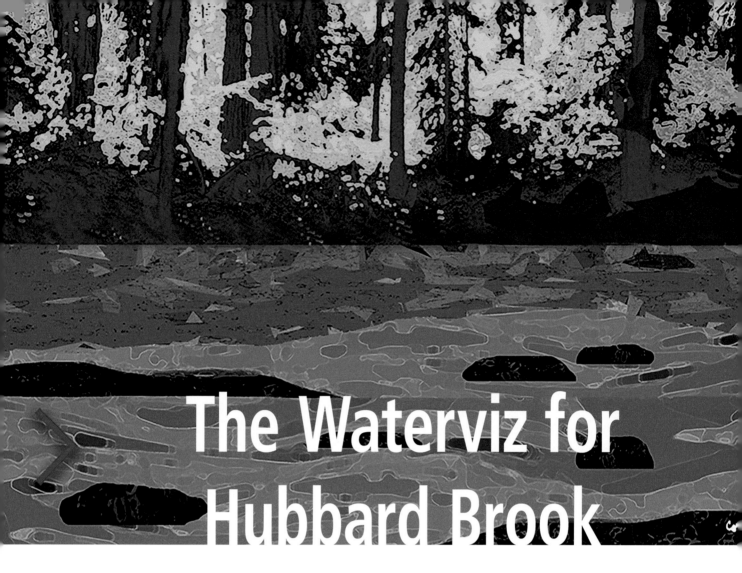

The Waterviz for Hubbard Brook

The Confluence of Science, Art, and Music at Long-Term Ecological Research Sites

Lindsey Rustad, Xavier Cortada, Marty Quinn, and Torrin Hallett

The Waterviz at Hubbard Brook is a new and transformative water cycle visualization and sonfication tool that lies at the nexus between hydrologic sciences, visual arts, music, and information design. For the Waterviz, water data, including precipitation, streamflow, soil water, and groundwater, are captured on an hourly time step from a small watershed at the world-renowned Hubbard Brook Experimental Forest in the White Mountains of New Hampshire using an array of digital environmental sensors. These high-frequency data are then transmitted wirelessly to the Internet and used to drive a computer model that calculates all components of the water cycle for the catchment *in real time*. These data, in turn, drive artistic visualizations and sonifications of the water cycle, accurately reflecting the hydrologic processes occurring at that moment in time. The Waterviz was inspired by a team of scientists led by Dr. Lindsey Rustad, the artist Xavier Cortada, and the musician/composer Marty Quinn, to integrate science, art, and music to create an opportunity for people anywhere in the world to intuitively experience the dynamic and ever-changing inputs, outputs, and storage of water in this small, upland forested watershed as they are occurring. In this chapter, we asked each of these three contributors to the project, as well as Waterviz student Torrin Hallett, to provide their insights on why they got involved, what they learned, surprises along the way, and reflections on the collaborative process.

Dr. Lindsey Rustad is a research ecologist for the U.S. Department of Agriculture (USDA) Forest Service Center for Research on Ecosystem Change in Durham, New Hampshire; co-director of the USDA Northeastern Hub for Risk Adaptation and Mitigation to Climate Change; a team leader for the Hubbard Brook Experimental Forest in NH; and a Fellow of the Soil Science Society of America. She received a BA in philosophy at Cornell University in 1980, an MS in forest science at the Yale School of Forestry and Environmental Sciences in 1983, and a PhD in plant science in 1988 at the University of Maine.

Xavier Cortada is an American artist. His work has been exhibited in museums, galleries, and cultural venues across the Americas, Europe, Asia, Antarctica, and Africa. Cortada studio is based at Pinecrest Gardens, where he serves as artist-in-residence. Since 2011, Cortada has based his engaged art-science practice at Florida International University, serving in the School of Environment, Arts and Society (SEAS) and the College of Communications, Architecture + The Arts (CARTA). A hallmark of his work is engaging scientists in his art-making and producing participatory science-art projects and exhibitions.

Marty Quinn is a composer/data scientist and founder of Design Rhythmics Sonification Research Lab. He has focused on the perception of data thru music and visualization for over 25 years. His many works include "The Climate Symphony," "Water Ice on Mars," CRaTER Live Internet Radio, "Walk on the Sun" interactive movement and image sonification for the visually impaired, and "Touch the Future: Hear the Climate Change." He has worked on the NASA IBEX and CRaTER instruments teams at the University of New Hampshire, where he holds an MS in computer science.

Torrin Hallett is a fifth-year student at Oberlin College and Conservatory pursuing bachelor's degrees in horn performance, composition, and mathematics. He has written music for a variety of solo, chamber, and large groups as well as for the 2014 PBS documentary *Consider the Conversation 2*. He is assistant conductor of the Northern Ohio Youth Orchestra's Wind Symphony and has also taught brass in Panama. In 2017, he performed with the Oberlin Sinfonietta at the Bang on a Can Marathon in Brooklyn.

Introduction

Water. We all care about water. Where it comes from, where it goes, how it gets there. Although scientists have studied water and its movements through the atmosphere/plant/soil continuum for centuries, a new generation of digital environmental sensors now provide terabytes of water data to the Internet in real time, allowing for an even closer look at these important hydrologic processes. In an era where providing clean and plentiful water to over seven billion people is an increasing global concern, it is critical to enable a better scientific and public understanding of the water cycle. The Waterviz is a novel data visualization and sonification tool that captures water data from digital environmental sensors at the U.S. Department of Agriculture (USDA) Forest Service's Hubbard Brook Experimental Forest,[1] New Hampshire, in real time. Integrating science, art, and music, the Waterviz is designed to simultaneously engage the reasoning, visual, and acoustical sensors of the brain, bringing to life pattern and process in water data.

In a nutshell, water cycle data (including precipitation, streamflow, soil water, and groundwater) are captured using an array of digital sensors on an hourly basis from a small watershed at the world-renowned Hubbard Brook Experimental Forest in the White Mountains of New Hampshire. These high frequency data are then transmitted wirelessly to the Internet and used to drive a computer model that calculates all components of the water cycle for the watershed *in real time*. These data, in turn, drive artistic visualizations and sonifications of the water cycle, accurately reflecting the hydrologic processes occurring at that moment in time. The Waterviz was inspired by a team of scientists led by Dr. Lindsey Rustad, the artist Xavier Cortada, and the musician/composer Marty Quinn, to integrate science, art, and music to create an opportunity for people anywhere in the world to intuitively experience the dynamic and ever-changing inputs, outputs, and storage of water in this small, upland forested watershed as they are occurring.

This chapter highlights our personal reflections on why we are involved, what we learned, surprises along the way, and reflections on the collaborative process. These are the views of just several of the many participants of this large, integrated project.

Voices

The Scientist's View Point, Lindsey Rustad

I believe in serendipity. The idea for the Waterviz arose as an intersection between my interests in digital data collection, a quest for new ways to

Title image, page 253:
Xavier Cortada, "H.J. Andrews Experimental Forest Water Visualization," digital art, 2016.

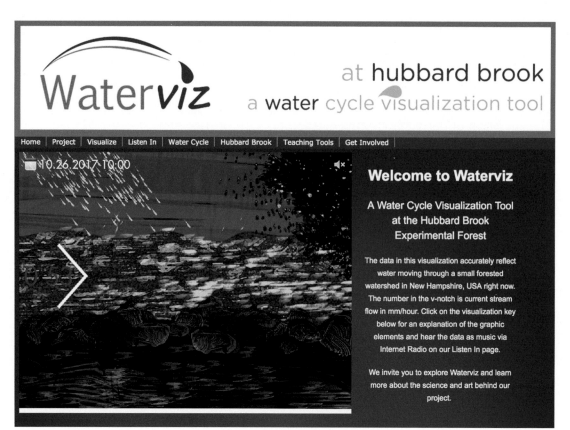

The homepage for the Waterviz for Hubbard Brook, a new water cycle visualization and sonification tool. https://www.waterviz.org.

understand and communicate Big Data in the environmental sciences, and a chance encounter with the Finnish Carbon Tree website. To step back, I have been intrigued by the potential for the automatic collection of environmental data using digital sensors connected to data loggers since the early 1990s, when a colleague, Dr. Steve Nodvin, started experimenting with these devices at a long-term research site: the Bear Brook Watershed in Maine. Since then, the technology has steadily improved and the costs have come down. More recently, starting around 2006, the USDA Forest Service team at the Hubbard Brook Experimental Forest, New Hampshire, began a major conversion from our historic analogue data collection, based on mechanical devices and strip chart recorders dating back to the mid 1950s, to a new digital environmental sensor platform. The new digital system produces over 100,000 data points per day from nine gaged watersheds, six automated weather stations, and three global climate change experiments. We are also actively collecting similar digital data from twelve other forests in the northeastern United States as part of the USDA Forest Service's Smart Forest initiative. The sheer magnitude of the data has been overwhelming, and has required the scientific community to rethink our approaches to data processing, analysis, and interpretation. The data are difficult for scientists to grasp, let alone the general public.

As I was grappling with these issues, I came across, by chance, the Finnish Carbon Tree website. The Carbon Tree represents a novel collaboration between scientists, artists, and educators with the aim to represent near-real-time ecosystem carbon cycle data in an artistic form that is intuitively understandable to any online viewer on the internet in the world. It was mesmerizing and I was hooked! The site uses weather data to drive a carbon cycle model in near real time, and then uses Adobe Flash Animation's moving particles to create the scintillating outline of a tree. The particles represent photosynthesis (i.e., carbon taken up by the virtual tree) and respiration (carbon released by the virtual tree). The number of particles are directly proportional to the data, and the outline of the shimmering tree expands and contracts as rates of photosynthesis increase and decrease (driven by temperature, moisture, and sunlight), and contracts and expands as leaf, stemwood, branch, and soil respiration cause the tree to lose carbon at different rates through respiration (driven by temperature and moisture). The balance between photosynthesis and respiration represents net primary productivity, and is illustrated by the size of the virtual tree, which can be seen over time, as the visualization cycles between night and day, and over the four seasons. The result is breathtakingly beautiful while also conveying fundamental information on how weather (and ultimately climate) drives the carbon cycle.

I instantly recognized the potential to use the same technology to communicate our new digital water cycle data from the Hubbard Brook Experimental Forest, and reached out to an initial team, including Mary Martin, our Hubbard Brook information manager, and Xavier Cortada, who had recently completed an artist-in-residence at the White Mountain National Forest, and with whom we had already engaged in an artistic collaboration. We were ambitious. Where the Carbon Tree had two fluxes (photosynthesis and respiration), we would have five (precipitation, streamflow, soil water, evaporation, and transpiration) and we would add an acoustical component. Although you can't "hear" photosynthesis and respiration, we recognized that the sound of water, at least precipitation and streamflow, is universally familiar to people worldwide. We originally imagined a recording of the sound of the stream, but Mary Martin introduced us to Marty Quinn at the University of New Hampshire (UNH), who was already turning real time cosmic data into music, and the collaborative opportunity to turn water cycle data into a forest symphony was launched.

So, how do you embark on such an ambitious project? We knew we needed to start by inviting Xavier Cortada back to New Hampshire and to Hubbard Brook to create the artwork. We also had reached out to the scientists running the Carbon Tree, who put us in touch with

Jussi Rasinmäki at Simisol in Finland who provided the digital "glue" to stitch the data and artwork together. Simisol was willing to work with us, and in fact contributed considerable time and talent to the project, but we needed a nominal fee to engage them more deeply. We wrote a one-page "invitation to collaborate," and floated it to a few organizations and individuals, who saw enough value in the concept to provide seed money. With this, we brought Xavier to Hubbard Brook, created a data pipeline with Simisol, and produced the first version of the Waterviz. To be honest, this first version looked much more like a very hungry caterpillar than the water cycle, but the particles moved, driven by the data, in the appropriate "masks" of the artwork, and we knew it would work. I recall being so excited with the first version that I charged into an Education and Outreach meeting at Hubbard Brook Experimental Forest to share our work, and received a lot of quizzical, skeptical looks. One fellow, who shall remain nameless, said "I don't know if you guys are brilliant … or just plain crazy …" The project is now one of our flagship education products at Hubbard Brook.

Carbon Tree. The Carbon Tree is a real-time simulation of the terrestrial carbon cycle in a Scotts Pine stand in Finland. The particles moving into and out of the virtual pine tree represent carbon uptake in photosynthesis and carbon release via respiration. The Carbon Tree is a cooperation between the Institute of Atmospheric Research and Earth System Science, Department of Forest Ecology of University of Helsinki, software company Simosol Oy, and artist Terike Haapoja. (See chapter "Carbon" in Repository Section).

This image was captured on November 25, 2017, from the website http://www.hiilipuu.fi/.

From these humble beginnings, we were able to fund a National Science Foundation Early-Concept Grant for Exploratory Research (EAGER) titled "The Waterviz at Hubbard Brook: The Confluence of Science, Art and Music at Long Term Ecological Research Sites." This allowed us to produce new, more sophisticated versions of the Waterviz for Hubbard Brook (see Xavier Cortada); develop the sonification as real-time Internet radio stations (see Marty Quinn); create a second Waterviz for the HJ Andrews Experimental Forest, an old-growth Douglas fir forest in Oregon; explore the basis of cognition of Big Data with neuroscientists from Dartmouth College; create lesson plans based on the Waterviz for middle and high school students; and provide funding for a summer undergraduate intern (see Torrin Hallett).

The biggest technical and artistic surprises were how difficult it is to make a compelling and beautiful visualization for five data components compared to the elegant two-component Carbon Tree, and how to effectively make visible the easily observed fluxes (precipitation and streamflow) as well as the unseen flows (soil water, ground water, evaporation, and transpiration). The biggest social surprise was how difficult it is to merge disciplines, and to truly integrate art and science. In the beginning, we were met with healthy skepticism, and frankly, a little derision. Some scientists, with raised eyebrows, asked the question of how can art (which some described as a two-dimensional painting on a wall) help us with our science. And artists wondered how to communicate with scientists. In the end, the project has engendered many thoughtful although sometimes difficult conversations between scientists and artists, has sparked new ways to look at and comprehend our data, and has allowed us to engage with a broader audience of "K through gray" learners. It is our hope, that through the Waterviz and related projects, it will be the norm and not the exception to have artists (writ large to include artists, musicians, writers, poets, philosophers, etc.) fully engaged with the scientific community, and to help solve some of the large and pressing socio-ecologic issues that face society today.

The Artist's View Point, Xavier Cortada

On the night of July 27, 2012, Dr. Lindsey Rustad walked me to the edge of a northern New Hampshire brook into which the water from nine watersheds flowed. The bright night stars wouldn't reveal the moving water. I could only hear it. I didn't have to see it to understand it. I could visualize it. It sounded magical. For decades Hubbard Brook Experimental Forest scientists have conducted intensive research on those nine watersheds to better understand the way forests work.

I was born in Albany, a four-hour drive from that spot—but I didn't grow up stepping across babbling brooks and hiking among the birches, the hemlocks, the maples. We moved to Miami when I was three and I have lived in the subtropics since. So, I grew up swimming around mangroves and snorkeling above corals.

I was in New England with my husband and composer Dr. Juan Carlos Espinosa at the invitation of the Arts Alliance of Northern New Hampshire.[2] We were immersed in the forest as part of a month-long artist this residency.[3] At 48, it was the first time I had ever spent any real time in a Northern forest. So, I reached out to USDA Forest Service personnel to help me explore the forest and better understand the natural and human history of the White Mountains.

Working with Forest Service hydrologists, I created an installation marking the impact of Hurricane Sandy on a brook it had rerouted.[4] Working with their biologists, I depicted the plight of an endangered bird, Bicknell's thrush (*Catharus bicknelli*), and the White Mountain butterfly (*Arctic Oeneis melissa semidea*), whose fragile Alpine habitats were being constricted by a warming climate.[5] Working with their archeologist, I took locals on a walk through the woods to find and dine in the ruins of a nineteenth century farmhouse consumed by wilderness.[6] The area was abandoned by the dinner party participants' ancestors when the government reclaimed it as forests to protect those living in communities downstream from agricultural and industrial runoff. During this residency, I also wanted to honor the scientists who study the forest to develop knowledge that has a global impact (and to rebut the growing sector of our society who deny science, especially climate science). So, I worked with Dr. Lindsey Rustad, a research ecologist with the USDA Forest Service, to develop four digital tapestries that portrayed Hubbard Brook's research on the forest's soils, water, vegetation, and wildlife. Her team's scientists participated in Wind Words[7]; they came to an overlook on the Kancamagus Highway, stood in front of their respective tapestries, and read from their published science articles. By having them speak their words to the wind, they were conceptually sharing their work beyond the pages of refereed research journals and across the Earth where their research also matters.

This was our first science-art piece. The collaboration was an essential first step that helped propel us to our second one. It allowed us to start to build trust, friendships, and an understanding of each other's disciplinary worldviews and methodologies.

Xavier Cortada with Lindsey Rustad reading excerpts from scientific papers to the four corners of the compass in a participatory art production of Wind Words at the Kancamagus Highway, July 26, 2012.

The second project began at the edge of the brook. While we sat there in the starlight, she asked me to help her with an idea she wanted to bring to life. She wanted to use Hubbard Brook data collected upstream from where we sat to visualize the water cycle. Learning about my art-science practice, the forest's lead scientist was engaging me as an artist to help increase that understanding.

I went back to Miami and in continuing our e-mail dialogue, she would write:

> The goal of the water cycle simulator is to use color, shapes, lines, movement and sound to produce a visualization of the water cycle, easily intuited based on an educated viewer's prior knowledge, but more deeply engaging of right hemisphere brain processes, such that the visualization will hold the viewer's attention long enough to instill the image and associated feelings from conscious memory (15–20 seconds) to longer term memory, and to invite the viewer to engage more deeply, for a longer period of time, and more often (invite to return) than a static, two dimensional, left brain hemisphere oriented chart, graph, or text.

During early May 2013, Dr. Espinosa and I returned to Hubbard Brook. He created a series of soundscapes and electronic pieces that were originally to serve as the score for our live stream data visualizations. We stayed on Mirror Lake (one of Earth's most researched lakes) and joined Hubbard Brook scientists as we designed ways of visualizing the live data collected onsite (e.g., precipitation input, stream water discharge, change in soil moisture, and evapotranspiration + groundwater).

I remember meeting scientists at their labs, in meetings. We would talk about their research. The science. I would try to gain insight on how they envisioned the water cycle. I would ask: What was the most important thing to capture? Which way would the water flow? What would we do with the ground water? How can we depict water in different states, doing different things, at different times?

We would talk, sketch, talk. Conceptualize. Answering our own questions, we would try to think about audiences. How could we engage them and spark further curiosity? We wondered about capturing and archiving extreme weather events so that they could be replayed. The team would also talk about keeping the piece dynamic and exciting. About logistics and technology. We discussed how to make the particles flow through the screen to look like stream water and how to make the particles go faster with higher stream flow. We figured out ways to make the leaves look like those of a Northern Forest. Everything was on the table.

Through a series of subsequent meetings and a myriad of Skype calls, e-mails and conference calls—all convened by our project leader, Dr. Rustad—the team eventually came up with what we called Waterviz 1.0.

On June 21, 2014, Dr. Espinosa and I received an e-mail from Dr. Rustad. It read: "I'll send a note out to the group tomorrow, but I wanted to let you know that WaterViz is LIVE!!!" It contained a link to the online site. We knew it wasn't our end product but were thrilled to click on it and see particles flow across a computer screen in real time. I responded:

SO HAPPY to see this!!!!
SO HONORED to be a part of this!
THANKS SO MUCH for your incredible leadership, vision and faith in our interdisciplinary process.
THRILLED, THRILLED, THRILLED!!!

In time, we brought in another composer, Marty Quinn, to help with the soundscape. As in the art, he would use actual live stream Hubbard Brook data to create the sonifications.

I returned to Hubbard Brook on November 4–7, 2015, with a larger team and broader focus, to continue the exploration, "Real-Time Data Visualizations: A 21st Century Confluence of Art, Music and Science at Ecological Research Sites." We were funded by a Waterviz EAGER grant that was to evaluate the efficacy of our work.

Together, we kept working and redesigning. Fine-tuning.

Other contemporaneous projects followed from this initial collaboration, including the creation of an art piece to celebrate the sixtieth anniversary of Hubbard Brook in 2015. The digital work included the names of every scientist who published work about the experimental forest.[8]

In another project, *Water Paintings*,[9] I placed nine pencil drawings and nine pieces of watercolor paper inside nylon mesh bags and tied them to a rope at each of the nine weirs at Hubbard Brook Experimental Forest for a period of sixteen weeks in 2016. The final work included the stained drawings, water samples, data collected over the same sixteen-week period from the same nine weirs, and residue in filters. I wanted audiences to see what the water "painted" as it flowed and transported materials down the stream.

In October 2016, I visited the H.J. Andrews Experimental Forest in Oregon and finally, in 2017, I returned to New Hampshire to complete a mini-residency at Hubbard Brook to work on an updated version (WaterViz 2.0). As an artist, these past five years have provided me with an incredible perch. A perch from which to see science in the making. A perch that allowed me to work with others to innovate, to find ways of better understanding, and communicating the science that explains our world, its ecosystems, and our relationship to one another.

I remember first coming to the White Mountains and reveling in the natural and human history of the place. I would see a forest as compilation

Xavier Cortada's *Water Painting* installed just below Weir 3 at the Hubbard Brook Experimental Forest, New Hampshire.

of discrete organisms (mostly plants and animals) interacting with one with the other through space and time. I wanted to create art pieces, processes, and installations that helped mark those moments and places (e.g., a tree as clock, calendar; a predator as prey, or as the sculptor of ecosystems), mostly so we could make better choices to support healthy forests.

I remember meeting with scientists who helped me see a forest through their eyes. Sharing the knowledge they developed by observing and experimenting on those systems, they let me also see forests as a transportation system, a conduit for things in motion. Chemicals— nitrogen, carbon, phosphorus—moving through the trees. Water moving through its cycle. Soils as storage. Natural processes responding to the environment.

We taught ourselves how to see things differently. Together, we came out better because of our interactions.

There was an interesting thing that happened on the last day of my mini-residency at H.J. Andrews Forest. I had dined with their poets-in-residence, walked through an old-growth forest, witnessed the decomposition of logs that are part of a 200-year longitudinal study and lectured at nearby Oregon State University on the engagement value of an art-science practice. I spent some time with wildlife biologist Eric Forsman and learned about how his research on the range of the spotted owl helped lawyers fight to save the forests of the Pacific Northwest from the timber industry. The residency had been cut short by a few days because of a freak Pacific Northwest typhoon that prevented me from doing my water paintings there. So driving down a winding road on my way to the airport that morning, a deer slowly jumped out of one side of the thickly vegetated road and gracefully disappeared into the other side. But, for the first time, I saw things differently. I wasn't really seeing a deer. Or vegetation.

I saw one organism: One piece of the forest simply walked across the paved road and reconnected with itself on the other side.

Like the sounds of the brook I couldn't see on that starry night, it was a new way of seeing things. And it was magical.

The Musician/Composer's View Point, Marty Quinn

The moon led me to the forest. I had been working on the IBEX and CRaTER space instrument teams at UNH as a data scientist when a colleague at UNH suggested I attend a conference at Hubbard Brook to discuss universal design for learning (i.e., an approach to curriculum development that gives all individuals equal opportunities to learn) and its applications to research dissemination at Hubbard Brook. My approach to turning cosmic ray data into music in CRaTER Live, an Internet-based radio station that created music out of six detectors on board the Lunar Reconnaissance Orbiter, appealed to many at the conference. My friendship and collaboration with Dr. Lindsey Rustad, the lead scientist on the Waterviz project, continues to evolve from that moment. It was a natural step for me to apply the many techniques and original software I had developed over 25 years to the data of Hubbard Brook, thereby providing a novel method for data expression heretofore unknown to the team.

After this first introduction, I became a member of the Waterviz team. I designed a musical world that would respond to various data inputs— inputs that had been gathered from digital environmental sensors at Hubbard Brook—to create music that evolved with every hour of new data. The result was a live Internet radio station called Hubbard Brook Listen Live that plays the last 96 hours of data (the last four days of eight variables) as polyphonic music. This design was also used to play and record the past five years of data archived on the web site as mp3 audio files. In addition, a second design was developed to accompany the visualization that focuses on the current hour of data only.

As I am also a data visualization specialist, in addition to the Waterviz sonification, I developed a new visualization of Hubbard Brook data in a form we titled "HydroScape". HydroScape was a data exploration tool written in Processing.js that presented nine variables in what I call a Multivariable Encoded Line Chart. In a sense it can be thought of a thick line chart, where the components of the thick multicolored line chart are subgraphic components that all fit together to expand and contract as the data of multiple variables changes. All the points of these subcomponents hang off the xy location of a focus variable, and that focus variable can be changed by the user, thus allowing for the exploration of how many other variables shape the outcome of a focus variable. For instance, if the focus variable is streamflow, we can easily see that when there is a precipitation event, it is followed by a short-term increase in streamflow. However, if we focus on soil water levels, we see that these same precipitation events cause the soil water to dramatically increase, then slowly decrease over time, as the trees "breathe" and suck the water out of the ground on a daily basis. This was completely new information that I learned from the music and the visualization. The music also helped to act in the role of data quality indicator, highlighting some issues we had at one point with Hubbard Brook data ingestion and

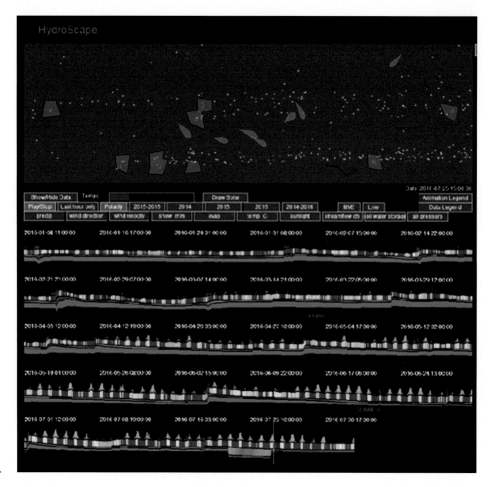

Marty Quinn's HydroScape, which combines data visualizations, musical sonfications, and Multivariable Encoded Line Charts.

processing. The music also effectively communicated that some years were more active than others.

Early on, the new visualizations brought to light a phenomenon called the mountain valley breeze effect, or more technically, the katabatic wind, that was apparent in the wind direction data. The breezes at Hubbard Brook often flow from west to east, or down the valley at night, and then reverse course, flowing east to west, and rising up the valley with solar warming during the day, bathing the forest in these gentle, and perhaps cleansing katabatic breezes. These sonic and artistic observations led to further interest and investigation from meteorologist Dr. Eric Kelsey and his students at Plymouth State University. These data were also used in a section of a dance performance titled "Life After Life," created by my wife and creative partner Dr. Wendy Quinn, PsyD, where the wind direction became the pitches played on a violin. By using our sonification technique of not playing the same note if the data does not change on subsequent data points, the music exuded both rhythm and great musical beauty, adding to the charm of the dance. Other pieces of data music from Hubbard Brook were also an integral part of this performance. A video recording of the performance and science/arts panel discussion at

3S Arts in Portsmouth, New Hampshire, were presented at a subsequent conference at Hubbard Brook. We were pleased to have Dr. Lindsey Rustad participate in the artist-scientist panel that we held for the public after the live performance.

During this entire creative process, we came to recognize more fully that the integration of artists and scientists requires sensitivity of all parties to when definite content creation versus artistic content creation are in play. Science is heavily weighted to definite content creation—for example, to buy or build a particular sensor that will work in a particular place and start recording data, analyze the results, and write a paper about the findings. Artistic content creation does not fall in the definite content world; it can be intangible and transient, and does not require hierarchical management organizational structure to be successful. Artistic creations also do not fit as well into the 5–15 minute talk frame so often allocated for scientific papers. And when not enough time is allocated for its proper experience, artistic projects may be short changed, or not understood, especially when presented alongside scientific papers. They are not the same thing as scientific projects but should be seen to be equally valid and requiring different sets of handling and perceptual accommodations.

The artistic creation may be both intangible and concrete at the same time as when data music may be perceived as music in and of itself, while simultaneously relaying to a person familiar with the sonification design, all the subtle details present in the scientific data. A unique quality of the arts—while integrating ideas of beauty, expression, balance, change, emotion, color, line, movement—is that they can communicate definite content as well as the infinite expression of the soul.

The Student's View Point, Torrin Hallett

I was brought onto the Waterviz project as a summer student intern at Hubbard Brook in the summer of 2016. As an undergraduate studying mathematics, music composition, and horn performance, this project presented a unique and exciting way for me to synthesize my areas of interest. I was also excited about the possibility of using art to communicate scientific data and their implications to a much larger audience than might be reached with the current prevailing methods of data representation. After observing Marty's existing sonification work, I began my own project, whose goal was to create a unique sonification of the water cycle data collected at Hubbard Brook for the year 2015. I used a computer program called MAX/Msp, which I had had some experience with, to create an aural graph of the data that allowed people to hear the data.

I worked with data for nine variables that had been collected at Hubbard Brook: streamflow, solar radiation, wind speed and direction, soil water storage, temperature, air pressure, humidity, and evapotranspiration. I had to decide which sound would represent each variable and also how that variable's data values would influence that sound. I created a patch in MAX/Msp that read in the data values for all of the variables in chronological sequence, performed mathematical operations to scale those values as necessary, and then used those scaled values to manipulate a variety of sonic parameters including volume, pitch, playback speed, and panning.

The largest continuous challenge throughout the process was ensuring that the piece was balanced—that all the variables could be heard, and that one variable's sound did not overpower the others'. Although I had already worked with MAX/Msp in school, there were a plethora of things that the program could do that I had not yet learned. I would frequently show my work to Marty and Lindsey to receive feedback on what I had done so far. Their constructive criticism usually outlined a vision of the ways they wanted my project to be improved, and then it was my job to find and piece together the tools available in MAX/Msp to come up with a solution. There were of course many solutions to every problem, and it was up to me to pick the one that I thought could be done most efficiently with the tools I had. This type of thinking seemed, to me, to be similar to the problem-solving techniques that I had encountered in my upper-level, proof-based math classes, where we were often given a starting point (some conditions), an end goal (the conclusion we had to prove), and various tools to get there (mathematical theorems). It was up to us to piece together a path from the beginning conditions to the final conclusions using the theorems at our disposal.

Throughout the entire process, I learned a lot about the data collection processes that occurred at Hubbard Brook. I would often accompany other scientists in the forest as they collected data or checked sensors, so that I could see hands-on how the data that I was using for my piece were being collected. Although, at the beginning of the project, I was nervous and unsure about what exactly I was going to be creating, I left the process feeling proud to be part of a greater team of scientists and artists who were working together to disseminate data to a wider audience. I was also excited to see how art and science, two simultaneous yet often independent passions of mine, could come together in a groundbreaking and exciting new project.

To Conclude

The Waterviz for Hubbard Brook is an artistic portrayal of water data, digitally streaming out of a small forested watershed in New Hampshire in real time, reinventing itself every hour in infinite variation and form, and based on algorithms developed collaboratively by artists, musicians, and scientists. The process to create the Waterviz was exciting, exhilarating and sometimes painful, as we sought to share worldviews, purpose, and methodological approaches, and to communicate our work to a diverse audience. Throughout the project, we learned about each other, our disciplines, and new truths and ways of understanding our data. Our unified goal was to make water data more intuitively available to a global audience, and to bridge the continuum from the scientist's world of facts and logical understanding of scientific data, to the artist's world of building empathy from that understanding, to eventually the policymaker and land manager's world of creating action from that empathy. In the end, we hope to better use scientific data and principles as a solid platform to protect our planet's water resources. It is a lofty socio-ecological goal, but one perhaps best tackled not by individuals working alone, but rather by multidisciplinary teams of scientists, artists, humanitarians, land managers, and policymakers seeking to share a common vision and voice for the future.

Endnotes

1. The Hubbard Brook Experimental Forest is owned and operated by the United States Department of Agriculture Forest Service. It was established in 1955 as a center for hydrologic research for the Northeastern United States, and continues as a major center for forest ecosystem and hydrologic research.

2. Xavier Cortada blogged about the artist's residency at New Hampshire's White Mountain National Forest in his White Mountain Trail Mix (https://whitemountaintrailmix.wordpress.com). The artist residency was a collaboration of the White Mountain National Forest and the Arts Alliance of Northern New Hampshire (http://aannh.org).

3. My first art-science residency was five years earlier: In 2007, I visited Antarctica as a National Science Foundation Antarctic Artist and Writer's Program fellow. I did so to help better understand the knowledge being developed in that continent around the science of climate change. Immersed in their workspace, I learned from scientists and was inspired to use their work to generate new art. Indeed, I melted the ice they gave me on paper adding paint and sediment to create new works. These *Ice Paintings* were made in Antarctica, about Antarctica, using Antarctica as the medium (provided to me by the very researchers who inform us about Antarctica). With the ice paintings, I wanted to melt the very ice that threatened to (melt and) drown my city (Miami). The work, beautiful and serene, would be a precursor of horrors to come (see http://cortada.com/2007/ice-paintings).

4. Xavier Cortada, *Surrender at Tunnel Brook*, White Mountain National Forest, 2012 (see http://cortada.com/2012/TunnelBrook/).

5. *Time-Space Species: Glacial Relics and the Thrush* exhibit at Patricia Ladd Gallery in Center Sandwich, New Hampshire (see https://whitemountaintrailmix.wordpress.com/2012/07/19/talk-exhibit-performance-at-patricia-ladd-carega-gallery/).

6. Xavier Cortada, *Hill Farm Dinner Party*, White Mountain National Forest, 2012 (see http://cortada.com/2012/HillFarm).

7. Xavier Cortada, *Wind Words*, White Mountain National Forest, 2012 (see http://cortada.com/2012/WindWords).

8. Xavier Cortada, *Hubbard Brook 60*, digital art, 2015 (see http://cortada.com/events/2015/60).

9. Xavier Cortada, *Water Paintings*, Hubbard Brook, 2016 (see http://cortada.com/clima2016/gallery/hubbardbrook).

Aesthetic Engineering

Giving Visual Credence to Restoration Processes

Bruce James, Daniel McCormick, and Mary O'Brien

Daniel McCormick and Mary O'Brien collaborate in the art practice Watershed Sculpture. They create environmental art installations that take on a remedial trajectory. They create sculptures that influence the ecological balance of compromised environments in watersheds across North America including those in Colorado, Michigan, North Carolina, New Orleans, Nevada, Los Angeles, and San Francisco. McCormick is an interdisciplinary artist and designer with skills in the fields of environmental design, sculptural installation,

Title image: *Bay Clay Oyster Reef*, 2013. This pilot project is a community-built remedial installation made from bay silt dredgings. It reframes scientific best practices and methodologies to restore the San Francisco Bay native oyster, and provide shoreline protection from sea level rise. 18″ × 60″ × 48″, 2013.

Photographer: Mary A. O'Brien. Reprinted with permission from Daniel McCormick and Mary O'Brien, 2015.

and ecological restoration. O'Brien has a background as a creative director, filmmaker, and sculptor. Both hold degrees from University of California, Berkeley.

Bruce James is a professor emeritus from the Environmental Science and Technology Department at the University of Maryland, College Park, where he taught and conducted research from 1986 to 2015 in the areas of soil chemistry and soil and civilization. He has an abiding interest in transdisciplinary links between soils and human societies, both in modern contexts and in ancient civilizations. He believes that an enhanced appreciation for soils as natural systems and resources will grow in the public conscientiousness if it is studied and viewed from the perspectives of the natural and social sciences; and the humanities, especially through visual appreciation of soil beauty.

Soil and landscape art can inform each other in ways that can enhance human appreciation for the land and its essential soil functions for ecological integrity, agricultural production, and human health. Art that restores natural soils and land functions can be combined in a transdisciplinary way with the study of soil in the laboratory. The combination of macroscale field projects and molecular-level laboratory studies leads to creative moments of intuition about the true nature of soils, their resilience, and how they function naturally and under the influence of human restoration. Soil chemist Bruce James and artists Daniel McCormick and Mary O'Brien conducted dialogue for this chapter with phone calls and e-mails, which accommodated their distance at opposite ends of the United States.

Why do you work with soil?

Bruce James: My appreciation of the soil comes from admiring the colors and other properties of soil profiles in the field and small samples in the lab, especially when such perceptions are informed by chemistry, microbiology, and physics. My reaction to your work and our dialogue in the context of this book is a recognition of the beauty, function, and purposeful human actions to restore soils in a way that is pleasing to the eye while trying to recreate natural processes and attributes of soils that will be able to sustain themselves through cycles of degradation and restoration.

How do you view the soil? What is your approach to combining beauty and function in your work as artists?

Daniel McCormick and Mary O'Brien: We take a watershed approach to working with soil. As environmental artists working in a remedial capacity, most of our work is concerned with protecting and restoring watersheds. We have experienced that even the most degraded soil can sustain some growth. That becomes a starting point for a project and the beginning of a restoration cycle. This often begins at a soil level, whether damaged or eroded, or as a foundation for other remedial actions.

We are compelled by the idea of using sculpture in a way that will allow the damaged areas of a watershed to reestablish themselves. Our work is intended to give advantage to natural systems. After a period of time, as the restoration process is established, the artists' presence diminishes. Moving away from an anthropocentric view of nature, we use a series of site-specific activities to initially give aesthetic weight to the restoration process. Then, through eventual succession into the environment, the restoration is established. The end result of our work is remedial. The art moves beyond the created work to become a catalyst for restoring specific ecotones in the watersheds where we work.

Place-Based Observations and Projects

Daniel McCormick and Mary O'Brien: A new appreciation for the importance of soil can be encouraged by scientists and artists alike and, through their efforts in tandem, bring the awareness and realization of the importance of soil protection. As artists working on the land, we take into consideration questions such as:

Thicket, Completed Weaving & At 6 Months, 2012. A woven sculpture of native willow and elderberry branches on Adobe Creek in Los Altos, California. Installation mitigates bank erosion due to invasive species removal and involved over 400 donated hours from 100 volunteers. Dimensions: 8′ × 52′ × 6′, 2012.

Photographer: Mary A. O'Brien. Reprinted with permission from Daniel McCormick and Mary O'Brien, 2012.

What is beyond the visual surface of the soil (microbiological, chemical)?

How does soil affect air and water quality?

How does soil affect place? Is it part of an aesthetic experience, or does it itself possess visual qualities worth considering, worth preserving?

Bruce James: While sampling soils in the field, one captures their visual and olfactory beauty. When those sensory perceptions are coupled to scientific knowledge of soil properties and behavior, we realize a refined appreciation of soils far different from traditional notions of soils as lowly dirt underfoot.

Soils are natural bodies in the landscape that would not exist if natural thermodynamics held sway, and all of the organic matter would react with oxygen and become carbon dioxide and water. Minerals would be weathered and wash out to sea. Soils are a testament to a lack of chemical equilibrium and to the importance of low entropy systems similar to living things. I know that these are technical terms from the lexicon of soil chemists, but they help us to understand the mystery, persistence, and resilience of soils.

Daniel McCormick and Mary O'Brien: In our work in different places across the United States, we found the soil to be as distinct as the landscapes to which it belongs. Addressing restoration issues in both urban and rural watersheds, we work in a conservation capacity, adjusting to the different regional textures and behaviors of the soil. Whether it's addressing erosion, urban runoff, flood plain disruptions, wind disturbances, logging, agricultural overgrazing, development, or water use issues, we address soil issues first and foremost.

In Charlotte, North Carolina, during heavy rains, urban runoff rips soil out from under traditional ground covers. Its bright red silt pours into streets, storm drains, and creeks during these high-water events. When we took on a project to mitigate storm water runoff in the riparian zone of the last remaining small bottomland hardwood forest that bordered Little Sugar Creek, hundreds of Charlotteans helped us create sculptures that helped spread and sink groundwater before the silt-laden flows reached the creek. An unexpected result was that native sedge returned to the forest and created a new microhabitat for raptors in search of small rodents was created. Two pairs of barred owls took up residence in the pocket forest.

In rural California, working on an active salmon spawning creek, much was needed to keep silt and agricultural nitrogen away from salmon spawning grounds. The National Park System, the local Resource

Urban Runoff Remediation, 2010. Woven Willow, Dogwood, and Elderberry over coir mat and compacted eradicated nonnative brush. One of three sculptures installed to mitigate storm water runoff from urban development into the largest flowing creek in Charlotte, North Carolina. Dimensions: 4' × 100' × 3.5', 2010.

Photographer: Mary A. O'Brien. Reprinted with permission from Daniel McCormick and Mary O'Briene, 2010.

Conservation District (RCD), local schools, and ranchers all worked with us on sculptures designed to control erosion adjacent to the creek. With research provided by a park service hydrologist and best management practices (BMPs) from the regional RCD, the sculptures were designed to filter runoff within the riparian zones along the creek. By slowing the seepage of nitrogen-laden agricultural wastes into the creek, the oxygen levels in the creek increased, improving the outcomes of the spawn. Willows that were live-staked into the sculptures, on the other hand, thrived with the excess nitrogen, and in the last decade have grown into mature trees.

What is soil?

Bruce James: I am strongly opposed to the use of the word "dirt" to describe soil. Dirt is soil out of place, often dried and pulverized for study in laboratories, and it has a popular connotation as being on the back of your neck, under the rug in your apartment, and as something to sweep or wash away. It is a four-letter work in the worst sense, but soil is also a four-letter word with a softer sound like love, hope, and good. The German word for dirt, *Dreck*, is equally harsh and unacceptable. Soil is not just a word used by university researchers and technical experts to sound authoritative. It is a word that encodes our history and linkages to the Earth, at least in English and many other languages.

Daniel McCormick and Mary O'Brien: Distinctions matter. Different soils have different behaviors, uses, and levels of regeneration, fragility, erosion, productivity. Soil is specific and placed-based. We've seen direct evidence how soil chemistry and geography can influence the aesthetics of soil.

Creating more and more impervious surfaces adds to the problem of soil erosion. Climate change affects soils. Forest fires expose soils to erosion. Prolonged droughts contribute to soil degradation. Chemical enhancement of soil changes its character. A few decades ago healthy soil was maintained biologically, simply, organically. It is our intention to advocate for "soft technologies," using many of the methods that originated out of necessity from the Dust Bowl era.

What are the challenges of conserving and preserving soils?

Bruce James: The intrinsic values of soils are not given consideration because instrumental value and cost-benefit analyses within an environmental or neoclassical economics system are the default foundations of North American environmental policy making and discussion. Even if your environmental ethics is principally informed by visual perception, beauty, mystery, and related human responses to soils, these are not given much credence in conservation and preservation arguments. It is noteworthy, however, that when a conservation group tries to convince its members to support an effort to save an ecosystem, species, or landscape, they use beautiful pictures to do so. The soil, however, is particularly challenging to represent as beautiful, without a refined knowledge of its inherent qualities.

Daniel McCormick and Mary O'Brien: Municipal governments throughout the United States are advocating for, and in some localities mandating, soil preservation through established best management practices (BMPs). Healthy soil practices could be adopted by citizens. Perhaps policy makers will galvanize behind protection of soils, as has happened with air and water.

We adopt a problem–solution approach to soil conservation. For example, addressing a 150-year history of erosion and siltification in California, we have utilized silt originating from the California Gold Rush that reached San Francisco Bay. Bay Clay Oyster Reef is made from this silt. The sculptures serve as reef substrates to recruit oysters along eroded bay shorelines. Oyster beds can become key to the restoration of tidal marshes protecting them from storm surge and eventual sea level rise.

Flood Control Wall, 2014. Woven fascines on the Carson River. A collaboration with The Nature Conservancy, Nevada Chapter, and the Nevada Museum of Art. Combines a biological trajectory with a remedial outcome to intercept floodwaters and enhance habitat areas. An inquiry into present-day relationships between community, land, and water involving over 1600 donated hours from local volunteers. 4′ × 3.5′ × 340′, 2014.

Photographer: Mary A. O'Brien. Reprinted with permission from Daniel McCormick and Mary O'Brien, 2014.

Daniel McCormick and Mary O'Brien: While projects such as these launch us into an arena of litigious issues, our goal is to inspire alternatives and aesthetic considerations from scientists, engineers, and policy makers. We also believe that the use of these "spoils" brings an awareness of the history of these issues. The artist-as-activist role can be important in conservation efforts.

What is the future of soil?

Bruce James: One of the biggest challenges to the health of soils is the loss of organic matter and the erosion of soils from their place of formation to surface waters. The soil profile embodies the wholeness and integrity of a soil, and when it is destroyed, a natural soil becomes "dirt," or soil out-of-place. We must seek to avoid the conversion of soil into dirt.

I envision the soil of the future as a whole, healthy, functioning body underlying every kind of terrestrial ecosystem. In the laboratory, I envision it as field-moist material that is biologically and chemically active. Through cycles of disturbances of intermediate severity that increase the entropy of soils temporarily (increasing their disarray), soils can recover and actually become increasingly lower entropy systems of greater stability. If humans can manage soils with a knowledge of how disturbance, recovery, and restoration proceed, soils will persist and indefinitely support diverse ecosystems and human cultures that are dependent on soils. I see this future for at least 5000 years, based on how we humans have lived in, on, and with soils since the beginning of agriculture 10,000 years ago and in the first cities roughly 5000 years ago. I am optimistic about soils, provided that we humans begin to extend the same concern and care for them as we are beginning to do with surface waters and air (both of which we take into our bodies for the maintenance of our health). We often talk about clean and refreshing qualities of water and air (especially when choosing a place to live or go on vacation), but we have yet to do so with soils.

The entropy of soil is mysterious, but it underlies the persistence of soils, much as entropy allows all living things, including human beings, to exist on Earth. We need to understand how entropy change affects the beauty and function of soils over myriad temporal and spatial scales.

Daniel McCormick and Mary O'Brien: We see soils of varying conditions that support riverine and terrestrial plants and animals. However, we have learned that soils need the proper conditions in order to thrive and support this life. We firmly believe that human communities can learn how to best use and conserve their soils to support all the uses of future generations.

The Aesthetic Bridge between Art and Soil Science

Bruce James: Transdisciplinarity connotes using the language and ways of learning from diverse disciplines, and in this sense, goes beyond inter-, cross-, or multidisciplinary studies that may simply be a collection of disciplinary perspectives cobbled together. The languages of the visual arts, natural sciences, and history (especially environmental history) can find common ground in the study of the soil. This book is a big step in that direction.

Daniel McCormick and Mary O'Brien: We cannot imagine a way to move forward with soil conservation, protection, and restoration without multiple collaborations at all levels, governments, scientists, artists, and citizens. Art and science can be so intertwined that one informs the other. We have experienced the power of these kinds of relationships. We envision projects between artists and scientists creating interdisciplinary collaborations. Land owners need knowledge, research, and methodologies that can help them conserve their soil. Soil scientists need to understand sociology, demographics, and policy. Aesthetics can be a bridge that illuminates both sides.

Storm Surge Mitigation, Mississippi Delta, New Orleans, LA, 2013. An environmental installation for species restoration and storm surge rehabilitation in Southern Louisiana. Recycled mixed-media with newly planted Bald Cypress seedlings. A one-acre island in a freshwater bayou in the Mississippi Delta, Venice Louisiana.

Photographer: Mary A. O'Brien. Reprinted with permission from Daniel McCormick and Mary O'Brien, 2013.

Rocks, Radishes, and Restoration

On the Relationships between Clean Water and Healthy Soil

Aviva Rahmani and Ray Weil

Aviva Rahmani's current project, The Blued Trees Symphony, has fellowships and support from the New York Foundation for the Arts, A Blade of Grass, and the Ethelwyn Doolittle Justice and Outreach Fund to replace toxic fossil fuel infrastructure with sonified biogeographic sculpture. Rahmani's work is internationally exhibited, written about and published. Her transdisciplinary PhD is from Plymouth University, United Kingdom, and her masters is from the California Institute of the Arts, where she was teaching assistant to Allan Kaprow and

Title image: Large portions of the Mississippi River and its tributaries are natural floodplains that have been developed for agricultural use and urbanization. Contemporary flooding, often caused by climate change now endangers people in those regions. *Fish Story* was designed to address some causes of that threat by proposing restoration to the Wolf River. *Tributaries #4; The Wolf River Spills its Boundaries*, encaustic paint on paper Google Earth satellite mapping 16″ × 20″, 2013.

Reprinted with permission from Aviva Rahmani.

Morton Sobotnick. She is an affiliate with INSTAAR, University of Colorado at Boulder, and has taught at Stony Brook University. Rahmani's *Trigger Points/Tipping Points*, film on global warming premiered at the 2007 Venice Biennale, as part of Gulf to Gulf, a NYFA sponsored webcast project accessed from 85 countries. *Ghost Nets* (1990–2000) restored a town dump to flourishing wetlands. *Blue Rocks* (2002–2005) helped initiate the U.S. Department of Agriculture's restoration of 26 acres of critical wetlands in Maine.

Ray Weil is professor of soil science in the Department of Environmental Science and Technology at the University of Maryland where he conducts research, teaches undergraduate and graduate courses, and conducts outreach education to the environmental and farming communities. He is a Fellow of both the Soil Science Society of America and the American Society of Agronomy, and has twice been awarded a Fulbright Fellowship to support his work in Africa. He is probably best known for his ecological approach to soil science in writing the 11th to 15th editions of the internationally most widely adopted, translated, and cited textbook in the field, *The Nature and Properties of Soils*. Weil is a leader in researching and promoting the adoption of sustainable agricultural systems in both industrial and developing countries. His research focuses on soil organic matter and plant management for enhanced soil health, ecosystem functions, and nutrient cycling for water quality and agricultural sustainability.

Ecological artist Aviva Rahmani and soils scientist Ray Weil discuss their various approaches to restoration work, concluding that a transdisciplinary approach is the most effective solution to degraded systems. They compare issues that they have encountered respectively in wetlands and fisheries degradation (Rahmani) and in agricultural ecosystems (Weil). Rahmani describes two projects, Ghost Nets (1990–2000) in Maine, and Fish Story (2012–2013) in the Gulf of Mexico and Memphis, Tennessee. Weil discusses how his use of daikon radishes helps wean agribusiness farmers from the use of heavy fertilizers. Together, they compare ecosystem conditions and discuss how their insights and discoveries can be applied to remediate problems at the local level, as well as large landscape problems such as eutrophication in the Gulf of Mexico. Their conversation reveals where art becomes science, and science becomes art.

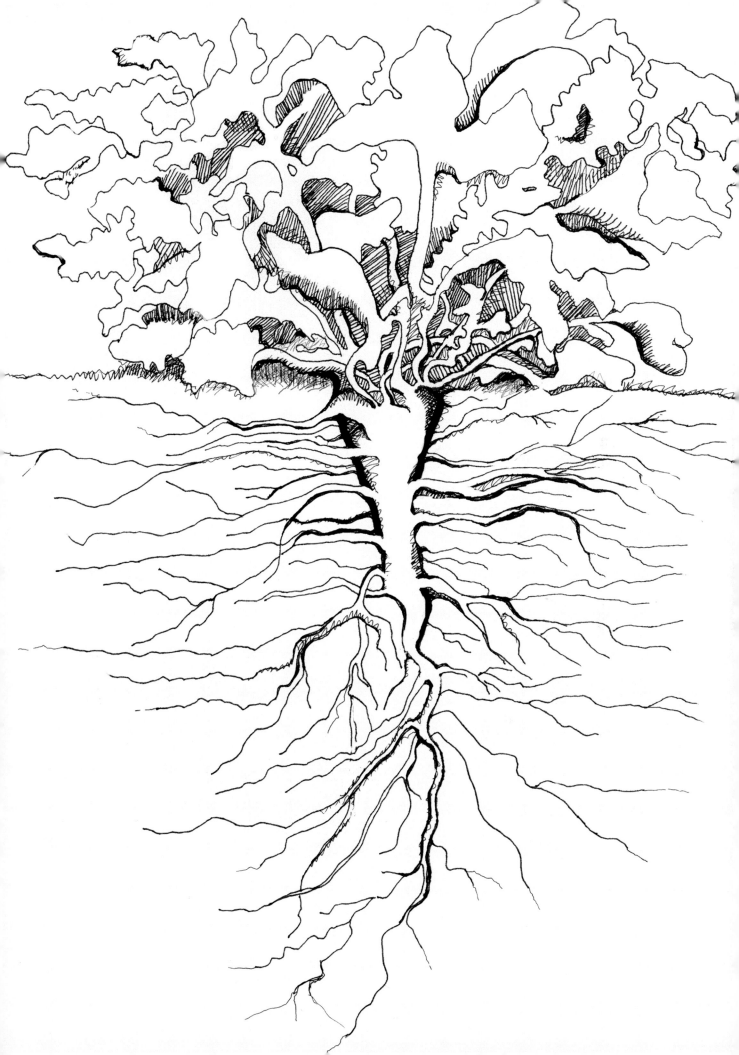

Previous page:

Daikon at Work, ink drawing on paper 11.5" × 9" 2017 (based on a conversation with soil scientist Ray Weil about no-tillage agriculture and soil health).

Reprinted with permission from Aviva Rahmani for Field to Palette 2018.

Ray Weil: Aviva, how are you interested in restoring natural systems? You've done a lot of land restoration. In *Ghost Nets*, you were doing some on-the-ground restoration in the estuary around your home, rebuilding the wetland up from the sea. You called it a *pocket salt marsh*?

Aviva Rahmani: Yes. I coined the term because the entire upper half estuarine system of the Gulf of Maine is characterized by small, rocky intertidal marshes rather than the flat expanses most people associate with salt marsh. Many of these small marshes have been filled in since settler times before we had a chance to know the unique functions of these tiny marshes compared to larger barrier estuarine systems. I restored a 1/3 acre salt marsh as a test model. It was part of the old town dump in Roberts Harbor, on a hill sloping into Penobscot Bay in the Gulf of Maine. The site started off as a pile of granite tailings (riprap) that were thrown into the estuary by the granite industry. That created land for a deep-water wharf to take finished stonework down the coast in schooners. When they stopped quarrying, it became the town's garbage dump. I bought the whole site (two and a half acres), built my home there, and restored the local upland riparian zone and the estuary. My intention was to provide local landowners with a regionally appropriate model for other landowners to replicate. I thought the restoration work could trigger a bioregional response. To make my case, I used Geographic Information Systems to map species distribution. It was very difficult to create conclusive proof of the immediate impact, but it did catalyze new research about the community interactions in these saltmarshes at the Wells National Research Reserve. The director of the Reserve, Michele Dionne, supported the idea.

There was no soil when I began in 1990, just rocky detritus. I began with soil creation, which any habitat restoration needs. I spent most of 10 years creating soil from green manure by planting legumes, encouraging leaf decomposition, composting, and hauling up seaweed that washed up on the shore. Finally, in 1997, I removed the granite debris that separated the tide from the land and opened up the coastline again.

Ray Weil: How did you happen to find the site?

Aviva Rahmani: I decided on the site because it looked promising on geological maps. It had many edges and lots of water. Local fishing was very rich. The Gulf Stream and other abiotic factors seemed to indicate the makings of a biological hot spot, a place where habitat complexity could provide rich biological diversity. And while I did that, I also organized the site into a series of stroll gardens and microhabitats for human visitors and other critters using the site for foraging or nesting. I selected plants that birds and small mammals might eat. Now the soil is thick with earthworms.

My goal in restoring the soil was to achieve a healthy aquatic system. Water depends on soil-making systems that result in fertility. The project seemed to be an equally psychological, creative, and practical commitment. I named it *Ghost Nets*, after the fishing drift nets that come loose and strip mine the ocean of marine life by trapping them in a wholesale extractive process. By focusing on the metaphoric, different observations emerged than would have emerged from quantitative analysis alone. *Ghost Nets* is an analogy for what I think the whole human race may have to embrace; accommodating our lifestyles to habitat reparations.

Ray Weil: How does your art, your paintings, or photographs, or objects, intervene in natural systems? And how can natural systems be seen as part of your artistic process?

Aviva Rahmani: The paintings and installations are visual formulas to intervene in conventional thinking … A sense of loss informs all my work. For example, in *Fish Story* (2012–2013), a participatory mapping project I did in Memphis, in collaboration with paleoecologist Jim White and wetlands biologist Eugene Turner, I was able to negotiate with stakeholders, especially big corporations who have developed the riverfront around Memphis and are focused on short-term financial profits and losses. The painting reminds me that I have to introduce that anguish without self-righteousness to engage their willingness to be part of the solution instead of the problem. It is more than evocative illustration. Contemplating the image of the river helps me think through the consequences of losing ecological structures and lets my mind drift into imagining new solutions.

In *Ghost Nets*, I think that the main artistic impact was creating a precedent to say: "You can do this yourself. Theoretically, you could buy a dump site. You could restore it. You can create soil where there was none. You could make these beautiful gardens, and it would benefit the region." If every degraded pocket marsh in the upper half of the Gulf of Maine were restored, it might have a larger biological impact. The difference between

what I did and creating any old backyard habitat for birds and butterflies is in how carefully I calculated where to position my efforts, my decision to begin with a dump site, and my need to share the restoration work with the public as a creative process. I took the environmental philosopher Robert Elliot's position on restoration seriously. Elliot claimed that restoration is fake because it's *artful*. I decided, Okay. If it's going to be art, then let's really make it art.

Ray Weil: We can look for the art in working with soil and the soil working in art … I use photography to interest people in my work with soil. You use natural phenomena to interest people in your art.

Aviva Rahmani: We both use visuals, albeit differently, even though our intentions might be similar. I draw attention to a site with art. The art is the middle of a process, not the end product. The first part is identifying *where* to pay attention. The middle is the restoration. The final part is monitoring how the system might change rhizomatically on an ecological and cultural level. My interest is in forging another approach to engage people differently in the science and art of restoring ecological systems.

1HOUR   PLANTCAM   JUL.25,11 08:00 AM

Still from *Earth Time*, a time-lapse film of the restored *Ghost Nets* site reviving across seasons 2010–2011.

Reprinted with permission from Aviva Rahmani.

The *Ghost Nets* site was developed from a former town dump to demonstrate how individual landowners might contribute to contiguous regional restoration and support fish habitat. *Ghost Nets Garden, West Quadrant.*

Photograph by Aviva Rahmani 2012, Reprinted with permission from the artist.

Ray Weil: You've talked about *Ghost Nets* and other works as *trigger points*. Did you borrow that term from acupuncture? Does it reflect relationships between the greater landscape and the soil underneath? In your work on the Wolf River, are you thinking about the whole watershed and the hydrology that impacts the city of Memphis in terms of certain sites?

Aviva Rahmani: Just as we activate meridians of the human body, it seemed like the biogeographic dynamics of Gaia might be activated for marine and coastal recovery. A similar approach originated in South America, called nucleation, for forest succession. (See, for example, Zahawi, R.A., Holl, K.D., Cole, R.J., and Reid, J.L., 2013, Testing applied nucleation as a strategy to facilitate tropical forest recovery, *Journal of Applied Ecology*, 50: 88–96. doi:10.1111/1365-2664.12014.) My process push-pulls between intuitive diagnoses of what needs to be done, on-site discovery, and learning about the scientific methods and strategies to create contiguous systems. My conceptual premise is what I call *trigger points*, locations in the landscape that indicate where dynamic confluences can catalyze a cascade of impacts—negative and positive. That is, the discreet trigger point can be identified

by layering data, but it is conditional on and interdependent with the much larger system. I elaborate more on this theory in my forthcoming book. In 2013, I did a series of works I collectively called *Fish Story*, including an installation at the Memphis College of Art, an international webcast, and a journal article. Part of the project was conceptual, part was the social interaction, and part was on site. Wetlands biologist Gene Turner and I canoed the Wolf River toward the Mississippi and noted what kinds of species had been taken out of the system. We were looking for a trigger point that might impact the Mississippi Water Basin and the Gulf of Mexico. For example, most of the canopy trees were gone, along with the beavers, which left the water dead relative to how it had been in preindustrialization times before 1930. We conjectured that if contiguity were reestablished between the Wolf River and the Mississippi, at exactly the point where the Army Corps of Engineers had interrupted the flow of water, ecological relationships might be restored and have wider consequences on regional sustainability.

Ray Weil: Following up on the idea of trigger points and environmental health, how much do you, or the people you work with, diagnose what's

Fish Story, cut and acrylic painted paper installation at Memphis Institute of Art, Memphis, Tennessee, 2013 for *Memphis Social* curated by Tom McGlynn. H: 132″ × W: 1728″

Photograph by Katie Maish. Reprinted with permission from the artist.

Wolf River Entrance Point #1, starting an exploration of a new, previously unmapped Lost Swamp Trail portion before the Ghost River section of the Wolf River by canoe with Aviva Rahmani, Gene Turner, and the Wolf River Conservancy from Bateman Bridge to Moscow, Tennessee.

Photograph by Aviva Rahmani 2013. Reprinted with permission from the artist.

wrong with the river and how that came to be? Is it a matter of toxic chemicals? It is a matter of clearing vegetation? Erosion from the uplands? Dredging? All of the above?

Aviva Rahmani: All of the above. In the case of the Wolf River, degradation has been exacerbated by an invasion of privet, which has out-competed the canopy trees that provided the habitat with shade. I think the most important thing is to reconnect the tributaries. Then fish populations can become indicators for success. As in the ocean, there's been dramatic overfishing. The rivers are also subject to lots of nitrate pollution. The question is whether the populations can recover if the linkages are restored. Meanwhile, a lot of the Mississippi River is being periodically, artificially restocked with fish for sport.

Ray Weil: The fundamental problem with the Mississippi Basin is dumping too much nitrogen in the rivers. That's coming mainly from large-scale farms. So, farmers are going to be the ones to solve that part of the problem. Waterfront development on the Wolf River is mainly about local development. You're going to have lots of local people involved. But in terms of the nutrients coming down the Wolf River, it's the farms in the watershed that have to change.

One of the best ways to control the loss of nitrates is to introduce cover crops … I've been studying cover crops for the last 10 or 15 years in the United States. These crops are grown to feed the soil, to help enrich it, and to improve the farming system.

I'm really excited because over the past 10 years or so I can see the use of this practice growing. In fact, in recent years I have been speaking at cover crop meetings and field days where hundreds of farmers come to exchange ideas and hear about cover cropping. I think it's pretty amazing and going to change the face of mainstream, large-scale American agriculture. Farmers are being

very innovative on thousands of acres of large farms producing our basic grains. My former grad student and I also made a website (www. notillveggies.org) focused on smaller farms producing vegetables. The farmers leading this movement see this as a win-win situation. There are indirect pathways by which adding cover crops into a system changes all kinds of things with benefits to the farmer *and* the environment. Growing cover crops is a very powerful way that farmers can enhance their soil's environmental filtration capacity and ability to transform excess nutrients into beneficial soil organic matter and a reduced need for fertilizer. In my research, I have seen that by using the right cover crop system, farmers can capture soluble nitrogen from as deep as 2 meters in the soil and prevent it from leaching to waterways.

Aviva Rahmani: I think these changes are a result of people demanding organic food, and conventional farmers working together with organic farmers. I was at a conference recently where a young woman representing small organic farms explained how they had convinced large grape growers in upstate New York to let weeds grow between the vine rows so they could graze their goats and sheep. It was a win-win for everybody because, of course, dung was added to the soil, the weed cover was kept down for the big farmers, and the small farmers had an inexpensive, clever way to feed their herds.

Ray Weil: Yes. There are lots of these synergies starting to come into play. Quite a few farmers are learning to integrate their livestock with crop production by using cover crops for both soil improvement and grazing. Weeds can be pretty useful too. I developed new, multipurpose cover cropping systems using a daikon-type radish cover crop that I had seen growing in Brazil. It grows so vigorously that even very business-minded farmers are quite impressed and excited about using it. Now there are a million acres of it planted for soil improvement.

Aviva Rahmani: Do radishes work like dandelions pulling up nutrients?

Ray Weil: Exactly … They pull up a lot of the nutrients that rainwater has washed down, even from several meters deep, and brings them to the surface. It loosens the soil and suppresses weeds. Some large-scale farmers have been able to skip their burn-down herbicide spray. The cover crop itself freeze-kills in the winter in cooler regions, so there is no problem dealing with it in spring. In the spring, it quickly decays and the nutrients are released into the soil. It provides big holes, which prevent runoff, and the radish attracts earthworms like a magnet. The worms love it because it's got a lot of sugars, which bacteria eat, and then the earthworms eat the bacteria.

If they have room to grow, these radishes can grow as big as baseball bats. This remarkable root impresses farmers into thinking, "Well, maybe this is doing something for me." And it is. It sells itself because it's so unique looking and grows so fast … More and more, I've been talking to groups

Portrait of a scientist and his radish cover crop.

Reprinted with permission from Edwin Remsberg, University of Maryland.

Mixed species radish cover crop.

Reprinted with permission from Ray Weil.

of large-scale farmers interested in soil health and biodiversity and how they can work to effect this change. It's an innovative, small percentage of farmers doing it, but you're finding it now in popular farm literature. All the agricultural journals and magazines have taken it up.

Aviva Rahmani: I'm thrilled to hear that. Maybe they could get to a point where they would even allow pigs and cattle to forage on agricultural lands.

Ray Weil: Cattle definitely. There are large-scale farmers with big equipment and credibility within the community, who are growing thousands of acres and doing it completely different now. They are integrating livestock and making more money. We find that the radish holes are like injecting fertilizer because it changes the soil structure and makes fields more fertile. When I show farmers plots or photos of my plots of radish cover crops, they see that there are virtually no weeds growing in them in early spring. It's absolutely weed-free! People think I sprayed herbicide on it, but it's the radish.

Aviva Rahmani: I'd love to see big farmers weaned off herbicides and pesticides.

Ray Weil: That's what they're trying to do—even bragging about how few, if any, herbicide and pesticide applications they need to make, and how little fertilizer they use. This means lower chemical load, which is relevant for the restoration of ecosystems like the Wolf River Watershed, and also lower production costs. The general American public is unaware of this revolution, but I think it's reaching a mainstream tipping point. Most U.S. farmland grows animal feed, although some farmers grow wheat for bread. I don't think most farmers are trying to get into the organic market. However, they are trying to improve their soil for the future, to make farming easier, more profitable, and preparing their farms for the coming large droughts and erratic weather from climate change. These farmers are putting food into the mainstream food system, and these new systems reduce their production costs.

Aviva Rahmani: One of the biggest problems though, and the reason why we're having this conversation now, is that we don't have 20 years any more to learn from each other. We need faster ways into changing these large systems.

We have to find ways for the general population to find their own points of entry into problems. It's one thing for us to talk about the methodologies of restoring soil or tributaries; it's another for this to happen in the lifetime of somebody whose dinner menu is going to change.

Ray Weil: Solutions take a while to work. Unfortunately, natural processes can take a long time to undo environmental damage, even when we change practices. Take the example in the Mississippi basin. Nitrogen already in the groundwater and soil profiles is going to take decades to flush out. It will take time for rivers to recover and for us to see change and impact in the dead zones—even if we started doing everything right overnight. This long delay in end-result makes it politically difficult.

Aviva Rahmani: I think there's also the issue of how we approach these problems. There are different models in different locations and for different problems. The real task is getting people to think about whole systems rather than disparate and disconnected sites. Even though the agents and relational dynamics might be different, approaches might be similar and overlap. We have to take all these approaches—the heuristic and the quantitative—and marry them. The problems are too large to rely on any one solution. We have to hook people's imagination to do things differently.

Ray Weil: Yes. There's no silver bullet, but we can use many little silver hammers to promote soil health and biodiversity—as scientists and artists, farmers, consumers, and politicians.

Dirt Dialogue

A conversation with and in memory of Jackie Brookner, compiled and completed by Maxine J. Levin

Born in 1945 in Providence, Rhode Island, **Jackie Brookner** received a BA in art history from Wellesley College in 1967 and completed all work for a PhD in art history at Harvard except for a dissertation before shifting her focus to sculpture in 1971. In her early career she worked in cast bronze and steel before shifting her focus to ecological art. Brookner collaborated extensively with ecologists, earth scientists, hydrologists, design professionals, communities, and policy makers, working to bring plant-based water remediation to parks, rivers, and wetlands together through habitat restoration, landscape sculpture, and active community collaboration. Her large-scale participatory remediation art projects create multifunctional public spaces where people can reconnect with healthier

Title image: Jackie Brookner, *Native Tongues*, installation of sound, sculpture and drawings at the Miro Foundation, Barcelona, Spain, 1997.

Image courtesy of the artist. http://jackiebrookner.com/project/native-tongues/

natural systems. Major projects are located in Salo, Finland (2009); San Jose, California (2008); Cincinnati, Ohio (2009); West Palm Beach, Florida (2005); and near Dresden, Germany (2002). Brookner received numerous awards including The National Endowment for the Arts, The New York Foundation for the Arts, ArtPlace, The Arts and Healing Network, The Nancy Gray Foundation for Art in the Environment, and The Trust for Mutual Understanding. *Urban Rain*, in San Jose, California, won several storm water management awards including Project of the Year from the American Public Works Association. Brookner was guest editor of the 1992 *Art Journal* issue "Art and Ecology." Brookner had teaching posts at Harvard University, University of Pennsylvania, and Parsons/New School for Design, where she taught from 1980 until her death in 2015.

Maxine J. Levin was a soil scientist, the national leader for Soil Interpretations with the National Soil Survey Center, Soil Science Division, U.S. Department of Agriculture (USDA) Natural Resources Conservation Service (NRCS) in Beltsville, Maryland; Lincoln, Nebraska; and Washington, DC and retired in 2018. She worked as a soil chemistry technician, soil survey mapper, soil correlator, and soil interpretations specialist with the National Cooperative Soil Survey (NCSS) in California, Maryland, New Jersey, Arizona, Florida, Tennessee, and Utah. She was the principal investigator and author for the City of Baltimore Soil Survey, the second all-urban soil survey of the United States after Washington, DC, which has led to continuous input since 1988 in urban-related soil projects throughout the United States and internationally. She also worked as a representative to the USDA department (Council for Sustainable Development) for soil survey and conservation operations; USDA representative to United Nations Rio Convention U.S. delegations in desertification, land degradation, biodiversity, wetlands, and global warming (conventions and field visits to Kenya, Argentina, Turkey, Spain); liaison to Smithsonian Museum of Natural History Museum for U.S. Soil Monolith Collection; U.S. Embassy Science Fellow, in Kigali Rwanda (2008) and Bangkok Thailand (2010); and lead for US NCSS conferences and field trips since 1998. Her degrees were in soils and plant nutrition, 1975, University of California, Berkeley, California and MS in agronomy, 2017, Iowa State University, Ames, IA.

Jackie Brookner, landscape and ecological conceptual artist, and Maxine Levin, soil scientist with the U.S. Department of Agriculture (USDA) Natural Resources Conservation Service, were introduced to each other by Alexandra R. Toland in early 2014 to talk about soils and art in anticipation of the International Year of Soils in 2015. Their introduction was one of many threads in a growing network of interdisciplinary soil stewards. The original concept of the meeting was to have a conversation; tape, transcribe, and edit it; and focus on wherever the conversation went about soils from the viewpoints of science and art combined. They had three conversations, two by phone and one in Brookner's studio. To initiate the conversation there were several e-mail exchanges facilitated by outgoing questions from Toland that the two women responded to individually and continued to make reference to in their larger dialogue. To set the tone of the conversation, the first three questions and answers are included here. The title of the chapter, "Dirt Dialogue," refers to the working title for the entire book, which was later abandoned after much protest against the "D" word, but remained in all of Levin's and Brookner's working documents and e-mail exchanges.

Hidden in the roots of our words we find what we seem to want to forget—that we are literally the same stuff as earth. My work explores this identity while undermining the assumptions and values that keep us from acknowledging it.

Jackie Brookner

Alexandra R. Toland: Describe your personal relationship to the soil.

Jackie Brookner: I'm mostly working with, rocks, water, soil, and plants, and whatever lives in them, in my artwork.

I've had to learn to be patient, to let things grow … It is a collaboration with ecosystems in many ways, because I'm learning so much all the time from whatever I'm working with, be it the water, the soil, the plants. But there are important differences. If I'm holding a stone or a rock, or I'm next to a mountain, the density of that is so different, compared to something soft, like plants or foliage, and that affects the body differently. So, I'm interested in personal relationships to individual places and ecosystems—it isn't the soil or the place. It's THIS soil in THIS place.

Maxine Levin: My focus is on the natural landscape and soils in the landscape. How do natural soils and soils formed by human (anthropogenic) changes work in the landscape? It is something that has fascinated me for over forty years (my entire adult life). I wonder at the wonder that I have felt in "digging deep," looking closely (either through a microscope, or a set of lab data, or a soil profile in a pit or full landscape in person or on a map) … how it all fits together and makes sense … and I wonder if others can see these connections that seem so obvious to me. Much of the time my ideas about the soil and landscape were purely intuitive, and from those intuitive ideas I was amazed when others confirmed those ideas with scientific data.

In the last ten years with my contacts with the Smithsonian Dig It! exhibition, I thought long and hard about sharing that wonder. I would like to explore more about art in the landscape versus natural landscapes. Are the feelings the same? Can sustainable art touch emotion the same way that a natural landscape does? It would be interesting to explain the need of watersheds and to use landscape art to share the concepts both scientifically and viscerally. I am interested in landscape art to change folks' ideas about land and soil protection for the International Year of Soils.

Alexandra R. Toland: Soil is often described in terms of its opposing qualities: "Healthy/Deteriorating," "Abundant/Scarce," "Fragile/Robust," "Alive/Inert," "Endangered/Secure" "Sacred/Mundane" … Many of these go beyond direct observation to individual perceptions and beliefs. How do you rate these qualities?

Maxine Levin: In science, we try to use quality descriptions such as these to address public concerns and put numbers (indexes) and measurements to them. These are all terms that we have asked the soil survey to quantify.

Jackie Brookner: It's hard to answer that with a single answer because the state of the soil depends on the context. I also don't believe in the dichotomy between sacred and mundane (or dichotomies in general). The mundane is also the sacred. The organization into binaries is often misleading.

Alexandra R. Toland: What other qualities come to mind when you talk or think about soil? What other beliefs do you hold about the soil?

Maxine Levin: I go back to my first years of mapping soil in Mendocino County, California, and intuitively feeling the connectivity of soil to life and physical functions of earth with water, air, plants and animals. Visually, soil is beautiful, though individually some soils as a stand-alone are not that impressive. Many of desert soils with pale color and low biological activity and carbon are relatively bland visually, especially with no structure or rock fragments. The beauty is in the contrast.

(left) Jackie Brookner, *Of Earth and Cotton*, mixed media installation of sculptures, cotton and soil, dimensions variable. 1994–1998. Exhibition view, *Of Earth and Cotton*, Denton, Texas, 1995. (right) Portrait of the feet of Elie Wilson, 1994.

Images courtesy of the artist.

Jackie Brookner: I think about how beautiful soil is, how diverse are its colors, textures, smells; how important it is for people to understand that healthy soil is richly alive, with complex structure; how long it takes for topsoil to develop, and how endangered it is by development, agriculture, deforestation, erosion, etc.

Maxine Levin: Jackie, talk me through some of your artwork with soils.

Jackie Brookner: Back in the '90s I was doing work, installations mostly, in museums that very much were about soil and regional history and how people and cultures and environments affect each other on a local level. One was called *Native Tongues*. It was created for an exhibition in Barcelona. Right now, Barcelona is going through tough times. When Franco was in power, he forbade the speaking of anything but Castilian because he didn't want anybody to be speaking regional languages. So, land, home land, and language in Spain, and particularly in Catalonia, are powerfully connected.

So, what I did was I recorded people speaking Catalan and Castilian as much as I could and then tried to imagine the shape of the tongues that would speak these languages based on the sounds. It's a very fanciful thing. Catalan, I don't know if you've ever heard it, but it's very gruff and kind of convoluted, and Castilian, of course it depends who's speaking it, tends to be more melodious. So I went around to different towns in Catalonia and in central Spain and collected soils. And they had the most amazing colors. And using the soil I made a giant wall drawing that was about 50 feet long and based on phonetic diagrams that show where we put the tongue when we make certain sounds.

What I loved about these diagrams is that they look like maps. And like a map, there was a kind of a key that I made that indicated which tongue came from which area. And there was also a sound installation of people talking, playing off this very rigidified notion of language, codified language versus the very chaotic way a party sounds or the din of a bunch of people talking at once. This reflects nuances and differences in the landscape. Getting back to one of your earlier comments, it's interesting that some of these soil colors by themselves may not be that interesting, like the desert soils you mentioned. But actually I find just about every soil beautiful (except the depleted and degraded ones) because, when you put them together and you compare them, you realize their textures and colors are so different and rich and speak to diversity on so many levels.

Maxine Levin: Yes. It's all about the contrast. That's what I experienced when I was doing the monoliths for the Dig It! exhibition. We had all 50 states represented and also Puerto Rico, and Virgin Islands, and Guam. Each state picked one soil profile that they felt spoke to them, and there was this incredible diversity represented. But go on about what you were saying …

Jackie Brookner: Well, I've done a similar "soil profiling" in the southern United States. After the *Native Tongues* project, I did a project called *Of Earth and Cotton* that traveled across the South from "94 to 98" and I followed the westward migration of the cotton belt in seven locations. This was actually a project that changed my life, and I'll tell you why in a second … So, at each of these seven locations, I spoke with people who farmed and picked cotton by hand in the 1930s and '40s, before mechanization. Part of what I wanted to know was what have we gained and what have we lost by not working the earth with our bodies anymore. It's not that nobody does it these days, but so few people do. These big combines machines with cabins cut you off, spew out enormous amounts of chemicals. So we have this relationship between the farmer and the soil that is much more distant in this country. So, I started talking to people; I collected soil in their backyards and in nearby cotton fields, or sometimes if I saw some fantastic soil on the road I would slam on the brakes and gather that too.

Maxine Levin: Well this is sort of interesting because that's the part that I loved about doing soil surveys—connecting with local people, place by place, and then really working out the essence of that county, that region, that place, and how it all went together.

Jackie Brookner: What do you mean when you say the essence?

Maxine Levin: Well I was thinking it was partially the people and working the land, or the reaction of the plants to the land, and the water to the land. Just working out that essence of how the land fit together as a big puzzle.

Jackie Brookner: That's fantastic and so interesting. What I found was that when I asked questions about "the land," they didn't know what I was talking about. But if I asked them a specific question about the soil, or how to grow cotton or something, they could go on forever because they had all this experience. This general idea land was an abstraction that meant nothing.

Maxine Levin: You said that project changed your life?

Jackie Brookner: Well, what happened in this project is that I realized how I had been very private in my work before that. I was working mostly in the studio and having shows, and then going back to the studio. But I had such amazing exchanges with people on all levels during that project that it actually kind of changed my life. I mean, some of the conversations were very spiritual and very personal, and with complete strangers. So I understand at that moment that I wanted to work with people and that if I was going to be really working with ecological issues, it had to be with people because we're the ones who are creating environmental problems and can be the only ones really who can come up with solutions to problems we've created. Unless we simply disappear and let things evolve very, very slowly over many years.

So, at that point I started a kind of long apprenticeship with myself to learn how to work with communities really working on environmental issues. I went and talked to people in Texas, in North Carolina and South Carolina, in Tennessee and Georgia as well. So, I had all these rich conversations and all these different kinds and colors of soils. And you could see the diversity and richness.

Maxine Levin: The soils in California that grow cotton are relatively boring visually. It's like a fine, sandy loam, not these black, black clays from Texas and the red clays from the Carolinas. Tell me about the feet on the soil.

Jackie Brookner: In the installations, there were foot sculptures that rested on 60 tons of soil and 2500 pounds of ginned cotton. They were all different colors and textures; the soil became a metaphor for the diversity of peoples who worked the land. I wasn't casting the feet into molds. I was literally sitting there making portraits of people's feet, of the people I met and spoke with. It gave me time to talk to people, and it opened up this space of intimacy which I didn't expect, because I was sitting on the ground below them in a very non-threatening position, and I was occupied doing something else so they could really open up. It kind of relaxed people, and then they got to see their own feet, portraits of themselves coming to life out of mud basically. So, people were fascinated by that and the human connection was just amazing. I recorded conversations about what a typical day was like, and then before you went into the installation there were 40 photographs from the 1930s of cotton farming and cotton picking that were taken for the government.

Maxine Levin: Like Dorothea Lange and Ben Shahn …

Jackie Brookner: Exactly. They're quite amazing. At the time, Roosevelt started to employ artists to document rural America; it was also how he raised sympathy for the poor.

Jackie Brookner: So, from that project I went on to create a series of *Tongue Chairs*.

Maxine Levin: Was that also part of the *Native Tongues* exhibition in Barcelona?

Jackie Brookner: No. These I made here … I started doing these chairs and some other work in the studio, and I wanted to give people a sense of how soil is nurturing and to create an intimate and very pleasurable and profound connection to it rather than this image of dirt that people revile or denigrate. There's a whole other discussion there about sex and shit and why that's denigrated in the first place, but basically, I wanted something people could feel with their whole bodies. Not just look at. And the tongue is a part of the body where sex and speech come together; mind and body. You can't talk without moving your tongue. So, speech is a bodily act, but of course it is also very logical, as a part of our intellectual functioning.

Maxine Levin: And the tongue is somewhat hidden most of the time, like the soil.

Jackie Brookner: Yes, it's hidden. I love that. Like the feet, in a way, these are threshold places. I think of the tongue as threshold where, if you stick it out, it's very meaningful.

Anyway, so I wanted you to see these because I do love these pieces. Some are rockers and they're very comfortable. This particular one, the tongue lounge, people can't believe how comfortable it is. It's got a real resonance when you sit in it.

Jackie Brookner,
Tongue Lounge, 1993,
42 × 31 × 55″ |
Earth + Wood.

Image courtesy of the artist.

Jackie Brookner: OK, so let me tell you more about what I'm doing now.

The interaction of soil and water come together best in my current project, The Fargo Project, in which an 18-acre single function storm water detention basin is being transformed into a multifunctional community commons. In addition to continuing to function to store storm water, the restored prairie habitat will enhance water quality and help restore the soils. During the excavation of the basins, the soils were largely removed. The grass, which has been established there, is growing directly into the parent material. We are working with soil, plant, and wetland scientists to assess and monitor the changes. And we have been and continue to engage diverse community members in the entire process from conception of the space to implementation and programming, in order to provide a place where people can have intimate experience with natural systems, other species, etc. The programming for prairie restoration and the community garden directly work with the soils and plants.

Maxine Levin: That sounds fascinating.

Jackie Brookner: So, there's this gorgeous storm water retention basin that's about 10 feet deep on one side, a little more on the other. It's got this concrete low flow channel going through it and the water in that channel is really disgusting. The channel arcs and then makes a Y and joins one of the northern sections and then you can see from the different colors of the grass where it's wetter and where it's drier. You can see the blues and the yellows.

Maxine Levin: So, is it sort of like a vernal pool, where it dries out in seasonally?

Jackie Brookner: Not exactly. It's engineered to drain within 72 hours, but there's not much monitoring so it ends up causing problems.

Maxine Levin: Right. I grew up with these kinds of storm water basins in Fresno, California. The few rains we had, because we only had like 10 inches of rain a year, would always cause a flood. All the streets would flood and then they got the idea to put in these storm water drains, and big basins. So then, when it rained, it would fill up with water and then slowly recede. Sometimes it would take a month.

Jackie Brookner: No, the ones in this work are engineered to drain within a few days. They're piped and they have valves and pumps and stuff like that.

Maxine Levin: It's sort of interesting, when I was in Bangkok for four months working as a U.S. Embassy science fellow, one of the things we were very interested in was sea level rise flash flooding. Is Bangkok sinking? I mean, that the whole city is sinking! That entire city used to be canals and now all those canals are streets. They completely fill up with water during the rainy season and people run around in big rubber boots. But one of the concepts they talk about with the Mekong is *monkey cheeks*.

Jackie Brookner: Monkey cheeks! What's that?

Maxine Levin: They were areas with big water basins along the Mekong that would fill up with water and hold the water in the system for a long time without flooding.

Jackie Brookner: That sounds neat. Natural basins.

Maxine Levin: So these basins would fill up, and they wanted to start reintroducing the idea and importance of water basins, and they called them monkey cheeks because they would fill up with water and then slowly let the water out. So, officially monkey cheeks are defined by the authorities there as "waste water improvement and flood management basins."

Jackie Brookner: I love that. It would be very interesting to check that out and find out more about those. Of course, it's such a different ecology in Fargo … One thing I also realized, doing what I'm doing up there, is that there is a huge focus on water quantity. They don't filter the water and most of the time, except for maybe 30 days a year at the most, they're just

these big, grass-covered basins and they divide neighborhoods throughout the city.

So, I wanted to create something that could be more ecologically and socially functional in a number of ways. It was important to restore the prairie, but I also wanted to do it with the community—and there are many diverse communities here. There are a lot of refugees from all over the world and also a poor Native American population and then the majority of people are of Scandinavian descent. So, we did a huge outreach project. And what came out of that was that people wanted a place to connect with enclosed nature in the city. They really loved the idea of a diverse prairie here. Some places would be wetter than others and some places would be drier than others because of the topography and the flow of the water.

So, it's largely a restored prairie that can store and filter water, but there's also going to be hiking trails throughout it. A natural play area. An orchard area. Some water features where the water comes in or out and we don't know yet what that's going to be because we want to see what the water does once we take out the concrete channel.

Maxine Levin: So, are you thinking about salinity and the flush of where the water goes up and down through the system?

Jackie Brookner: Yes, we're working with some soil scientists and we're working with some wetlands experts, and soon we're starting a kind of eco-lab to get some monitoring done before everything starts getting torn up. There are so many unknowns at this point. It turns out that there's a sand band that goes through it, but we don't know exactly where it goes. The drainage situation there is generally very poor. It's mostly clay in the area that has very poor drainage, but then there's this sand band going through it … So, we've got tons to learn just from watching and getting to know the site under these developing conditions. And then we're trying to engage people as much as we can in the restoration processes and subsequent maintenance.

Maxine Levin: Well, it's so interesting. You've picked this landscape that superficially doesn't look like much, but has a lot of potential to be quite beautiful. And there's so many processes going on through there. I know that with saline pools it depends on whether the water is draining through or coming up, and then you have to think about how the water is flowing through the system and where the salts come from.

Then the thought of the vernal pools, essentially that's what they are, are also tied to animal behavior and habitat. For instance, ducks like to live in pairs. They'll want that little pond just for themselves. In fact, they'll settle on a tiny one as long as it's just theirs, just their honeymoon suite. They don't like to share, and we keep on wanting to make them share. So, we've destroyed many a little vernal pool, and the smaller they are the more likely it will be destroyed.

Jackie Brookner: Right. Well, we've got geese out there too. For me, restoring these drainage and storage functions are also about resorting habitat.

Maxine Levin: And for me, the beauty is in the function. This is something I'm quite emotional about, in fact. When I did the soil survey, when I was out in the field, that was what I was the most interested in—the combination of the landscape and the function and looking at that true essence of the land and then the people, of course, who live on that land.

Jackie Brookner: Part of the really intriguing part for me is the question of how do we communicate function. What does it mean and then how can

the joy and pleasure that it evokes be conveyed? That is an aesthetic question. And aesthetics for me isn't just what something looks like or sounds like, it's about there's a sense of human connection to the world we live in and to each other.

At the beginning, you were talking about the wonder that you have—of digging deep and understanding how soils form and change and then wondering if others can see those connections ... to open up that kind of scientific understanding to a level of wonder brings you to a powerful, sacred place, really, which certainly aesthetics is a part of.

Maxine Levin: Yes, it's stuff like this that does keep me going.

Jackie Brookner: In the modernist view of aesthetics, there was this idea of form following function, but it was a very formal thing with physical symmetry. I think today we need a much deeper sense of function—not only when the landscape is functioning for humans and for human exploitation, management, agriculture, etcetera, but rather function for the sake of ecosystem functioning, of which we are of course a part.

That aesthetics in a larger sense of the word. And to experience aesthetically, in this larger understanding, it's not just about the senses. It's about the senses integrated with all our other facilities. To me, ultimately, that does take us to the sacred. But how do we engage on that deeper level of landscape functioning and our place in it, without just using it? You really need to understand the bigger picture to understand your place within it. It's not about just seeing the formal thing before you, the picture of what is, but about seeing the relationships which are, in a way, invisible.

Jackie Brookner, *Fargo Project*, Pilot Project at Rabanus Park Stormwater Basin, Fargo, North Dakota, 2010. Community working together at WeDesign/Charette to develop a schematic design with the workshop ideas.

Image courtesy of the artist.

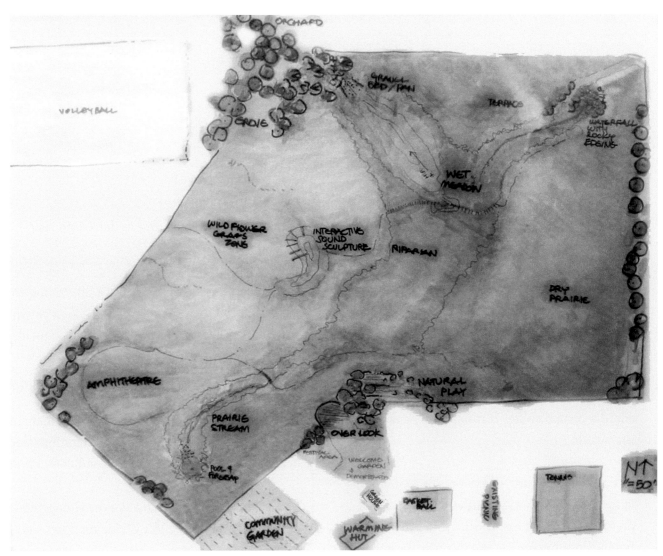

Jackie Brookner, *Fargo Project*, Pilot Project at Rabanus Park Stormwater Basin, Fargo, North Dakota, 2010. Panoramic view of the storm water basin at Rabanus Park.

Image courtesy of the artist.

Maxine Levin: Yes! And that's how I have approached the science of soils and mapping the landscape.

It's interesting. Now that I'm in the field of soil interpretations, the whole point is to delve out of the database we've created, to zoom out of these incremental descriptions of soils throughout the landscape in order to describe the function. How could a certain soil or certain place be used and managed?

In daily practice, the suitability for septic tanks is the dull version of it. But we've also started to do interpretations for wildlife (like desert tortoise habitat) or mycological organisms (like valley fever habitat). So, we're starting to delve into that, but it's really peripheral.

Jackie Brookner: The thing that's interesting to me is that the information already exists in what you've done. It's now just a question of how it's used and what you're actually looking for …

Because, I mean, the landscape is still the same landscape. You can take the same body of knowledge and get opposite interpretations from it depending on what your desires and context are.

Maxine Levin: Totally. This is something that's always frustrated me. We divide mappers into lumpers and splitters.

Jackie Brookner: Just like in taxonomy.

Maxine Levin: Yes. Taxonomy … the same thing. "Lumpers" and "splitters." You need both. You need the splitters to get the detail, but then the splitters can't map. I used to be so frustrated with them. They wanted to get it right and they wanted to incrementally measure everything in the site, but you never got at the function when you did that. They'd get lost in the minutiae and could never see the big picture of how the landscape was functioning. I would go in as what we call a "correlator" and try to tease it out of them. They'd just see—here's the cup, here's the saucer, here's the spoon, here's the plate … But they couldn't figure out how to drink and eat a meal.

Jackie Brookner: I understand. A number of years ago, I went on a field trip to Venezuela with a group of botanists to look at the rainforests in the Orinoco. I had never been there and I thought, "My god, this is going to be amazing."

Maxine Levin: And you got a bunch of splitters?

Jackie Brookner: Yeah. The whole time all they did was count sepals and blades of grass. And here we were in this gorgeous rainforest. There was only one "correlator" guy who had the bigger picture in mind. I thought, "My god, this is excruciating—they're not seeing the (rain)forest for the trees." So, I know exactly what you're talking about.

One idea, and I'm just thinking selfishly because it would be so interesting to me …

I would love to develop this conversation with you over time. It's not going to be in the next hour. But one thought I had, and I don't know when I could do this or if you would even want to, but it would be really fun to go on some landscape excursion or field work together. It would be interesting to just talk through what you're seeing and, of course, I would respond to how I see it. That could be an amazing thing. Maybe not a whole survey, but just a little piece of the landscape. Would that interest you?

Maxine Levin: Totally! That would be really fun.

I do these soil education tours. I try to work with people on these tours that I set up, to help them understand landscape functions by fluctuating from detail to the bigger picture and back to detail until they get it. What they don't realize is I'm trying to light a fire in the field—to illuminate the bigger picture by showing all the little details.

Jackie Brookner: Where I go with that immediately is that we humans are the little picture and that there's this much bigger picture out there that we have to oscillate back and forth between.

Maxine Levin: The two, yes … When we're sampling soils, we will have several pits that we're going to sample, maybe eight or nine in different sections of the landscape (in the region of investigation). We'll start out with the little picture. We'll dig the pit, describe the profile and tease out the meaning, observing the important physical soil properties—all the "little" aspects of it. Then we'll take soil samples back to the laboratory for an even teenier look at things. But in the selection of the pits, we have to look at the big picture and think about what are the most critical soils to sample that will tell the story of the bigger picture.

Jackie Brookner: That, itself, is very interesting. How do you make those choices?

Maxine Levin at a field site in West Virginia helping to dig a pit to collect a soil sample for a Soil Monolith, 2010.

Courtesy of the author.

Maxine Levin: That, to me, has always been the fun part. We're in the business these days of statistical analysis, where we can actually decide on the location of the pit based on statistical variability. But we'll never get at the essence of what we're doing that way. Statistically, we can't take enough samples to properly portray nature … so, we start doing it in our heads.

Jackie Brookner: That's the intuitive part. But it always comes back to the question of how we ascertain the big picture, the big stories we're telling ourselves. The stories we tell ourselves depend completely on what we're choosing to focus on. Unfortunately, we're telling ourselves some pretty bad stories these days. So, we need to ask, What stories do we need to be telling ourselves now?

Maxine Levin: Yes. At the same time, when we do these surveys, I can see the value of the splitter point of view. It's important to spend a period of time doing the random selection, getting the statistics right, because maybe it will point out stuff that we were blind to. The big picture can sometimes be our blind spot.

Jackie Brookner: Absolutely … If you're only looking at the big picture, it's not going to tell you the details. And if you're only looking at the details, it's not going to give you the big picture. So, we need to be able to be flexible and move back and forth …

Anyway, it has been really fun talking with you again and I think it would be amazing if we could continue our dialogue out in the landscape. I wish we had all day to talk, but unfortunately, I have to go out of town later on.

Maxine Levin: I know and I gave a promise that I would be back at 11 o'clock.

****

It has been since 2014 that I (Levin) traveled to New York City and visited Jackie Brookner in her studio home in SoHo. It's hard to remember details versus general impressions. I rang the bell buzzer on the street and was allowed into a large freight style elevator and taken to the top floor, maybe eight floors up.

Jackie greeted me at the elevator door and there I was in the top floor of the building with a large open area with floor to ceiling windows, the equivalent of two stories high, rough floors with unfinished wood like a warehouse. There was her table and a couch and a few of her iconic tongue chairs that I recognized from her website. Her simple kitchen was free-form with an old enamel gas stove and salvaged wood countertop shelves and cast-iron sink in the corner. She offered me a cup of tea. I thought at the time that you only see this kind of setting in the movies. It was straight out an avant-garde film from the 1960–1970s. She had been living and working in this studio apartment from right after college—for almost 40 years. Originally, the studio was all that she could afford, which at the time was an open warehouse space in the middle of a rundown slum. There were no services—barely functioning electrical and water, with salvaged furniture and appliances from the street. She slept and worked in the same place. The landlords updated the freight elevator just for her about 20 years prior. Then, SoHo grew up around her with multi-million-dollar spaces and inevitable gentrification. The NYC rent control protected her, so that she could keep the warehouse space for her original bare warehouse '70s rent level. The owners were now offering her a few million dollars to give up her lease, and she was tempted given a recent cancer diagnosis and the impacts that it may have on her.

She gave me a tour of her studio which included many of the art pieces that I had seen on the website from the museum exhibitions. Some were in pieces on wooden flats, and some were on the wall, which were living sculptures—biosculptures she called them. There was stone and sculpted pumice in the shapes of tongues with moving water and green, small-leafed leguminous, orchid and air plant and cactus vegetation to illustrate self-sustaining systems that clean the water and feed small fish in connected pools below. It is hard to visualize now, but I recall it feeling like a dream with the filtered north facing sun coming through the windows. Though I am no longer sure what was real or my imagination or my mind's eye memory.

We talked and laughed and drank some more tea and continued our dialogue for an hour and a half.

Our dialogues had been so stimulating for both of us. We fantasized that when this little project was over that we might try our own book on the philosophical connections between soil, culture, and art. She worried about what would happen to her art pieces if she moved, and whether an agency or organization might be interested in any permanent installations at no cost for the future to artfully demonstrate sustainability in action. I went home and started to edit the two dialogues that I recorded earlier. We exchanged a few e-mails, but I assumed that she was busy when I didn't get an immediate response. She was a very busy person with lots of appointments and connections and multiple ongoing projects all over the world. Jackie Brookner died about nine months later with our manuscript unfinished. Her last site-specific installation was her own natural burial in which she was wrapped in a shroud on a simple plank of wood. She had become one with the earth that she loved so much.

Jackie Brookner, ecological artist, 1945–2015.

Photo credit: Terry Iacuzzo.

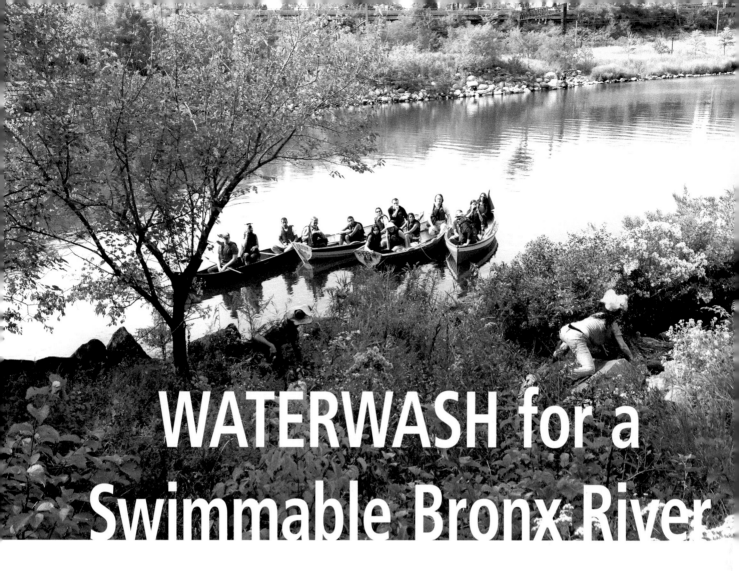

WATERWASH for a Swimmable Bronx River

Edward R. Landa and Lillian Ball

Lillian Ball is an ecological artist/activist working on water issues with a multidisciplinary background in anthropology, ethnographic film, and sculpture. She has exhibited and lectured internationally, recently at the Anchorage Museum, Seville Biennial, and Reina Sofia Museum. Awards include: New York State Foundation for the Arts Fellowships, John-Simon Guggenheim Foundation Fellowship, and a National Endowment Sculpture Grant. She was named 2012 Environmentalist of the Year by the North Fork Environmental Council and awarded a citation by the New York State Assembly for WATERWASH Bronx River. Since 2006, Ball has served Southold Town's Land Preservation Committee, advising her maritime municipality on land use/stewardship.

Title image: A 2013 performance by Rev. Billy and the Stop Shopping Choir included "Extinct Golden Toads" while the audience looks on from RTB boats, WATERWASH Bronx River, 2013.

Reprinted with permission from Lillian Ball.) (Previously published in Public Art Review, Issue 49, Fall/Winter 2013).

Edward R. Landa is a soil scientist and adjunct professor in the Department of Environmental Science and Technology at the University of Maryland-College Park. His research has focused on the sources, fate, and transport of trace elements and radionuclides in soils and soil-like materials. Recent work has examined tire wear particles as a source of zinc to the environment, the impact of such particles on the development of tadpoles in simulated storm water retention ponds, and the effects of tire leachate on mosquito development. His recent historical research has examined early twentieth century debates within the United States on soil fertility, and on bridges between the Russian and American soil science communities during this same era.

In a series of e-mails and phone calls, Ed Landa and Lillian Ball discuss community activism, the public and private challenges of watershed restoration, and the imagery and language of soil remediation. The conversation focuses on the ongoing WATERWASH® wetland restoration project initiated by Ball in 2008 along the Bronx not far away from where Landa grew up. Funded by the LIS Futures Fund, the project successfully transformed a boat ramp parking lot into a renaturalized biotope. WATERWASH Bronx River (2009–ongoing) is a collaborative green infrastructure solution to runoff pollution in the Bronx River. Apprentices from Rocking the Boat, a nonprofit doing environmental river work with local youth, planted 8000 native plants. WATERWASH Bronx River filters commercial parking lot runoff before it enters the river, opens private property to pubic use, and was sponsored by the New York State Attorney General's Office.

Paloma Mcgregor's modern Dance 2013 performance of "Building a Better Fish Trap" at WATERWASH Bronx River, 2013.

Reprinted with permission from Lillian Ball.

Edward R. Landa: Your WATERWASH Project on the shore of the Bronx River immediately caught my attention, as I grew up nearby. How did this transformation from neglected, anonymous space come about?

Lillian Ball: The National Fish and Wildlife Foundation introduced me to Adam Green, the founder/director of Rocking the Boat (RTB), a Bronx-based boatbuilding and environmental education and youth-development organization. Adam and I hit it off right away and went about researching and visiting various sites along the Bronx River. The NYC Parks Department had suggested I look at Concrete Plant Park, a newly finished Bronx River Park that needed some interpretive signage and wetland rethinking. Since the park was basically done, it didn't seem very challenging. However, the overgrown landfill across the river and behind the parking lot of ABC Carpet's outlet store had storm water runoff going directly into the river and presented a real opportunity to improve the water quality.

We are both familiar with the kinds of nasty runoff that can be treated by wetland filtering, giving more bang for the buck than simply keeping runoff out of the sewer system. At the location of the WATERWASH Bronx River parking lot, runoff was previously piped directly into the river. We diverted it through a pipeline and a flume to reduce turbidity, and worked with Drexel University environmental engineering faculty and students to monitor input and releases. Our wetland project also enhances the native habitat for birds and insects, as well as humans seeking experience in nature.

Edward R. Landa: My work has included investigations of vehicle-derived contaminants such as tire-wear particles and brake pad dust. So, parking lots are more than inert surfaces with painted lines to me. And the soils that receive their runoff are more than adjacent sponges. They are filters, absorptive barriers, habitats, and bioreactors. I am pleased to see that the U.S. Geological Survey—where I spent

36 years—began pilot-scale testing in 2013 on storm water–associated contaminants coming off a major Bronx highway, and their retention by a remediation strategy similar to that at WATERWASH Bronx River. Engineered, soil-filled "pop-up wetlands" constructed on a pier in the Harlem River near Yankee Stadium intercept and treat the highway runoff before it reaches the river.

Lillian Ball: And you grew up there? Did you swim in the river as a kid? Paul Chapman, the vice president of ABC Carpet, evidently did too, which was one of the reasons he was interested in allowing the WATERWASH Project to be built on the commercial property. Technically, the Bronx River has not been swimmable or fishable for decades due to the several combined sewer overflows (CSOs) that occur with storm events (Stormwater Infrastructure Matters (SWIM) Coalition, http://www.swimmablenyc.info).

Edward R. Landa: Let me put the location of my childhood home in the Bronx in a time and space perspective.

For a kid in the 1950s, during the Mickey Mantle era, it was about three miles north of Yankee Stadium. Teaching would grind to a halt, and radio broadcasts of key games of the World Series would be piped into the classrooms over the PA system. I count Mel Allen and Red Barber, along with my junior high school earth science teacher, as early mentors. I grew up overlooking the Harlem River. It was an industrial zone, not to be swimmed in, and hard to access. So, it's really encouraging to see the rebirth of the shoreline here, with wetland restorations at new public parks such as Sherman Park and Muscota Marsh.

Despite its largely urban character, the Bronx has a special place in homegrown natural history studies with the Bronx County Bird Club and its members from the 1920s, such as artist/ornithologist Roger Tory Peterson, and wildlife biologist Joe Hickey, who grew up near the Bronx River and went on to fill Aldo Leopold's position at the University of Wisconsin. I see the RTB kids and their exploration of the soil-plant community within the WATERWASH Project as part of that proud tradition.

Lillian Ball: Yes, there is an impressive naturalist history in the Bronx. And it has been moving, as well as frustrating, to work with the young adults from RTB. The energy is incredible when it's focused, and that's our job to see what gets them excited. They are remarkably savvy about the pollution issues on the river and do tours with school groups where they eloquently lecture about water quality. Their life circumstances don't always give them many options, so exposure to this experience might offer alternatives. I just try to cheer them on.

As an artist concerned with water issues, the WATERWASH concept merges art with wetland restoration and community activism. WATERWASH Bronx River is a wetland/grassland park that filters runoff from a 30,000 sq ft commercial parking lot, and opens up private property to public use. The project was awarded a special citation from the New York State Assembly for innovative cooperation between the artist, the community, and a local business. Job skills apprentices from RTB were paid to plant over 8000 native plants and carry out ongoing maintenance.

Edward R. Landa: I love the fact that the community here is not just called in after the fact to comment and sign-off on somebody else's efforts. They are hands-on, hands-in players! They are not the traditional "stakeholders," rounded-up *pro forma* to satisfy some bureaucratic requirement.

They are stakeholders, shovel wielders, and grass planters—literally the "boots on the ground," digging in the soil and delving into its complexity on a visceral level.

Lillian Ball: I am so glad that comes across to you, since I love that "boots on the ground" part too. For me, that is one of the most substantial reasons the WATERWASH Project is art. It is a very different approach than making work to exhibit in a gallery or museum context, where you only affect a rarified audience. Terms used to describe this type of work in art theory include, for example, relational aesthetics and social practice. It's artwork that has broader, perhaps more ambitious humanist goals than simply making objects. My work depends on a web of relationships literally embedded in the soil, which encourage stewardship long after I am gone. The people involved become part of the creation and have a sense of ownership. With the Bronx River WATERWASH I realized how crucial it is to have that process begin before the design and continue after construction. This kind of local stakeholder participation with public ecological art helps focus on local solutions to challenges like climate change. Participants who are involved with building such a project have much more of an investment in its long-term success. And the process is never ending. The Zen Buddhist master Suzuki Roshi has said "a garden is never finished."[1] I say, WATERWASH projects are never *done*. They are ongoing and open ended. Nothing is permanent, or ever executed exactly as it is designed.

Edward R. Landa: For me, WATERWASH operates on two levels:

The first is primarily aesthetic. Even when we are sampling soils for highly technical purposes, I think all soil scientists revel in the aesthetic experience—the unique smells, the color changes with depth, the glint of a mica flake, and the emergence of an earthworm. You and the team have created an unexpected, uniquely tranquil place, away from the concrete and asphalt and the noise of the city—a place that eases you from the anchor of the land to the flow of the river.

The second level is the way it reveals the active earth, which is no small task in a paved-over, urban terrain. In an area where "constructed" means cinder blocks and roofing tar, a constructed wetland is an invitation to explore the soil as *terra nova*—to experience it as a medium with metabolism, architecture, and memory.

Lillian Ball: The aesthetic experience is important to make people care about the site. A shared sense of investment and accomplishment comes by making the community part of the visioning and creative process as well as involving them in the planting and stewardship. I consider "place making" one of my primary tasks as an artist. The visual aspects are one of the ways to draw people in. That goes beyond the actual wetland design and construction to devising events, which draw in participants both before and after construction. Before construction at WATERWASH Bronx River, we had some teenage girls from Soundview School visit to help pot some of the native plants like *Solidago sempervirens* (salt-marsh goldenrod). When I gave them the job of putting natural rocks into pots, they were confused by the request. There seemed to be no distinction for them between brick and concrete and natural stone. I realized how little opportunity they had to dig in the dirt! Two years after construction, for example, during a climate event funded by Franklin Furnace, Reverend Billy, the Stop Shopping Choir, and the extinct Golden Toads performed for visitors who ate a lunch grown and cooked by Bronx Wildcat Academy students. For the final 2013 event, "Bronx River WATERWASH" a 50-minute documentary of the construction process by ReOrient Films was shown at the Bronx Museum, followed by a Q & A panel highlighting RTB participation.

Meanwhile, my own experience with the soil on the Bronx River site was initially perplexing. When we began excavation, the situation became obvious—it was essentially a landfill. Oil tanks, sinks, and porcelain emerged with tires and old pieces of metal. But to my amazement, there was soil there too, in the spaces between the junk. Then, while doing the research about the kind of mixture we wanted for the wetland grassland, I kept coming across suggestions for planting with 2/3 sand and 1/3 peat moss. Evidently, the consensus is that native plants from the Long Island Sound area don't need high nutrients, but the weeds and invasive species do. So, using a mostly sandy mix makes it less likely to sustain the undesirables. Eighteen inches of sand and peat moss in the wetland and six inches in the grassland worked well to effectively filter storm water runoff during the time it was retained in the wetland.

Edward R. Landa: As a soil scientist that seems counterintuitive at first. High sand equates to low fertility. But maybe that's what was needed for the site. I think that the basic strategies that soil scientists use in their exploration of soil properties and functions are not unlike those that you use as an artist.

We dissect with sieving, and demarcate with horizon designations. We interrogate, doing percolation tests to see how fast water can enter the soil. We extract DNA to identify microbes present; we measure responses of plants and animals to nutrients and contaminants … I love being a soil scientist. Several years ago, a colleague passed on to me the wonderful view that "soil science is everything a soil scientist is interested in." That is a profound truth that reflects the expansive nature of our field, and how it touches

so many aspects of our existence—be it the Bronx River at WATERWASH, the compost applied to an urban garden, or an exploration of biodiversity in Central Park. I count a urologist, an endocrinologist, a historian, a mosquito ecologist, and a soil microbiologist among my present research collaborators. As my wife would say: "Don't we live in the best neighborhood!"

Lillian Ball: I think the creative process is similar for both scientists and artists; we just have different toolboxes. I like the agency of being an artist in the same way you appreciate the wide range of methodological options in your work. Some might say, "Art is everything an artist is interested in," but then you might have to qualify that statement. Usually, these kinds of constructed-wetland projects are led by scientists or government agencies. The inclusive, collaborative nature of the artist-led WATERWASH projects are challenging for some viewers, since the aesthetic built environment is distinct from the underlying ecological issue of the work. For me, that's the kind of ambiguity that makes good, thought-provoking art.

Edward R. Landa: I find the same attraction in imagery and language.

In terms of images—please, no more cupped hands with dark, rich soil and the optional seedling.

Gives us more to chew on!

And for language, give me stuff like "probes the subsoil ecology of the modern self." That's the description of a new book on fossil fuels as a force in shaping the American society.[2] It is wonderful phrasing unto itself, but for me as soil scientist, it speaks to a different plane of abstraction—one somewhat less metaphorical and more reality based. *Ecology … and subsoil* are words that made me think about the mystery and complexity of

Early construction of the wetland at WATERWASH Bronx River, 2011.

Reprinted with permission from Lillian Ball.

Job skills apprentices from Rocking the Boat celebrate the completion of planting, WATERWASH Bronx River, 2011.

Reprinted with permission from Lillian Ball.

Overall view of the completed wetland, WATERWASH Bronx River, 2011, dimensions variable.

Reprinted with permission from Lillian Ball.

soil, but are both within my comfort zone. And then the stretch to *the modern self* ... Now that's food for thought! That's what I seek in art.

Lillian Ball: I like the probing part of that quote. It's applicable to both of our disciplines. Another one of the things we share is this concept that culture is inherent in our work. My background in anthropology and ethnographic film is not overtly used in WATERWASH Projects, but is a part of how I see the world. The humanist connection to the land is something that must be built into soil science.

Is there such a thing as a soil sociologist?

Edward R. Landa: We have an active environmental sociology program here at the University of Maryland. So, I think there is movement toward more dialogue at both the natural science and social science ends of the spectrum. As to soil sociologists, it's not one of our top four specialties— soil chemistry, soil physics, soil microbiology and biochemistry, and pedology. Johan Bouma, professor emeritus of soil science at Wageningen University in The Netherlands, has been a tireless voice for broader thinking about the relationships between soil science and society. He has urged us on the inside to "avoid the trap of making an exclusive analysis from our own perspective without trying to really comprehend and integrate visions by others, be it colleague scientists, citizens, stakeholders, planners or politicians. This inward-looking attitude does not create mutual understanding and often leads to a sterile debate. Better ways have to be explored. We need help here and philosophy and sociology can offer valuable insights into the processes involved and their context."[3]

Science is about focus for me, while art is more effective if it's a bit unfocused—abstraction, blurring of borders—these are things I like. There's a lot of to think about when we look at a soil. It truly is the "excited skin" of the Earth.[4]

Lillian Ball: Bouma's warning is one that fine art culture can listen to as well. In my experience, interdisciplinary collaboration between the arts and sciences strengthens both perspectives.

It's so urgent now with climate change and sea level rise challenging the natural world. This moment in time needs all the active solutions we can collectively dig up.

Endnotes

1. Chadwick, D. 1999. *Crooked Cucumber: The Life and Zen Teaching of Shunryu Suzuki.* Broadway Books, New York.

2. Book jacket quote from Stephanie Le Menager, Moore Endowed Professor of English at the University of Oregon, in *Carbon Nation: Fossil Fuels in the Making of American Culture* by Bob Johnson, University Press of Kansas 2014; http://www.kansaspress.ku.edu/johcar.html.

3. Bouma, J. 2004. "Facing future challenges for soil science!" Available at http://www.regional.org.au/au/asssi/ supersoil2004/keynote/bouma.htm.

4. From Nikiforoff, C. C. 1959. Reappraisal of the soil. *Science* 129: 186–196. Constantin Constantinovich Nikiforoff, a Russian émigré with a doctorate from the University of St. Petersburg, was a soil scientist with the University of Minnesota and later the U.S. Department of Agriculture's Soil Survey Division. A resident of Hyattsville, Maryland, and an avid walker, he had a flowing white mustache and goatee, and was a frequent presence in the soil science group on our campus during his retirement years (1957–1979) when this poetic and, in all senses, visceral descriptor was written (personal communication, Del Fanning, professor emeritus, University of Maryland, Department of Environmental Science and Technology).

Backyard Portals

A Solutions-Oriented Approach to Understanding and Valuing Soil

Stacy Levy in conversation with Patrick Drohan

Stacy Levy graduated from Yale University with a BA in art and a minor in forestry. She earned her MFA from Tyler School of Art, Temple University. She is a recipient of the Pew Fellowship in the Arts. Levy's works are often realized outdoors in urban nature and industrial landscapes, often as artist–scientist collaborations. Her work registers the changes in nature over the course of a day, a season, or a several years. Many of her recent projects

Title image: The AMD&ART team worked closely with the community through a series of neighborhood meetings and field days, planting trees with the Boy Scouts, horticulture groups, community service groups, and prisoners. All of the soil on the site was an engineered mix of reused materials from dredging, recycled paper, and fly ash.

Reprinted with courtesy of the artist.

utilize storm water runoff to make rainwater an asset to the site. Projects like *Dendritic Decay Garden, Spiral Wetland, Springside Rain Wall & Garden*, and AMD&ART are less of a destination than an unexpected invitation to look, wonder, and know. Levy's works aim to create social encounters that transform a passive gaze into an active understanding. Reinforcing the active gaze is the volunteer-based construction model of many of these projects, a model wherein a community comes into direct contact with the problem. By knowing it, they can begin to rethink it.

Patrick Drohan is associate professor of pedology at Penn State's College of Agricultural Sciences. His research examines people's use of landscapes and the accompanying changes in soil function across the larger ecosystem the soil supports. His research addresses basic science questions, but also demonstrates how this new knowledge can be used in applied research to improve land management and ecosystem stability. Drohan has taught over twenty-four different courses in the natural and physical sciences, and currently teaches urban soils; soil genesis and classification; field interpretation of soil properties; and a study abroad course in Western Europe on land and society's co-evolution through time. His involvement in the Dig It! exhibition at the Smithsonian Museum of Natural History created a social space for encountering and engaging with soils for tens of thousands of museum visitors.

We met for the first time on a misty, early Winter's morning in Pennsylvania at Stacy's barn-studio. Over coffee, and surrounded by Stacy's whirlwind of projects, we realized a commonality where profession and life-practice merged and empowering others to see the wonder of nature was a life-goal. Scientists and artists are translators of the natural world. Our work is to investigate and then to express the voices of sites and systems. The right balance of evocation and information allows us to "translate" our understanding of nature for others. It is this shared conviction that unites us in our thinking and in the direction of our work.

Interaction with the soil is at an all-time low. We walk on concrete that covers the soil, and we eat food grown in soil three thousand miles away from where we live. Our direct contact with soil as a complex, life-sustaining system is minimal. In a given room full of people, it is unlikely than anyone has touched the soil that day, week, month, or even season. Without knowledge of, and contact with, natural systems, there is a disconnection. We lose the sense that we could repair damages. Slowly we begin to believe that the task of reversing environmental harms may be unattainable.

But a new form of confidence is evolving. This new (or, *renewed*) commitment to environmental responsibility relies on the sense that the answers to our problems will not come as a centralized, monoengineered package. Rather, the solutions will come in smaller and more scattered packages, and may have more to do with our backyards than with federal programs or large-scale engineering projects. The notion of living within the solution is very interesting, as *solutions* engage negatives and positives. By its nature, a solution acknowledges the existence of a problem; it also liberates and empowers us to make visible action and change. Wildlife ecologists suggest that we can be instrumental in linking fragments of habitant by creating native habitats right in our own backyards. Doug Tallamy,[1] for example, has proposed that we are *responsible* for creating backyard landscapes that provide "ecosystem services to the natural world of insects and birds."

The growing eco-art movement could be seen as an indicator of this solutions-based ethos. When environmental artworks first began to emerge in the early 1970s, some artists (such as Michael Heiser and Robert Smithson) were partly responsible for acts of degradation, while others (such as Newton and Helen Mayer Harrison and Betsy Damon) turned toward a solutions-oriented approach. Artists began to collaborate with scientists and engineers and a "third party" emerged from the engagement of multiple minds and fields in developing solutions for ongoing, site-specific issues. The solution has become a way for the artist to both remediate problems and to call viewers' attention to the issues at play. This two-fold approach is part of a larger design movement that changes how we think about the relationship between people and engineering. Rather than see people as passive recipients of engineering solutions, we are beginning to think about those solutions as invitations to awareness and agency.

Shared passions can emerge when scientists and artists meet. Our desires are often kindred: to show how the world works and to find ways to express those workings to people who have not yet felt awed by nature. As an artist, I, Stacy, may have more leeway; my investigations are not tied to data collection and hypothesis. I work in a more flexible system. As a scientist, I, Patrick, mind these constraints, but also have the power of facts and rationale to support my expressions. For a variety of historical, social, and economic reasons, the language of science is currently imbued with a lot of power; it's a language that many people find persuasive when it comes to understanding how things work. But both scientists and artists are translators of the natural world. Our work is to investigate and then to express the voices of sites and systems. The right balance of evocation and information allows us to "translate" our understanding of nature for others. It is this shared conviction that unites us in our thinking and in the direction of our work.

In a conversation about solutions-oriented art and science, we spoke about how understanding nature is related to value systems. Knowing something is the first step to valuing it, and how we value nature is fundamentally related to all our other values. We guessed that a rural upbringing could give you a very different value system than growing up in the city. Yet we both grew up in largely urban settings where tiny portions of the natural landscape served as portals to understanding broader systems of nature. We wandered these little remnants with a great sense of wonder and excitement. We were both curious about trees and insects and birds, and these fragments of nature in our urban settings were very valuable to us. We asked ourselves: "How do you invite and encourage others to be curious? How do you create a portal to understanding? How do you get a child to look at the soil and begin to wonder about the life forms between its grains?"

AMD&ART Project at Mine Number Six in Vintondale, public park and water treatment facility, Pennsylvania, 1995–2005. In collaboration with Julie Bargmann, landscape architect; Robert Deason, hydrogeologist; and T. Allan Comp, historian. AMD&Art Project for Vintondale is a completely new method of designing a passive water treatment solution for acid mine drainage (AMD), a nasty cocktail of heavy metals which continually seeps out the abandoned mines.

Reprinted with courtesy of the artist.

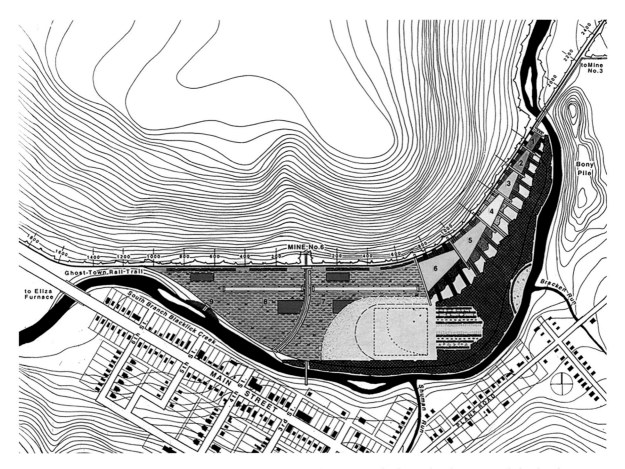

Rather than a typical engineered solution, in the AMD&ART project we were both *treating* the water and *showing* the process. The Litmus Gardens, hedgerows of native trees and shrubs, vivify the process of the water treatment, reflecting the color of the water as it progresses throughout the treatment basins from deep orange, to yellow, and then to pea green.

Reprinted with courtesy of the artist.

AMD&ART Project at Mine Number Six in Vintondale, Pennsylvania, 1995–2005. The project seen from the bony pile in the first snow after excavation. The basins had not yet filled with solution. Acid mine drainage pollutes hundreds of miles of streams in Pennsylvania. At Mine Number Six in Vintondale, in the coal mining region of south central Pennsylvania, artists, landscape architects, scientists, and historians collaborated on ways to treat AMD while interpreting the coal mining history and the passive treatment processes.

Reprinted with courtesy of the artist.

Fostering wonder is the soul-work of our allied fields. After our short exchange, we were reminded of the imperative of creating social encounters with nature. Thinking back to those childhood rambles, we are both interested in creating visual portals for people to begin to understand the natural world. We spoke at length about how our choices and our values are interlinked, about our tendency to treat the soil the way we treat each other: with or without respect, responsibly or irresponsibly.

The possibility of using our own agency to answer environmental needs suggests a new angle when it comes to creating social encounters with natural systems. In our work as scientists and artists, we have both sought to illuminate the ecological power of backyards and local landscapes, encouraging people to see that many of the fixes we need may come directly from changes in how we envision, use, and live in these familiar spaces.

Confronting Acid Mine Drainage

The AMD&ART Project introduced a completely new method of designing a passive water treatment solution through the employment of art and the engagement an entire community. Acid mine drainage (AMD) is a nasty cocktail of heavy metals that continually seeps out the abandoned mines. Typically, passive and active solutions for acid mine drainage are constructed out of view, enclosed by a chain-link fence. The project's initiator, historian Allan Comp, began to feel strongly that the cleanup process for industrial water pollution should be a visible part of everyday experience. In 1997, he recruited a team to come up with a new vision. As the environmental artist on that team, I joined landscape architect Julie Bargmann and hydrogeologist Bob Deason to examine engineering solutions for acid mine drainage. We envisioned a treatment system that would engage the landscape by creating a park that celebrated the treatment of water, rather than render that process invisible.

Vintondale, Pennsylvania, the site of the project, had been degraded by years of industrial coal extraction, and the local community was tired of large-scale fixes that resulted in the loss of landscape. The coal-mining industry had left the area with little in exchange for polluted water and man-made mountains of slag. Bargmann, Deason, and I wanted to do two things: to give the town of Vintondale a chance to reclaim its own backyard, and to preserve the paradox inherent in the cohabitation of natural beauty with the industrial past. We did not want to create a sanitized landscape devoid of the dirty elements of coal mining, but rather to highlight strange juxtapositions: resilient nature growing around a mixed palette of residual pollutants like "yellow boy," "red dog," and "boney."

The project was twofold in its aim to create a living solution that framed the incongruities of nature and industrial toxins: to treat the water and to show the process. Our team was fascinated by the moving parts of the treatment chain and wanted to share this visual activity with the community. Seeing and enjoying the system energized the neighborhood; residents developed a sense of ownership and excitement about the project. Boy Scouts, horticulture groups, community service groups, and prisoners were some of the members of the community to get involved through a series of neighborhood meetings and field days, tree planting, and working activities to build site furnishings and create wildlife habitat. A bike trail, baseball field, picnic grounds, and bird watching were also integrated into the design.

Hedgerows of native trees and shrubs complement The Litmus Garden, a series of treatment ponds that clean acid mine drainage through a sequential alkaline processing system. The color of the water is animated by the plants as it progresses through each of the treatment basins from deep orange, to yellow, and then to pea green. The design of the water treatment wetlands brings the massive scale of the mining operation back to the site. We raised plinths of soil demarcating the footprints of the original mine buildings, including fallen coke ovens and a washery, and planted them with red maples. The final rinse of the water is through a series of wetlands allowing the water to slowly seep back into the Black Lick Creek.

Springside Rain Wall, 2008, PVC pipe, glass pipe, native perennials, rainwater, the Springside School, Philadelphia, Pennsylvania. In collaboration with the Philadelphia Water Department and the Pennsylvania Horticultural Society.

Reprinted with courtesy of the artist.

Dendritic Decay. Cored concrete, removed asphalt, soil, and native plants. Washington Avenue Green at Pier 53 on the Delaware River, Philadelphia, Pennsylvania, 2010. In collaboration with Biohabitats Inc. Supported by the Delaware River Waterfront Corporation. This part of the garden is 1200 square feet. Overall the site is 250 yards × 140 yards.

Reprinted with courtesy of the artist.

Close-up of girl interacting with plants and soil of *Dendritic Decay.*

Reprinted with courtesy of the artist.

Making the Watershed Visible

Several other recent works develop the solutions-oriented, participation-based approach to celebrating the neighborhood watershed that was so central to the AMD&ART project. For the *Springside Rain Wall & Garden* project, rainwater was carried from the roof gutter of the Springside School in Philadelphia through a series of curving blue PVC and glass drainpipes to depict the local watershed. Portions of the pipe are clear glass to verify the passage of rainwater through the system. The rainwater from each drain splashes into a stone or concrete runnel and flows down a graded bioswale filled with native irises, creating a trail of blue flowers in the late

Spiral Wetland, 2013. Project for the Artosphere, Walton Art Center Lake Fayette, Fayetteville, Arkansas. 300 linear feet long, 4 feet wide floating on the water. Closed-cell foam mat, 7000 *Juncus effusus* native plants, rope, anchors.

Reprinted with courtesy of the artist.

spring. The water then flows into a planted infiltration basin and soaks into the ground. The swale slows the passage of the storm water, while the basin allows the water to seep into the ground. This slow infiltration keeps the rainwater moving deeply through the soil, rather than running off the surface and flowing directly into the storm sewers. The project handles all of the runoff from the science and art wing of the building, allowing this part of the campus structure "to drink its own water."

In collaboration with the Philadelphia Water Department and the Pennsylvania Horticultural Society, the native plants in the garden were planted by kindergarten through 12th grade students of the Springside School. The space was once an ignored pocket of eroded lawn, but is now an area for watching blossoms and seed heads and studying the ecosystem of a native wet meadow. It has created natural habitat for butterflies and other insects. Both the watershed pipes and the rain garden are highly visible from the road, giving a graphic connection of rain to landscape. Rather than acting invisibly, this project legibly celebrates the connection of rainwater and watersheds and makes a rain an event to look forward to.

In a subsequent project at Pier 53 on the Delaware River in Philadelphia, I removed bits of asphalt in the design of the local watershed. In collaboration with Biohabitats Inc. and supported by the Delaware River Waterfront Corporation, this part of the garden takes up 1200 square feet of the overall site of 315 square yards. The design consists of the existing cored concrete and the spaces in between—removed asphalt, soil, and native plants. The destructive power of plant roots is harnessed to create different ways to break down the remnant industrial hardscape of the park. Because the removal of the entire surface of the concrete and asphalt would have used up the entire design and construction budget for the park, we used the freeze and thaw cycles and the power of roots to do the job of decaying the concrete and asphalt over time, while conveying storm water into a rain garden. The patterns of hardscape removal and planting depicts the watershed of the site.

While the *Springside Rain Wall* and *Dendritic Decay Garden* projects render local waterways visible by merging unusual mapping techniques and public sculpture, the project *Spiral Wetland* uses closed-cell foam mats, 7000 native rushes (Juncus effuses), rope, and anchors in the construction of a floating sculpture. Inspired by the *Spiral Jetty* by Robert Smithson, one of the most famous and enduring of the land art pieces, *Spiral Wetland* reaches back to the beginning of eco-art but envisions the next stage of our thinking: to heal and transform the environment for better. Floating wetlands have been used recently for water treatment in New Zealand, Australia, China, Singapore, Canada, and the United States. They are a form of biomimicry: re-creating the natural processes at work in a typical wetland. These constructed wetlands help to remove excess nutrients from water by exposing the water to microbial processes facilitated by the plants and organic matter of the soil. They work to improve water quality and produce much needed wetland habitats for fish and other water creatures, while reminding us of the roots of land art.

Soil for the Masses

Engaging a college student about soil in a classroom or laboratory is very different from engaging a member of the public who may not even be aware that they should know anything about soil. The college student has chosen, via a course they enroll in, to learn about soil. While how much they learn is due to their effort at understanding and the quality of the lesson, the experience is a chosen one. How does a scientist capture the attention of the general public? *Dig It!* was a chance to entice the public's interest in soil during their visit to the Smithsonian Museum of Natural History in Washington, DC.[2]

Visitors to the Smithsonian have a variety of reasons for attending: to see a particular exhibit; general interest to educate themselves or their children; simple curiosity. Exhibits in the museum somewhat compete for attention, thus designers want to grab the attention of someone wandering by an exhibit. The *Dig It!* entrance tried to do this too. *Dig It!* used a play on scale at the entrance in part to attract attention but also to draw one's thinking down to the scale of the elements that make up the soil. This unique perspective change began to challenge previously held assumptions about the soil and simply wow people by making them think about the world from a very different literal view.

Soils are colorful and *Dig It!* took advantage of this natural feature to show-off the beauty of soils from around the United States using "monoliths". Monoliths are like faces of a soil's profile that are lifted from an excavated hole. These "faces" of the earth were collected from all U.S. states and territories. Little interpretation of each monolith was provided, with exception of its name (what soil scientists would call a series). If someone wanted more specific information they could look this name up on the Internet. What was powerful about this section of the exhibit was the bold display of soil diversity in the United States. This was evident in the colors, visible texture or roughness of a soil's face, the presence

(top left) First exhibit rendering based on discussions from the 2002 Smithsonian Institution exhibit workshop; and 2008 *Dig It!* opening (top right) State monoliths display, (bottom left) One of the entrances (note the similarity to the first rendering in part a), and (bottom right) Soil water movement and particle size display. Photos courtesy of the Smithsonian National Museum of Natural History.

Image originally appeared in *Soil Science Society of America Journal, 74*(3), 697–705. Reprinted here with permission from the author.

of rocks, roots, and the various shapes of particles and structures. The theme of texture and shape was highlighted many times in various exhibit components in order to teach important concepts tied to a soil's capability to provide ecosystem services. These services may include the capability to hold water and gases, play a critical role in chemistry and thus fertility, and finally in showing how soil supports our life every day via food and commodity production, engineering, water filtration, oxygen production from plants, and so forth.

The success of *Dig It!* happened because scientists turned to artists to help release a new creative energy that in the end enhanced the communication of the science. Personally, this effort helped me (Patrick Drohan) think about better ways to communicate soil science by using a visual language that people naturally gravitate to. I know the experience of helping make *Dig It!* a reality has helped me see the importance of art in science as a critical component of the educational experience; the integration of art and science achieves a better, more enjoyable educational experience.

Endnotes

1. *Bringing Nature Home* by Douglas W. Tallamy, Timber Press, 2009.

2. For more information on *Dig It!*, see Drohan, P.J., Havlin, J.L., Megonigal, J. P., and Cheng, H.H., 2010, The "Dig It!" Smithsonian soils exhibition: Lessons learned and goals for the future, *Soil Science Society of America Journal*, *74*(3), 697–705; and Megonigal, J.P., Stauffer, B., Starrs, S., Pekarik, A., Drohan, P., and Havlin, J., 2010, "Dig It!": How an exhibit breathed life into soils education, *Soil Science Society of America Journal*, *74*(3), 706–716.

Don't Worry, It's Only Mud

Patricia Watts and Amy Lipton in conversation with Mel Chin

Mel Chin is known for the broad range of approaches in his art, including works that require multidisciplinary, collaborative teamwork and works that conjoin cross-cultural aesthetics with complex ideas. In 1989, he developed "Revival Field," a pioneer project in the field of "green remediation," the use of plants to remove toxic, heavy metals from the soil. From 1995 to 1998 Chin formed a collective to produce *In the Name of the Place*, a conceptual public art project conducted on the popular prime-time TV series, *Melrose Place*. At Knowmad, Chin worked with software engineers to create a video game based on rug patterns of nomadic people facing cultural disappearance, and his hand-drawn, 24-minute film *9–11/9–11* won the "Oscar" of Chile for best animation in 2007. A current project, *Fundred Dollar*

Title image: *Revival Field*, 1993 Harvest, Pig's Eye Landfill, St. Paul, Minnesota.

Photo credit: David Schneider.

Bill, focuses on national awareness and prevention of childhood lead poisoning through art making. Chin's work was documented in the popular PBS program, *Art of the 21st Century*. Chin has received numerous awards and grants from organizations such as the National Endowment for the Arts, New York State Council for the Arts, Art Matters, Creative Capital, and the Penny McCall, Pollock/Krasner, Joan Mitchell, Rockefeller and Louis Comfort Tiffany Foundations, among others. A multivenue exhibition of Chin's work titled *All Over the Place* will be presented in New York City in the spring of 2018.

Patricia Watts and **Amy Lipton** have researched and worked with hundreds of artists addressing environmental issues for over twenty years. Through the years, they both have taken a special interest in farming and food production. Watts founded ecoartspace in 1997 while living in Los Angeles and partnered with Lipton in 1999 out of New York City to become the first bicoastal platform exclusively presenting ecological art. Separately and together they have curated dozens of exhibitions and programs in the United States for museums, university galleries, nature centers, and sculpture parks. They have also participated on panels, given lectures internationally, and often write for books and art publications. ecoartspace is currently focused on developing a media archive of video interviews with pioneering ecological artists and is also working on a series of ACTION Guides of replicable social practice public art projects for educators. Lipton gave Chin his first solo show in New York in 1987, The Operation of the Sun Through the Cult of the Hand. Watts invited Chin to do this interview while staying at the Christordora House in the East Village, April 2015.

Patricia Watts: Everyone has their own definition of what soil is, so what does soil mean to you?

Mel Chin: Not being raised Christian, I don't think it has anything to do with creation, but it is everything about creating. Maybe it's not what it means, but what you associate it with ... My first association was in college, where I had to take a pottery class. I didn't want to take it. I wanted to be an artist. Painting and sculpture and all that. I remember my first critique with an instructor. His name was Bob. He said, "Who did this?" and his finger was pressing on it, and he pressed it so hard it broke. He just picked it up and threw it in the slip bucket and said, "If you're going to make an ashtray, make a damn good one." Then he said, "Don't worry, it's only mud." And that unleashed it for me. Once you get into clay, you get into soil and all its capacity.

That association is also to me the earthiness of the '70s. There was a collective creation in it for me, something out of the earth and not out of a tube. And the people that were doing clay then were mostly all women. I became so obsessed with it I almost think I became psycho-ceramic. I'd be there 24/7. I'd quit everything to be amongst the smell of it and the mix of it. If anything, you have to become part of the dirt to liberate yourself. I got so heavy into it. That's where the origins of my standing in conceptual art began—from clay. If you look at Chinese Song pottery, or the stuff from 3000 BC Japan, you see strange behavior. When you think about soil, you learn about its origins. To me it's a point of origin for ideas, not just material.

Patricia Watts: Do you think your early experience with clay in some way informed your later works like *Revival Field* and the project you're doing now in New Orleans?

Mel Chin: After the Hirschhorn show in DC, I'm in the elevator and I hear these voices, distinctly

questioning, "What do you love, man?" I said, "I love making these objects, using these materials. I love researching and creating these things. I think I could do it my whole life." And a voice said, "Stop." I felt like that was exactly what I needed to hear. So, I came back to New York and I didn't pick up a hammer or paint, no art, nothing. I wandered the streets and I'd go to the library. I socialized very little. It was 1989. I was looking at anything but art, just forgetting about it for a while. Then I came upon Terrence McKenna, a botanist in a hippie magazine, *The Whole Earth Review*. He spoke about Jimson Weed, *Datura*, being able to pull toxins from the earth. That's how the idea started ... I was interested in the idea that the soil could be reborn through a process. I think the poetry of the idea that plants could do this sounded great. But at the time it was layered with an incredible degree of anxiety because it meant stepping out of the gallery and dealing with the environmental reality of what people were going through. People's lives and their relationships were being compromised by something in the soil! Something that was originally cool and sustaining could inordinately bring about death and toxicity.

Amy Lipton: How did you find Rufus Cheney?

Mel Chin: Well that was the quest. I met Terrence McKenna as well. But the thing about Terrence's claim was that I needed corroboration. Was it true? Because I started studying the plant and Datura does have certain kinds of alkaloid compounds that can sicken you as much as it can do anything else, so I asked, "Is this what I want?" I just wanted to understand the biomechanics of it.

Amy Lipton: It's hallucinogenic as well.

Mel Chin: Oh yeah, and it has unpredictable kinds of results. The poetry became layered with responsibility. Do I really wanna do this?

Would it really work? So that's how I spent my time. Looking for any article that would tell me Terrence was correct, that it could do this. Nothing. There was nothing from months of research. So, I finally called a landfill expert in Texas. He said he didn't know of any papers based on Datura. But, he told me about a guy named Rufus Chaney, who was a senior research agronomist at the USDA [U.S. Department of Agriculture] at that time. So, I called him. I asked, "Dr. Chaney do you know about Datura and its ability to pick up heavy metals?" He said, "Well, Datura gets you high, but it will not pick up any heavy metals." But he also said, "If you want something that will, this is what I'm trying to research." It was interesting. When he first heard I was an artist he was all, "Oh I can tell you about the chemistry that can make a tree go yellow or red for painting." I told him I wasn't interested in painting, that I wanted to know about this metal absorbing thing in plants. It became apparent that I needed to learn what he knew.

Dr. Chaney was working on sewage sludge, as an expert in transfers of heavy metals in humans, plants, and animals and their pathogenic effects. His research in plants was not supported. So that's where *Revival Field* was born. I said, "Ok, tell me what I gotta read," and it was *Biological Methods of Prospecting for Minerals* by Robert Richard Brooks (1983), which is a great little book that describes all the methods of prospecting with plants. It talks about the origins and understanding of it. It goes back to Africa to look at the copper hunters, who were looking for bioindicator plants, "If you dig here, you'll find copper." It's field knowledge. And when he told me there had been no replicated field tests, I felt that we should start this. To draft *Revival Field* seemed very important.

The NEA [National Endowment for the Arts] Inter-Arts grants supported collaboration between visual artist with someone like a dancer or a musician, but I chose a scientist, Dr. Chaney. Advancing the first replicated field test in the world would be a breakthrough. The history of the rejection of the grant by the chair of the NEA and the subsequent reversal is another history. While I was embroiled in the strategy to respond to the NEA, Dr. Chaney and I had conversations about different species … It began when we planted the first tiny seeds at the laboratory. The idea was to go to a place that had been toxified by cadmium (Cd). We located a site closed because thirty acres were contaminated with sewage sludge ash with significant levels of cadmium. That was our first *Revival Field* site … So, back to my association of dirt—this time loaded the possibility of harm waiting for a "poetic" method of transforming it.

Patricia Watts: So, *Revival Field* was successful? The plants did absorb the cadmium?

Robert Richard Brooks, *Biological Methods of Prospecting for Minerals* (New York: John Wiley & Sons, Inc., 1983).

Photo credit: Severn Eaton.

Species difference in uptake of Cd and Zn on revival field-St. Paul, Minnesota, 1993.

Species	Zn in Shoots		Cd in Shoots	
	−S	+S	−S	+S
	-------mg/kg dry weight-------			
Thlaspi	2350.0	1330.0	10.0	9.5
Silene	33.0	43.0	1.3	2.5
Lettuce	60.8	80.5	4.9	7.6
Corn	41.3	31.5	3.1	2.8
Red Fescue	28.5	36.2	1.2	1.6

Mel Chin: Yes. In the second year, lab analysis confirmed cadmium and zinc uptake. In fact, it proved hyperaccumulation. It became a three-year project, which confirmed the science. That was its most important contribution, a confirmation of a scientific hypothesis.

Amy Lipton: Did Dr. Chaney continue with that work?

Mel Chin: He did. We did a *Revival Field* in St. Paul and we started others. In fact, five post-ops have come out of *Revival Field*. You can learn what we did in Palmerton and in Germany. You can start to understand the relationships between bacteria and fungus in the soil. You can really understand it as a living entity that can be compromised deeply by industrial/postindustrial processes—how it deadens it, making it incapable of sustaining life. I mean zinc is good for us. Eat some oysters, whatever. But too much can be terribly toxic for plants. So, you have these combinations of zinc and cadmium. So, my association of soil is mostly from a toxic angle—from my visits to places like smelter sites or mountainsides that had been laid bare.

Patricia Watts: Describe the success or failure of the collaboration. What was the most challenging or rewarding thing for you?

Mel Chin: Well, collaboration can take different forms. It can begin as an initiator or a catalyst. But it's not just one thing … I don't think there was any failure to this in the sense that it's still a process. Now we can say we're collaborating, but at the time it was a cooperation. In this case, it was a cooperation between two identical dreams. It's almost like going up to somebody and saying, "Hey, what's your dream?" His dream was to make it real. My dream was to make it real. And, I said, "Well we both need this. How do we do this together?" I could phrase it one way, he could phrase it another. If there is cooperation or collaboration, then it should be done with the same rigor from both disciplines. If you're coming from the humanities, then how dedicated are you to its outcome? It's not about success or failure, it's about applying the rigor to determine if these cooperations are for real … And in some cases, I felt that I should submerge, because I felt the biggest obstacle would be art. The term. For political or social reasons. But, that's how the conceptual art evolved, with the association that we could do this. We used the application of critical thinking for aesthetics, so to speak, so we could engage things from formal structures. I tried to push forward a new aesthetic that was forming—the aesthetics of our existence.

Back then with *Revival Fields*, there were a couple things that were very freeing. The first freedom was beginning to imagine a project that may be completed after you're dead. You are freed by that constraint of time. So, there's that first liberation. The second liberation was to understand art as a catalytic structure. It's not necessarily a language that forms, but a platform on which another language can emerge. You start standing on that level, and you can approach almost anything. And it's not about chasing down dirty spots, or badness. It's about approaching the world and being compelled into action.

Patricia Watts: *Revival Field* was probably the first artwork of its kind to be a test site for a scientific process to provide data for future usage.

Mel Chin: Yes, and it propelled the new site. The Palmerton superfund site was started two years later on public lands … We had the most data for *Revival Field #1*, and *#2* was the Palmerton site in Pennsylvania, which was the site of a zinc-smelting factory. I think the Palmerton site was probably one of the most important ones, with some of the leading people in the field of

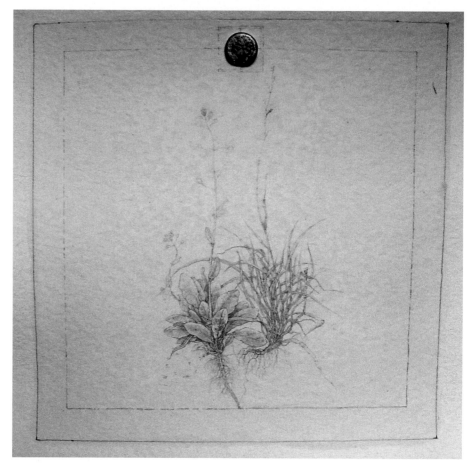

Photo and drawing of *Thlaspi caerulescens* by Mel Chin.

Reprinted with courtesy of the Artist.

bioremediation. It lasted for quite a while. I remember it was like a fan shape—to show that it was a formal configuration to appease the art people. And that had some significance for me, to make it more visually interesting from my point of view. But the science was pretty demanding … There were 96 different conditions in *Revival Field* that we had to follow. There were certain types of fertilizer, certain types of amendments, and all that data had to go back to Chaney. Not all the plants survived in the wild. You need three years to eventually get it right.

Patricia Watts: Can you list some of the plants you used?

Mel Chin: A variety of dwarf corn, romaine lettuce, red Merlin fescue, *Silene cucabalis*, to name a few. But it was *Thlaspi caerulescens* (Alpine penny-cress), a known hyperaccumulator, that we had to absorb cadmium and zinc. As far as I understand from my collaboration with Chaney and book research, Thlaspi can survive in unheard of levels of zinc pollution. Nothing can compete with it.

Patricia Watts: Can you talk about the *Operation Paydirt* project in New Orleans, which began only months after Katrina?

Mel Chin: Here we looked not only at the physical damage, but also at the physiological and psychological impact of the storm. It was intense. I went down there and was challenged by a tragic situation of such great magnitude, my first thought was "maybe my conceptual art was inadequate to respond."

Something that got me started was the NRDC [Natural Resources Defense Council] report mentioning the toxic environment in the soils that Katrina had brought in. The other thing that got me going was that the EPA [Environmental Protection Agency] said that it didn't get any worse. So, you had two camps: one is horrible, and one it's not. So, I wanted to find out. That's when I called Chaney again, who referred me to Dr. Howard Mielke, with whom I had consulted with during *Revival Field*. He had spent twenty-five years studying the soils of New Orleans. He was in the School of Pharmacology at the time at Xavier and was closing down his lab when I met him. He said the EPA was correct, but they didn't say how bad it was before the storm. He produced a map that showed levels of lead in the soil, and showed concentrations in neighborhoods that were way above the 400 parts per million limit. So, I asked, "So what does this mean as far as people." He said, "This lead map correlates with 30% to 50% of the inner-city childhood population, poisoned or compromised with lead—*before the storm*." And that's where the project born, something had to give voice to these kids, who had no say in what threatens their future.

At the time, there was almost no economic support in the way of addressing the issue. Someone who saw me speak at a panel at MIT [Massachusetts Institute of Technology] came up to me and said, "All these people are applying what you taught us in *Revival Field* in New Orleans to clean the soil. What do you say to that"? I said, "The first thing I'd tell them is cease and desist." If you go into a place that is so hurt, so damaged, so disrespected, why would you want to bring in unverified science to people who have already lost so much? So, I went down armed with this curiosity and the knowledge from *Revival Field*, that no plant is known to accumulate lead (Pb). I discovered Dr. Mielke's lead map and his plan to use soil from the Mississippi that was clean and washed from the heartland, reclaiming it from some of the deposits to cover the toxic ground in an effective way. He had tested some applications and the lead levels dropped every time it was done. And, the thing about Katrina is that it brought about a plan for lead remediation in New Orleans by bringing in uncontaminated soil. It made it safer. That's the only silver lining of Katrina. It reduced the levels of lead in the soil to some extent.

There's controversy over how much lead is in the soil, or how much is bioavailable, and how much is in the interior of the house itself. So, if you're living in a poverty-stricken area most of the time, and if the soil has high levels of lead and the house has high levels of lead, then children could be accumulating vast quantities, as hand-mouth relationships are a continual thing for kids. By the time they're three years old they could be totally damaged. So, *Operation Paydirt* begins with this knowledge. It's not a soil experiment. It's more of an experiment in democratic engagement about values, and using the 100 Dollar Bill drawing as a way of giving value to someone's contribution against a problem … *Operation Paydirt /The Fundred Dollar Bill Project* was to immediately teach a child and a parent or grandparent what they can do to be safer. That original lesson has value and after 450,000 drawings of currency, we still maintain that "kids are worth it."

The project has evolved, it started with how to deal with contaminated soil but it was always about the lead in the blood, brain and heart of a child. It's about contributing to bring 100% child lead poisoning prevention to America. Because the CDC [Centers for Disease Control and Prevention] and EPA say is 100% preventable. So, my question is, How do we make that real? Why is it still here? Why is it in Houston? Why is it in Saint Louis? Why is it in New York, in the Bronx? Why is it in New Orleans? Why, if it's 100% preventable? We're dedicating these *Fundred* funds from the people of America to catalyze the first process with all the coalitions we've partnered with. We're combining all these efforts to figure out which solution to begin with. That's what our first funding should support. What's the most pragmatic thing we can do to start

the process of 100% preventable childhood lead poisoning? And this started, again, with dirt. So, if you ask me about my relationship with soil, then I'd say it is the catalytic potential for these projects …

Patricia Watts: Tell me if I'm wrong, but I think I've heard you say a few times in the past … that when it comes to the science, leave it to the scientists. That the artist's role is that of creating a connection with the community—the social practice, but not the science. Is that right?

Mel Chin: We can be the conveyer of process and of science. We can do that. We can put it in a form where it can be understood. We can recompose data, not to make it more aesthetic but to make it more understandable. Traditionally, how art is used is we wear our ideas on our sleeves and we wanna be transparent, and it's because we have a voice in society, OK? Yeah, you can market something and make millions of dollars; you can be an activist and have everyone know your name. But if you're doing a collaborative project, let's make it collaborative. It belongs to the people. And, I don't see it any other way. So, you bring the people's will, and you exchange it for the value of whatever that might be to deliver the results back to the people. They should be represented, and we're trying to do it in cities with methodologies that we can show have been effective.

Patricia Watts: So finally, in your perspective what are the biggest challenges facing the health of the soil? Are you optimistic or pessimistic about the future of the planet's soils?

Mel Chin: OK. The biggest threat to soil is people. Human beings. The track record is terrible. I mean, everything is associated. The loss of our ecosystem in terms of plant and animal forms is devastating in the smallest amount of time in the history of the earth. So, am I pessimistic?

Map indicating ppm lead (Pb) levels in pre-Hurricane New Orleans soil.

Courtesy of Howard W. Mielke, PhD, Department of Pharmacology Tulane University School of Medicine.

Absolutely. I don't believe there's any real hope as long as *we* exist. Soil will go its way and it may recuperate. If you wanna believe in the Lovelockian hypothesis, that's cool. But maybe not on our watch. We're incapable of that because our psychological makeup is based in very short temporal spans. We don't have the foresight to do that. So why do we make art, engage in social practice, when we live within such a community of greedy, self-serving losers? The alternatives are just less boring. The options are exciting and the possibilities make it worthwhile. So, I believe that the actions that we do are in some cases remedial, or regenerative. It's part of our jobs as creatives. It's what we do. So, I'm OK with that. There's no cross to bear here, it's just boring, complacent, a conspiracy with degradation. You have that choice.

The Safehouse for Fundred Dollar Bills in New Orleans' 8th Ward, 2008–2010.

Photo Credit: Mel Chin. Courtesy of the artist.

The Fundred Reserve and Fundred Armored Truck temporary takeover of the Corcoran School of Art, George Washington University.

Photo credit: Fundred.

The Art of Decay

Soil Decomposition Explored through the Visual Arts

Farrah Fatemi and Laura Fatemi

Farrah Fatemi has a doctorate in Ecology and Environmental Sciences from the University of Maine, specializing in soil nutrient cycling. She works on collaborative projects that explore art as a tool for science communication and advocacy, and conducts educational outreach in the state of Oregon.

Laura Fatemi is former museum associate director and curator for the DePaul Art Museum in Chicago where she organized numerous exhibitions over the past fifteen years. Some of her notable projects focused on environmental issues such as climate change and soil

Title image: Jane Fulton Alt (American, born 1951), *Burn # 86*, 2009, inkjet print.

Reprinted with courtesy of the artist.

degradation. Fatemi retired in 2016, is a practicing artist and holds an MFA in drawing from Kansas State University. She currently lives in Chicago.

We are fundamentally connected to the Earth's nutrient cycles through the processes of death and decomposition. Yet we often fail to recognize the life-affirming power of these processes, whereby old energy is harnessed for new growth and beauty. Here, the works of four artists included in DePaul University's 2015 exhibit *Rooted in Soil* are explored: Sally Mann, Jae Rhim Lee, Jane Fulton Alt, and Claire Pentecost. Together, these artists explore the connection of the body to the soil and urge viewers to reassess their own place in the natural world. Take Sally Mann's haunting black-and-white photos of withering bodies in the *Body Farm* that engage viewers in the raw reality of decay in the afterlife. Or Jane Fulton Alt's ethereal photographs in *The Burn* that evoke a reverence for the destructive power of fire, but also a lament for the loss of what once was. Both Mann and Fulton Alt draw our attention to the delicate balance of decay versus renewal. In contrast, Jae Rhim Lee and Claire Pentecost take a stance of activism and advocacy, speaking up for the soil substrate which has no voice. For instance, Rhim Lee, using innovative research on mycoremediation, identifies new pathways to sustainability through green burial. And Pentecost reminds us that soil is a living and breathing substrate that can lose its integrity and life-affirming quality with misuse and degradation. Together, these four artists, using radically different approaches and mediums, give us the space and framework to reexamine our position in the natural world and to cultivate respect for the life-affirming qualities of soil.

As humans, we are ghosts of former beings—an amalgamation of energy and molecules that were once bound in the dead and lifeless. Kinetic energy, the energy of motion, is the currency of life. Yet this currency is only made available to us by an invisible set of lowly organized creatures through the process of decomposition. These creatures in the soil liberate the energy of the dead so that it can be used for birth and renewal. Decomposition is so crucial for life that without the recycling of nutrients and energy by this process, humans and all higher living beings would cease to exist. Even our own bodies will eventually serve as a delicious feast for a rich and diverse set of soil organisms that will provide fuel to form new life. We are intimately connected to the Earth's nutrient cycles through the process of decomposition.

The nature of mortality, decay, and renewal are subjects that have long been explored by artists. But now, some artists are probing more deeply and directly into the ephemeral nature of life cycles, and our connection to the soil environment. Here, four artists are discussed that were included in the exhibition *Rooted in Soil* at the DePaul Art Museum in Chicago

Sally Mann (American, born 1951), *Untitled,* (*The Body Farm*), 2000, silver gelatin print from wet collodion negative.

Copyright Sally Mann, courtesy of Gagosian Gallery, New York.

SALLY MANN *Untitled,* 2000 Gelatin silver print, 30 x 38 inches, (76.2 x 96.5 cm) Edition of 3

GAGOSIAN GALLERY

from January to April 2015. *Rooted in Soil* emphasized the ecological and cultural values of soil. Within the exhibition, works from Sally Mann, Jane Fulton Alt, Jae Rhim Lee, and Claire Pentecost formed a nexus on the topic of death and decay. Through captivating and innovative work, these artists challenge us to ponder our ephemeral existence and decomposition as a life-affirming process of renewal.

Virginia-based photographer, Sally Mann, says that she has "long been afflicted with the metaphysical question of death: what does remain? What becomes of us, of our being?"[1] In her series of photographs titled *Body Farm*, Mann takes us to the Forensic Anthropology Center (FAC) in Knoxville, Tennessee, to bear witness to the decomposition of human remains. The FAC is a woodland site where corpses lie exposed to the elements for analysis by forensic researchers. Photographs in the *Body Farm* series capture corpses in various stages of decomposition. Some of the photographs were taken close-up, others from a distance. The result is disarming but quietly resonant. Mann's photographs allow us to take a moment to contemplate mortality in its purest form in nature and the soil. "We don't talk much about what happens when we die. Years ago, sex was the unmentionable thing; now it's death … I thought it was an important part of the project to actually confront human death. I am curious about why people are bothered by a simple fact of nature. I like to make people a little uncomfortable. It encourages them to examine who they are and why they think the way they do," the artist commented on her work.[2]

Mann's photographic process also lends a unique vantage point from which to observe the bodies at FAC. She uses a wet-plate collodion process developed in the 1850s, and frequently shoots with an antique camera. This approach allows Mann to integrate the inherent imperfections of the medium with the subject matter. Blurred, marked, and blemished traces obscure some of the gruesome reality so that the viewer is drawn into the image. Mann describes one photograph of a decaying body, which she referred to as *Tunnel Mann*: "Since the day was cloudy, the exposure was a long one. The furiously churning maggots over the necessary six-second exposure gave *Tunnel Man* a beautiful diaphanous veil over his ruined features."[1] The photographs in *Body Farm* call attention to the grotesque, but also illuminate the beauty of our human connection to nature.

A quiet contemplation of our ephemeral existence can be observed in much of Sally Mann's work over the past several decades, from the eeriness of Southern landscapes, to her more recent work exploring the ghosts of slavery. Personal grievance and loss have surely influenced the *Body Farm* works—Mann discusses losing her parents and grappling with her husband's affliction with muscular dystrophy in her 2015 memoir *Hold Still*. As an artist, she intuitively understands what ecologists do about our place in the ecosystem: "In a sense, *Tunnel Man* had more life in him; life was feeding on him, the beetles and worms making inroads and leaving behind soil into which stray seeds would sink their fibrous roots. Surely the people is grass."[1] It is this kind of unabashed examination of mortality that reinforces our deep connection to the soil, to the land, and to other beings.

While Mann has a poetic and deeply contemplative approach to mortality, Korean-born, U.S.-based artist Jae Rhim Lee takes a business-model activism approach to dealing with the dead. In her work, Jae Rhim Lee confronts burial practices and addresses the toxic funeral industry. "Somehow death acceptance is needed for environmental stewardship. All the industrial toxins we emit into the atmosphere and the soil become part of our bodies. That is difficult to accept because it means we are also physical beings, animals, who will die and decay," says the artist.[3] Over the course of our lives, our bodies accumulate hundreds of synthetic compounds, many of which can act as toxins. These chemicals can contaminate the soil and connected waterways. Common embalming chemicals, including formaldehyde (a carcinogen), only further exacerbate the problem. Unfortunately, cremation is not necessarily a better environmental option compared with embalming and coffin burial. Combustion of a human body requires extremely high temperatures, and this energy-intensive process can emit pollutants such as carbon and volatile mercury. In some countries, high-tech human composting methods allow for low environmental impact upon burial. In Sweden, for instance, bodies are frozen in liquid nitrogen, shattered to small fragments, and placed in a compostable box. In recent decades, a growing number of people have been advocating for green burial methods, where the body is reconnected with the earth without traditional embalming and coffin encasing. Green burial cemeteries are increasingly becoming popular in Europe—seen as a way to exercise one's right to be one with the Earth. However, green burial is catching on less quickly in the United States, where a culture of religion and a multibillion dollar funeral industry cling tightly to traditional burial methods.

Jae Rhim Lee has developed a green burial suit designed as an ecofriendly alternative to traditional burial practices. The burial suit is inoculated with fungal species that are efficient at breaking down toxins in our bodies and in the soil, a process called mycoremediation. She began her experimental mycoremediation research by feeding her own hair, skin, and nail clippings to different types of mushrooms. Fungi are some of the most efficient decomposers on the planet, having evolved a sophisticated set of enzymes that they spew out to digest organic matter. Through the process of organic matter decay through enzymatic breakdown, many of the compounds that would be considered toxic to animals or plants are broken down into less harmful forms. With her work, Lee bridges the disciplines of the arts, science, and business; burial suits are now available for purchase for both humans and pets. More important, the *Infinity Burial Project* reminds us of our innate connection to fungal networks in the soil, and increases social and cultural awareness of the benefits of green burial.

Renewal from decay is also a recurrent theme for Chicago-based photographer Jane Fulton Alt in her series *The Burn*. Fulton Alt began the photographic journey of *The Burn* in 2007 in prairies and woodland areas around Lake Forest, Illinois. She was mesmerized by the ethereal qualities and the power of controlled burns, which are simultaneously destructive and regenerative. "To witness and photograph a controlled burn is to place oneself in the presence of a certain 'terrible beauty.' I attempt to

Jane Fulton Alt (American, born 1951), *Burn # 49*, 2009, inkjet print.

Reprinted with courtesy of the artist.

capture the ephemeral moment when life and death are not opposed but are harmonized as a single process to be embraced as one," says the artist.[4] These controlled burns not only accelerate decomposition in prairie soils, but they help liberate nutrients and energy that would otherwise remain locked away, unable to fuel new life. Furthermore, the gases produced in fires, including ethylene and ammonia, as well as the smoke itself, act to scarify seeds and promote germination. The full turning of the life cycle can be observed over the course of one burn and its aftermath. Once the dry grass that has been stunted by drought or the cold of winter has been combusted, it can then make way for new growth.

The photographs in *The Burn* are captivating—resplendent in the powerful beauty of the layered smoke, a liminal space that Fulton Alt has captured. "Fire and Smoke are my equivalents: abstract manifestations of an inner state where the unknowable resides. Smoke both conceals and reveals. While flames leap from the earth, the densely layered landscape is enveloped in veils that are alternately transparent, translucent, and opaque. The foreground melts into the background in the quickly changing terrain, altering one's sense of scale and space,"[3] writes the artist. *The Burn* is also a meditation on the transition between life and death. When Fulton Alt began her project, her sister just started chemotherapy treatments and her first grandchild was born. The images capture a transitory state, when one shape or form of energy is transformed to another. In this way, Fulton Alt's photographs result in a subliminal contemplation of the ecological connections in nature, and the way that combustion and decomposition necessarily bridge the realms of decay and regrowth.

In a final example from the *Rooted in Soil* exhibition, social consciousness is at the forefront of a work by the Chicago-based interdisciplinary artist Claire Pentecost. The deconstruction of the "old" to make way for the "new" in soil decomposition is a transformative power, the result of which can be observed firsthand in Pentecost's piece *Proposal for a New American Agriculture*. For this piece, Pentecost buried an American flag in a compost bin. The result is a disfigured, discolored flag, turned to a dark brown hue, frayed with gaping holes. Here, Pentecost is contemplating the unraveling of an unsustainable industrial agricultural system. Pentecost sees a clear connection between soil health and the health of our bodies: "Our bodies have co-evolved with millions of beneficial microbes, many on a continuum with the beneficial microbes in the soil. Unfortunately, we have been destroying many of these microscopic symbionts with misuse of antibiotics and overuse of microbial cleansers. Similarly in the soil, pesticides are destroying the healthy biology that makes farming without poisonous chemicals possible."[5]

Proposal for a New American Agriculture was featured at the Salina Arts Center in Kansas in the fall of 2015, in conjunction with the Land Institute's Prairie Festival. The Land Institute is a forerunner in sustainable agriculture research in the Midwestern United States. With *Proposal for a New American Agriculture*, Pentecost asks us to think more holistically about the soil ecosystem, as they do in their work at the Land Institute. For the *Rooted in Soil* exhibition in Chicago, Pentecost designed an interactive installation touching on themes of agricultural and human health. For *Our bodies our soils*, at the DePaul Art Museum, Pentecost displayed soil and compost from different locations under glass jars; visitors could pick up the jars and smell the aroma of each of the different soils. Lori Waxman, writer for the Chicago *Tribune*, commented on the experience: "I scented dirt from Kilbourn Park, the Bowmanville community garden, the University of Illinois at Chicago facilities compost and a Monsanto Roundup Ready cornfield in Three Oaks Township, Michigan … My nose isn't as keen as my eyes—I'm an art critic, not a perfumer—but even I found the Monsanto cornfield's lack of odor alarming …"[6] With the interactive installation, Pentecost provided a space for viewers to connect physically with the soil, and think about the linkages between soil and human health.

The work by Pentecost and other artists in *Rooted in Soil* are potent reminders of environmental problems and cultural attitudes that threaten soil ecosystems. In *Rooted in Soil*, a common thread emerged—one that emphasized symbioses and connection in the natural world. The message of ecological synergy is more subtle in the eerily beautiful works of Sally Mann and Jane Fulton Alt, and more direct in the activist-minded works of Jae Rhim Lee and Claire Pentecost. Yet imbedded in their work is an acknowledgment of the complexity of biological processes working together in the life cycle. Environmental problems arise when we fail to preserve the integrity of these processes. For example, in the industrial agricultural system that Pentecost critiques, fertilizers and pesticides have reduced the ability of the soil to sustain the life-affirming process of nutrient recycling.

In death, some of our own matter and energy will find a home in the soil—perhaps in a muddy microbial matrix or in an atom buried deep in the regolith. But in life, we are as only as healthy as our soils are. Practices that compromise soil health also compromise human health. Artists can play an important role in building awareness through visual and visceral experiences. They remind audiences of the renewal brought by the life-affirming process of soil decomposition, and the need for innovative solutions to nurture the integrity of our ecosystems.

Claire Pentecost (American, born, 1956), *Our Bodies Our Soils*, 2015, installation.

Reprinted with courtesy of the artist.

Claire Pentecost (American, born 1956), *Proposal for a New American Agriculture*, composted cotton flag.

Reprinted with courtesy of the artist.

Endnotes

1. Mann, Sally. 2015. *Hold Still: A Memoir with Photographs.* New York, NY.

2. Mann, Sally. 2015. *Hold Still: A Memoir with Photographs.* Boston: Little, Brown & Company.

3. George, A. 2011, July 20. Designing a mushroom death suit. *New Scientist.* Retrieved October 31, 2017, from https://www.newscientist.com/blogs/culturelab/2011/07/designing-a-mushroom-death-suit.html.

4. Fulton Alt, Jane. 2013. *The Burn.* Heidelberg: Kehrer Verlag.

5. Schoenberger, E. 2015, September 28. Conversation with Claire Pentecost. Retrieved from http://elisa-shoenberger.squarespace.com/blog/2015/9/28/conversation-with-claire-pentecost.

6. Waxman, Lori. 2015, March 4. Dirt takes center stage at DePaul Art Museum. *Chicago Tribune.* Retrieved from http://www.chicagotribune.com/entertainment/ct-ent-0305-rooted-in-soil-review-20150304-story.html.

Function 4

HOME: Soil as habitat, biological hotspot, and gene pool

Home

The focus shifts from the physical, chemical, and hydrological movements within the soil to the biological functions of the soil in the fourth section of the book. A single teaspoon of healthy soil can hold over a billion individual organisms, representing tens of thousands of individual species of bacteria and fungi. More than a quarter of the planet's known species live in the soil. The thriving layer below the surface of the Earth is actually responsible for almost all life above ground. Ideally, the pore space of the soil can make up 50% of its total volume, creating a porous web of interconnected tunnels and caverns that soil macro- and microorganisms call "home." In this section, protagonists of bio art and bio design, artistic microscopy, and film come together to investigate the lives of the world below in an attempt to better understand the kind of world we as humans want to live in above. The function of the soil as biological hotspot, habitat, and gene pool is examined in this section through the lenses of biohacking, Afrofuturism, and postcapitalist sculptural intervention.

In chapters by Peter Haff, Wanuri Kahiu, and Carlina Rossée, and by Ellen Kandeler and Georg Dietzler, habitat is seen as a merging of the technosphere with the pedophere, a shifting place where biosensoric technology can enhance human well-being. Biopolitics and the aesthetic visualization of the microscopic are key focal points in chapters by Simon Park, by Natsai Audrey Chieza and John Ward, and in an interview with Suzanne Anker by Regine Rapp and Christian de Lutz. Meanwhile questions of microbial consciousness, genetic biodiversity, and soil macrocosms are brought up in chapters by Daro Montag and Diana Wall; Amy Franceschini and Régine Debatty; and Alejandro Meitin, Claire Pentecost, Brian Holmes, and Ela Spalding.

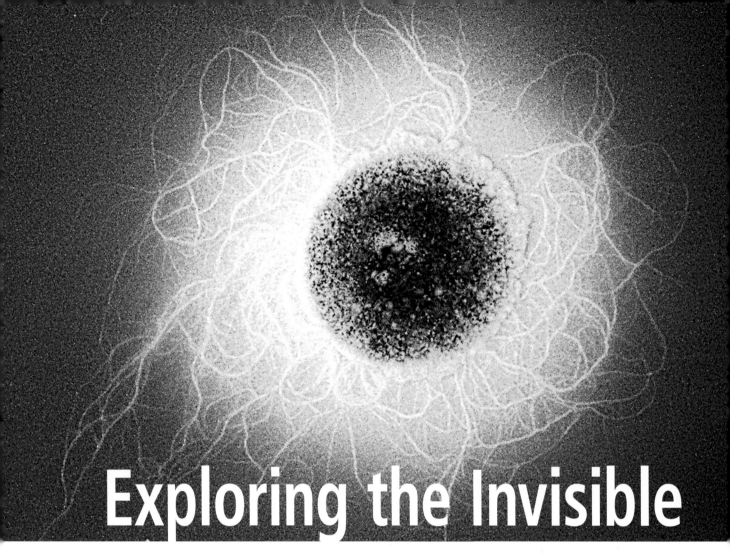

Exploring the Invisible

The Exemplary Life of Soil

Simon F. Park

Dr. Simon F. Park is a senior teaching fellow at the University of Surrey, where he teaches microbiology and molecular biology. An internationally recognized molecular microbiologist, he has published two textbooks and over sixty research papers in refereed journals, books, and other periodicals. For nearly ten years now, he has also worked at the fertile intersection between art and science, where his practice has been inspired by the aesthetics and processes of the usually invisible microbiological world. He collaborates with artists and also produces his own work. The outcomes of these projects have been widely disseminated and have been featured at such venues as The Natural History Museum, The Science Museum, The Royal

Title image: *Solaglyphs*, visualizing the soil microbiota. A complex and multispecies glyph made by the community of soil bacteria as they emerge from a droplet of a soil suspension to colonize the bacteriological agar. The cleavage of a chromogen in the agar, by the bacteria, generates fluorescence, revealing the bacteria via their own molecular activity.

Reprinted with permission from Simon F. Park.

Institution, The Science Gallery (Dublin), GV Art (London), The Wellcome Collection, and The Eden Project. He recently won the Peter Wildy Prize for his outstanding outreach work in microbiology.

For the section of this book on the pedo-microbiome and the habitat and gene pool functions of the soil, Park reflects on the different scales of life in the soil, and how their functioning fulfills a range of vital processes that can be visually brought to light with the help of microscopic photography, laboratory protocol, and a little collaboration with the soil organisms themselves. Park regards members of the soil biota as independent agents, and thus as cocreators in the aesthetic process, so that each work is generated by the soil itself rather than simply being a representation of it.

Soil is the matrix upon which all terrestrial life depends, and soil organisms are mostly responsible for performing its many vital functions. Despite this, the soil biota and its essential activities are often overlooked. This living soil component and its collective biodiversity is an important but poorly understood aspect of terrestrial ecosystems. The life-forms that live here carry out a huge range of vital processes that are important for soil health and fertility. This applies not only to natural ecosystems but also to the agricultural systems that underpin our civilization. Within the soil, the manifold activities of its many life-forms interact in a complex food web. Beyond this, soil can now be considered to be a superorganism, that is, a self-regulating ensemble of living organisms that are closely integrated with their immediate material environment and with the whole soil system behaving as a recognizable entity. Microbiological, near-microscopic organisms in the soil represent the largest fraction of terrestrial biodiversity. Yet these life-forms and the essential activities they preform are all too often overlooked in favor of larger life-forms, such as trees and forest animals.

The works described in this chapter, as well as many of my explorations that I have pursued over the last ten years, have sought to reveal the inherent creativity of the soil's disregarded life, to reveal its subtle and usually hidden narratives, and to bring them to light. In this sense, the work is generated by the soil itself rather than simply being a representation of it. In the following I give an overview of the range of organisms I have worked with. The simplest and most widely used system for classifying soil organisms uses body size as its basis, with all life falling into one of three categories: macrobiota (generally >2 mm in diameter), mesobiota (0.1–2 mm in diameter), and microbiota (<0.1 mm in diameter). As detailed next, I have worked with all three classes of soil biota.

At the largest end of the spectrum of scale, I have made works that have recorded the activities of soil macrobiota, which include larger organisms such as ants, millipedes, and centipedes. In order to do this, I carefully coated glass slides with a thin layer of carbon and gently pressed these into the soil next to an ant's nest. In the example here, a single ant walked over the extremely sensitive layer of carbon and lifted it where it touched the surface with its feet to reveal footfall as white specks of transparency.

At a smaller scale, soil organisms include microbiota such as nematodes and smaller mites, and mesobiota such as springtails and other microarthropods and larger mites. Such organisms are so small that their activity does not leave a visible trace on the carbon-based devices described earlier. Consequently, drawing on the theme of revealing invisible

Next page:

Complex and multispecies glyph.

Reprinted with permission from Simon F. Park.

processes, I developed a process using bioluminescent bacteria, which naturally emit a blue light and are used as markers in biotechnology. This process reveals the often-overlooked activity of these minute soil creatures. In these works, bioluminescent bacteria are inoculated onto the surface of a device containing agar so that they cover the entire circumference of the medium. The device is then implanted into the top layer of soil and observed over a period of up to four days.

When the microbiota and mesobiota enter the device, they inevitably walk over the bioluminescent bacteria and inadvertently collect them on their feet (or other bodily parts depending on their mechanism of locomotion). Then, as they continue their journey and walk over the uninoculated agar surface, they leave behind a trail of bacteria in their footsteps. Because of the microscopic nature of the footprints and the small numbers of bacteria at the beginning of the process, these tracks are at first invisible. However, after a day or so the bacteria grow into visible points of light (colonies) that now reveal the otherwise invisible tracks that the organisms leave behind. These tiny tracks provide insight into the vastness and complexity of bacterial mobility. Here the bacterially generated light acts as an amplification process that uniquely reveals activities that would otherwise, and perpetually, remain invisible. In the example here, the complex activity tracks were made after the device was implanted into the soil of an ancient woodland (Old Down Wood, Hampshire, United Kingdom) and left overnight on a warm summer's evening. When a similar device was implanted into a neighboring field of wheat, at the same time, the pattern of activity was far less complex and so the process gives an excellent visual representation of the biodiversity of a soil. In his painting *The Exemplary Life of Soil*, Jean Dubuffet similarly sought to give an impression of the soil as "teeming matter, alive and sparkling." His portrayal was the inspiration for this work, and this is reflected in its title as it reveals the traces of, and the activity of, his living and teeming matter.

Bacteria are single-celled microorganisms that are usually only a few micrometers in length. They are the smallest life-forms of the soil biota. Despite their diminutive size, bacteria contribute the highest amount of biomass to the soil (around a ton per acre). A teaspoon of healthy and productive soil usually contains between 100 million and 1 billion bacteria. Bacteria also perform many vital functions in the soil. Many are decomposers that recycle organic matter and are thus essential for the carbon cycle. Some are mutualists that form symbiotic and protective associations with plants (the most well-known of these are the nitrogen-fixing bacteria), and others, the lithotrophs, are involved in the recycling of inorganic compounds.

 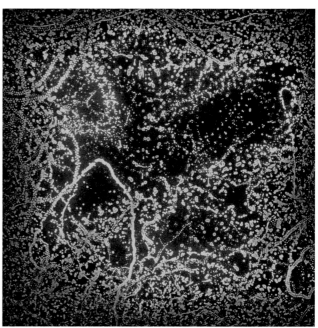

The Exemplary Life of Soil, visualizing the activity of the soil mesobiota and microbiota. Tracks made by minute soil creatures revealed by the growth of bioluminescent bacteria. The tracks after (left) one-day incubation and then (right) after four days.

Reprinted with permission from Simon F. Park.

Jean Dubuffet, *The Exemplary Life of Soil* (1958). Vie exemplaire du sol (Texturologie LXIII), oil paint on canvas, 1360 × 1681 × 65 mm.

Copyright 2017 Artists Rights Society (ARS), New York / ADAGP, Paris

I've worked as a bacteriologist now for over thirty years, and during this time our understanding of bacteria has radically changed. When I began my microbiology undergraduate degree, bacteria were widely regarded as rather simple, ultimately selfish, and very uncommunicative life-forms. However, the idea that bacteria were simple solitary creatures stemmed from decades of laboratory experiments in which they were grown under artificial conditions in the laboratory. We know now that in the natural environment bacteria are able to communicate with each other using a process called quorum sensing, which enables bacterial cells to coordinate their behavior, and which also imbues them, like ants and bees, with a form of social intelligence. In this context, I see the soil bacteria that I use here in my work very much as independent agents and thus as cocreators in the aesthetic process. My role is to understand the bacteria, their characteristics, and capabilities, and to provide the necessary environments that allow them to grow and to flourish. The works are thus generated by autogenic processes that better reflect the reality of the natural world and the soil ecosystem.

The final works here explore the aforementioned concept of the soil superorganism. To make these I take small samples of soil from various locations, which represent biopsies taken from the global soil superorganism. These samples are then spotted onto a rich bacteriological agar allowing the "tissue samples" to grow in the laboratory *in vitro*. In this way, the vital but usually invisible bacterial component of the soil superorganism emerges onto, and colonizes, the agar as a complex and unique (for each different sample) living multispecies glyph. When the same soil samples are diluted prior to plating onto the agar, the individual bacterial soil microbiota within the sample are resolved as a complex of individual, yet still connected colonies. Like Daro Montag's bioglyphs (see Daro Montag's dialogue chapter with Diana Wall (Chapter 32) for more on bioglyphs.), in which soil bacteria interact with and modify photographic film to generate images, in the bacterially generated glyphs here the creativity of the works does not reside solely with the scientist or artist.

To add an additional aesthetic to these works and to reveal the molecular activity of the complex bacterial soil community, I next incorporated a chromogen into the agar. This chemical compound is colorless and nonfluorescent until the enzymes present in the soil bacteria interact with it. These proteinaceous and biological catalysts cleave the chromogen to release a fluorescent compound. This compound is normally invisible, but when it is exposed to ultraviolet light in a dark room it absorbs this energy and reemits it at a different wavelength so that it glows an ethereal blue color. In this way, the bacterial component of the soil microfauna reveals itself through its own molecular activity.

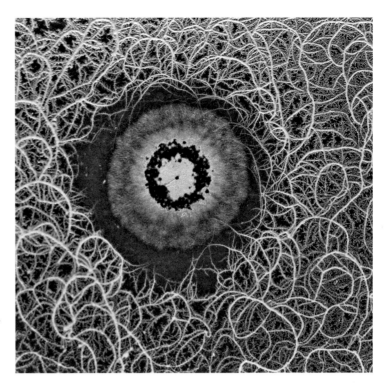

Solaglyphs, visualizing the soil microbiota. A complex and multispecies glyph made by the community of soil bacteria as they emerge from a droplet of a soil suspension to colonize the bacteriological agar. When the soil sample is diluted, and then plated onto the agar, individual bacterial colonies from the soil microbiota emerge.

Reprinted with permission from Simon F. Park.

The Small Turf, visualizing the soil microbiota. A complex and multispecies glyph made by the community of soil bacteria as they emerge from the soil supporting a small clod of turf to colonize the bacteriological agar. The cleavage of a chromogen in the agar, by the bacteria, generates fluorescence, revealing the bacteria via their own molecular activity. The turf observed in (top) daylight and (bottom) in the dark under exposure to ultraviolet light.

Reprinted with permission from Simon F. Park.

Albrecht Dürer, *The Large Piece of Turf* (1503). Watercolor, pen, and ink; dimensions 40.3 cm × 31.1 cm; location: Albertina, Vienna.

Public domain of the Google Art Project.

In the final aspect of this work, I drew inspiration from Albrecht Dürer's watercolor painting *The Great Piece of Turf*. Based on Dürer's meticulous observation of nature, the painting is a representation of a group of wild plants and also a snapshot of a living and chaotic undergrowth. Moreover, the painting does not visually isolate nor distinguish the various plants but represents them, together with the soil, as a distinct and singular thing, much like a superorganism. When a clod of turf is placed onto the chromogenic agar, described earlier, it appears to bloom from its underside, as the complex bacterial community present in its underpinning soil colonizes the medium and reveals itself via its own enzymatic activity. Dürer completed his painting in 1503, long before the advent of the science of microbiology, and would have been unaware of the role played by soil bacteria. And so it is my hope that the work here reveals this invisible and overlooked, yet vital, aspect of soil biology.

Soiled

Reflecting a Natural Body through Socioaesthetical and Biopolitical Viewpoints

An Interview with Suzanne Anker by Regine Rapp and Christian de Lutz

Suzanne Anker works at the intersection of art and the biological sciences. Her variety of mediums range from digital sculpture and installation to large-scale photography to plants grown by LED lights. Her work has been shown at the ZKM Center for Art and Media, Karlsruhe; The J. Paul Getty Museum, California; the Pera Museum in Istanbul; and the International Biennial of Contemporary Art of Cartagena de Indias, Colombia. Her books include *The Molecular Gaze: Art in the Genetic Age*, published in 2004 by Cold Spring Harbor Laboratory Press, and *Visual Culture and Bioscience*, copublished by University of Maryland and the National Academy of Sciences in Washington, DC.

Title image: Suzanne Anker, *Astroculture (Shelf Life)* 18, 2009, Archival inkjet print, 24″ × 36″.

Reprinted with permission from the artist.

Christian de Lutz is a curator and visual artist, originally from New York. Cofounder and codirector of Art Laboratory Berlin (ALB), he has curated over forty exhibitions, including the series *Time and Technology*, *Synaesthesia*, and *[macro]biologies and [micro]biologies*. His curatorial work focuses on the interface of art, science, and technology in the twenty-first century. He has published numerous articles in journals and books. His new publication *[macro]biologies and [micro]biologies: Art and the Biological Sublime in the 21st Century* theoretically reflects on ALB's 2013–15 program.

Regine Rapp is an art historian and curator, specializing in twentieth and twenty-first century art. She worked as assistant professor for art history at the Burg Giebichenstein Art Academy Halle until 2013. She is cofounder and codirector of Art Laboratory Berlin (ALB) and has curated over forty shows and published several books, most recently *[macro]biologies and [micro]biologies: Art and the Biological Sublime in the 21st Century* (2015). Together with her team from ALB, Rapp has coordinated several international conferences, including the Sol LeWitt Symposium in 2011, the Synaesthesia Symposium: Discussing a Phenomenon in the Arts, Humanities and (Neuro-)Sciences in 2013, and most recently the conference Nonhuman Agents in Art, Culture and Theory in 2017.

The American artist and theoretician Suzanne Anker has been one of the most important figures working at the intersection of art and biology. Her work over the last four decades combines inquiry into science and the newest technologies with a strong aesthetic sense. She is also chair of the BFA Fine Arts Department at the School of Visual Arts (SVA) in New York, which features a cutting-edge bioart laboratory. Under her guidance, a mixed group of art and science students took part in the 2015 iGEM (International Genetic Engineered Machine) competition with the project *Soiled*, which won a gold medal for Best Art and Design Project. Informed by the tests experts perform on soil samples, *Soiled* applies color as a metric for analysis. Furthermore, the group sought to design a device that would minimize or do away with the toxic reagents used by most tests. The project included taking soil tests from over fifty locations in New York City.

In 2014 Suzanne Anker was invited by Christian de Lutz and Regine Rapp to exhibit her work in an exhibition titled *[macro]biologies II: organisms* at Art Laboratory Berlin in Germany. This interview is a follow-up concerning the issues emerging from the installation of *Astroculture (Shelf Life)* (2014). The significant issues that materialized were international soil transport, indoor urban gardening, and the effects of light and color on plant growth.

Soiled. Fieldwork SVA student Andrew Cziraki uses an auger to collect soil sample at a desired depth of 6–8 inches in East New York on July 17, 2015.

Photo by Darya Warner.

Soiled. Exhibition view of *Soiled: iGEM International Competition* (February 27 to March 12, 2016), SVA Chelsea Gallery, New York City.

Photo by Raul Valverde.

Reprinted with permission from Suzanne Anker.

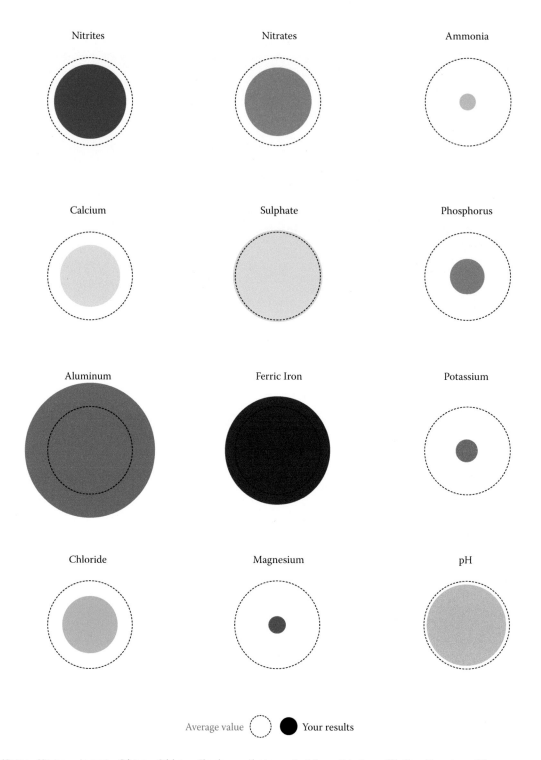

Average value ⭘ ● Your results

Nitrites	Nitrates	Ammonia	Calcium	Sulphate	Phosphorus	Aluminum	Ferric Iron	Potassium	Chloride	Magnesium	PH	
1.2	6.7	1.2	433.8	147.5	0.588	79.5	3.4	67.2	0.473	19.7	5.8	PPM on location
1.6	9.7	7.3	696	161	1.656	59	3.1	298.8	0.828	112	7.11	PPM Average

Soiled. Diagram of elements' concentration (PPM) in soil sample location #2. Sample collected on July 11, 2015, at 1 pm in Sunnyside, New York. Location was a small backyard outlined with a wall of shrubs and plenty of vegetation. Auger reached a depth of 7 inches and extracted a dark brown sample that was moist and moderately coarse.

Design by Raul Valverde. Reprinted with permission from Suzanne Anker.

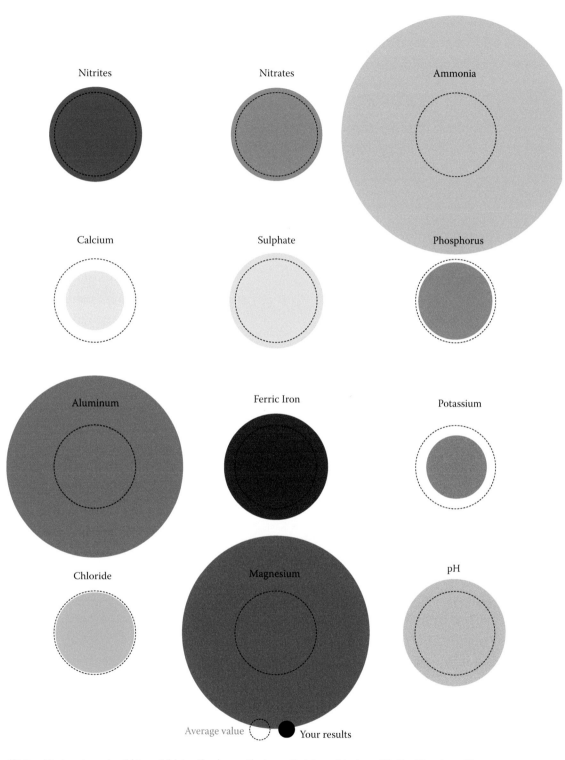

Nitrites	Nitrates	Ammonia	Calcium	Sulphate	Phosphorus	Aluminum	Ferric Iron	Potassium	Chloride	Magnesium	PH	
1.6	9.5	18.4	436.4	162.2	1.350	112.5	3.5	199.4	0.700	227.5	8	PPM on location
1.6	9.7	7.3	696	161	1.656	59	3.1	298.8	0.828	112	7.11	PPM Average

Soiled. Diagram of elements' concentration (PPM) in soil sample location #17. Sample collected on July 18, 2015, at 12 pm in Park Slope, New York. Location was a communal outdoor space outlined with chain fence and two trees. Auger reached a depth of 8 inches and extracted a grayish brown sample that was damp and mildly coarse.

Design by Raul Valverde. Reprinted with permission from Suzanne Anker.

Regine Rapp and Christian de Lutz: There is a long history of using color as a metric in soil science. Field scientists today still depend on Albert Munsell's color charts for determining iron and humus content and guiding taxonomic classification based on color changes in horizons. pH charts also use color as a metric. What is different or innovative about your use of color classification?

Suzanne Anker: The Munsell classification system was applied to the actual color of the soil itself, whereas *Soiled* utilized the resulting color of chemical reactions with the soil extract to indicate the level of nutrient present. Unlike common pH tests that require the user to make a qualitative assessment based on the color they visualize, we used a method to generate more precise quantitative results. Each sample was processed in a spectrophotometer, which is an apparatus that beams an assigned wavelength of light through a solution and measures how much light is absorbed. However, there is a notable difference in the use of color as a metric in the *Soiled* project and color as an experience in the *Astroculture* project. The former being a seemingly objective way to read nutrient composition in the soil, while the latter becomes a manifestation of sensual/emotional phenomena through the disorienting use of violet lighting. Color perception, however, is rife with subjectivity, even in Munsell's classification card samples. Our microfluidic chip attempts to turn color into a quantitative entity.

Regine Rapp and Christian de Lutz: SVA's Bio Art program is unique, bringing together students of fine arts with visiting scientists for projects such as *Soiled*. What were the dynamics within the team in deciding on this project and where it would go?

Suzanne Anker: As a team, we were concerned with working on a project that reflected NYC's interest in urban farming. Many rooftop and backyard gardens are being developed in addition to the already extant community gardens. At first, we thought we could sample soil from parks (and even cemeteries) to ascertain toxic elements existing in community spaces. After further consideration and investigation, it was brought to our attention that soil samples cannot legally be gathered from New York City's public parks. Soil samples from cemeteries pose similar problems. At that point, we decided we would contact students, faculty, staff, and alumni from SVA who expressed interest in having their soils sampled, particularly if they had a backyard or access to a green patch. We also altered the project's scope from toxicity to nutrition. Real estate landlords can be quite contentious in NYC. Thus, any legal ramifications outside of our work were also outside of our scope. We chose soil for a number of reasons, the first of which is that soil is a living substance. We felt that many people would appreciate knowing which nutrients may be missing from their patch of green. Qualities of soil are in many ways dependent on pH factors, because conditions that are too alkaline or acidic affect the solubility of nutrients. Ideally soil should register a pH of neutral, which is seven. The behavior of certain nutrients is affected by changes in pH levels. For example, phosphorous, potassium, and nitrogen do not do well in acidic soil because the nutrients cannot properly dissolve and be available for uptake by a plant's roots. However, the character of soil can also be adjusted by supplementing the proper nutrients.

Regine Rapp and Christian de Lutz: How has your previous work with soil influenced the project?

Suzanne Anker: I am interested in the political and structural implications of soil in that there are specific rules for its transportation among countries. In order to bring soil from one country into another, special documents are necessary. It's almost akin to getting a passport for a plant.

Even when I constructed *Astroculture* at Art Laboratory Berlin, I was prevented from bringing peat pellets into Germany from America. Soil can carry disease since it is a form of life. Working with peat pellets has been inspiring since they begin as dehydrated discs and turn into little pods that resemble pin cushions when hydrated. The emerging shoot wiggles its way out of the darkness. The discs are wrapped in a small biodegradable membrane holding them together. Sometimes the roots comingle with other nearby plants forming a symbiotic relationship, which increases their size. They are acquired as sterilized pellets for starting seeds made from Canadian sphagnum peat moss, a type of soil harvested from bogs. They allow for the proper amount of aeration and drainage for the embryonic seed to sprout. Soil has provided civilization with its complex properties: part microbe, part insulator, and a reservoir for nutrients, water, and oxygen.

Regine Rapp and Christian de Lutz: There has been a lot of hype about "urban gardening" and "urban agriculture," yet the development of cities—especially during the Industrial Revolution—has left a lot of toxic substances in the soil. How was this reflected in test results from different parts of New York?

Suzanne Anker: There is a plethora of toxicity in New York's soil, particularly lead. There are also free soil-testing programs in New York conducted by the soil chemistry department at Cornell University in collaboration with the New York City Department of Parks and Recreation. Our lab is not equipped with the high-end apparatuses for such testing, nor do we have the ability to engage with many reagents since we are a Level I lab. Although we did not test for toxicity we did discover an enormous amount of differentiation regarding nutrients. Some samples were in the range of a balanced set of nutrients, while others showed heavy doses of one nutrient over another (see Data Charts on pages 358 and 359). (The data was visualized by comparing individual results to the average of all the samples processed, so maybe a better word for balanced would be "common.")

The NYC Soil Survey from 2014 was a very cohesive overview of the physical characteristics and patterns of urban soil, but did not include much data on the chemical compositions. While our students in the field did record information about the sites characteristics, we have yet to cross-reference that information with the nutrient profiles, but the project is still evolving. The initial goal for the iGEM competition was to redesign the outdated industry standard for testing soil, so the students primarily focused on conceptualizing a device that would allow the backyard gardener to accurately and safely generate data for a soil profile.

Regine Rapp and Christian de Lutz: Genspace in Brooklyn has been a vibrant meeting point for the worlds of art and science. It is also a very community-oriented project. What role did Genspace play in the project?

Suzanne Anker: We are very collaborative with Genspace, sharing students, instructors, and ideas. Ellen Jorgensen and Oliver Medvedik have even been scientists-in-residents in our lab. For the *Soiled* project Genspace helped us with their social media access, which was our participant base. Daniel Grushkin at Genspace encouraged us to compete in iGEM as well.

Regine Rapp and Christian de Lutz: Beyond bringing in an aesthetic design component into soil testing, what could the *Soiled* project offer an urban gardener? We were particularly intrigued with the "open source" style of the website, which sought, rather effectively, to demystify the whole process.

Suzanne Anker: Our project raised awareness to New York communities. We advanced the idea that soil is a living substance and requires a balanced environment to create healthy produce. Simply because a New York City resident is growing organic produce, unbalanced soil could inflect unhealthy results. Our project provided them with data for further research.

Reaching out to the historical geography of their sites while contacting Cornell University's soil testing programs, and conversing with peers and colleagues concerning their own experiences brought a rich and dynamic extension to urban farming.

Regine Rapp and Christian de Lutz: Is SVA or any of the participating students/advisers planning to continue this project? Will your developed prototype be soon (mass) produced for wider use?

Suzanne Anker: We are in the process of investigating this angle. We are also interested in developing open-source technologies. All of our projects are ongoing, as we continue to learn more about our materials and processes. Our latest project, the *Myo Tomato*, replaces proteins generally found in meat and reprogrammed into a tomato. It serves as a pun on the famous beefsteak tomato. Recently we were also part of the Biodesign challenge and presented our work, along with nine other schools, at the Museum of Modern Art in New York City.

Regine Rapp and Christian de Lutz: What role can you envision art and science collaborations in the future, regarding the complex issues of urban soil?

Suzanne Anker: I can envision a project involving the microbiomes inherent in soil. As a living substance chock-full of microbes, what role do they play? What are they? How do they react or suppress nutrients and even toxicity? Perhaps we

will examine the history of the sites more fully. The types of produce grown and some of the socioeconomic conditions underlying this project are thus augmented and enhanced. Clearly, bioethnicity also has a role as well in soil distribution and the kinds of botanical specimens produced.

Regine Rapp and Christian de Lutz: Was there any interest from your side, while developing the artwork *Soiled*, to reflect on the particular type of soils used?

Suzanne Anker: Because the soil samples we tested were of various types and arrived from diverse locations in New York City, our project reflected the varieties of this cultivating medium and its results. We will continue to research the shifting aspects of soil and what is superior for growing healthy produce. Some plants require a sandy soil mixture while others require a denser variety. Some need a working drainage component while others do not. This is a whole project in itself. This summer, I am testing the soil in various patches in my gardens in East Hampton, New York. Since particular plants favor particular nutrients and conditions, I anticipate seeing what sprouts and when. In addition to the soil autopsies, I plan on putting together a small aeroponic system in the Bio Art Lab at SVA as a control to our tissue culturing projects. Other research will entail light frequencies that encourage photosynthesis and how, and if that alters gene expression in the plants.

Regine Rapp and Christian de Lutz: Talking about soil: Another interesting art project of yours, also linked to soil, is the installation piece *Astroculture (Shelf Life)* from 2009. We at Art Laboratory Berlin had the pleasure to present it in the show *[macro] biologies: organisms* in 2014. And we remember well how ideal the conditions were, as the exhibition space was dark around the opening hours and therefore the seeds in the earth did not have so much real sunlight, but as suggested, the artificial

Suzanne Anker, *Astroculture (Eternal Return)*, 2015. Vegetable-producing plants grown from seed using LED lights. Galvanized steel cubes, plastic, red and blue LED lights, plants, water, soil and no pesticides. 42 × 14 × 14 in (106.65 × 35.65 × 35.65 cm) each set. Installation view at *The Value of Food*, 2015, The Cathedral Church of Saint John the Divine, New York City.

Photo by Raul Valverde.

Reprinted with permission from the artist.

LED-lights, and they did grow very fast. With the work title you refer to the term "astroculture," deriving from NASA's Space Development Program, whose apt trademark label is "Advanced Astroculture™" (ADVASC). NASA had experimented with different forms of gravity and extreme conditions for plant growth in space since 2001. Would you say that NASA has an interest in using a particular soil in their experiments?

Suzanne Anker: NASA has experimented with a number of substrates in order to grow vegetables in space. What is currently favored is the employment of aeroponics, which eliminates soil and sun completely. Since plants survive with the aid of water and nutrients, it is their rooting system, which is being watched extremely

closely in this method. Although hydroponic agriculture has been very popular for eliminating the requirement for soil, it is aeroponics, which employs water in a very light mist form. Under this system roots stay healthier and less prone to disease. As agriculture moves indoors and soil becomes increasingly toxic due to pollution from all sources, it is difficult to accept that soil may not be the most favored substrate for agricultural plants in the future.

Regine Rapp and Christian de Lutz: Are there any other aspects on soil in this or other art works you want to mention here?

Suzanne Anker: Yes, the experiments that NASA is conducting are uncanny in keeping up with the recent movie *The Martian*. NASA has conducted research for growing potatoes with soil from Pampas de la Joya in the Atacama Desert in Peru. The soil, harvested and delivered to the lab, is much like the soil detected on Mars. As interest evolves regarding life on other planets, potatoes emerge as a perfect sustenance for human survival. Are we ready for extraterrestrial farming? Soil may be our research hope. Most recently, I fabricated a large-scale version of *Astroculture (Eternal Return)* in one of the chapels at NYC's Cathedral of St. John the Divine. The piece was composed of thirty-one cubes and grew tomatoes, space lettuce (the same lettuce the astronauts grew), kale, beans, peas, and herbs. Many of the visitors did not think that the vegetables were real because of the fuchsia-colored emanations in the room, which was perfect for a chapel. Priests, schoolchildren, international visitors, and community citizens had a first-hand experience of what can be done in the dark and indoors with plant life.

Nematode State of Mind

Daro Montag in dialogue with Diana Wall

Daro Montag is associate professor of art and environment at Falmouth University, United Kingdom, where he was chair of the Research in Art, Nature, and Environment group (RANE) from 2004 to 2015. His work has been exhibited at galleries in the United Kingdom, Europe, United States, Australia, and the United Arab Emirates, and published in a number of catalogs. In 2002, he was awarded a prestigious art-science prize in Tokyo, and in 2007 he was commissioned to develop an art-science exhibition, which resulted in a project titled *This Earth*. In collaboration with the Centre for Contemporary Art and the Natural World (CCANW) he coordinated a series of exhibitions, conferences, and other events from 2013 to 2016 on soils in the environment. The resulting publication, *Soil Culture: Bringing the Arts Down to Earth*, provides a survey of historical and current artworks dealing with soil.

Diana Wall, University Distinguished Professor at Colorado State University, was appointed as the founding director of the School of Global Environmental Sustainability in 2008. A professor in the Department of Biology and senior scientist at the Natural

Title image: Beacon (film buried on Holy Island), 1995.

Resource Ecology Laboratory, Wall is responsible for helping faculty and students contribute to progress toward a sustainable future. A soil ecologist and environmental scientist, Wall's research explores how life in soil (microbial and invertebrate diversity) contributes to healthy, fertile, and productive soils and thus to society, and the consequences of human activities on soil globally. Her research on soil biota, particularly soil nematodes, extends from agroecosystems to arid ecosystems. Wall has spent more than twenty-five seasons in the Antarctic Dry Valleys examining how global changes impact soil biodiversity, ecosystem processes and ecosystem services. She currently serves as science chair for the Global Soil Biodiversity Initiative.

Dijon, France, was a wonderful setting for the first Global Soil Biodiversity Conference. The conference brought Montag and Wall together with about 700 scientists and others fascinated by the unique organisms that live in soils. From that meeting, Montag and Wall started an e-mail conversation on the spectacular differences of nematodes and pondered whether these and other soil organisms might be aware of their surroundings and could they be more sophisticated than we thought. The points of view from a scientist and an artist were surprisingly overlapping. In the end, both dialogue partners felt enlivened by their different perspectives.

Daro Montag: Having recently met you at the Global Soil Biodiversity conference in Dijon, I'd like to start this conversation by finding out about how you became interested in soil biodiversity. Is it something that you have always been interested in or did you arrive at it through a broader interest in soil?

Diana Wall: As an undergraduate at the University of Kentucky in Lexington, I took a class in microbiology, and as a graduate student there I studied plant parasitic nematodes. I was fascinated that something so small and which seemingly all look the same on first glance through a microscope, basically tiny worms pointed at each end, could on closer resolution show morphological characteristics and look very different. And, it turned out that the farmers and scientists studying these nematodes also recognized that the individual plant parasitic species caused very distinct physiological reactions in crop plants. I thought that it was quite astounding that something so small could be recognized by a farmer looking at a crop plant. That was the beginning of my interest in soil habitats and the variability of soil biodiversity.

Daro Montag: I have also been fascinated by nematodes. Indeed, they have contributed to some of the artworks that I have been making for over twenty years. Could you explain to someone who has not had a scientific training, a little bit more about nematodes? I know that under a microscope they look like very tiny worms, but they have very different roles in the soil ecosystem compared to earthworms that most people are familiar with.

Diana Wall: When I first studied nematodes, I was amazed at how I could see differences from one species to another under a microscope, much like recognizing differences between bird species. I thought it was fairly amazing that these nearly invisible animals, not even the size of an eyelash, could have so many of their own identifiable characteristics. And there are many species. Some are worm shaped, just smaller and tapered slightly at the ends. But much like a science fiction movie, you can see all their body parts—they are transparent and you can watch them moving, their intestine churning, the esophagus pumping, and for females, you can count the eggs in their ovaries. Some species have ornate cuticular "crowns" on their "heads," and there are distinctions between nematodes that feed as parasites on plants versus those that feed on bacteria or are predators. They are also quite beautiful! The plant parasites all have a rigid-looking stylet (protrusion) in their anterior region, and when they jab it into plants, much like a hypodermic needle, the plant can respond quickly by changing water transport or carbon flow aboveground. When this happens, the plant becomes stressed and perhaps its fruit size decreases. On the other hand, predators can be equipped with a tooth or a spear and eat other nematodes or soil organisms. There are many different species all preferentially feeding on one of five main food sources (microbes, plants, other

Next pages:

Daro Montag, *Bioglyph Radiance 2*, 1994. By burying strips of preprocessed film in different soil profiles, Montag allows soil microorganisms to eat away the gelatin surface and leave behind colorful traces. Bioglyphs are a method of recording microbial action in the soil, which varies according to depth, pH value, bulk density, and moisture content. Rather than isolating individual species, the artist records biological activity in concert.

Reprinted with permission from Daro Montag.

The omnivorous nematode *Eudorylaimus antarcticus* from the McMurdo Dry Valley region, Antarctica. Omnivorous nematodes can feed on algae, protists, other nematodes, and when these primary food sources are unavailable, switch to feeding on fungi and bacteria.

Photo: Jon Eisenbach. Reprinted with permission from Diana Wall's lab.

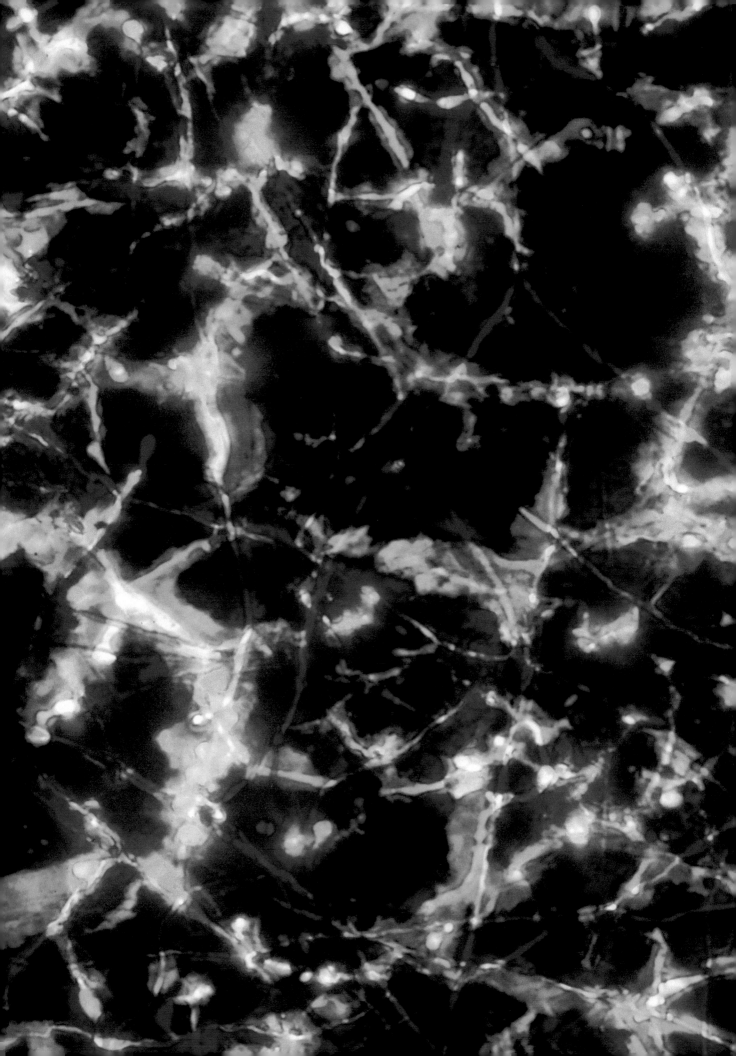

animals, fungi, or feeding as omnivores). And because nematodes are so abundant in soils, they occupy a central role in soil food webs and in the transfer of carbon and nitrogen. For example, as bacterial feeding nematodes graze on their prey, bacteria, they excrete excess nitrogen and release CO_2, thus regulating the populations of bacteria involved in decomposition.

Daro Montag: And yet despite the diversity of nematodes, they are very different to bird species. Birds show a huge range of diversity in plumage, size, and habitat, and of course most have the freedom of the air. But perhaps more importantly, birds also have a relationship with humans; we can relate to them through ours stories, art, and myths. There are very few cultural works that feature nematodes! Despite their vast numbers and significance in the soil, they simply don't feature in our songs or stories. Indeed, whilst there are a number of charities set up for the protection of birds, I think you'd struggle to establish a society for the protection of nematodes.

Perhaps this lack of cultural representation of soil microbes is due more to their miniscule size than the fact that they appear so alien. It is probably because nematodes are microscopic that we have difficulty crediting them with any real identity or personality. If, on the other hand, we get to see them alive, and greatly enlarged through a microscope, it is much easier to appreciate them as living entities, going about their business. So, although they are alien and appear to be the inspiration for a number of science fiction films, I think we struggle to relate to them simply because they are beyond our range of vision. If we were able to see them more readily, I suspect we would understand that they are doing what most animals do—searching for food, seeking a mate, avoiding unpleasant experiences, and generally "making a living." For me, it was only when I was able to see soil microbes through a microscope that I really began to appreciate them as individuals in their own right. This was one of the driving motivations behind my developing a creative method for making their activities visible through art.

Diana Wall: I have to agree that it is hard to tout a microscopic species, whether in soil or air, to the public as an icon for conservation. But we are working on it! It is exciting to think of how we as scientists who admire soil organisms might share what we see with the public. It is also fascinating to share the knowledge of these superstar organisms in soil. Every minute we sleep and beneath every footstep we take there are many types of organisms working together, much like an assembly line.

And I think there are a diversity of efforts, including yours as an artist, that are important to increasing public knowledge and appreciation of soil species over time. There are microscopes with big screen monitors so everyone can see what a scientist sees. There are other scientific tools that allow us to see soil and the many active animals moving through and across roots. These pictures and videos are being circulated on social media to a ready audience that can watch a tardigrade (water bear) moving its claws or a mite lumbering over a soil particle. Pictures that can be downloaded become images that can once again ignite curiosity and exchange of information about soil biodiversity. What does that funny little nematode do and why could that organism be important? I think we can introduce the organisms and how each kind is different—some crawl, some wiggle, and they all vary in shape and size. And the ways to present all this knowledge will just increase. The new Global Soil Biodiversity Atlas,[1] for example, presents images with information on different taxonomic groups, along with maps to show hotspots of biodiversity. Another example is a set of playing cards that was developed and is now online.[2] That cards are unique—one for an earwig, another for a collembolan … With advances forthcoming in scientific tools and technology and the Internet, there will be thousands of creative ways to share an image to the public, from social media campaigns to games and calendars and art. Perhaps we will soon have a nematode as a charismatic soil icon!

Daro Montag: One thing that intrigues me about these tiny creatures, in fact all creatures living in a handful of soil, is the question of consciousness. Whilst most people have little difficulty attributing the notion of mind to fellow mammals and larger, more visible species, I suspect the idea of "nematode mind" is not debated much amongst biologists—or maybe it is and I'm just not aware of it. But when you know these animals so well and look closely at them, do you ever see some level of consciousness at work? It seems that, just like larger animals, they are making decisions as they swim through the dampness. They must have a rudimentary sense of their surroundings and have some sort of inner experience. I'm not suggesting they possess a degree of self-awareness that is equivalent to higher organisms, but in some small way they have a form of sentience. And if this is the case, it significantly changes how we think about the soil and its community of inhabitants. Is this something that is discussed amongst scientists?

Diana Wall: I can't specifically remember any discussions on "nematode mind." However, I do know that many parasitic soil nematodes wouldn't survive without their insect hosts, and use sensory cues to find them. Whether they actually have sense of self-awareness that is equivalent to higher organisms is something to investigate further. But it is pretty exciting, for example, to know that microscopic parasitic nematodes have chemosensory, thermosensory, and other sensory abilities. Some species of soil entomopathogenic

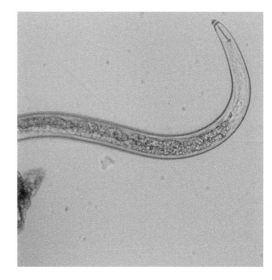

A plant parasitic nematode, *Helicotylenchus* spp., inserts its stylet into plant roots to feed and can damage the root and alter root architecture and plant metabolism and growth.

Reprinted with permission from Diana Wall's lab.

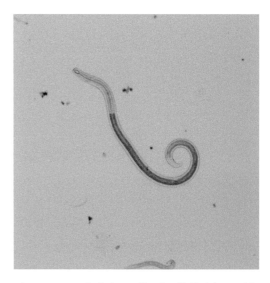

This predaceous nematode has a tiny tooth that is used to feed on protozoa and other nematodes and small mesofauna. Predaceous nematodes, like other predators, are at the top of nematode food web.

Reprinted with permission from Diana Wall's lab.

nematodes (nematodes that parasitize an insect larvae and kill it) are attracted to hosts by olfactory cues such as CO_2 and other volatiles. They can almost smell their way to their hosts. Only recently it was shown that besides being attracted by CO_2, individual nematode species differ in their response to odors from their insect

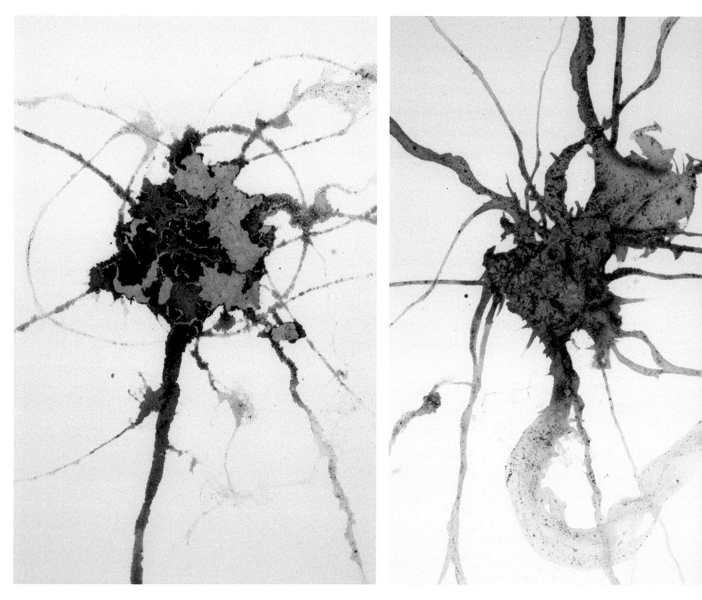

Daro Montag, *Worm Squirms 1–4,* 2010 (detail). Worm drawings with blackberry juice.
Reprinted with permission from Daro Montag.

hosts. So very specific interactions have evolved, which we might want to think of as one parasitic nematode species being more aware of a nearby host and seeking it out than the other parasitic nematode species. That isn't quite what I would call self-awareness, but I certainly would say the interactions in soils are more highly evolved than we appreciate!

Daro Montag: That is fascinating. I wasn't aware of how sophisticated nematodes are when it comes to perceiving their surroundings and

responding to that perception. We tend to think of perception as something that happens through the types of organs that we are familiar with—the eyes, ears, nose, and skin—but it's worth remembering that other creatures have evolved different sensorial abilities that are appropriate to their surroundings. Although I suspect we'll never get inside the mind of a nematode (or even another human being for that matter), it is incredible to imagine each individual nematode making rudimentary decisions based on an internal experience.

Daro Montag, *This Earth*, installation view. In: This Earth, Sherborne House Gallery, Dorset, 2006.

Perhaps not an awareness of "self" as we might understand it but certainly a form of awareness. What I find so appealing about this knowledge is that it gives me a way to connect with a creature that is, in all other respects, totally foreign. It is this thought that propels much of my art making, which is often done in collaboration with the nonhuman. By enabling other species such as earthworms and nematodes in the actual production of the artworks, I try to raise questions about creativity, consciousness, and what is meant by those terms. I tend to believe that creativity is not purely a human attribute but an inherent property of the living world. I'm not suggesting that other creatures make works of art that can be measured against our own cultural standards. I'm not even suggesting that "art" exists outside of the human world. But creativity is a separate matter to art, and that intrigues me. If creativity can be understood as a natural propensity of living matter, then this idea connects us to nature, and in this case the living biome of soil habitats, in a very deep way.

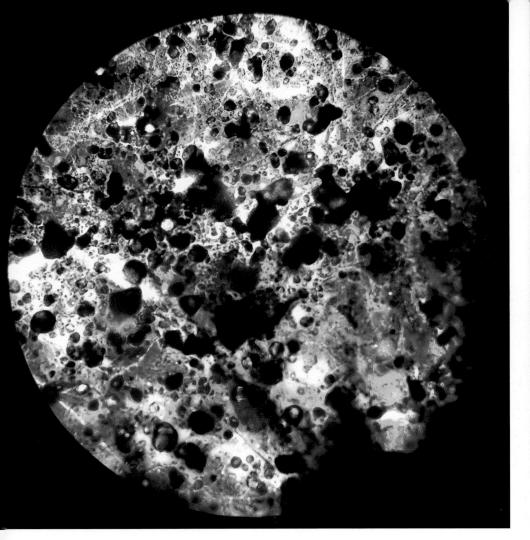
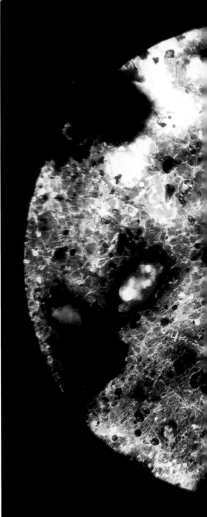

Daro Montag, *This Earth No. 2, No. 9, No. 10,* 2007. Bioglyphs made by burying film in soil.
Reprinted with permission from Daro Montag.

Diana Wall: I talked about this aspect with my great post doc and executive director of GSBI, Elizabeth Bach. As scientists, we believe that items that may appear beautiful or unique to our human eye may not necessarily reflect conscious creativity on the part of other organisms. Nevertheless, we appreciate the beauty and can marvel at it. Humans typically connect to the natural world through beauty (e.g., sacred landscapes and features, charismatic animals, intricate flowers) and we are learning day by day that beauty occurs in the organisms in the microscopic world of soil. Soil organisms can be creative in a literal sense, creating burrows, mounds, and other physical features in the soil, but these transformations are entirely functional, or a by-product of some essential activity that has evolved in that group of organisms—termites and earthworms come to mind. I do not suggest that soil organisms create for the sake of satisfaction or enjoyment!

Daro Montag: Oh, I agree. I would be very surprised to learn that soil organisms gained enjoyment from their toil. Nor do I think we should confuse beauty with art; art has long since separated itself from conceptions of beauty. Instead what I am trying to suggest is that creativity is a special attribute of living matter. In this sense, it is perhaps more akin to problem

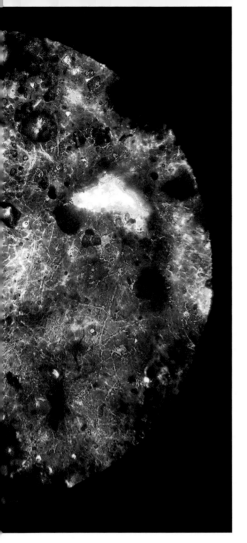
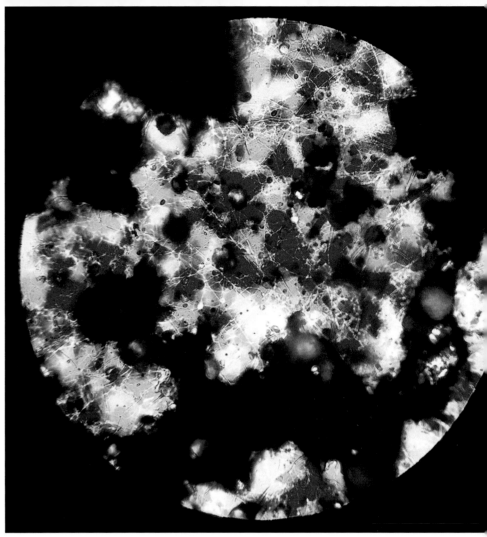

solving, that is, encountering a problem in the (physical or mental) world and devising a solution. Each action is essentially unpredictably open, and each organism might respond in a subtly different way as a result of its own internal "thinking," which is informed by a lifetime of experience. And although soil organisms may not be driven to create for the sake of satisfaction, who knows? There may be some very small satisfaction for a nematode as it swims along a trail of scent cautiously avoiding dangers along the way … Perhaps we should end with the idea that by focusing on the microscopic, and appreciating these tiny creatures as entities in their own right, we can gain a renewed respect for the soil and the miracle of life that inhabits it.

Endnotes

1. A hard copy of the Global Soil Biodiversity Atlas can be bought at https://publications.europa.eu/en/publication-detail/-/publication/c54ece8e-1e4d-11e6-ba9a-01aa75ed71a1. Web version of chapters can be downloaded from http://www.gessol.fr/game-hidden-life-soils.

2. Links for playing cards. Original version in French for GESSOL research program, set up by the French Ministry for Ecology: http://www.gessol.fr/English. English and Portuguese translations available from GESSOL website: http://www.gessol.fr/game-hidden-life-soils. The GSBI website has a download available at https://globalsoilbiodiversity.org/education.

On Color Hunting

The Rhizosphere Pigment Lab

Natsai Audrey Chieza and John Ward

Natsai Audrey Chieza is founder of Faber Futures, a creative research and development lab working at the intersection of design and biotechnology. Its mission is to leverage biotechnology for sustainable material futures by embedding design thinking into the emerging bioeconomy. Chieza's *TED Talk* highlights the new approach to material fabrication through biotechnology and how this interacts with contemporary challenges of resource scarcity and sustainability. Chieza holds an MA (Hons) in architectural design from University of Edinburgh, and an MA in textile futures from Central Saint Martins (University of the Arts London).

Title image: The rhizosphere is the section of soil that is directly influenced by root secretions and associated soil microorganisms. The notion that laboratory protocol can be reappropriated as a craft-led design by designers and the concept of material provenance from rhizosphere microecologies reveals the complex systems designers will need to engage with as they seek to drive narratives about how biology could be best used as technology for sustainable material futures.

John Ward is a professor of synthetic biology for bioprocessing in the Department of Biochemical Engineering, University College London (UCL). He has international recognition in synthetic biology and microbial biotechnology. He has a degree in biochemistry and a PhD in microbiology from Bristol University and did postdoctoral research in Manchester before joining UCL in 1983. His current research interests are using enzymes in biocatalysis, cell engineering for better bioprocesses, and using phages and proteins for biomaterials.

Chieza and Ward have a longstanding collaboration that aims to combine design thinking with biology to develop textile dye protocols from pigment secreted by *S. coelicolor* bacteria. They've established a set of reliable interventions that permanently transfer colorfast biopigment without the use of chemical fixatives, using a fraction of the water resource ordinarily required in industry. The Rhizosphere Pigment Lab is an iteration of the project that explores a wider context for scientific method: how the protocols for bioprospecting for pigment-producing actinomycetes interplay with similar concepts in design and how biology is beginning to inform a new way of thinking about the design process. The outcome of this project was an installation commissioned by Science Gallery Dublin for *Grow Your Own Life After Nature*, 2013.

In a quest for colors found in nature that would inform Issey Miyake's spring/summer 2009 collection, Dai Fujiwara and his creative team took to the Amazon to match 3000 color samples with the authentic hues found within the rivers' surrounding habitat. *Colour Hunting in the Amazon* documents their expedition on land and water. Vibrant textile color swatches were matched to the mosses growing on the trunks of fallen trees on the forest floor, while fresh flowers and leaves added bright interjections of magenta in contrast with wild glossy chartreuse greens. Leaves at different stages of decay gave an array of earthier shades and muted tonalities. In witnessing the plethora of color smattered on screen throughout this short film, the notion of ever describing a dustcoat as just "brown" is made to the viewer as carelessly absurd.

The sound of a paddle softly slapping the water's surface tells us the source material has changed: this time, colored strips are hung like bunting across narrow waterways, and when they disappear or blend into the fluid background of the river, a color is matched. As the river widens, so too does the mode of transport change. A speedboat decisively slices unambiguous flesh tones from this body of water. Fujiwara and his team eventually edit the color samples to just eight for a fashion collection that can be mapped back to a particular line of geographical and cultural provenance.

This kind of sensitive exploration into materiality is typical of design research methods that merge research-driven problem solving with an instinctive intuition fueled by creativity. Fujiwara abstracts the scientific act of bioprospecting, relocating it in the emotiveness of the craft processes that mediate the existence and interpretation of designed objects. In this instance, bioprospecting at the continental scale of the Amazon—being able to access authentic qualities of nature in a radically proactive way—provides a tangible materiality. This active engagement with natural systems forms part of an evolving trajectory of humans giving form to matter and embedding meaning at different scales.

A more precise understanding of bioprospecting originates from the scientific community. Here, to bioprospect is to search and identify and document those substances made by living organisms that may be of medicinal or commercial value. In this context, unlike in Fujiwara's expedition, not all of nature's commodities (both material and immaterial) are immediately visible. In the natural world, bacteria tend to be invisible to the naked eye. Knowing which habitats they occupy, how to isolate them from those environments, and what they can do for you in a sanitized laboratory takes highly specialized training and practice in biology. How then does a designer whose knowledge system is imbedded in the macroscales of culture, society, economics, and philosophy access the invisible potential of a microorganism to create new materials and processes that can, like Fujiwara's, be applied to textile design and thus have both commercial and cultural value?

Provenance is a powerful component of design thinking and making. Knowing the origins of a material helps to determine its value, cultural capital, and natural history. It tells us how it can be fashioned and therefore what it can become. Provenance enables new narratives to form around an object's place in the past, present, and future. As we leap into a future driven by the fourth industrial revolution, synthetic biology will drive the biofabrication of consumer products, ushering in a new context from which to understand materiality. At a DNA level (micro), how we engineer a protein will have direct consequences on a global ecosystem level (macro). And since this new age of living technology is emerging in parallel with anthropocentric narratives of resource scarcity in a world of megapollution abundance, there is urgent need for wider transparency and engagement on the imminent redesign of biological systems for our material environments.

The Rhizosphere

From the Greek "pertaining to roots," the rhizosphere is the soil zone immediately surrounding the roots of plants and trees. It is rich in microorganisms, both bacteria and fungi, which are often in symbiotic relationships with the plants whose rhizosphere they inhabit. Many plants exude sugars, amino acids, and other compounds from their roots, which encourage the growth of specific species of bacteria and fungi in the rhizosphere. The bacteria in turn sequester molecules such as iron from the soil and pass these on to plant roots. Young plants can secrete into the rhizosphere up to 40% of the sugars that they build in their leaves using sunlight, water, and CO_2 so that this microbially rich zone around plant roots is a rich hunting ground for microbiologists looking for novel bacteria.

Many of these bacteria are defined as plant growth promoting bacteria (PGPD) or plant growth promoting rhizobacteria (PGPR). They help plants by fixing nitrogen from the atmosphere and passing it on to plants in the form of easily assimilable nitrogen compounds. They also kill or prevent the growth of pathogens by secreting antibiotics and binding iron from the surrounding soil and providing this to the plant root for transport up to the leaves.

The Rhizosphere Pigment Lab was commissioned by Daisy Ginsberg for Grow Your Own Life After Nature at Science Gallery Dublin 2013. The exhibition invited dialogue and debate on the emerging applications of engineering life through synthetic biology from engineers, designers, scientists, artists, and biohackers. The curatorial team mirrored the multidisciplinarity of this emerging field: artist and designer Alexandra Daisy Ginsberg, Anthony Dunne (Royal College of Art), Paul Freemont (Imperial College), Cathal Garvey (biohacker), and Michael John Gorman (Science Gallery).

This work articulates the unique protocol involved in determining what color a tarragon plant might provide within its microecology that a rosemary or mint plant cannot, and vice versa. What kind of tools would a bioprospector need to recontextualize sourcing raw materials of life for their application in design? A three-stage experiment illustrates: the botanical provenance of the bacteria, the evaluation of pigment produced by the microbial colonies from each plant, and the corresponding selected samples that actively dye silk scarves *in vitro*.

The bacteria in the rhizosphere surrounding several plants were isolated using the solid agar media R5 containing nystatin at 40 micrograms/ml, which we have successfully used to stimulate the pigments from the soil bacterium *Streptomyces coelicolor*. Soil (0.2 g) from the roots of mint, tarragon, basil, oregano, and rosemary plants grown in 9 cm pots was suspended in 10 mL of sterile water. Dilutions of the soil suspension were made by taking 100 microliters of the first suspension and placing in 10 mL of sterile water and mixing. This was repeated three times to give a 10^{-6} dilution of the original suspension. Then 100 microliter amounts of each dilution were spread onto the surface of R5 agar plates containing nystatin in 9 cm petri dishes. The plates were incubated at 30°C. Bacterial colonies were picked off with a sterile inoculating loop and restreaked onto R5 agar with nystatin.

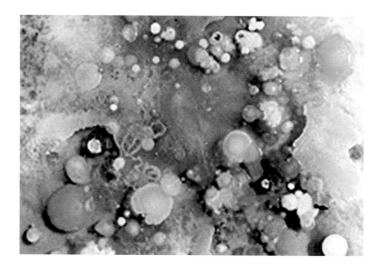

Bacterial groups such as *Actinomycetes*, *Pseudomonas*, *Bacillus*, *Rhizobium*, *Agrobacterium*, *Xanthomonas*, and *Burkholderia* are among the bacterial groupings found in association with plant root systems. Of the *Actinomycetes*, *Streptomyces coelicolor* is one such example. *S. coelicolor* produces an antibiotic called actinohodin, which changes color from red to blue depending on the acidity of its environment. The name *coelicolor* comes from the Latin "sky color."

The authors have collaborated on experiments with *S. coelicolor* to better understand its natural provenance and how its pigment can be applied to textiles and function as a dye with reduced water use and the elimination of toxic substances. The experimentation seeks to develop and refine protocols to mediate replicable growth conditions of *S. coelicolor*.

The project is an early example of the new terms of engagement of nonscientists working with innovative processes for materials design in the field of biotechnology. If biology is now technology, meaningful engagement with its tools requires both designers and scientists to pick up new tools and become fluent in each other's vocabulary of thinking, making, and validating.

The Rhizosphere Pigment Lab invites the audience to witness the alchemy of the "unseen" emerge through a unique collection of biologically colored and patterned silk scarves. While charting the progress of this live experiment, these fluid fabric forms illustrate how research, science, and design are defining new craft processes with living systems.

Part of the designer's training in the laboratories involved learning and implementing protocols on bioprospecting from various herb plants cultivated. Being able to examine part of the microbial ecosystem of a root structure unique to a species of plant, and then isolating pigment-producing bacteria, fungi, and yeasts from those samples in a quantifiable manner was a key moment of discovery.

If a universal protocol for the bioprospecting of organisms from the soil as a site of provenance exists, how might it inform the development of fermentation processes of cultured dyes from *S. coelicolor*? And how, if the methods can be easily adopted and distributed, will the textiles produced reflect the culture of the people who choose to work with it in the production of their own textiles? To successfully translate isolated bacteria into the textile dye production protocol, to ascertain color pallet, and then to transform this into material artifacts is to understand the kinds of methodologies designers will need to explore as they interface with biological systems as collaborators, tools, and platforms.

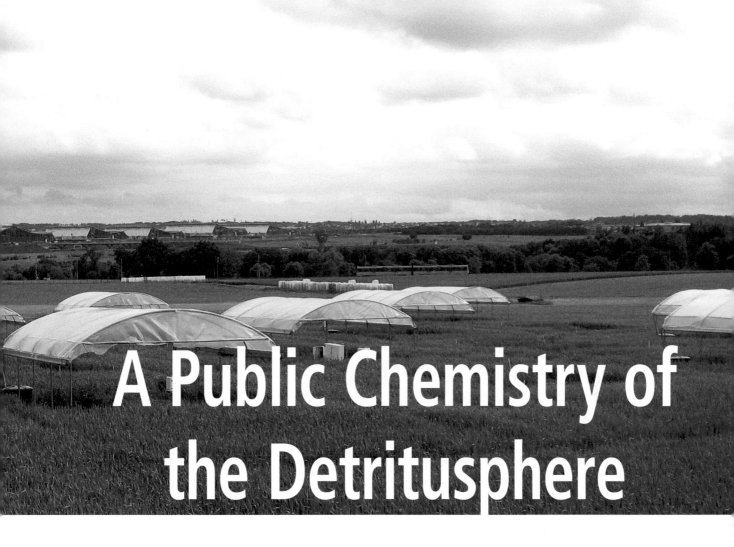

A Public Chemistry of the Detritusphere

Ellen Kandeler and Georg Dietzler

Georg Dietzler is a Cologne-based artist-curator recognized for cross-disciplinary cultural projects, conferences and seminars. Working internationally as a social, environmental and political artist; he integrates research and practice to develop future scenarios. Dietzler is recognized for experimental artwork on landscape issues, nature-culture interrelationship, natural studies by landscape meditation and the use of mushrooms in bioremediation.

After a position as a researcher at the Federal Institute for Soil Management, Vienna, **Ellen Kandeler** moved to the University of Hohenheim, Stuttgart, Germany, as a professor of soil biology. She is author or coauthor of over 200 scientific journal publications. At Hohenheim, she maintains an active teaching program focusing on soil biology and soil science. Her current research program focuses on the small-scale distribution and activity of microorganisms in soils as well as on microbial responses to environmental change (climate, management, heavy metals).

Title image, previous page: Hohenheim Climate Change (HoCC) experiment (Heidfeldhof, University of Hohenheim, Stuttgart, Germany).

This station was established in the summer 2008 to manipulate both soil temperature and precipitation on an arable field. This field experiment allows investigating carbon cycling in an agricultural ecosystem adapting to both future climate scenarios.

Photo: Christian Poll.

Ellen Kandeler and Georg Dietzler in front of the University of Hohenheim, Stuttgart, Germany. Reprinted with permission of the authors.

On a sunny October day on the campus green at the University of Hohenheim, Ellen Kandeler and Georg Dietzler embarked on a tour of the collection of soil profiles, soil maps, and the laboratories at the Institute of Soil Science and Land Evaluation, and chatted extensively about their backgrounds, personal history, research interests, and the driving forces behind their work. Dietzler related how his fascination in research-based arts and scientific topics is deeply rooted in his childhood growing up in the Rhineland-Palatinate countryside, and spending time with foresters, farmers, and hunters. When he moved to Dortmund he found himself disconnected him from nature while living in a region famous for its heavy steel and coal industry. He became aware of the contaminated industrial fallows and began making artwork as a way of reconnecting with nature, beginning his interest in the detritusphere.

Kandeler's research on soil and fungi pays more attention to rural, agricultural, and forested sites, but she talked about what she had learned through a research project on urban soils in Stuttgart. Railway station soils and some soils on the outskirts of Stuttgart were tested to observe whether microbes were still alive or largely reduced. She examined railway stations where many organic pollutants are present and revealed that bacteria as well as fungi are heavily affected by different chemical compounds. Her study did not find much adaptation of soil microorganisms in these polluted sites. This kind of research is always accompanied by a close description of abiotic soil properties. For example, it is important to know whether a pollutant is mobile and can be transported within a soil profile or whether the pollutant is tightly adhered to material surfaces and is consequently not accessible to soil microorganisms.

Despite the largely urban and industrial focus, Dietzler has also worked in rural contexts, such as a project called ZERO EMISSION (2006) in South Korea. Oyster mushrooms were used to transform crop and organic waste into highly fertile soil. The regional cropping systems are used for food production as an additional income source for farmers. Therefore, organic wastes from food plantations, such as rice, coffee, cocoa, and sugar, among others, can be used for bioremediation. By growing regional edible mushrooms, most decomposed remnants of plant crop waste can be transformed into nutrient-rich food for animals (e.g., chickens and pigs) or into highly fertile soil in a process that takes more time. Nutrient-rich mushrooms feed the community or offer additional income.

Kandeler and Dietzler then talked about exploring aspects of the natural world that are normally invisible, exploring formats for sharing knowledge on approaches and strategies in the arts and sciences. They discussed aspects of human experience and emotion that usually fall outside of language: limits of understanding, collectivity and conflict as tools, and the transformative potential of science, art, and culture to imagine the world differently. Kandeler's scientific and Dietzler's artistic research focus on showing, introducing, and stimulating discourse on manageable options, and on solutions that matter for improving environmental impacts of various kinds.

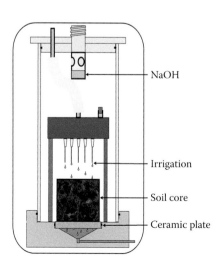

Soil biologists at the University of Hohenheim use microcosms to investigate the response of microorganisms in the detritusphere to environmental changes. A small rain machine provides irrigation; suction at the bottom allows collection of leachate. Litter on top of the soil core stimulates microbial growth and improves microbial degradation of organic pollutants.

Reprinted from Poll, C., Pagel, H., Devers, M., Martin-Laurent, F., Ingwersen, J., Streck, T., and Kandeler, E., 2010, Regulation of bacterial and fungal MCPA degradation at the soil–litter interface, *Soil Biology and Biochemistry* 42, 1879–1887.

During their tour, Kandeler showed Dietzler some research tools, such as basic soil cores to sample undisturbed soils and complex experimental setups in the lab. She then presented one of her simulated microcosms, which allows one to study the activity of soil microorganisms in the detritusphere under controlled lab conditions. She explained that her group used these microcosms to investigate the decomposition of herbicides at the soil–litter interface. Kandeler described in detail her research on biogeochemical interfaces, her research on the detritusphere. She explained that the detritusphere is defined as one of the most important hot spots in soils where up to 90% of decomposition of litter material and other organic material takes place. Consequently, this microhabitat is colonized by a high biodiversity of bacteria and fungi responsible for the degradation of any organic material. This sphere is also predestined to contribute largely to the detoxification of organic pollutants. Dietzler described his self-decomposing laboratories as closed-circle systems that are more related to green architecture, earth ships, and complex systems. For example, the mushrooms he has used in several installations, *Pleurotus ostreatus (oyster mushroom),* have the unique ability to cause an "organic blasting" of highly toxic PCBs into nontoxic substances without absorbing toxic agents into the gourmet mushroom's fruit body. Mushrooms on such sites can still be harvested and eaten.

When it comes to field experiments Kandeler and Dietzler both involve teams of various specialists. For the art projects, Dietzler seeks local or at least regional environmental agencies and laboratories as consulting engineers for monitoring the degradation of PCBs in various climatic zones. In the end, both dialogue partners concluded that they are looking at the same topic from two different perspectives. While Kandeler tries to understand why organic pollutants are decomposed to a higher extent by fungi when a second organic substrate is available, Dietzler hopes to investigate through his art that fungal decomposition is aesthetic as well as functional and highly relevant for the general public. After their tour of the campus, they finally settled into Kandeler's office to start recording their dialogue.

Georg Dietzler, Oyster mushroom, *Self-Decomposing Laboratories*. Architectural sculptures are used as living machines for site remediation in various climate zones. Straw, clay, lime, wood, oyster mushroom cultures, nutrient solutions, and PCB-contaminated soil, among others are used as time-based sculpture materials of alternative architecture. Setup are controlled experiments and monitored by local or regional environmental labs due to regional health and safety regulations. PCB-contaminated soils are processed by regional oyster mushrooms as a method for ecological restoration. Could this knowledge be transferred to temporary site-specific artworks designed for industrial sites and former army bases as a method of ecological restoration? What is needed for large-scale projects? Since 1992 Dietzler has been researching such possibilities under various climatic conditions. Based on a scientific research project by the University of Göttingen Faculty of Forest Sciences and Forest Ecology, Germany, 1988/89.

Georg Dietzler, *ZERO EMISSION*. Oyster mushrooms transform crop and organic waste into highly fertile soil and can be used as regional cropping systems for food production and as additional income sources for farmers. Agricultural monocropping, like rice, coffee, cocoa, and sugar, produce tons of unused crop waste. By growing regional edible mushrooms, most decomposed remnants of plant crop waste can be transformed into nutrient-rich food for animals (e.g., chicken and pigs) and, in a longer lasting process, into highly fertile soil. Nutrient-rich mushrooms feed the community or offer additional income. Images from ZERO EMISSION, South Korea, 2006. Decomposing architectural sculpture, materials: Wood (logs and branches from the mountain), crop waste (rice straw, rice hulls as nutrient medium) from rice fields/mills of the Gongju region, jute textile, and regional oyster mushroom cultures.

Research and Art Outreach to the General Public

Ellen Kandeler: I think we'd probably both like to live in a better world. I am interested in understanding why a toxic system might change into a nontoxic system, and you seem to be making work to make these things visible. Therefore, from this perspective we have similar interests. How do people know you are an artist and not just a biotechnologist trying to improve soil systems? I think artists are always interested in the reception of their art. They need feedback from a larger number of people. This is a big difference between artists and scientists. On the other hand, scientists are happy to go to conferences with a small group of maybe the fifty best soil scientists worldwide. If they agree with your results, then you are happy. Is it true that you also need feedback from other artists about your system or your approach, or are you happy with a broader feedback from the general public? And what is your real aim? Do you want to change the world and improve the environment, or do you want to be accepted as an artist who produces very interesting sculptures?

Georg Dietzler: My approach might not be so different from yours. As an artist, it is possible to introduce some ideas to a wider community. I don't know how scientists communicate their research, as it seems less accessible to the general public. Publications of work and presentations at conferences are important for a scientist. For me as an artist, it is also important that my work is published. Currently, I consider it more important to have my work recognized by the scientific community than to be exhibited in large museums.

In Finland, my colleague Tuula Nikulainen and I had an amazing experience working on a funding application for the science-orientated Nessling Foundation in Helsinki in 2006. It was for a process-based public artwork in Salo by the NY-based artist Jackie Brookner. She would focus on cleaning up a sewage pond using bioremediation in collaboration with the Scottish Findhorn Eco-Institute. When the grant had been accepted, they called us up, it was the first time that they would fund a public art project! They were curious about the topics and issues, the values that we introduced into the work. The art questions were not the same as the science questions they would normally consider.

Ellen Kandeler: I think it's a good idea that art stimulates questions for science. Science and art can be complementary—we are both looking deeply into mechanisms to understand the biology and chemistry of the decomposition of organic pollutants and you are able to visualize these processes.

Georg Dietzler: Last week I had a really interesting experience in Belgium. Some months ago, I was asked if I could imagine a seminar for Flemish politicians on how to run pilot projects between arts and environmental practitioners. Some members of the Flemish government wanted to start pilot projects. The idea was to go to an investor from an industrial site to develop the site for housing. In the beginning, investors, artists, architects, and urban planners would collaborate with each other. Scientists are still missing from the conversation, but bringing these different people together provides a completely different way of thinking about the site and reusing the site. From the investor's side this is quite a new approach. I hosted an all-day workshop there and, you know, in the beginning the administration was wondering what will happen if we include people who have different ideas. If you want to connect with people in the region, you have to listen to them. You have to come into dialogue with them. If you tell people, "well, we are going about it this way because we are absolutely sure that this is the right way," you are not offering people the opportunity to become active in the process. Get people interested in taking up a responsible stewardship. So, there are different possibilities when reaching out to a wider audience. Is it possible to consider a system of stewardship? It might not be so easy because stewardship needs a completely different time scale and amount of attention.

Ellen Kandeler: It is a very difficult task trying to make a rather complex matter understandable for society without losing the truth. It has become very popular to exchange knowledge with different members of the society. For many years now the University of Hohenheim has invited children to learn more about specific research topics. For example, Iris Lewandowsky explained to children how plants could save the world. The university chooses a variety of topics: Do we need money in our society or not? What does fertilization mean? How does food grow?

In the future, we should try to work on direct interplay between science, art, and society. Currently, direct contacts between soil scientists and art are rare. You just have to look at the structure of universities. You will never find the institute of soil science in the same building as arts and music departments. Therefore, I like this project of writing a book. It is a way to broaden our views and to get the importance of soils into society.

Georg Dietzler: For me, it is important to sow ideas and let them grow. This means that the people I am working with also have to be active, which is not always possible, but opens up new possibilities for people.

Ellen Kandeler: I see teaching as another way of trying to stimulate ideas. This morning I met a student in my office that had hundreds of different ideas. Now is the time for me to support this person in order to select the best idea. So, with teaching one can say it is a sowing of ideas, but it is also guiding someone through their ideas with expertise in certain aspects.

Georg Dietzler: Having had the tour through all the different labs in your department, the first thing that comes to mind is how much computer technology is involved in your work. In my art projects, timelines and observation of phenomena are essential. Hasn't this also been an important principle of scientific research, its findings, and discoveries? How much time do you have to do research without computer technology? Nowadays, which role does spending time to observe phenomena have?

Ellen Kandeler: I think time for direct observation was much more important in the past. Still I think this is the basis for all kinds of research. We have to look at things in detail. If you are interested, for example, in soil structure, you did it by micromorphological analyses. This can now be done within several minutes by using computer tomography. The methods changed, but still the aim of the research is similar: You want to know the distribution of clay minerals, organic matter, and earthworm burrows.

The great advantage of computer technology is the fast exchange of information between researchers. In earlier times, we really had to visit our colleagues and travel by train for several hours and back again. Back then I sent manuscripts by the postal system. Nowadays up to twenty people can work on one manuscript at the same time. Therefore, science is totally different in comparison to twenty or thirty years ago. Years ago, one person mainly had one idea, did a nice experiment to cover one nice story written slowly within one year. He or she was able to cover all the steps from the experiment, going to the lab by him- or herself, doing all the measurements, making all the figures by hand, doing the statistics, writing the draft of the manuscript, and finally sending it to a journal. Writing a manuscript has changed a lot due to the fast exchange of drafts between different researchers, but I still like both concepts that a researcher can work on his/her own story or that interdisciplinary groups can develop a joint manuscript.

Modern computer technology also allows us to search for new topics within a very short time period. In former days one had to go and ask the library for a specific book, and if it was not available, one had to go to another library. To get into a new topic it probably took half a year. For example, once I decided that I was interested in copper pollution in vineyards, and having no knowledge of this topic, I would have to spend half a year starting with literature research and then start an experiment. But now, I can promise you, that within one week I am a specialist in any topic you can suggest. It could be copper in vineyards, it could be the distribution of a certain microorganism to a depth of 10 m in subsoils, and it could be whether microbes exist on a different planet or something else. Nevertheless, we also recognize some disadvantages of these new information technologies. Everybody expects a quick answer to every question. This phenomenon yields quick answers, sometimes at the expense of quality.

Georg Dietzler: The way you are describing sharing information has benefits, but also creates a lot of time pressure when someone is not fast enough. Thinking from the perspective of the arts, I need to have time to ask questions, which is surprising to me. Is this still possible in science?

Ellen Kandeler: In science, there are only rare moments when a group of scientists really have time to develop and discuss new ideas under very quiet conditions. We have to shut down our computers, retreat to a small island or a silent place. Our project team did this several times. Being in a silent place for several days, thinking and discussing how we could develop future research. For example, colleagues from different disciplines like soil science, bioinformatics, and ecology met and worked on a joint data set. It is amazing how fast a small group made progress in understanding a certain topic. These retreats also stimulate a much better work–life balance. The historical concept that humans need a balance between working and relaxation brings us back to the old Greek and Roman times where people met to walk and think. If we choose today to reserve an afternoon just to talk and listen to each other, this would make more sense than trying to be faster on the computer. I am sure that scientists and artists will return to this really old-fashioned system of just spending time together in a relaxed atmosphere and having good discussions.

Georg Dietzler: Also in the arts, you need this sort of balance, letting ideas flow like they do today. Residencies are a good way of letting ideas unfold. And traveling by train, reading something on the train, or looking at the passing countryside is important. Much more important, related to this book, is what you mentioned: taking time for conversation between scientists and artists. How do you find the right person to work with? Who is open and curious enough? Often enough the reply is it "sounds interesting, but unfortunately, I'm too busy doing my own research." One question that regularly comes up among international artists working on environmental projects is, Is it really possible to collaborate with scientists? There is not just one answer. It is very much about the way people can connect; it is more than just doing research from different points of view.

Ellen Kandeler: I am sure that transdisciplinary dialogues will make great progress in the future. We need colleagues that have an open mind to stimulate these kinds of discussions between the arts and soil science. I think a development of something new is taking place, which emerges from the close interdisciplinary approach between art and soil science. My wish for the future is that the collaboration between science and art can develop in a peaceful, creative, and inspiring environment. A prerequisite will be that scientists who are normally trained to focus on very specific topics in their specialist field of interest will have the chance to broaden their view toward a better understanding of artistic practice.

Georg Dietzler: It's not too much of a surprise that I agree with you, Ellen. It would be wonderful if artists and scientists together with writers/journalists form inspiring, engaging growing communities, with many others joining in to become open-minded, curious citizens interested in our surrounding natural world. Each sharing numberless tiny findings, more than

phenomenological observations, any changes through science, arts, or whatever. Stimulated by one's own free will, to act responsibly toward nature, raising awareness and responsibility for environmental, social, and political issues linked to independent, interdisciplinary international networks: cooperation of artists, scientists, media, economy, etcetera, by using each other's potential of creativity to shape a future. The essentials are peace, harmony, wholeness, completeness, prosperity, welfare, and tranquility referring to the well-being, welfare, or safety of an individual or a group of individuals. Recently I read sort of a biographical science novel *The Invention of Nature* by Andrea Wulf. It's about Alexander von Humboldt, who worked closely with artists, even in the eighteenth–nineteenth century; still an inspiring figure and in many ways relevant nowadays.

Acknowledgment

We would like to thank Barbara Rosenhart for transcribing the record of our interview and Katie Mackie for English editing.

Soil Macrocosms

Microbes, People, and Our Cumulative Effects

Alejandro Meitin, Claire Pentecost, and Brian Holmes in conversation
with Ela Spalding

This interview took place online, traversing three time zones and two languages. Due to the need to translate between questions and answers, most questions were posed to both parties, except for some cases in which I (Ela Spalding) followed up specifically with Claire Pentecost or Alejandro Meitin. Since Brian Holmes is a link between all of us and knows both interviewees quite well, he was invited to send some questions that are also included here. The conversation traveled from the microcosm to the macrocosm, through perceptions on time and different cosmovisions, along noticeable parallels in approaches to soil as a place for regeneration and shifting paradigms about how we relate to ourselves, each other, other life-forms, as well as the environment and economy. Both Pentecost and Meitin are artists literally making soil, one landscape and one exhibition at a time, to rewrite the narrative of how people live in the world today.

Claire Pentecost is a Chicago-based artist and writer who researches life and food systems, agriculture, and bioengineering. She advocates for the role of the amateur in the production and interpretation of knowledge, while her longstanding interest in nature and artificiality predicates her recent responses to anthropogenic climate change. Since 2006 she has worked with Brian Holmes, 16Beaver, and many others organizing Continental Drift, a series of seminars to articulate the interlocking scales of our existence in the logic of globalization. Pentecost has shown her work in venues such as dOCUMENTA(13), Whitechapel Gallery, the 13th Istanbul Biennial, Nottingham Contemporary, the DePaul Art Museum, and the Third Mongolian Land Art Biennial. She is represented by Higher Pictures, New York, and is professor and chair of the Department of Photography at the School of the Art Institute of Chicago. Her website is www.publicamateur.org.

Alejandro Meitin is an artist, lawyer, and founder of the art collective Ala Plástica, based in the city of La Plata, Argentina. He has participated in the research, development, and implementation of collaborative art practices working with residents, youths, farmers, artists, activists, architects, landscape architects, local authorities, and pollution control experts. His collaborations and proposals have involved regional, national, and international rivers and water resources. Ala Plástica (https://alaplastica.wixsite.com/alaplastica) is an art and environmental organization that works on the rhizomatic linking of ecological, social, and artistic methodologies, combining direct interventions and precisely defined concepts to a parallel universe without giving up the symbolic potential of art. Since 1991, Ala Plástica has developed a range of nonconventional artworks focused on local and regional problems in the social and environmental realm. Ala Plástica works bioregionally within Argentina as well as internationally in collaboration with other transformative arts practitioners, scientists, and environmental groups.

Brian Holmes is a researcher, activist, and essayist based in Chicago. He has been involved with issues of globalization and social change over the last twenty years, and is known for his writing on the intersections of artistic and political practice. He is the author of a blog titled "Continental Drift or The Other Side of Neoliberal Globalization" (https://brianholmes.wordpress.com/).

Ela Spalding is an artist and founder of Estudio Nuboso (www.estudionuboso.org), a platform for art and ecology in Panama and beyond. Her work focuses on the human–environment relationship through art, culture and facilitating exchanges across disciplines and communities. In 2014, she ran a pilot multidisciplinary residency called Suelo, aiming to articulate the natural and cultural value of place through multiple ways of relating to and appreciating soil. It was during this residency that she got to know Claire Pentecost and Brian Holmes, who participated as guest facilitators. She shares her time between Panama and Berlin.

Title image, previous page: *Offer to Pachamama*, Camino de la Sal, 2005. Ala Plastica.

Image credit: Matilde Zúccaro.

Ela Spalding: For the sake of giving our readers a richer introduction to both of you and your works, let's start with a question from Brian [Holmes]: How did you each come to perceive soil, and to care about it? Specifically looking at your particular geographical contexts, do your prairies, Claire, and your pampas, Alejandro, have anything to do with your concerns about soil, or did you come to the question of soil along very different avenues than the ones suggested by your regional history?

Claire Pentecost: I live in the Midwest, which some people call the Heartland or the Breadbasket. It's actually a toxic zone though because of the extent of industrial farming that goes on here. It's like a constant background noise, this practice that happens all around me that disregards everything that good healthy soil has to offer. But it was kind of a coincidence that I moved to the Midwest from New York in '97. Around the same time, I was invited to make a proposal about biotechnology for Creative Time. When I started researching, I found that the place where biotechnology was already in operation was agriculture. In the U.S., we were eating transgenic agriculture and no one was talking about it. So that got me looking at food systems, which initiated a twenty-year investigation on the subject. For many years, I was looking at the industrial global food system where the magic of soil is replaced by pesticides and synthetic nitrogen. Soil is not really a conversation point in industrial agriculture, which is kind of amazing to me now. When I started looking at alternative kinds of farming that don't use toxins or artificial fertilizers. I realized that soil is probably the most critical factor, because everything happens there—it's a fantastic ecosystem of which plants are only a part.

Alejandro Meitin: In the first years, our work was concerned with urban dynamics. In 1995, we started to take a more regional perspective, leaving urban art aside to focus on greater ecosystems and territories, particularly in the area of the Parana River delta and La Plata River estuary, where our main activities take place. That year we started the task of restoring a degraded ecosystem on the coast of La Plata River by planting a series of rhizomes in a performative manner. We worked with the Botany and Applied Botany Departments of the Natural Science Museum of La Plata University, where we learned techniques for restoring coastal ecosystems, particularly soil restoration, using semiaquatic emergent plants. That's how we started focusing our art work on ecosystem recovery, which in turn allowed for a cultural and economic recovery of communities that depended largely on natural resources that were degraded due to the urban expansion of the City of Buenos Aires. This work is ongoing in various exercises that we have continued to develop throughout the region.

Beyond soil restoration by means of plants, we have also developed a special interest in the industrial agriculture practiced in the La Plata watershed, which stretches over five countries: Brazil, Uruguay, Argentina, Paraguay, and Bolivia. This is an incredibly large area, almost the size of the European Union, that we have been transiting for twenty-five years. We are particularly interested in the ecosystem destruction caused by the expansion of the agricultural frontier for soybean cultivation, which not only destroys soils, but entire communities who depend on the production of this transgenic crop. In this sense, I recently collaborated with Brian Holmes on the creation of a dynamic interactive website showing the problems due to industrial agriculture in the territories of La Plata and Mississippi watersheds.[1] This was for the exhibition *The Earth Will Not Abide*[2] in which Claire also participated along with other artists.

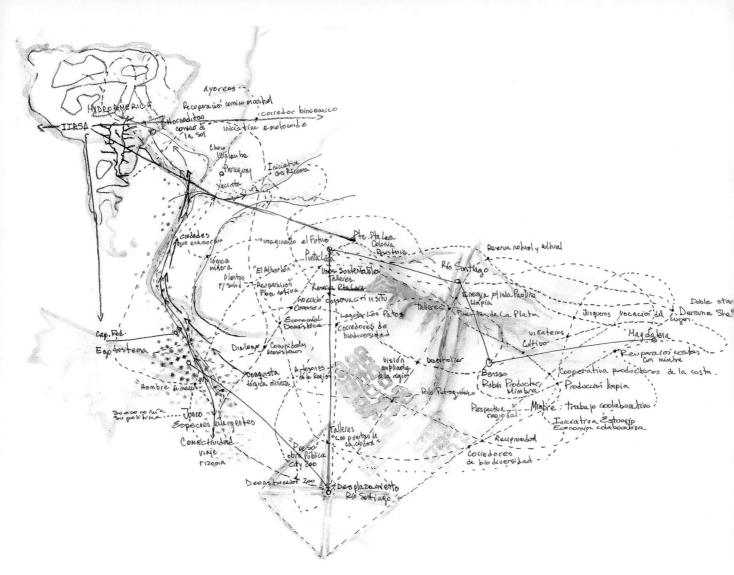

Connectivity. Drawing of Ala Plastica's network of rhizomatic relationships in the La Plata watershed.
Image credit: Silvina Babich.

Ela Spalding: I've noticed that in both your cases knowledge of place and knowledge exchange are key. How do your artistic practices related to the knowledge of soil and place serve as a starting point or as a way to articulate social action?

Alejandro Meitin: I have worked for a long time with the concept of "vocación del lugar," which has to do with how communities identify with systems that are environmentally recognizable through a comprehensive totality definable as a place's vocation or calling. This integration of the place's symbolic role and the forms built in the

natural landscape have been represented by art in most cultures.

Currently there is advanced research on immunodeficiency that recognizes that the human body is connected to a network of neurochemical communications with our surroundings, which determines greatly the state of our health and well-being. This neurochemical relationship or this natural understanding becomes altered when the environment is redefined by the division of areas for economic interests, commercial means of communication, and financial institutions, generating just a few common spaces of shiny modernity. This alienation produces a hostile reality in which humans start to perceive the environment as an object outside of themselves, leading to situations of great toxicity, which affect the quality of life of ecosystems, from the soil to the overall sanity of society. These effects are felt not only in terms of consumer society, but society as a whole—from the microcosm to the macrocosm.

Our practice focuses on the integration of humans and their surroundings, to strengthen the vocation of a place through communication, dialogue, and the recovery of "poder hacer," or the power or agency of communities. For us it is very important to work beyond the art context to address socioenvironmental problems, exploring noninstitutional and intercultural models of working in the social sphere.

Claire Pentecost: That is very beautiful. There is a lot there. One thing I want to say is that soil, from my point of view, is not static. It's not a thing whose qualities are frozen in time. If it's alive, it's changing, evolving, and responding to how we treat it. The vegetation and geology of a place have to do with the quality of the soil. In any given place, soil has a character and a history, and is more or less unique. For this reason, I've been emphasizing a practice called soil chromatography, where I'm making portraits of soils. I want to convey to people the idea that soils are distinct; they are part of what gives any place its specific character.

In terms of health, I love what Alejandro is saying. I hadn't heard about this research on a neurochemical reaction of people to a place. What I've looked at along the same lines is the fact that our bodies are inhabited by microbes with which we have coevolved for hundreds of millennia. Our physical and mental systems are interdependent with whole populations of microbes that live in and on our bodies. In the last few generations our relationship to the environment has been to eradicate microbes, seeing microbial life as pathogens, germs, contamination, or something that is a pest, or a disease. While all those things can be true, that's only a part of the story. We have created so many chemicals to destroy microbes—our approach is one of war and, like war, it wipes out the good with the bad. You can look at the scale of this from the pesticides, herbicides, fungicides, insecticides that we put into our soil all the way to the antibiotics that we put in our bodies.

Antibiotics are undoubtedly a miracle; they save lives, including my own. But we have come to use them so irresponsibly that we have destroyed a lot of the beneficial microbes that live in our bodies and are passed specifically from mother to child. The fetus has no microbes living within the amniotic sack, but when the fetus passes through the birth canal it is washed in microbes that are part of its whole apparatus for survival. With every generation of antibiotics, as well as biocidal soaps, cleansers, and insecticides, the mother is passing on fewer and less diverse populations of microbes. So, in affluent countries we have new health disorders that are largely autoimmune disorders. We have found that the microbes in the child's body teach the immune system to regulate itself; if those microbes aren't there, the immune system doesn't know how to self-regulate and it overreacts, etc. These disorders are not seen in less affluent countries where the war on microbes has not been so intense. I'm fascinated that many of the medical specialists working on these matters have decided that children need to play in dirt! And need to have contact with other animals. This is a way of saying that there is a continuity between the microbial life living in the soil and that in our bodies.

Claire Pentecost, *old friends and unloved others*, 2013. "Old friends" is the term coined by Dr. Graham Rook to refer to the many, mostly microbial, life-forms who live in and on our bodies. "Unloved others" refers to an issue of *Environmental Philosophy* edited by Deborah Bird Rose and Thom van Dooren (2011) focusing on the biological and social ecosystems that are often ignored, feared, scorned, or despised. This silverpoint drawing on panel of Da Vinci's perfectly proportioned human (man) with microbes were first shown at *The Spirit of Utopia* exhibition at Whitechapel Gallery in London, 2013. A later version was exhibited as part of the "Our Bodies Our Soils" installation featured at the exhibition *Rooted in Soil*, DePaul Art Museum, 2015. The drawing in London used soil from London and the version in Chicago used local soils.

Image courtesy of the artist.

As in so many things, we need to develop a more and more precise appreciation of how we are not isolated organisms. We are part of larger ecosystems, consisting of the microbes in my body and all around me, in the soil, on the leaves, and in decaying matter. So, it's hard to say what the effects of so many assaults on microscopic life are. One is fragmentation, of landscape, of lives, of our relationship to any given place.

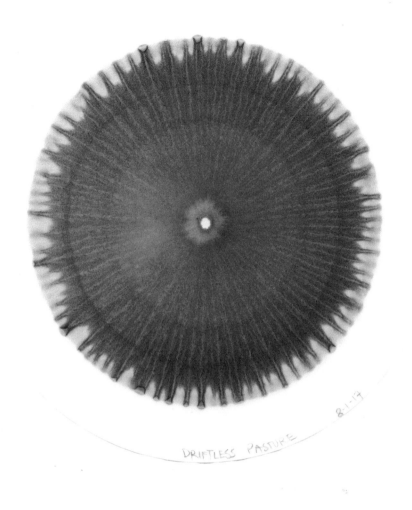

Claire Pentecost, *Driftless Pasture #18*, soil chromatography, 2017. The "Driftless" refers to an ancient land that escaped the passage of the last Ice Age, some 11,000 years ago. Unlike the rest of the Upper Midwest, the topography of this region has neither been scraped smooth nor buried in glacial drift. In summer 2017, Claire Pentecost and Brian Holmes traveled through the Kickapoo Valley watershed collecting soil samples to use in an experimental cartographic project. Soil chromatography is a rare technique that derives an image of the soil from the application of a carefully prepared solution to a piece of light-sensitive filter paper.

Image courtesy of the artist.

Soil Chromatography

To make these I (Claire Pentecost) prepare a soil sample with a dilute solution of sodium hydroxide which gently breaks down the components of the soil—the mineral and organic matters. Then I prepare a piece of filter paper with a 0.5% solution of silver nitrate. After that is dry I infuse it with the soil sample to get a unique, holistic portrait of a given soil.

Chromatography is used to examine all kinds of things and can be used to study leaves, fruits, roots, and so on, but this specific soil process was developed by Erenfried Pfeifer, an associate of Rudolf Steiner, in order to formalize a simple, inexpensive process to grasp the condition of a given soil. It is a qualitative, not quantitative, process.

Brian Holmes: Alejandro, have you learned anything from indigenous people about the way they see soil? In the Indo-European family of languages, there is a common etymological root linking humus and humans. Is there a similar trait in Amerindian cosmologies, or maybe something of a similar importance?

Alejandro Meitin: Yes, we are very interested in linking the scientific world with ancestral knowledge and the basis of this knowledge, which often reaches the same conclusions but represents them in different ways with respect toward the Earth. The cosmologies of the area in which we move are very present in the communities we work with. For instance, in the 2006, we were invited by the Kolla Ecotourism Council of the upper Andean communities of the Puna mountain region, in northwestern Argentina, to work together on the recovery of an ancestral path: Camino de la Sal. There we were able to deepen our understanding of the concept of Pachamama or Mama Pacha, a divinity known as Mother Earth. She represents the Earth, but not only soil or geological earth, or nature alone, but all of its totality. She is an immediate, everyday deity, whose presence evokes permanent dialogue. This dialectic relationship between soil, nature, and (her) presence is a form of neurochemical relation that is not only creative, but protective and provisional, favoring life, fertility, etc. Coincidentally, today August 1, is Mother Earth Day in our region—a very important celebration for many people. It is a day to open windows to new possibilities, to clean the house, to burn cleansing herbs, to drink aguardiente with rue to cure the organism and prepare for winter.

I think it's important to acknowledge how different the cosmovisions of South American cultures are compared to those of European cultures; otherwise this can be read as simple customs or something of lesser validity. The comparative advantage we have regarding this worldview is that these cultures, with their spiritual connection to the land, are very much alive here, while in most developed countries they disappeared a long time ago, with almost all of their ancestral knowledge.

Brian Holmes: It seems like modernism is more interested in the stars than the soil. Claire, I think you once used a phrase about the cosmos underground. Have you gone on thinking about that? Why are you attracted to microcosmology of the underground? What can that cosmology do for the way that people see or feel the world?

Claire Pentecost: Science and the specialization of our knowledge has been an amazing force. It has been prodigious in terms of teaching us a series of perspectives on how things work together in the biosphere, but it's been very slow to understand the macrocosmos—the idea of the Earth as a macroorganism, whether you call it Pacha Mama or Gaia or Mother Earth. I have been focused on what I would call the cosmos underground, just because it's so relatively unknown, overlooked, unnoticed, and undervalued. Why is soil so undervalued? I think partly because in Judeo-Christian culture the earth is associated with dirt, disease, and death and even hell. In general, it definitely is where things decompose, get recycled, and become new life, so it's a highly charged place. However, if we're only seeing it as a place that harbors the dead, we're only seeing half the picture.

Studying microbiology has really changed my sense of time, partly because microbes so predate us. Bacteria, the very early single-celled life-forms on this planet changed the atmosphere and made it one that supports organisms like ours. The other thing that has changed my sense of time is the Anthropocene. The idea that, as small a proportion of the Earth's history as our presence represents, our impact is significant enough to justify naming a whole new geological era.

Ela Spalding: Regarding the element of time, it's interesting that both of you started this line of work in the '90s and have continued for over twenty years. I find these long time lines inspiring and realistic, in the sense of how long it takes for new ideas to sink in, or to see change in a landscape or a person or a community.

Claire Pentecost: Yeah, I see it in myself. In terms of looking at food and the way I relate to the earth and to soil, it's taken me years to change my behavior. It's important to understand that most of us don't change our behavior overnight. At the same time, when I started researching food in 1998 no one wanted to talk about it. People would say, "Oh, you're gonna tell me all this depressing stuff and there is no alternative and you know, I still have to eat!" But in twenty years the whole social landscape has changed. We're still having to reckon with big corporate agriculture and its destructiveness, but the awareness has totally changed and people are starting to demand not only healthier food but better treatment of animals and better treatment of the Earth. Twenty years is actually very short to see so much social change; it gives me some hope.

Alejandro Meitin: We have been inspired by the world of plants and how they make their path in the long term, which gives time for growth. In the art world and I think in the scientific world too, immediacy prevails. Everything starts and ends and has to have a measurable result. That is why we don't like talking about "projects." We feel this word implies that we need to have a premeditated ending to what we are proposing. We prefer to talk about "initiatives" that lead to "exercises."

For instance, the Bioregional Initiative started with the exercise The Reed (Junco/Especies Emergentes). Junco (*Schoenoplectus californicus*) is a semiaquatic species that spreads subterraneously via rhizomes, generating a body along a territory and expanding in time. These plants have various aptitudes that one can read. One is that their cilia hold sediment, creating new soil for new plants to take root. Another is that their underground rhizomes allow for the purification of water as it passes slowly through them. We found a metaphor in the consolidation of soil for the creation of new territory, which at one level is in reference to Deleuze and Guattari's study of rhizomes.[3] With this metaphor we started the ecosystemic recovery on the coast of the La Plata River and simultaneously took this rhizomaticity to our work with the territory. As the soil recovered, we expanded on the plane of community work. We later worked with willow (*Salix viminalis*) growers in the delta region, with whom we also started a process of crop recovery, using the junco and wicker fibers to refresh the skills and livelihood of the local communities. This is how our work, having started in the southern coast of the La Plata River in '95, has extended along the entire Plata watershed until today.

Ela Spalding: We've touched on belief systems, analogies, and metaphors related to soil. If soil health is directly related to the health of a community, soil wealth (rich, biodiverse soil) should lead to the wealth of a community. However, since the current economy's idea of wealth is based on extraction, do you think it is possible for the world to perceive wealth as abundance of biodiversity? Can we change the narrative of our current economy from one of extraction to an economy of restoration?

Claire Pentecost: In the Soil Erg project, I envisioned soil—and when I say soil I mean

Next pages:
Walkers, Camino de la Sal, 2005.

Image credit: Alejandro Meitin.

Transformation. Image of the recovery of a degraded landscape on the coast of the La Plata River, from seedling to induced growth using reed rhizomes and other native species, 1995–2011.

Image credit: Alejandro Meitin. Junco—Especies Emergentes. 1995–2011 Desembocadura arroya La Guardia, Punta Lara, Ensenada, costa del Río de La Plata.

living healthy soil—as the basis of an economy. An economy is a set of social relations mediated by certain promises and/or substances or goods, but the important thing is really the relations. It's also a set of relations with the earth around us that we depend on. I wanted to propose that soil can also be a set of social relations, because soil must be maintained, replenished at the rate we take from it. When I say social I am including other beings beyond the human. In the larger ecology, there are active relationships that have consequences and in restoration we are working with, not against creatures that inhabit the soil. For about 8000 years most humans have been dependent on agriculture, which can be seen as an extractive practice. That's the way we've done it so far, taking from the soil and not nurturing it. I was asking, what if our economic system demanded that we engage in a restorative practice in addition to an extractive practice? What if that was just part of the workings of our economy, that we were constantly restoring the life in the soil? You can do that through compost. I was just in England studying with a Colombian man, Jairo Restrepo,[4] who teaches restorative agriculture, which has a way

of remineralizing the soil. There are a lot of practices by which we can contribute to life and the environment.

Ela Spalding: And if we actively engaged in these we could potentially flip the script about the Anthropocene and our actions would stop destroying landscapes and fragmenting our relationship to nature, but actually improve it.

Claire Pentecost: I think that at this point, we have to accept that the Anthropocene is a kind of mirror, a reflection that the human is a biological and now even a geological agent. We have to accept that we have an outsized effect on our environment. The question is, Is it going to continue to be destructive or is it going to be something that starts with observation and appreciation of delicate systems to which we can contribute? Of course, now it's a question of improving our environment because we have caused so much damage. I don't think originally it was about that, because I think the way the biological system works can hardly be improved upon. It's just such a beautiful system. Think about

Evidence: Oil Spill in the Río De La Plata, Citizen Culture: Artists and Architects Shape Policy. Santa Monica Museum of Art, 2014. Image credit: Alejandro Meitin.

it, life on Earth can capture energy from a star and recycle its own waste. It's amazing! Right? There is no waste in nature. We're the ones who create these wasteful substances that are a problem. The big question right now is how can we participate in this Earth system without destroying it.

Ela Spalding: Alejandro, you have done a lot of work on reimagining geopolitical, environmental, and economic landscapes. Have you noticed a change in the economy, visions, or narratives of the communities with which you are involved?

Alejandro Meitin: Yes, there are many examples. But one needs to have the ability to appreciate what emerges from these slow processes. Oftentimes we work in a territorial context and the vision of the people with whom we work is one of immediacy. They ask "What is the result? What does this practice do?" This logic is very close to how the market and consumer society

function, and it reproduces the system that is leading us to ruin. So how can we enable the creation of a new world, leaving the ocular-centric and short-term results aside and become able to acknowledge other kinds of emergents? We like to talk about reading the effects of meaning. For us the "meaning" of a given work is not centered in the physical locus of the object, or in the imaginative capacity of a single viewer. Rather, it is dispersed through multiple registers. The "work" is constituted as an ensemble of effects and forces, which operate on numerous levels of signification and discursive interaction. We propose that a new pedagogy for long-term processes is required to make an accurate reading of their effects, which have nothing to do with spectacle, yet are often what create new territory, new soil. This reeducation is imperative in times of social change.

Ela Spalding: This makes me think of a saying about how we cannot solve our problems with the same thinking we used when we created them. So,

Claire Pentecost, *Soil Erg*, 2012. Multipart work created for dOCUMENTA13 consisting of vertical container gardens around the city of Kassel, stacked ingots, and oversized coins made of pressed compost, a series of graphite and soil drawings in the form of giant dollar bills, and appropriated museum pieces that all revolved around a central theme—the soil as postcapitalist currency. The bill/drawing shown here is a portrait of the biologist and evolutionary theorist Lynn Margulis.

Image courtesy of the artist.

as we shift our practices, we also have to shift our thinking, ways of seeing and the entire systems around them.

Claire Pentecost: Alejandro is right. These efforts should be evaluated under different terms than the ones we've been given. It demands of us that we be very well tuned to emergent changes, to things that may not be so obvious. As a cultural producer, I feel that what I'm trying to do is produce cultural changes so that what we value is different. For instance, the culture we have now values convenience. We are willing to sacrifice all kinds of things for convenience. This is something that has to be completely rethought. There is a philosophical question that Plato asked in the Western tradition: What is a life worth living? That's part of what I hope artists like Alejandro, Ala Plástica, and myself (and the other artists in this book) are contributing to—a different idea of what constitutes a good life for a human on Earth. I propose it's a life that has a sense of shared fate with other living things; the idea that it's not just about me or my tribe or my species; rather, how can I contribute to a robust life for diverse life-forms? I actually think that we have a lot of the technology and knowledge we need to solve our problems, it's just that we need the will. What's going to produce that? We need a vision that acknowledges that it's actually preferable to live as though other species matter as much as our species.

Ela Spalding: Alejandro, what is your take on this?

Alejandro Meitin: I feel that while the self-destructive values that rule today continue to be sustained, the societal structure is limited by a small percentage of the population who wield power and benefit from destroying the soil, the air, and the water. It will be very complicated to see a change, unless it stems from direct connection and reciprocity among communities. Many of the changes that have occurred in the last 200 years have been forced on individuals through the use of weapons. So, it is not only through a philosophical movement that one can change the world today. However, we are all doing what is possible for things to improve.

Claire Pentecost: Sin duda, es muy complejo.

Alejandro Meitin: Pero bueno, allí estamos.

Endnotes

1. http://ecotopia.today/riosvivos/mapa.html.

2. The exhibition *The Earth Will Not Abide* poses questions about the ecological and social viability of industrial agriculture and extractive land use. These artist projects, spanning video, creative mapping, paintings, and installation, comprise an investigation of and aesthetic response to these questions, employing a variety of media and methods. Artists Ryan Griffis and Sarah Ross, Brian Holmes and Alejandro Meitin, Sarah Lewison and duskin drum, and Claire Pentecost draw connections between lands in the United States, Brazil, Argentina, and China that have been—and are currently being—engineered to support an agricultural economy based on monoculture.

3. Deleuze, G., and Guattari, F. (2015). *A Thousand Plateaus: Capitalism and Schizophrenia*. London: Bloomsbury.

4. https://blog.permaculture.org.uk/articles/harvesting-sun-jairo-restrepo.

Soil Procession

A Seed Journey to Preserve Genetic Diversity

Régine Debatty in conversation with Amy Franceschini

Amy Franceschini received her BFA from San Francisco State University in photography and her MFA from Stanford University. She has taught in the visual arts graduate programs at California College of the Arts in San Francisco and Stanford University. Amy is a 2009 Guggenheim fellow and has received grants from the Cultural Innovation Fund, Creative Work Fund, and the Graham Foundation. She is the founder of the international arts collective Futurefarmers. Founded in 1994, Futurefarmers is a rotating group of artists, activists, farmers, and architects who share a common curiosity around the intersection of nature and culture. Futurefarmers use various media to destabilize logics of "certainty." They often start with preexisting forms or systems, for example, food policies, public transportation systems, and rural farming rituals. They see these as "ready-mades" that they carefully deconstruct

Title image: Flatbread Society soil procession, 2016.

Photo by Monica Loevdahl.

as a means to visualize and understand their intrinsic logics. Through this disassembly new narratives emerge, whereby reconfigurations of the principles that once dominated these systems offer new agencies.

Régine Debatty is a writer, curator, critic, and founder of the award-winning blog "we make money not art" (http://we-make-money-not-art.com/). Debatty is known for her writings on art, science, technology, and social issues. She writes and lectures internationally about the way in which artists, hackers, and designers use technology as a medium for critical discussion. She also created A.I.L. (Artists in Laboratories), a weekly radio program about the connections and collaborations between art and science for Resonance 104.4 FM in London (2012–2014), and is the coauthor of the "sprint book" *New Art/Science Affinities*, published by Carnegie Mellon University in 2011.

Seed varieties have declined significantly since the beginning of time. First, with plant domestication and now, increasingly, through homogenization, industrialization, privatization, and commodification of our seed stock. Independent groups are currently working as private protectors of genetic diversity[1] by cultivating endangered varieties in their home gardens, sharing seeds with other seed savers, but also lobbying the European Union to make sure that a new proposal for seed marketing regulation[2] will promote agricultural biodiversity, small-farmers' rights, global food security, and consumer choices.

The need for a robust and vibrant culture of seed diversity was one of the motivations that led Amy Franceschini and Futurefarmers to establish the Flatbread Society—a collective of farmers, artists, activists, scientists, and other people involved in urban food production and preservation of the commons. Since 2012, the group has been working in a permanent "common" area on the waterfront development of Bjørvika in Oslo, Norway. They built an urban farm, an allotment community, an ancient grain field, and a bakehouse.

Last year, a delegation of the Flatbread Society embarked on a yearlong sailing expedition that took them from Oslo to Istanbul. Onboard was a rotating crew of artists, sailors, anthropologists, activists, writers, and ecologists. As for the cargo, it consisted mostly of grain seeds that had been lost or forgotten. Along their journey to the Middle East, where the cultivated grains originated, the members of the crew stopped in harbors to meet artisan bakers and farmers, make flatbread, and collect and exchange seeds, as well as document and retrace the journey that the seeds made thousands of years ago. Régine Debatty talked with Amy Franceschini about the Flatbread Society's extraordinary sailing adventure and about their efforts to raise awareness around the need for the development of plant genetic diversity.

Flatbread Society, Futurefarmers'
canoe oven, 2013.

Photo: Max McClure.

Flatbread Society seed collection, 2014.

Photo: Futurefarmers.

Flatbread Society, 2016.

Photo by Monica Loevdahl.

Régine Debatty: Amy, I'm curious about the seeds you've decided to take on this "reverse journey" to Turkey. Which varieties of grains did you select exactly?

Amy Franceschini: We started with a Finnish rye. We came to this rye when searching for someone farming "ancient" grains in Oslo. We have been working on a public artwork in the former port of Oslo for the last five years. The centerpiece of this work is a grain field featuring ancient grains that have been rescued from interesting places.

For example, this Finnish rye was found between two boards in an old sauna used by the Forest Finns in the early 1900s to dry their grains. This grain was thought to be lost, but an amateur archaeologist rediscovered it.[3]

Our project in Oslo is located on a "commons," a piece of land set aside within the waterfront development that should be accessible to all. We took the tradition of Norwegian commons to heart in this project and wanted these ancient grains to symbolize the biological commons, which is currently at risk due to the privatization and commodification of our seed stock. Michael Taussig,[4] the Seed Journey onboard ethnographer, says that "the return of ancient seeds is like reverse engineering, taking apart this long history fold by fold. This voyage is an allegory, one forever open to chance. Our participation from afar breathes wind into the sails of the future."

Régine Debatty: Will you be collecting other seeds along the way?

Amy Franceschini: Yes. We collect and share at each stop. Our mothership, RS10 Christiania, carries an ingeniously crafted miniboat that holds all of the seeds we collect "like a chalice." It contains small amounts of old wheat and rye seeds collected along the journey. These seeds are like jewels. The disproportion in size between the small chalice-boat and the mother vessel carrying it symbolizes the preciousness of the cargo as does the very idea of a prolonged voyage using wind and sail as the means of propulsion.

Régine Debatty: Why did you choose to travel with these particular seeds?

Amy Franceschini: Each of these seeds has a particular story of rescue associated with them. And through the planting and exchanging of them comes an awakening. For example, a variety

RS-10 Christiania, 2010.
Photo: Martin Hoy.

of barley that we have with us came by way of St. Petersburg. Nikolai Vavilov collected more seeds from around the world than any other person in history. He was one of the first scientists to really listen to traditional farmers, peasant farmers, and ask why they felt seed diversity was important in their fields. During the siege of Leningrad in 1941, Vavilov was imprisoned by Stalin and starved to death. He became the main opponent of Stalin's favored scientist Trofim Lysenko for his defense of Mendelian theory. Just a few blocks away in the Vavilov Research Institute of Plant Industry, Vavilov's staff scientists locked themselves in the seed bank to diligently protect his seeds. Over half a million people starved to death during the twenty-eight-month siege while these twelve scientists filled their pockets with grains so that future generations would be able to grow food. When the Allied troops arrived at the seed bank they found the emaciated bodies of the botanists lying next to untouched sacks of wheat and other edible seeds—a genetic legacy for which they paid with their lives.

Another wheat we have is called Brueghel. During an archaeological dig in a church in Pajottenland, Belgium, charred rye and wheat seeds were found. The archaeologists and a group of local farmers call these seeds Bruegelseeds because they date from the time of Bruegel, the old Belgian painter known for landscapes and peasant scenes. A group of young farmers want to make a "Bruegelbread," a bread made in the landscape of Bruegel. For the moment, all of the grain for human consumption in this region comes from abroad and they would like to reinstall a local chain from the Bruegelseed to make the Bruegelbread. On April 1, we summoned the ancient Bruegel grain to imagine what Bruegel would paint or make in 2017. Futurefarmers hosted a Seed Ceremony at the Heetveldemolen–Heetveld Windmill. On this day, many farmers gathered to share their unique and locally cultivated grains. A handful of Bruegel grains was launched onto the Futurefarmers Canoe Oven and rowed from canal, to river, and into the Schelde to Christiania.

Régine Debatty: What do you mean when you say that the seeds have been "rescued"? Rescued from what or whom? And how? To what purpose?

Amy Franceschini: The Latin root of the word "rescue" is to return. The seeds we choose to carry with us are seeds that were either lost or fell out of production and then found again, like the stories I referred to earlier, or they are seeds that farmers have taken out of gene banks that have not been grown for thirty to eighty years. These farmers are trying to return these seeds to the ground and into production rather than sitting dormant in gene banks. Our collaborator, the Norwegian farmer Johan Swärd, is busy collecting grains out of gene banks and getting them into the soil. He says, "We don't need a museum to conserve varieties, what we want is to grow them."

Many of these seeds fell out of production before the green revolution, so they have not been homogenized. But if they are not grown each year as a landrace, they do not have the opportunity to adapt to their growing environment; to the soil, local climate, and socially adapted and acquired taste.

Régine Debatty: Why is it important to use heritage grains?

Amy Franceschini: To see an agribusiness supplanted locally adapted seeds with fewer varieties of seeds: until the late nineteenth century, most plants existed as highly heterogeneous landraces. Over the past century of modern breeding, attempts to produce cultivars that meet the advanced agriculture demands of an increasing population has resulted in the landraces being almost wholly displaced by genetically uniform cultivars. The result has been a narrowing genetic base that puts these plants and the future of food production at serious risk.

Régine Debatty: And isn't this complicated by the trend of patenting grains?

Amy Franceschini: It is said that wheat and rye were domesticated in Kurdistan and through gift and trade, not to mention wind, birds, and animals, made their way north to Europe to become the "staff of life." These "old" seeds come loaded with a forgotten history at once social and biological. The domestication of plants involved a long march through trial and error, not to mention chance, whereby certain varieties became reliable foodstuff. It was a revolution in world history, ushering in what is called the Neolithic period with tremendous consequences, one of which, of course, was deforestation. Another was the birth of the state and private property. Patenting grain is yet another moment along that trajectory. We sail along the cusp of many contradictions.

Cultivators in each and every microclimate developed their own varieties of seed stock from their harvests to the present time when, all of a sudden, such practices have been declared illegal. Another revolution is afoot. Farmers who continue with old stocks are at risk of arrest. They too are now an endangered species. In referring to themselves as such they cement an alliance—biological and political—with the plant world, which is what Flatbread Society is doing as well.

At the moment, they are in such a small market and the farmers who are bringing these lost seeds back into production are not interested in profit as such, but sustainability. The ideal scenario for these seeds would be to stay small in scale in terms of production and very local, so that they are adapted to a local biotope, ecology, taste and climate. This would enable a very local, durable,

and resilient economy, but not one based on surplus or growth per se.

But yes, these seeds can be in danger of being collected by large companies, patented, and homogenized. For example, some Tanzanian farmers now face heavy prison sentences if they continue their traditional seed exchange.[5]

Régine Debatty: Is there anything in the story or design of the Seed Journey project that particularly connects with the history and intentions of Flatbread Society?

Amy Franceschini: Flatbread Society is a durational public art project in Oslo, Norway, which includes a grain field, a bakehouse, and ten years–plus of artistic programing. The grain field connects Norway's agricultural heritage to the present, extending the metaphor of cultivation to larger ideas of self-determination and the foregrounding of organic processes in the development of land use, social relations, and cultural forms. The presence of this grain field against the backdrop of the city of Oslo and the "Barcode" buildings—its openness and fluidity—stand in stark contrast, culturally and physically, to the rational and rigid urban development in the surrounding areas of Oslo.

In 2015 a group of seventy-five people, swarms of bees, and a colony of airborne and soil-based microorganisms gathered in a location now called Losæter, a museum without walls, where an expanding inventory of ancient grains are growing. Since then, a selection of seven grains have been planted on this new common area in Oslo. Each variety has been "rescued" from various locations in the Northern Hemisphere—from the formally registered (e.g., seeds saved during the Siege of Leningrad from the Vavilov Institute seed bank) to the informal (e.g., experimental archaeologists discovering Finnish rye between two wooden boards in an abandoned sauna in Hamar, Norway). Together with local farmers Johan Swård and Anders Naes, these seeds and the knowledge of how to grow, harvest, mill, and bake them have become embedded in the project.

Régine Debatty: The Seed Journey emphasizes the preservation of agricultural biodiversity. Could you talk about how the project also encourages soil biodiversity? You mentioned, for example, that soil-based microorganisms were part of the

Soil Procession in Losæter. How do these nearly invisible, uncultivated species find a place in your work?

Amy Franceschini: A few things. The Svedjerug (Finnish rye) we are growing is a perennial, whereby the first year-one seed produces twelve tufts of grass. When this grass dies out, it goes back into the soil producing its own nutrients for the coming year when the seeds become twelve stocks of tall rye. This particular grain is a testament to the species being able to prosper

Flatbread Society, 2016.
Photo by Monica Loevdahl.

Horse plough in Losæter.
Photo by Vibeke Hermanrud.

without chemical inputs and tending to its organisms in the soil.

In terms of the microbes as collaborators in the project, when we built up the soil on site, we drew attention to our companions by dedicating the site to them and their health. We did this through a Soil Ceremony and a renaming of this land to Losæter (meaning a common in the city). This was all formalized in a Declaration of Land Use signed by farmers, artists, and politicians during the Soil Ceremony. Soil was the ultimate witness. Our proposal to guests and participants in this project was to consider the microbial world in every action as a collaborator and to acknowledge and tend to its presence—caring, nurturing, or celebrating it. Through the Soil Ceremony, we wished to raise the cultural understanding of our companion species by bringing it upon the declaration document and claiming it as the true "culture" of this new development. Alongside the National Opera, The National Library, and the Munch Museum being built next door, this new microbial area is now considered a vital institution contributing to the diversity of cultural production represented along the waterfront.

Régine Debatty: Ever since I've followed your work, I've been amazed by the way you connect with audiences outside of the traditional art venues. So how do you communicate this project and the issues behind it (politics of food production, role of grains in the economy, environmental challenges, knowledge sharing, etc.) to the people you encounter along the way?

Amy Franceschini: We still depend on arts institutions as our main support. They are a very important amplifier of our work. They have a much wider media reach than we do. Through them, we have the capacity to bring farmers onto a cultural stage, which in some cases validates their work as seminally cultural. For example,

when we exhibited at the National Museum in Cardiff for the Artes Mundi shortlist exhibition,[6] we were able to host a Seed Ceremony/Exchange with the Welsh Grain forum.[7] We hosted this in the National Library inside the National Museum. Since we had access to this space our project became more legitimate and we were able to extend this legitimization to the farmers by inviting them. The farmers' efforts became legitimatized as a cultural practice and was documented by the BBC, which gives their work a voice.

As Seed Journey moves along its route, word gets out and when we arrive in small or large harbors we are often welcomed by the town mayor, the local newspaper, and passersby. We also try to align with local harvest and sowing festivals, and with the maritime community. Our boat has an allure in itself, so a few times when we arrive at a marina, the harbormaster is quite proud to have us dock and calls the local media. For us, the message must move beyond the art venue, but the art venue is a valuable collaborator.

Régine Debatty: Could you tell us a few words about the people who accompany you on this journey? What is their role and how did you select them?

Amy Franceschini: The project inherited an imaginary dimension early on. Each time we speak of this journey, it fills one's mind with joy, hope, and wonder as well as being viewed as a critical project that needs to be happening right now. There is an absurdity and persistence to the project that captures people.

Many people enlisted themselves and soon enough we had an incredible crew of artists, anthropologists, ecologists, farmers, and sailors. The core crew was born out of a conversation in Ghent, Belgium, over two years ago, whereby,

Soil Ceremony, 2015.
Video still: Futurefarmers.

Soil Ceremony, 2015.
Video Still: Futurefarmers.

we asked each other, "If you had to be on a boat with this mission, who would need to be on this journey?" At this point it was more of a fantasy, but we wrote down many names, and most of them are now formal crewmembers.

The idea is that a rotating crew of artists', scientists', writers', and farmers' research interests influence the journey, but the grains ultimately guide the route. Seed Journey maps not only space, but also time and phylogeny. While the more familiar space yields a cartographic map, time yields history and phylogeny yields a picture of networks of relationships between and among living beings—relationships between cultural groups, but also between human and nonhuman

Declaration of Land Use Signing, 2015, Futurefarmers.

living forms such as seeds, sea life, and the terrestrial species from the various places and times we will traverse.

Régine Debatty: What do you hope will be the impact of this reverse journey?

Amy Franceschini: We try not to think in terms of "impact." This is a work in progress and it still has a lot to tell us and to discover. We hope to protect this space without the terms of impact, outcomes, etc. But of course, a basic drive for the project is to raise the status of the small farmers' work, validate this work, connect farmers in various locations so as to strengthen the network that is working to protect farmers' rights, and most importantly to keep the seeds in the hands of many rather than a few.

Endnotes

1. The ARCHE NOAH Seed Bank, https://www.arche-noah.at/english/about-arche-noah/the-arche-noah-seed-bank.

2. Plant Reproductive Material, EU marketing requirements, https://ec.europa.eu/food/plant/plant_propagation_material/legislation/eu_marketing_requirements_en.

3. See "The Rediscovered Rye," http://www.flatbreadsociety.net/stories/14/the-rediscovered-rye.

4. Michael T. Taussig, http://anthropology.columbia.edu/people/profile/376.

5. Ebe Daems and Kweli Ukwethembeka Iqinsio, 2017, "Tanzanian farmers are facing heavy prison sentences if they continue their traditional seed exchange," https://www.grain.org/fr/bulletin_board/entries/5633-tanzanian-farmers-are-facing-heavy-prison-sentences-if-they-continue-their-traditional-seed-exchange.

6. Amy Franceschini/FutureFarmers, September 17, 2015, Back to the exhibition National Museum Cardiff, Gallery, http://www.artesmundi.org/en/artists/amy-franceschinifuturefarmers

7. Welsh Grain Forum, http://welshgrainforum.co.uk/.

Future Worlds

Intelligent Soil, Technospheric Colonization, and a Habitat of Emotional Particles

Carlina Rossée in conversation with Wanuri Kahiu and Peter K. Haff

Peter K. Haff is professor emeritus of geology and civil and environmental engineering in the Division of Earth and Ocean Sciences of the Nicholas School of the Environment at Duke University, Durham, North Carolina. His research focuses on the role of technology in the Anthropocene and he is part of the Anthropocene Working Group, a body of scientists, social scientists, and other experts constituted under the International Union of Geological Sciences to develop a proposal for recognizing the Anthropocene as a formal unit of geological time. Over the last years, he participated in The Anthropocene Project (2013–2014), the subsequent project Technosphere (2015–2018), and the Anthropocene Curriculum (since

Title image: On the set of Pumzi, actress Kudzani Moswela and director Wanuri Kahiu in the Virtual Natural Museum.

Photo: Mark Wessels.

Printed with permission of Wanuri Kahiu.

2013) at Haus der Kulturen der Welt (HKW), Berlin. His concept of the technosphere as a geologic phenomenon suggests that the human–technological nexus has emerged as new Earth sphere, analogous to the classical spheres of air, water, rock, and biology. Like these ancient spheres, the technosphere operates largely beyond direct human control. In his article "The Far Future of Soil" (2014), he relates this concept to soil, drawing on four scenarios for its future: logistic soils, picture postcard soils, geoengineered soils, and smart soils.

Wanuri Kahiu is an internationally acclaimed storyteller and filmmaker from Kenya. She is the cofounder of AFROBUBBLEGUM, an artistic platform that "supports, creates and commissions fun, fierce and frivolous African art." In her film *Pumzi* (2009), which is Swahili for "breath," she makes the interconnectedness between the spheres of humans and nature graspable in powerful images. The film was often labeled as a work of Afrofuturism, envisioning a postapocalyptic dystopia, "35 years after World War III, the Water War," where the human sphere is separated from its natural environment as a self-sustaining system on a desertified planet. Sealed off from contact with nature, human society lives in a hermetic technospheric regime of control and function governed by the Maitu Community Council. Out of this fictional world and the dialogue with Peter Haff, Kahiu further developed his smart soil scenario toward the idea of intelligent soil.

Carlina Rossée is coordinator of the project Anthropocene Curriculum (anthropocene-curriculum.org) at Haus der Kulturen der Welt (HKW), Berlin. Since 2013, this onsite and online platform for experimental modes of knowledge production and pedagogy apt to the challenges of the Anthropocene has been developing toward engaged collaborations between partnering initiatives worldwide. Trained in comparative literature and philosophy, Rossée's research interests span transdisciplinarily across fields around cosmologies and concurring narratives of nature and the human, contingency, causality and determinism, religion, and science in different knowledge systems.

At the edge of a humanly caused environmental catastrophe in which an evermore resource-absorbing technosphere has spiraled out of control, which speculative futures of inhabitable soil can we possibly imagine? How can the Earth's stories, its granularities, and microclimates be heard and told within a complex entanglement of spheres, time scales, and knowledge systems? This virtual conversation with filmmaker Wanuri Kahiu and geologist Peter Haff was inspired by two of their pioneering works—Kahiu's film *Pumzi* and Haff's text "The Far Future of Soils." (Haff, Peter, 2014, The Far Future of Soil, in G. Jock Churchman and Edward R. Landa (eds.), *The Soil Underfoot: Infinite Possibilities for a Finite Resource*, pp. 61–72, CRC Press, Boca Raton, FL.) From a mutual fascination, common visions unfolded—on an agency of soil, Earthly awareness, institutions as spaces to interact with the future, storytelling between science and the artistic imaginary, and transdisciplinary dialogues as a ground for resistance.

Maitu Community, 35 years after WWIII, the Water War …

The Maitu community has successfully launched the Maitu "intelligent soil" program. To date the Maitu Community has created 7.5 tonnes of intelligent soil with the use of nano technology. Unlike the other "intelligent soil" samples on the market, the Maitu Intelligent Soil (MIS) mimics the fertility of the historic and geographic soils of the area with the ability to adapt to an indoor environment. The MIS has grown three strains of coffee plants and one strain of tea plants. The soil scientists are currently working with the seed mothers to increase the efficiency of the plant and the soil's ability to help it reproduce. Within five years, MIS soil will be available for interplanetary export.

The Maitu Community regularly sends explorers to the outside world to collect soil samples. Over the past five years we have seen the soil in the surrounding area continue to lose nitrogen, phosphorous, grow in acidity, and become increasingly coarse and hard. Tests also show that the soil samples continue to contain high levels of pesticides that break down at a very slow rate. The Maitu Community is working with a handful of scientists to experiment on different sections of soil and help return them to optimal health. So far, their findings are inconclusive.

The Maitu virtual natural museum is the oldest museum in the Eastern Africa territory. It was opened in 2022 to commemorate the closure of the Nairobi National Park. The Museum contains thousands of hours of archival footage from the National Park and its environs. The installations include projections of the largest forests in the region. The Museum was relocated to the second floor in the East wing of the first tower in Maitu Community. The museum contains a permanent collection of the National Park and also houses different exhibitions donated by partnering museums across the Africa territory. The Maitu Natural Virtual design was based on the style and architecture of the Nairobi National Museum and Archives. Its mission is to collect and curate archival footage of extinct natural landscapes, wildlife, flora, and aquamarine life for future generations.

The Maitu Community practices the "community first" article as defined by the 2020 Global Environmental Policy. The community first policy is philosophically based on the ideals that the goals of the individual should not outweigh the goals of the community. Where once the individual, was considered the pioneer, the collective is now hailed for its progress, adaptability, alignment, and responsiveness to the natural world around it. History has shown that societies are made stronger by their disruptors, like Prof. Wangari Maathai, leader of the Pan-African Green Belt Movement. However, in the Maitu Community, a streamlined living and working environment for the optimum micro-climate for humans to live peacefully, dissidents like the fictional inhabitant Asha put the delicate eco-system at risk. Accountability must remain with the Community and be governed by the Council.

Next pages:
Asha in the outside world. Film still from *Pumzi*.

Carlina Rossée: Peter, your essay "The Far Future of Soil" (Haff 2014) centers on the quite radical concept of the technosphere as an increasingly autonomous geological force beyond human control, in which you draw out four future scenarios for the soil. You start out by describing different functions of the soil, suggesting that the function of the soil as habitat for plants and microbes (and organisms of all kinds) also implies a cultural function, that soil is an existential medium of "play"—children's play, artistic practice, and ritualistic traditions. Could you elaborate a little more on this particular function, as well as discuss the ambivalent interdependency of the technosphere and the pedosphere in the global habitat we call Earth?

Peter Haff: The operation of the technosphere depends on other spheres, such as the atmosphere, hydrosphere, or pedosphere (the soil component of the lithosphere), to provide goods and services that are not usually fully represented in a standard economic accounting. For example, the atmosphere provides oxygen and a medium of travel for pollinating insects, the hydrosphere offers rain and streams, and the pedosphere provides filtration of water and nutrients for crops, all of which are necessary for the technosphere to continue to exist and are generally provided "free" of cost. The source of these necessary natural goods and services is called "natural capital." Natural capital evidently summarizes the physical interdependency of the technosphere with other spheres, but has a cultural dimension as well. Thus, natural spheres are a repository of cultural natural capital: as attractions for recreation provided by a beach; for kite-flying by the air; for sailing by a lake; or for worship and other rituals on holy ground and sacred places provided by the soil. Both cultural and physical natural capital are undergoing a process of diminution as the technosphere continues to remake and modify the natural world.

The other theme here is abstraction of natural capital into the technosphere itself; for example, one could argue that agricultural soil, especially at industrial scale, is not natural capital provided by the pedosphere, but rather a resource that has been captured by and become part of the technosphere itself. Continuing this idea to its logical conclusion leads to the notion that humans (as well as domesticated plants and animals) are no longer parts of the biosphere, but now function in essence as parts of the technosphere, i.e., that they are technological artifacts. An interesting question to consider is whether the technosphere could ever develop to a stage where it was completely self-contained, i.e., where all natural capital had finally come under the controlling dynamics of technology. It seems to me that this will not happen if it requires human control, because the system is too complex to be understood by humans, much less controlled by them.

Carlina Rossée: Wanuri, in your film *Pumzi* you envision a postapocalyptic dystopia, "35 years after World War III, the Water War," where the human sphere is separated from a lost natural environment. Human society is functioning as a self-sustaining system on the desertified planet, rigidly sealed off from contact with nature in a hermetic technospheric "inside" world—the Maitu Community. Little water is left, or rather it is recycled to an extreme, all fertile soil is gone and with it all plants and nonhuman living organisms—except for a single archived seed and a mysterious glass of soil. Could you elaborate on your understanding of the soil as an essential basis of life on this planet?

Wanuri Kahiu: The Maitu Community was created to be a thriving, vibrant community in what was once considered sub-Saharan Africa. In it, humans have reduced pollution and increased their ability to be self-sustainable. In the bigger world of *Pumzi* the reason that everybody lives on the inside is because they think the outside is dead. The land has been completely preserved and protected from humans by forcing them inside with the aim of saving the planet. The Maitu Council intended to help the soil recover and return to what Haff describes as "postcard picture" soil.

When the film starts, the Maitu Community has inhabitants living in the residences for three to four generations. The cornerstones of the community are built on hard work, community service, and limited social interaction. In order to survive, the Maitu Community creates hydroponic soil materials to grow genetically modified food. Most of the agricultural processes are automated with computers and involve little input from humans. Meanwhile, the Maitu Council itself is using the more fertile areas outside the contained community (shown at the end of the film) to grow forests and create a thriving eco-climate. In order to save the trees, the Council has created the "inside world" to keep people away from nature. In this environment, the need for soil is limited. The general perception of soil is as a substrate beneath the buildings and toxic and unfertile in the surrounding landscape. The once fecund land surrounding the inside Maitu Community has turned to desert and all the trees have fallen. Fertile soil is highly sought, it would mean life on the outside and the people trapped inside would be able to leave.

Peter, you say that we would have to reduce our energy intake in order for that threat to be countered. But what does that mean for human interaction with the land? If, as you say, even ploughing up the soil is intrusive, are you thinking of sort of a permaculture layering? What role do humans play in your future scenarios?

Peter Haff: My scenarios envision ways in which one could imagine technology transforming landscape as it appears to us. This leaves open the question that is more fundamental in your film of how do people exist and function within that framework. In developing the idea of the technosphere I tried to adhere to a basic principle—to think as much as possible about how the modern world works without specifically invoking the human element. The key question is, what needs to be physically true regardless of how humans function in that framework? Human existence occurs within some physical setting, and that setting can either be "natural," i.e., the result of planetary and biological evolution, or humanly determined. The modern landscape is mostly not accidental, but reflects the effects of increasing human population, industrial activity, agriculture, and other technologies.

Carlina Rossée: Peter, your fourth and most far-reaching scenario shows parallels with Wanuri's film insofar as the technosphere is becoming an evermore dominant sphere. You imagine a self-controlling system of soil permeated by digitally autonomous nanoparticles. This system is detached from any environmental influences and surrounded by a future "picture postcard" landscape (prettily picturesque, like the scenes typically shown on postcards) in which artificial plants simulate nature. How does this scenario of "smart soil" relate to an understanding of soil as a natural system and habitat? Is this a feasible prediction of the future of precision agriculture?

Peter Haff: The smart soil scenario is more like a cartoon. The idea illustrated there is fanciful even if it is imaginable and is meant to suggest how the continued technologization of nature might look if carried to its logical extreme. Perhaps if technology were powerful enough to completely computerize soil, the basic assumptions of large-scale agriculture would need to be rethought. Or perhaps smart soil will be by-passed entirely. For example, at what level might food be "synthetically" created without the use of soil, which substance is mainly at present a convenient medium to hold plants in place and provide an environment that can easily supply water and nutrients directly to the biologic plant "factory"? But maybe all that work could by then be outsourced to industrial factories, or even to individual homes, e.g., artificial meat from 3D-printers, etc.

Wanuri Kahiu: Peter, what about the concept of soil intelligence? Is this different from your idea of smart soil?

Peter Haff: The scenario of the smart soil envisions each grain as a tiny programmable computer. The idea of soil intelligence is taking this one stage further where the soil isn't just preprogrammed but has a mind of its own. This idea was inspired by Stanislaw Lem, the Polish science fiction writer, in his novel *Solaris*. In Lem's book a space station orbits the planet Solaris, which is covered by an ocean possessing a kind of emotional intelligence. The ocean has the capacity of mental telepathy and can influence people's thinking on the space station. The extraterrestrial ocean can incarnate human thoughts and memories. Back on Earth, even natural soil can in some circumstances evoke human emotions, so how powerful might be the emotional influences of a programmed, not to mention an intelligent soil?

Wanuri Kahiu: What kind of characteristics would soil have when imagined in an analogue way as a kind of emotional organism?

Peter Haff: Perhaps its emotional influence might tighten the bond between humans and the Earth, moving us beyond our understanding of soil's utilitarian qualities to a deeper, felt appreciation of how we as humans animate soil within ourselves during the moment we are alive.

Carlina Rossée: Wanuri, maybe you could add something here about the meaning of place. For instance, why did you choose to place your story where you did, in a not-so-distant future, and to thereby call upon the tradition of science fiction? I was also hoping you could talk about the genre of Afrofuturism (which you have often critiqued) as an alternative narrative to the dominant environmental catastrophe fictions of Hollywood, which in many ways reinforce colonizing and exploitative modes that have led us to this moment in the first place and still seem to guide visual culture?

Wanuri Kahiu: Afrofuturism is science fiction, speculative fiction, fantasy, and magical realism made by people of African descent. However, the use of science fiction, or speculative fiction, has always existed in Africa. It is present in our creation myths, legends, stories, naming, etc., and remains so. While I appreciate the need to contextualize work, one must be cognizant of the different types of work, their genesis, and evolution. While the "West" may view the work as an alternative narrative, I believe it has been a primary narrative on the continent and the West is the alternative. Social scientists have suggested that we learn our norms and values from the bedtime stories we are told, therefore science fiction, fantasy, etc., may be considered essential building blocks of many different cultures around Africa. As artists, we create what we know or imagine; it is rooted in where we live and who we are and in our history. The films I create are a result of this and examine humankind in its entirety and how we all contributed to our current state. While colonization severely disrupted the lives of many and impaired the freedom of generations (the effects of which continue to be felt), traditional methods of life within Africa have also created inequalities that we continue to address. The work we create is in response to the conflict we find in Africa today and not in reaction to the Western narrative.

Asha in the Virtual Natural Museum. Film still from *Pumzi*.

Copyright Awali Entertainment, 2009.

Printed with permission of Wanuri Kahiu.

Asha by a dead tree. Film still from *Pumzi*.

Copyright Awali Entertainment, 2009.

Printed with permission of Wanuri Kahiu.

Carlina Rossée: Where do you see your own role as an artist addressing that current state of conflict or crisis? In our contemporary condition, where humans are only beginning to understand their planetary entanglement and responsibility as driving forces of planetary change (and I do believe that there are many cultures that have always seen and been aware of that), how do you see your role as a filmmaker and storyteller? What role do artistic practices and aesthetics—in a broader understanding of the perception of the senses—play in this process of blending past, present, and future? How can storytelling inspire different behavior toward each other and toward the planet?

Wanuri Kahiu: The traditional role of the storyteller in Kenya is to be a seer and a historian, that is, to present possible futures of the world while maintaining a firm grasp on our different histories. The seers would forecast possible environmental changes and migration patterns, while the historians would remember weather patterns and lineage. There was as firm a link to the future as to the past. The role of modern-day storytellers (in film, theatre, art, music, etc.) is to continue to be the thread between the two worlds so that others may traverse freely into their chosen multiverse.

Pumzi stimulated important conversations about environmental change and water conservation. The images in the opening sequence of the film were excerpts of real stories about water scarcity. After screenings, audiences shared stories about their own experiences and talked about the ways they conserve or recycle water in their homes. Over the years, the questions that have come up were not only about the environment, but about social structures, conditioning, tradition, and the role of gender and race. Female scientists of African descent were glad to see Asha in the

role, while the tech world was more interested in futuristic forms of communication.

We are wired to analyze a constant flood of images of people and places and sort them into rational representations of our world. When similar images of people based on gender, race, sexuality begin to emerge in media and art, we start to form an idea about those people and places based on the history of what we have seen. Artistic practices should disrupt the repetitive pattern of stereotypes with new perspectives and projections of the world. Through art we are able to present many different worlds and possibilities, and allow the audience to think of new ways to interact with the themes and the story. Art is meant to inspire change, or at least inspire questioning.

In *Pumzi* we used science fiction to imagine the consequences of our current relationship with the environment. The world was created by researching current technology used to recycle kinetic energy, while Asha's costume was based on ideas around wearable technology. Science was integral in creating a logical and conceivable world, and scientific ideas were put to the test in a stimulated, imagined world. Making *Pumzi* illustrated the possible feedback mechanisms that can exist between science and science fiction. Work in one field is able to push the other to question, investigate, and imagine different scenarios and test hypotheses. Both science and science fiction should be able to ask questions about where and how we live and turn to the other for answers.

Carlina Rossée: Peter, where once humans were the colonizers, you suggest that now technology has become a more aggressive form of global colonization that affects and indeed dictates our basic human interactions. Technology is more likely to create a far more lasting effect on the human psyche, faster degradation of resources, and massive social inequality than colonizers of the past. It may be the truest antagonist of our days and more stories may change to speak of the struggle of man against the machine. So how far does your plea for developing scenarios go to address this? What can we learn from scenarios—a term originating from the theatre—as opposed to the limited capacity of scientific prediction?

Peter Haff: We usually think of humans as the colonizers and exploiters, which is true as far as it goes, but the technospheric perspective also makes the point that technology (as part of a global networked system) is in some ways the prime colonizer and exploiter in the world, and it colonizes (or incorporates) and exploits nature as well as humans. Colonization is moving from a social phenomenon to a technological process.

Science often has difficulty making reliable predictions about the future behavior of complex systems, like societies or economies, and even in the case of physical systems like climate or soil ecosystems. Invoking scenarios is one way to approach the problem of a limited time-horizon in scientific prediction in the face of the need to make critical decisions that will affect our future. Scenarios are imaginary, but if grounded in what we know about the physical, biological, and social world, they can help focus attention on impending challenges to human well-being and suggest possibilities for effective action.

Carlina Rossée: Does this plea for scenarios point toward new epistemologies that might question the primacy of science in Western modernity?

Peter Haff: Science, and more to the point, networked-technology, is so far an unstoppable force that, although in many ways was spawned in the West, seems to have no trouble extending its imperium to all corners of the globe. The problem for the West in moving to such a new and different dynamical state, as I would call it, is that most Western populations could literally not survive without the resources (for example, food, fuel, and shelter) provided by the soils of the world, and have come to depend on a functioning technosphere with its scientific base. Non-Western populations of course recognize the challenge to their traditions and local hegemony from this process, but resistance is difficult. Technology has an effective method of extending its sway far and wide—it acts to give people what they (think they) want, usually in the form of more "efficient" modes of communication, transportation, medical treatment, and so on. At some point these realized wants, which began as optional, tend to become necessities that society cannot live without.

Carlina Rossée: Why do you see these speculative narratives as so crucial in a time when humans are no longer actors in front of a static scenery called nature?

Peter Haff: Because technology is accelerating, but human processes (thinking, decision making, etc.) are not. At some point, relatively soon, the clash of these time scales will fail our expectations and cause a rupture in the system that currently sustains the well-being of much of the world population. We cannot reliably predict any details of how this will occur, but science can tell us that it will occur in the absence of some change in how humans respond to the offers of efficiency that technology makes to us every day. Scenarios and storytelling à la Wanuri are ways to raise awareness of the changes that humans are facing, as well as possible responses, even though we cannot scientifically predict what the future will be. This is the value of scenarios and storytelling.

Carlina Rossée: As a geologist involved with the Anthropocene Working Group, the group tasked to formally define and identify the Anthropocene as our current geological epoch, how do you see the Anthropocene as a concept that can lead toward new forms of understanding our fragile contemporary condition?

Peter Haff: I distinguish the social Anthropocene from the geological Anthropocene. These are entirely different concepts whose distinction is often unrecognized, thus sometimes leading to confusion and unnecessary conflict. The "social Anthropocene" refers to actions of modern humans in the political, economic, and humanistic realm—it is indeed a political concept—whereas the "geological Anthropocene" refers to the actual time in Earth history that we occupy today. The scientific task of establishing the geological Anthropocene as a formal unit of geological time is a purely technical/scientific endeavor. In the end, the success or failure of the formalization process is determined only by the professional assessment by the geologic community of the stratigraphic evidence (coming from observation and measurement of strata in the field).

The geological Anthropocene doesn't directly say anything about our contemporary condition, fragile or otherwise, except that there has been a (geologically) recent and notable change in Earth stratigraphy that is (argued to be) similar in its meaning to changes that have been previously recognized (and now enshrined in the Geological Time Scale) as marking demonstrable and significant changes in Earth history.

Carlina Rossée: How do you see your own role as a scientist facing this situation?

Peter Haff: To try to develop an Earth-based (rather than human-based) picture of how the world works. The concept of the technosphere was intended to remove explicit reference to humans in our treatment of the Earth system, thus hopefully reducing my own role, as well as the role of everyone else, to "simply" one more kind of phenomenon presently active in the current state of Earth functioning. Whatever residue of understanding of our current condition may result is, as much as is possible, independent of any human bias (which, on its own, tends to inflate the importance of humans as geological actors).

Carlina Rossée: Which restraints do you deal with, especially under the political auspices of the current government, which negates climate science and blatantly disregards efforts of global environmental governance as set out by the Paris agreement?

Peter Haff: No restraints, although the political turmoil in the U.S. is perhaps a symptom of an ever-faster technosphere that allows less and less time for the slow response of humans to changes which many of them are suddenly forced to deal with, e.g., the challenge of automation. The phenomenon is like fluid turbulence, where effects generated far away suddenly sweep through a previously quiescent region.

Carlina Rossée: Wanuri, could you talk a little bit about the role of the individual as global actor against the backdrop of institutional or governmental power? The main character and heroine of the film, Asha, is curator of the Virtual Natural History Museum in Maitu Community. When she leaves the museum and then the indoor community housing unit, she takes on another role—that of pioneer. Here, the individual is posited against the institution, demonstrating the power of one person to take action to change the world (or at least a piece of it), as in the case of Wangari Maathai. Does this portrayal relocate accountability from institutions to the individual?

Wanuri Kahiu: The Maitu Council believes it has a higher duty to preserve nature over individuals, however, what they seek to save will only benefit a few. However, the more vulnerable individuals in society (in this case, the majority) do not benefit from that good. When Asha sees the potential to save the people in the inside world, she seeks it out. The role of the individual here is to scrutinize, to question, and to investigate the institutions.

Carlina Rossée: The hypothesis of the Anthropocene has often been criticized as a "Western" concept, reestablishing reductive categories of "the human" and implying a dichotomy between nature and culture that is culturally specific and not at all universal. Would you agree to this critique, Wanuri? And, if so, what alternative concepts would you suggest as more encouraging and inclusive for society to ethically and collectively act upon our situation toward a viable global ecopolitics?

Wanuri Kahiu: I am curious in creating and curating narratives that are not so polarized. So far, the norm has been stories from the global North versus stories from the rest of the world. We are beginning to see the flood of stories from other places that include other theories, ways of learning, and techniques for understanding our place in the world. Perhaps by actively curating the stories that already exist, from places other than the global North, we may find new ways to consider society and ecopolitics. There are far too few stories about cultures that continue to live in symbiotic relationships with nature. Adding more stories of the global South to our collective consciousness may help address the current imbalance. The conflict between people and nature is a human experience and we must draw on our collective wealth of knowledge to address it.

Carlina Rossée: Coming from a cultural institution that is experimenting on how to facilitate new forms of collaboration beyond traditional disciplinary divides, I was wondering how different modes of scientific and artistic knowledge production become intertwined. How can new alliances between the sciences and the arts be organized and established toward a form of "Earthbound knowledge," as Bruno Latour calls it, that can help address pressing issues of climate change, water and soil insecurity, and biodiversity loss?

Wanuri, you spend a considerable amount of film time in the Virtual Natural History Museum. What role does the institution play in your film? Does the nature and function of the museum critically reference contemporary attempts to archive biodiversity and preserve objects of nature, from colonial models of natural history museums and botanical gardens to the Svalbard Global Seed Vault in the remote ice of Norway?

Wanuri Kahiu: While creating the virtual natural museum, we looked at many different "preservation" institutes like the Svalbard Global Seed Bank. The film not only references the need to safeguard natural seeds from around the world, but also references our growing apathy for these institutions. The museum like the one in Nairobi is old, rarely updated, and empty. The Council in Maitu similarly does not consider the virtual museum important or relevant. Instead it is an attempt to keep up the rouse that the outside is dead. The virtual museum is a critique on society's interest in preserving things once they are dead or near extinction, while limited effort is made to keep them alive before they perish.

Meanwhile, the role of the curator is becoming increasingly vital. With the growth of content across all platforms it becomes more important now to be able to sift through everything and collect it into similar themes, thoughts, and ideas. The role of the institute now is not only to preserve but to create spaces where people can interact with the work. Institutions should include the feedback and narratives of their audiences in their collections. Conversations by many can allow for different perspectives and perhaps encourage us to learn from each other or to question our own social norms. Institutions like museums should not only preserve the past but include ways to look, think, and interact with the future. They should be places to present ideas, experiment, fail, and try again.

Carlina Rossée: Finally, Peter, where do you see necessary transformations in academic institutions? How can universities and research institutes rise to meet the challenges of the Anthropocene?

Peter Haff: Universities tend to be too conservative, encouraging work in areas that are in the news, so to speak, and so where scientific work will likely generate attention and resources. This is not unexpected, since, as part of the technosphere, the university dynamics is configured, like that of any other institution, toward its own (short-term) survival. In my opinion, there is often too much pressure in academia to work toward direct solutions of identified problems. As E.O. Wilson has advised young scientists, "If you are looking for a scientific problem to work on, and hear the sound of drums and bugles in the distance, make

sure to march in the other direction." At least in areas like environmental science and policy, universities still need more faculty and students who will challenge the terms of debate. But to that end, they need more philosophers, visionaries, and storytellers.

Smart Soil, grain-sized microcomputers drawing by Peter Haff. The sketch on graph paper suggests a rectangular perfection underlying the regularity that technology is trying to impose on the world.

Reprinted with permission from Peter Haff.

Function 5
HERITAGE: Soil as embodiment of cultural memory, identity, and spirit

Heritage

The fifth section focuses on the cultural heritage and archive functions of the soil. These functions encompass the soil's capacity to provide aesthetic pleasure, recreational enjoyment, cognitive development, and spiritual or religious enlightenment. These have proven difficult to evaluate, monitor, or protect with the same standards as other soil functions because there are no "indicators" or other metrics with which to measure the cultural functions of the soil. Philippe Baveye et al. argue that "from a scientist's perspective, it makes sense to advocate strongly that the assignment of values to soil functions/services should rely heavily on their quantification, in actuality … for a number of cultural, aesthetic, and spiritual services, quantification based on fundamental scientific principles would be entirely meaningless, and valuation is where the story starts." (Baveye et al., "Soil 'Ecosystem' Services and Natural Capital," 22.) The contributors in this section take up that story of valuation. They take as points of departure singular sites, material samples, personal encounters, and cultural histories to create works in celebration of the complexity and diversity of the soil as embodiment of collective identity and spiritual ancestry.

The cultural heritage of the soil is evoked as a visual ethnography in space and time in a conversation with Ekkaland Götze and Winfried Blum; as an educational opportunity in chapters by Ken Van Rees and Symeon van Donkelaar, and Nance Klehm and Akilah Martin; a narrative of shifting land-use practices and beliefs in the chapter by Ruttikorn Vuttikorn, Myriel Milicevic, and Prasert Trakansuphakon; and as a medium of "deep time" used for advocating indigenous rights in a chapter by Mandy Martin, Libby Robin, Guy Fitzhardinge, and Mike Smith, and in an interview with Cannupa Hanska Luger by Alexandra R. Toland. While E. Christian Wells and Marlena Antonucci outline ideas about cultural soilscapes and soil memory, Linda Weintraub offers advice for becoming a soil connoisseur, and Beth Stephens, Annie Sprinkle, and Fred L. Kirschenmann discuss the image of soil as lover (rather than mother) as new ways of conceiving soil heritage.

Reframing Heritage

Cultural Soilscapes and Soil Memory

E. Christian Wells and Marlena Antonucci

E. Christian Wells is professor of anthropology and director of the Center for Brownfields Research at the University of South Florida, where he has served previously as the founding director of the Office of Sustainability and as deputy director of the School of Global Sustainability. Wells is an applied environmental anthropologist whose research focuses on environmental justice issues involving soil and water contamination in underserved communities. Over the past twenty years, he has undertaken research throughout the United States, Central America, and the Caribbean, resulting in over one hundred books, articles, and essays. He was recently awarded the Global Achievement Award for Outstanding Global

Title image: Soil Kitchen. Future Farmers, 2012.

180 W Girard Ave. 19123-1660 Philadelphia, PA 39° 58' 7.8096" N, 75° 8' 20.508" W.

A vacant building rehabbed with a large working windmill made of wood and a reused car axle, workshop spaces, pop up kitchen, and EPA soil testing van. In this chapter Future Farmers' Soil Kitchen is used as a case study in the interpretation of cultural soilscapes.

Student Success from the University of South Florida and the Black Bear Award by the Sierra Club of Tampa Bay.

Marlena Antonucci is the community engagement coordinator for the Rio Grande Farm Park in Alamosa, Colorado, an innovative and multiuse park along the Rio Grande corridor that fosters an equitable, local food system. Antonucci earned her MA in art history from the University of South Florida, focusing on the use of soil as an artistic medium and how brownfield sites in particular might be reimagined by contemporary artists as meaningful places. She is currently working with San Luis Valley artists to install public artworks in the Farm Park that draw attention to the preservation of the valley's agricultural heritage, water resources, and open public spaces. Her greater goal is to transform visitors' interactions with the land by creating spaces for contemplation so they may observe the complexity and beauty of natural systems.

Cultural soilscapes denote anthropogenic landforms, such as farms, fields, and forests, which were created by a combination of natural and cultural processes. Since these places are imbued with perceptions and values by the people that created or experience them, they are subject to interpretive processes of meaning making. Central to these processes is the concept of soil memory, or the capacity of a soil to embody the impacts of human behaviors over time. As legacies of social-ecological dynamics on the land, soilscapes are integral to cultural heritage. In this chapter, we discuss the ideas of cultural soilscapes and soil memory with the aim of broadening understandings of heritage. We also explore how one contemporary artist collective transforms the cultural soilscape of an industrialized urban space into a meaningful place for community residents to reengage with local cultural heritage.

Why is dirt "dirty"? When something is "dirty," why do we say that it is "soiled"? Generally speaking, how do our perceptions—anecdotal and scientific—of dirt and soil influence and sometimes determine how we interact with these substances? Over ten years ago, one of us proposed the concept of "cultural soilscapes" ("a given area of the Earth's surface that is the result of spatially and temporally variable geomorphic, pedogenic, and cultural processes"[1]) to help answer these questions and to develop an analytical vocabulary with which to study and write about the values and meanings of soil. Since that time, some provocative scholarship has emerged that deals either directly with the concept[2] or with similar ideas.[3] Together, these contributions demonstrate how soils form an important part of cultural heritage, that is, the legacy of artistic expressions, customs, practices, values, and beliefs inherited from prior generations. Since the arts are integral to both tangible and intangible cultural heritage,[4] in this chapter we—an environmental anthropologist (Wells) and an art historian (Antonucci)—extend the conversation to consider one case study, Soil Kitchen, in which cultural soilscapes are imagined and visualized by the contemporary arts collective Future Farmers.

Cultural Soilscapes

Contemporary artists and scientists studying soils from various perspectives have increasingly been engaged in conversations about the value of soil beyond the environmental services it provides and its capacity to produce sustenance.[5] Some of these conversations have centered on cultural values and the idea that, for many communities on the planet, soil is an intrinsic part of their heritage.[6] Jon Sandor and colleagues,[7] for instance, describe the heritage of soil knowledge among different cultural groups across the world. Drawing from the rapidly expanding field of ethnopedology (folk soil taxonomy), which examines the overlaps and divergences between Western soil science and local ecological knowledge,[8] they argue that soil knowledge from other cultures through time is of great value because it is long term and time tested, and because it is holistically integrated within other aspects of society, including politics, economy, and religion.

Wells and Mihok,[9] for example, demonstrate how ancient Maya worldview that legitimated Maya kingship was materialized in religious cults of spirits associated with agricultural production. In this case, soil was conceived as a gift from the ancestors that had to be repaid with human blood. Certain types of soil were often treated as sacred substances, which supported its long-term care and conservation. Traditional farmers today in certain parts of the Maya region believe that soil conveys *itz*, a holy substance that gives life to plants.[10] Wells and Mihok[11] show how these ideas were expressed through the design of hieroglyphic inscriptions on monuments and in codices, ancient bark-paper pictorial manuscripts that served as almanacs for agrarian rituals.

In the case of the Maya and in many other examples across the world,[12] cultural soilscapes can be a useful way to understand soil formation as both an outcome of human behaviors and the chemical, biological, and geological processes that produce and change soils over time. The concept is also useful for understanding the cultural heritage of soils more broadly because it refers to perceived spaces whose boundaries are delimited, not by Western scientific understandings of the unique edaphic characteristics of anthrosols (i.e., any soil impacted by human activity), but by the individual viewer's lived experience. In this way, cultural soilscapes are constituted by our worldviews, values, and beliefs that we use to make sense of our experiences as well as our behaviors that stem from them.

This relational view of cultural soilscapes, which situates them as social constructions of reality, recognizes that people and soils are mutually transformative[13] such that "modifying the medium of expression, in this case the soil, alters the expression itself."[14] As a result, to use this analytical concept for understanding cultural heritage and contemporary soil artworks, we need to recognize a broader set of attributes of soil. For example, in addition to studying soil's color, texture, moisture, acidity, and cation exchange capacity, we also need to study its social value, cultural meanings, heritage, affordances, and the legacies it presents to us from previous generations.[15] In other words, we need to interrogate a soil's memory of past human activities.

Soil Memory

Cultural soilscapes mediate the relationship between past human behaviors that impacted soils and present/future practices that have to contend with this heritage. Wells[16] describes this property as "soil memory," or "how soils encode the physical, biological, and chemical effects of different human behaviors" over time. Archaeologists have employed this concept to detect and study past land-use practices by, for example, characterizing the physical and chemical changes to soil caused by different human activities in the past, from hundreds to thousands of years ago.[17] Yet, even more recent impacts to cultural soilscapes, such as the formation of brownfields (polluted or contaminated spaces) in contemporary urban settings, can form part of a soil's memory.[18]

Brownfields as cultural soilscapes, in particular, have provided soil artists with a great deal of inspiration in recent times.[19] For example, for one week in April 2011, the artist collective Futurefarmers[20] showed the relational role of cultural soilscapes by inviting residents of the city of Philadelphia, Pennsylvania, to bring samples of soil from their yards to an abandoned brownfield they called *Soil Kitchen*[21] for a soil-health screening. The brownfield site, occupied by an abandoned building, was reimagined as a social space, a soup kitchen, and a soil lab powered by a working windmill that was erected on its roof.

The city of Philadelphia was originally designed in the seventeenth century as a green town with greenbelts, parks, and public squares, with residences planned as detached houses with ample space for kitchen gardens. Over the centuries, Philadelphia grew into a thriving industrial metropolis with a peak population of over 2 million. Over the past 50 years, however, quality of life shifted as infrastructure was unable to keep up with the population boom, resulting in factory closings and a mass exodus of over half of a million people. In 2011 there were 26,000 vacant homes, 31,000 vacant lots, and 54,000 vacant industrial and commercial buildings in the city. Some of the various types of properties that have become brownfields in the city include mine-scarred lands, former meth labs, and abandoned factories. These soilscapes thus represent a palimpsest of previous land uses, from agriculture to industry. In planning *Soil Kitchen*, Futurefarmers questioned how it could use the soil's memory to remind the city, confronted with a postindustrial condition, of its agricultural heritage. By inhabiting one of these derelict spaces, the artists crafted a space to question, disrupt, and draw attention to a community issue—the myriad deserted spaces left over from a postindustrial state.

Philadelphia's Office of Arts, Culture, and Creative Economy commissioned *Soil Kitchen* to coincide with the 2011 National Brownfield Conference organized by the U.S. Environmental Protection Agency. Embracing the mission of the conference, Futurefarmers proposed a potential outcome of a community with remediated and nourished soils. Preliminary illustrations of *Soil Kitchen* show how the space was to function. In one illustration, the flow of the multiuse space was demonstrated. The participant would walk through the entrance into the "Soil Area" where soil is collected and tested. There are test tubes hanging in the window above workspaces. A map of Philadelphia's neighborhoods is hung facing the door. The middle of the space, called the "Soup Area," contained bowed eating benches in a circular configuration and a soup station near the back wall. The "Workshop/Screening Area" is located at the far end of the room where chairs and a projector are set up for meetings and lectures.

During the installation in 2011, thrift-store bowls and mismatched spoons were arranged next to a cafeteria cart parked along the center of the back wall of the *Soil Kitchen* site. There was a sign above the cart that read, "SOIL * SOUP * LEARN." Every day of the week a local business, Cosmic Catering, prepared two different varieties of soup sourced from local, organic farms. *Soil Kitchen* exchanged about 50 quarts of soup per day. Functioning under a barter economy, participants were able to imagine and taste soup made from vegetables grown in their community. Their soil was then tested for heavy metals by the Environmental Protection Agency (EPA), providing individuals with knowledge about the historical and present condition of their land, and providing the EPA with general and anonymous data for areas of concern in the city. *Soil Kitchen* also provided participants with the tools to make decisions about what can be done with their soil, through lectures and workshops on topics ranging from composting to cooking lessons.

Many anonymous people submitted soil samples; over 350 samples were tested and 300 bowls of soup served each day of *Soil Kitchen*. The samples

SOIL KITCHEN

SOIL MAP/ARCHIVE

WORK/DEMO TABLE

Servers

Soup TABLE

FOLDING CHAIRS

EATING AREA

① ② ③

←ENTRANCE

←GIRARD→

① SOIL AREA
work/demo table
- SOIL COLLECTION
- TESTING
- INFO GATHERING

② SOUP AREA
- EATING BENCHES
- SERVERS

③ WORKSHOP/
SCREENING AREA
- TURBINE WORKSHOP

Future Farmers
SCALE: 1/8"=1'0"

A
B
C
ONE PLEASE

TO WIND TURBINE WORKSHOP

② VIEW OF SOUP SERVICE
AND GENERAL SEATING

③ VIEW OF PROJECTION/
WORKSHOP SPACE.
FROM EXTERIOR

SOIL KITCHEN

Window

B

A

① View of SOIL AREA from Exterior
Soil Kitchen /Philadelphia 2011

2ND Street

Soil Kitchen. Illustrations by Amy Franceschini and Dan
Allende/Futurefarmers 2010. *Soil Kitchen* view from
inside, 2011.

collected were numbered and the individuals marked their number on an illustrated map hung on the wall of the space. The hand-drawn map demarcated neighborhoods, indicating citywide involvement. When the project was completed, it charted a communal identity characterized by a broad concern for soil health and heritage. The map visually expanded the impact of the work far beyond the boundaries of the brownfield site because citizens left with a different perception of their community. The soil's memory reminded them about how the soilscape had changed over time. At the same, the *Soil Kitchen* map became a tangible material object that reflected a collective consciousness. Offering a temporary space to reflect and critically engage with soil history resulted in a shared sense of communal responsibility.

Futurefarmers created spaces of heightened awareness and appreciation for daily social actions in an unusual setting. They engaged sociality in order to reveal a collective interest in the community's soil. *Soil Kitchen* project manager, Dan Allende, noted that the space provided a platform for generations of local farmers to come together and share their practice for the first time, a meeting that he said would continue to take place after the project concluded.[22] In this way, Futurefarmers employed soil as the subject, which then shaped the participants' relations. If cultural soilscapes act as "reservoirs of shared ecological knowledge and its manifestation in the soil record,"[23] then the City of Philadelphia was reimagined as a cultural soilscape through *Soil Kitchen* because it manifests shared values and beliefs and also organized the designed environment in which Philadelphians live.[24] Applying the concepts of cultural soilscapes and soil memory to soil artworks provokes a new engagement with the material by showing that culture, heritage, and soil are not isolated but mutually interdependent. In this way, soil artworks such as *Soil Kitchen* can empower their viewers to decenter their worldview and understand soilscapes from a more holistic perspective.

Endnotes

1. Wells, E. C. 2006. Cultural soilscapes. In *Function of Soils for Human Societies and the Environment*, edited by E. Frossard, W.E.H. Blum, and B.P. Warkentin, p. 125. Geological Society of London, London.

2. Butler, D. H., 2011, Exploring soilscapes and places inside Labrador Inuit winter dwellings, *Canadian Journal of Archaeology* 35(1):55–85; Pauknerová, K., R.B. Salisbury, and M. Baumanová, 2013, Human-landscape interaction in prehistoric Central Europe: Analysis of natural and built environments, *Anthropologie* 51(2):131–142; Roos, C.I., and E. Christian Wells, 2017, Geoarchaeology of ritual behavior and sacred places: An introduction, *Archaeological and Anthropological Sciences* 9(6):1001–1004; Salisbury, R.B., 2012, Engaging with soil, past and present, *Journal of Material Culture* 17(1):23–41; Toland, A.R., 2015, Soil Art—Transdisciplinary approaches to protection, doctoral thesis, Department of Soil

Protection, Technische Universität, Berlin, Germany; Vittori Antisari, L., S. Cremonini, P. Desantis, C. Calastri, and G. Vianello, 2013, Chemical characterisation of anthro-technosols from Bronze to Middle Age in Bologna (Italy), *Journal of Archaeological Science* 40(10):3660–3671; Wells, E.C., and L.D. Mihok, 2010, Ancient Maya perceptions of soil, land, and earth. In *Soil and Culture*, edited by E.R. Landa and C. Feller, pp. 311–327, Springer, New York.

3. Conesa, F.C., A. Lobo, J. Alcaina, A. Balbo, M. Madella, S.V. Rajesh, and P. Ajithprasad, 2016, Multi-proxy survey of open-air surface scatters in drylands: Archaeological and physico-chemical characterisation of fossilized dunes in North Gujarat (India), *Quaternary International* 436:57–75; Contreras, D.A., 2017, (Re)constructing the sacred: Landscape geoarchaeology at Chavín de Huántar, Peru, *Archaeological and Anthropological Sciences* 9(6):1045–1057; McAnany, P.A., and I. Hodder, 2009, Thinking about stratigraphic sequence in social terms, *Archaeological Dialogues* 16(1):1–22; Montgomery, D.R., 2007, *Dirt: The Erosion of Civilizations*, University of California Press, Berkeley.

4. Ruggles, D.F., and H. Silverman (eds.), 2009, *Intangible Heritage Embodied*, Springer-Verlag, New York.

5. Adams, C., and D. Montag, 2015, *Soil Culture: Bringing the Arts Down to Earth*, Centre for Contemporary Art and the Natural World and Falmouth Art Gallery, Dartington, England; Boivin, N., 2004, Geoarchaeology and the goddess Laksmi: Rajasthani insights into geoarchaeological methods and prehistoric soil use, in *Soils, Stones, and Symbols: Cultural Perceptions of the Mineral World*, edited by N. Boivin and M.A. Owoc, pp. 165–186, UCL Press, London. Cox, R., R. George, R.H. Horne, R. Nagle, E. Pisani, B. Ralph, and V. Smith, *The Filthy Realty of Everyday Life: Dirt*, Profile Books, London; Landa, E.R., and C. Feller, 2010, *Soil and Culture*, Springer, Dordrecht; McNeill, J.R., and V. Winiwarter (eds.), 2010, *Soils and Societies: Perspectives from Environmental History*, The White Horse Press, Cambridge, UK.

6. Boivin, N., and M. Ann Owoc, 2004, *Soils, Stones, and Symbols: Cultural Perceptions of the Mineral World*, UCL Press, London; Winiwarter, V., and W.E.H. Blum, 2006, Souls and soils: A survey of worldviews, in *Footprints in the Soil: People and Ideas in Soil History*, edited by B.P. Warkentin, pp. 107–122, Elsevier, Amsterdam, The Netherlands.

7. Sandor, J.A., A.M.G.A. WinklerPrins, N. Barrera-Bassols, and J. Alfred Zinck, 2006, The heritage of soil knowledge among the world's cultures, in *Footprints in the Soil: People and Ideas in Soil History*, edited by B. P. Warkentin, pp. 43–84, Elsevier, Amsterdam, The Netherlands.

8. Barrera-Bassols, N., and J. Alfred Zinck, 2000, *Ethnopedology in a Worldwide Perspective: An Annotated Bibliography*, International Institute for Aerospace Survey and Earth Sciences, Enschede, The Netherlands; Krasilnikov, P.V., and J.A. Tabor, 2003, Perspectives on utilitarian ethnopedology, *Geoderma* 111(3–4):197–215.

9. Wells and Mihok, "Ancient Maya perceptions."

10. Dunning, N.P., and T. Beach, 2004, Fruit of the *lu'um*: Lowland Maya soil knowledge and agricultural practices, *Mono y Conejo* 2:10.

11. Wells and Mihok, "Ancient Maya perceptions," 317–318, 321–322.

12. Warkentin, B.P., 2006, *Footprints in the Soil: People and Ideas in Soil History*, Elsevier, Amsterdam, The Netherlands.

13. Ingold, T., 2000, *The Perception of the Environment: Essays on Livelihood, Dwelling and Skill*, Routledge, London.

14. Salisbury, "Engaging with soil," 26.

15. Wells, E. C., K.L. Davis-Salazar, and D.D. Kuehn, 2013, Soilscape legacies: Historical and emerging consequences of socioecological interactions in Honduras, in *Soils, Climate, and Society: Archaeological Investigations in Ancient America*, edited by J.D. Wingard and S.E. Hayes, pp. 21–59, University Press of Colorado, Boulder.

16. Wells, "Cultural soilscapes," 126.

17. Wells, E.C., and R.E. Terry (eds.), 2007a, Advances in geoarchaeological approaches to anthrosol chemistry, Part I: Agriculture, *Special issue of Geoarchaeology: An International Journal* 22(3); Wells, E.C., and R.E. Terry (eds.), 2007b, Advances in geoarchaeological approaches to anthrosol chemistry, Part II: Activity area analysis, *Special issue of Geoarchaeology: An International Journal* 22(4); Wingard, J.D., and S.E. Hayes (eds.), 2013, *Soils, Climate, and Society: Archaeological Investigations in Ancient America*, University Press of Colorado, Boulder.

18. Meuser, H., 2010, *Contaminated Urban Soils*, Springer Science and Business Media, Dordrecht, The Netherlands.

19. Antonucci, M., 2016, Cultural soilscapes: Land art in the expanded field, MA qualifying paper, School of Art and Art History, University of South Florida.

20. Artists/collaborators: Amy Franceschini, Dan Allende, Lode Vranken, and Ian Cox.

21. www.futurefarmers.com/soilkitchen

22. Wexler, D., 2011, Soil Kitchen combines art, environmentalism, and food, *Philadelphia Inquirer*, April 4.

23. Wells, "Cultural soilscapes," 126.

24. Antonucci, "Cultural soilscapes."

The Weapon is Sharing

Cannupa Hanska Luger in conversation with Alexandra R. Toland

Soil cultural heritage (see previous chapter by E. Christian Wells and Marlena Antonucci) refers to the legacy of cultural artifacts embedded in the soil by a specific group or multiple groups of people, inherited from past generations, maintained in the present, and preserved for the future. The International Soil Reference and Information Centre (ISRIC) includes "physical and cultural heritage" in its list of soil functions, but fails to offer guidance on how to protect them. Existing measures to safeguard these functions are vague at best, relying on a kind of "piggyback" protection. Where filter and buffering functions are protected in regulatory procedures and on-the-ground practice, so are archive and heritage functions. To fully understand the deeper meaning of soil heritage, as well as approaches to

Title image: *The Weapon is Sharing (This Machine Kills Fascists)*, 2017. Ceramic objects, variable dimensions. The phone objects are recreated out of clay in a state of vitrification. They may last forever and relay urgent recognition of our current environment. Forcing antiquity on our modern technology creates a way for Indigenous perspectives to immediately penetrate the historical record. The objects increase the permanence of our contemporary stories as told through society's current throw-away technology.

Image courtesy of artist.

its protection, we must listen to the groups of people whose cultures produced such arti-facts. Indigenous groups the world over offer practical wisdom for protecting soil heritage. To protect the soil is to hear their voices and respect their values and practices. In July 2017 I (Toland) spoke with the acclaimed ceramicist and multimedia artist Cannupa Hanska Luger about his views on soil heritage and soil memory, the challenges and joys of working with clay, dominant cultural influences, indigenous visual technologies, and connection to place. What started out as a list of questions about soil heritage developed into a much longer and richer conversation about the wider goals of re-indigenizing contemporary society. In reference to one of his more well-known works, *The Weapon is Sharing*, Luger ends the con-versation by suggesting that the strongest defense we have in protecting soil heritage—and future generations of humans—is sharing.

Born in North Dakota on the Standing Rock Reservation, multidisciplinary artist **Cannupa Hanska Luger** comes from Mandan, Hidatsa, Arikara, Lakota, Austrian, and Norwegian descent. His work communicates stories of complex indigenous identities coming up against twenty-first century challenges, including human alienation from and destruction of the land to which we all belong. Luger is known for his ceramic innovations and a practice that combines critical cultural analysis with dedication and respect for the diverse materials, environments, and communities. He tells stories using fiber, steel, cut-paper, video, sound, performance, monumental sculpture, land art installation, and social collaboration.

Luger holds a BFA in studio ceramics from the Institute of American Indian Arts. He was recipient of the 2016 Native Arts & Cultures Foundation National Artist Fellowship Award and has participated in artist residencies and institution lectures throughout the nation. He maintains a studio practice in New Mexico. His work is collected and exhibited nationally and internationally, including The Denver Art Museum Denver, Colorado; The Museum of Contemporary Native Arts Santa Fe, New Mexico; Radiator Gallery New York, New York; La Biennale di Venezia, Verona, Italy; Art Mur Montreal, Quebec; Museum of Northern Arizona, Flagstaff, Arizona; Rochester Art Center, Rochester, Minnesota; Navy Pier, Chicago, Illinois; and the National Center for Civil and Human Rights, Atlanta, Georgia.

Alexandra R. Toland: Since this is a book about soil and art, I want to begin our conversation with clay. Clay is an amazing substance. The presence of colloidal clay crystals, with their charged electrons, can attract toxic particles in the soil in a dynamic process of cationic exchange. Chemically bound to their clay partner, dangerous heavy metals and organic pollutants may be rendered immobile in drier areas. The molecular biologist Alexander Graham Cairns-Smith even proposed that clays could be a proto-organic template for biological replication, a mineral precursor to DNA.[1] As an artistic medium, clay is one of the oldest and most diverse. This book aims to explore different forms of knowledge about the soil, so I was hoping you could talk about your knowledge of clay from the point of view of a master ceramist. What have you learned about clay that you could share with us here?

Cannupa Hanska Luger: What I know about clay is still a learning process. I always consider clay my greatest teacher of working with clay. Really, what I see is a relationship that has developed over time and I feel like that relationship is ongoing. So what I know versus the process of gaining that knowledge are two entirely different things. My people, Mandan and Hidatsa, are clay people. They built objects and vessels and stuff out of clay, and that knowledge is gone now due to the historical practices in North America, or the United States. Our clan system got broken up in the 1940s and early 1950s, and that separation affected the knowledge that we have around clay, among other things. So, I don't have many elders that I can talk to about their knowledge and relationship to clay. Honestly, I have never actually seen any Mandan pottery. I've seen it in books. I've seen it online, but I've never physically held a ceramic object made by my own people. And the irony is, the clan that I'm from is Awa

Xe, which translates from Siouan languages to "Dripping Earth." We were the builders for our community and we lived in earth lodges that we would line with clay. You know when you make sand castles and drip the sand onto itself and it kind of makes these bubbly shapes? Well, that same thing would happen down the inside of our earth lodges, so that's what "Awa Xe" means.

Alexandra R. Toland: So, with that traditional knowledge of the Awa Xe lost, how did you come to use clay? When did you first come in contact with clay?

Cannupa Hanska Luger: I've only been working with clay for probably around 10 years now. I received my first lump of clay from Roxanne Swentzell and her daughter Rose. Roxanne is a Santa Clara Pueblo artist and she runs a ceramics studio and gallery and permaculture institute here in Sante Fe and is much more connected to this environment of New Mexico than I am, coming from North Dakota. She's amazing as a person, and art-maker, and elder.[2] I used to work with Roxanne and Rose at their permaculture institute. I learned to build with adobe there, which is an old technology that is still prevalent. Roxanne was like, "You're pretty good at building with adobe. You can keep your walls straight." So she encouraged me, and we would work on projects together. For example, Roxanne developed a program where they would go to all the different pueblos down here and build *Pantes*, which are traditional adobe bread ovens.[3] She would organize a group of people from the pueblo to help build a Pante. It was a process of teaching how to build, versus building something for them. It was their responsibility to maintain it and replaster the walls every year. So, those experiences were really my first introduction to clay.

Lake Sakakawea 2016, film still.

Image Credit: Dylan McLaughlin, Cannupa Hanska Luger.

One day, I came to work and Roxanne and Rose were trying to get some sculptures done for some art projects they were working on. I guess I was kind of antsy and waiting to work on something else and Roxanne just chopped off a lump of clay and handed it to me and was like, "You're driving us crazy pacing around like that. Have some clay and make something." So that was when I built my first sculptural form using clay, and that changed my whole focus. At that point, I was still in school at the Institute of American Indian Arts here in Santa Fe. I was a painter primarily, but I had come to school with painting as a preconceived idea. I was self-taught and it was really difficult for me to shift from what I taught myself to what the institution wanted to teach me about painting. So, I decided to start from zero and started taking classes from Karita Coffey, who is Comanche. She recognized that I was going to produce a lot of work, so she said, "I'll give you a table in the back of the classroom. Go to town. Just play with the clay, learn, figure it out. Have the conversations that need to be had with it, share your work, and we'll critique it." So, I began a kind of self-paced education in ceramics, which was very freeing. I would use tools improperly and my ceramics teacher would come over and shake her head and say, "That's not what that tool is used

for, but you're making it work. So I'm not going to stop you."

Through that whole process, because I didn't know much about the material, I was able to push my work past the expectations of our curriculum. And it was really fun. There were tons of failures in that process, which was important. The clay itself would be like, "Nope! I'm not going to do that. That's the point that I break." So it's really been a process of trial and error. And that's the part of my relationship to the material that I really like. I would end up building pieces that had broken before and would try to build them

again but knew where the weak points were so I could reinforce those areas. So I learned ways to engineer the clay in its weaker points. But I feel like the clay itself taught me all of that. It taught me necessities of engineering that I needed for the technology of ceramics. And now, ten years later, I sometimes feel as though I'm still in the same place. It's always difficult for me to consider myself anything more than a student of the material.

Alexandra R. Toland: Did you ever make pottery as your Mandan ancestors did, or were you strictly interested in making art? What's your view as a ceramic artist on the relationships between form

Aquatint of a Mandan Village by Karl Bodmer, 1839, from the book *"Maximilian, Prince of Wied's, Travels in the Interior of North America, during the years 1832–1834"* by Prince Maximilian of Wied (Ackermann & Co., 1839).

and function, art and craftsmanship? I ask this in reaction to some of the things you mentioned about "engineering" the clay. There are similar distinctions between science and engineering, as there are with art and design or arts and crafts in terms of functionality and purpose. But I think clay is a really interesting medium because it pushes us to think beyond such distinctions.

Cannupa Hanska Luger: I guess I've made functional objects now and again, but even my functional objects are produced in the context of art making, because that's what I wanted to engage with. But once again, working with clay kind of shifted my philosophy. I always knew I wanted to be an artist when I was younger, but as I worked and developed my relationship with the material itself, I recognized that we're all artists and it is the material itself that wants us to hone our craft. And the more I learned about the material itself, the more interested I became in craftsmanship. I try not to fire things that I'm not proud of. And I've done a lot of work

where I don't fire the work, where I just leave it unfired and let it remain clay because I'm aware of its molecular transformation. Once it becomes ceramic, it can never become clay again. Clay has endless potential. Ceramic is frozen. So I think there's a responsibility as a ceramic maker to know that you are ending the potential of this clay to be that thing from now on. So act accordingly.

At this point in my career, I consider everything I make, these sculptures or objects, to be a vessel. They all hold something, whether it is an idea, or a prayer, or conversation. But before any form is made out of clay, it is already a vessel—it holds water. The material itself, the clay, holds water molecules in its mineral structure. Then, through the process of firing it, you burn the water out. All the water, all the organic material, everything burns out in the clay and what you're left with is an echo of that vessel. I believe that echo, that void, wants to be filled with something always. The material itself always wants to carry something. Throughout all human experience of engaging with ceramics, the clay has always kept that potential. For everybody who makes anything with clay, the material itself wants to hold, wants to be that vessel. It wants to carry nourishment of all kinds, whether it be water, food, ideas, conversations, inspiration.

And that makes me want to be more responsible with it. I recognize that clay itself is endless potential. The material has the ability to travel around the world, and has done so. The kaolin has found its way around the entire globe with the shifting of tectonic plates. It's picked up little bits and pieces of its travels. It reminds me so much of our experience as a living beings. We developed in one place and yet we travel around the world. Through that process, we grated the surface and absorbed bits and pieces of our landscape. It's developed and changed our very physiology. The clay itself has done that same trip. I think that's why we have this intimate, ancient relationship with it. Even to this day, ceramics is one of those technologies that stays true, honest, and absolutely useful even within our advanced technological society. We still use ceramics in our electronics. Our space shuttles are covered in ceramic tiles. And through all of those boundaries that we've pushed within our human experience, we've taken clay with us. Perhaps also vice versa. Maybe it's the clay that has taken us along with it.

Alexandra R. Toland: What a beautiful thought … That reminds me of a concept introduced by E. Christian Wells—the idea of *soil memory*, or the ways in which "soils encode the physical, biological and chemical effects of different human activities."[4] Soil memory can be relived through chemical compounds archived in the pedosphere. These traces are indicators of prosperity and resilience, population rise and fall, human habit, ritual, diet, transport, technology, and environmental change. Early soil scientists have even described the soil as an environmental

"body," implying a living entity that perhaps at some level has the capacity to remember, or subjectively experience its own past. I wonder if you think that the soil can remember.

Cannupa Hanska Luger: That question is really interesting to me … I think what we could do, rather than extract history from the soil record, we could recognize the conversation that it has carried over the years. In talking about memory, it's not about transporting knowledge from the past to the future. In my view, it's about realizing that the soil is part of a larger cycle. So the question is framed according to a very human idea of time. We view time in this very linear pattern and yet I feel like it's more of a continuum, or something cyclical, or spherical like the Earth itself. So, rather than having the conversation around the linear memory of the soil, and rendering that story into human terms, I think we should begin trying to render our human story in terms of the Earth, to recognize how we are a part of this place, a part of this landscape …

In terms of clay, I definitely think it has memory. I've been doing a lot of slip casting lately—building ceramic forms, then making plaster molds out of those ceramic forms, and then casting them. And I swear, it is the hardest thing to try to get that casting line out of the material. You can carve it away, sand it down, and wet it. Then you throw it in the kiln, fire it, and you can see it again. The clay itself is just like, "Hey, this is where I was put together." It's almost like a scar tissue. It's smooth and it's pretty close to the same thing, but it's different. And if you're aware, you can see that line.

Alexandra R. Toland: That's in the studio. What about clay memory in situ? Could you talk about your experience with clays as they occur in their natural state, as part of the landscape? Do you ever dig your own clays?

Garrison Dam construction site circa 1950. Image from the Gus Sorlie Photograph Collection, Photographer unknown.

Photo Credit: State Historical Society of North Dakota, 2014-P-038-00062.

Cannupa Hanska Luger: Yeah, definitely. I go back to North Dakota every summer, and this is a very crazy but true story. They built a dam in 1948 on the Missouri River and it flooded out our lowlands. That's where we all lived. It was this beautiful river valley. There was a town there called Elbowoods. It was about two miles from our oldest village—or rather our youngest village prior to contact. That village was called Like A Fishhook. They called it Elbowoods and Like A Fishhook because the Missouri River did a 90-degree turn there that created this really lush and powerful river valley. So when the Army Corps of Engineers built the Garrison Dam, the whole area flooded.[5] Forty-two percent of my reservation is underneath Lake Sakakawea, which is strange to me because all of that happened before I was born and I don't know firsthand what it must have been like before the flood. But every time I go back to North Dakota, I notice how the high water mark had eroded the banks of the hills that ran through there and it exposed many different clay veins along the river valley. There are hillsides that have been sheared off completely from the high water mark. So harvesting clay from there has been easy because of that event.

Meanwhile, back here in New Mexico where I currently live, there is a rich ceramic tradition.

However it's also a visual language that I don't have the right to use. Pueblo pottery is not a tradition of my people. And yet the industry and economy here are primed for that kind of work. So I end up making objects that aren't tied to my culture specifically, because I recognize how the art itself has a kind of vulture's approach to culture. It consumes little bits and pieces of something that are perceived as dead in order to keep itself alive. I have a record that my culture still exists because I've harvested little bits and pieces of it. I'm from a specific cultural group, where many of my cultural practices are absolutely taboo to the people down here in the Southwest. Our cultures are so different. But here I am and I need to figure out a way to integrate my experience from the plains. So I'll go back to harvest clay whenever my material gets low. And what I'll do is, I'll wedge material from North Dakota into the clay that I'm using for my sculptures, so there is always a connection to the place I come from. Even if that visual language isn't directly cued into the objects themselves, it comes from that same place that I do. I have a little bucket here with clay from North Dakota. I let it dry and then I pulverize it and then I wedge this clay into the clay body that I'm working with so there is a resonance of place, which is important. Especially because I am not from this region, I like the idea of traveling the material with me into this place and into this artistic practice.

Alexandra R. Toland: Timothy Morton has used the example of Doctor Who's time machine as a way of explaining things as being bigger on the inside.[6] On the outside, the time machine is a police call box from the 1950s. On the inside, it's rooms and corridors and doors and closets and other worlds in space and time. I wonder what worlds lie beneath the surface of your objects and if there is a certain intended relationship between the earth materials you use and the final objects that result from that transformation of clay into ceramics. I'm especially interested in the *The Weapon is Sharing/This Machine Kills Fascists*. It's a clay cell phone transmitting still images of a volatile present state. Is it some kind of a time machine?

Cannupa Hanska Luger: I like the idea of sharing a philosophy rather than a physical object. But the object does have its place, mainly as a vessel for communicating conversation. The whole idea for *The Weapon is Sharing* came out of what I've experienced as our objects being consumed by anthropologists and institutions and sitting in museums. It's very difficult for us to get those objects back because we don't have hermetically sealed institutions to house them from now until the end of time. But that was never the function of those objects anyway. Their function was to be a part of the culture, to disintegrate, to fall apart, to be repaired, to be maintained, and to change.

Culture is the material that shifts and changes and becomes what we need, when we need it, and allows every generation to put their little bit of experience into that conversation so that it can be understood on a longer continuum of time. That's how you maintain culture. Freezing it, putting it in a museum, and isolating it from the people that it's supposed to benefit, I think, removes its meaning. And I would love to be able to get some of those

objects back, not to hold on to them forever, but to learn from them, apply them to the culture of now, and allow them to transition. We have to apply these objects to our time as we see fit. That is how cultural practice finds its way into the objects that I make. I'm trying to have a conversation with these vessels. I recognize the object as a byproduct. It holds the story that needs to be translated to people outside of our gnostic experience of culture. For lack of a better term, I'm trying to re-indigenize Western thought into its relationship to place and its relationship to the objects that are derived from a place.

Alexandra R. Toland: I like that term. It seems to offer more room for action and imagination than corresponding theories of decolonizing nature and culture.[7] Your idea of re-indigenization turns ideas of agency inside out. Not only are the descendants of colonizing bodies called upon to reconnect in new ways to places they were not culturally connected to in the first place, but the land itself is allowed to shape those bodies, as we shape it. The layers of cultural complexity involved, especially in a vacuum of Native knowledge, present a challenge to the project of re-indigenization. We, as a jumbled, heterogeneous, hybridized (and in many cases urban and queer) society are tasked with not only decoupling ourselves from deeply embedded, colonially constructed models of governance, economics, and culture, but we are also asked to seek out elders in everyday experience and embrace place—wherever that might be—in order to make meaning of our lives on this planet. It could be tricky to get this thing right!

Cannupa Hanska Luger: Yeah. The problem I ran into with decolonizing is that the emphasis of change is put on the people affected by the colonizer.[8] That's fucked up! How can you come in here, mess up my house, and have me clean it up? Rather, how about you come here, you mess up my house, and I teach you how to clean it, how to take care of it, how not to make the mess in the first place? I think that's a stronger conversation. What I also realize is that I have a lot of sympathy for Europeans because they've been colonized longer than anybody else. Their experience with colonization is so old that they forgot why they love this place, that they belong to this place. In every culture, if you go back far enough, you can see that we worshipped the Earth. I see that in the United States. Look at Mount Rushmore, look at the renaming of rivers and mountains. All of that is a process of trying to relate to place.

As human beings, we have an innate desire to belong. I have the incredible privilege of being from North America from time immemorial. And the plains have literally sculpted my physique. I wasn't really aware of that until I moved to the Southwest because I'm 6-foot-3 and I am not tall where I come from. Here in the Southwest, I am tall, I mean really tall. Native people from around here are relatively short, but I see the benefit of that in the context of the land. Being tall in an arid, deserty kind of environment is *not* an advantage. Moisture is wicked from my body here because I have so much more surface area and I am higher up in the wind. It's not advantageous to be tall here. But in the Great Plains when the grass is 4 feet tall, being 6 feet tall is a benefit. Every foot you gain in height gives you another 25 miles of vision in the Plains. There's benefits to that. So I've recognized that the landscape has sculpted us. If you think of it that way, this is not our land but rather we are its people. We belong to it and it has sculpted us more than we've sculpted it. It's like clay in a way.

So, going into this idea of re-indigenizing, let's figure out how we can have this conversation and recognize that we all belong to a place more so than it belongs to us. That's the conversation that we need to be having right now. In the United States, there's so much political upheaval and unrest and protest, but I see all of that as engaging with symptoms of the problem rather than the problem itself. The root of the problem, I think, is that we have convinced ourselves that we are separate from the landscape, that we've separated ourselves from the Earth. That's tragic, but if we can heal that rift, we won't have to go to the frontlines. We won't have to engage with patriarchy and colonial dominance and all of these other symptoms of not being a part of our planet.

Alexandra R. Toland: Symbolic allusion to soil— and clay—as a life-giving, spiritual force is anchored in many cultural traditions, from the ancient Roman worship of the goddess Terra to the Abrahamic religions conception of Adam, the first human, which literally means "of the earth" in Hebrew.[9] In your work, clay could be understood as a powerful symbolic medium of the spirit world. I'm thinking about your allegorical portraits of mythical creatures, bears, buffalos, and humans in the *Regalia Series*. Could you talk about deification of earth materials and how that might influence the figurative objects and sculptures you create? Are these works in any way representative of a spirit world?

Cannupa Hanska Luger: No. None of the sculptures I make represent any cultural deities whatsoever. I shy away from that tremendously because I don't think that consumers of art have the right to engage with deities. That's not cultural knowledge for me to share. And every spiritual experience that I've had is a personal exchange between me and creation. If I try to render that into an object, I am transforming something ecstatic, something that comes from outside *this* world, and cheapening that experience. I don't want to transform something beautiful and spiritual into an anecdote. And that's what would happen to it. There's no way for me to describe that. So, the objects I make are pulled from my own experience. From me, Cannupa, and everything I've experienced from birth until this time—from my childhood in the plains, to anime cartoons, to surgery MTV, to everything that I've experienced in this world.

So the figures in the *Regalia Series* aren't deities. They are ideas. I wanted these animal forms to complete the human figure. I wanted to make the point that regalia is not something you wear, it's something that completes you. It's a visual practice. I built that whole series at the same time here in the U.S., and actually all over the world, when the war bonnet was suddenly in vogue in different cultural contexts, being worn by celebrities and hipsters going to Coachella, and models at fashion shows. I wanted to emphasize that the war bonnet is not a *hat*. It's not just something that you put on your head. It's like a purple heart in military terms, a medal of honor that comes at great sacrifice to yourself to help your people. You have to earn the right to wear it. The war bonnet comes from *my* people, from the Plains Indians, and yet it's distilled into an icon for all Native American people. So, I'm afraid of it losing its function because, you know, I don't even have the right to wear it. I don't even know if I want the responsibility to have the right to wear it. And I sure as hell don't want to see any random celebrity or hipster at a music festival just throw it on because it looks cool. Because, you know, it *does* look cool. It's awesome. Not awesome like a pair of socks or something, but awesome like the stars, awesome like the depth of space. And you're removing that awesomeness in that moment of misappropriation.

It was never in vogue for Native people to talk about themselves. It was shameful to do that. It was shameful to walk into a place and talk about how great you are. However, our regalia carried on that conversation. You get to wear certain objects that show the prowess of your experience so that you never have to talk about it. You can just walk into a space and people can identify you through the visual cues that you are wearing on your person. Somebody from another tribe may

not speak the same language, or use the same visual vocabulary, but at the same time they'll be like, "OK, you're dressed to the nines. I can tell you have great accomplishments. I can respect you without you having to say anything." And I think there's something really beautiful about that. In Western thought we would consider that visual regalia art or fashion, or something along those lines, but it's much deeper than that. There's something in that technology, in that practice that we can learn from and apply to our current culture. Using regalia correctly, respectfully, can create a stronger connection to culture and engagement with community.

Alexandra R. Toland: Examples of regalia in Western contemporary societies seem to almost always be a hybrid of something else … Maybe these blatant examples of cultural misappropriation are the downside or failure of the greater project of re-indigenization. What do you think is at the root of that carelessness and disrespect, and what can we learn from such mistakes?

Cannupa Hanska Luger: It's all about the dominance of consumer capitalism in our culture. We still recognize clothing as identifiers. I dress a certain way and so I am lumped in with this certain group of people. I'm trying to shift ideas about our consumer culture by using moments of that very consumer culture. I'm trying to transform it from the inside rather than standing on the outside and pointing fingers. And that's what a lot of my work does in the context of "Native art." I want to engage with that conversation because there is no such thing as Native art. In fact, we don't even have a word for art in our language. It's just a part of our everyday life and our way of being. So, I'm not really interested in selling my culture back in the form of objects, but I do find the industry and economics of Native art fascinating and I like making objects that address that industry. In the end, there is no

Opposite page from left to right and top to bottom:

Regalia Series, 2013, ceramic and mixed media, dimensions variable. *Bear, Coyote, Wolf, Winter Fox, Elk, Owl, Lynx.*

Images courtesy of artist.

Regalia Series, 2013, *Lion*, ceramic and mixed media, dimensions 12″ × 10″ × 8″.

Image courtesy of artist.

Native art because there is no "Native American." That's a bubble definition with no real meaning behind it, another superbroad ethnic label like "Asian" or "European" or "African" that feeds into ideas of otherness and domination.

I've been thinking along those lines about concepts of equity in the larger conversation on consumer culture. There's this idea that there is not enough equality in the world, that there's an incredible struggle in the name of equity. And we've done incredible things in that struggle, but the idea is kind of frightening to me because it concedes that some groups are less than others. And I'm not ready to concede that I am less—not

because I don't have a war bonnet, but because I don't have a Maserati, or a tower in the sky, or all the things that the larger dominant culture perceives as wealth. It's a continued idea that the dominant culture is going to raise up third world cultures to their level of civilization. But I'm watching that larger culture fall apart, watching it cripple under its own weight. And when *this* fails, it's not going to be knowledge from laboratories and extractive processes that we'll need, but rather the knowledge from those people with less—the homeless, the poor, the third world societies who we've looked down upon. That's something that I personally recognize, growing up poor on the reservation. I'm recession-proof.

I'm depression-proof. I'm apocalypse-proof. I'm already living in a postapocalyptic world right now. The United States of America is like Mad Max for my people. And I was born into that. I was born into turmoil. So I see what's missing in this dominant consumer culture and a lot of it has to do with meaning and oral tradition and maintaining stories and intention and using it rather than logging it away as "heritage" or recording it on external hard drives.

Alexandra R. Toland: Thank you for bringing up this point. Sometimes the focus on inequality outshines the capacity for resilience and adaptability, both in human and nonhuman communities.[10] Examples of resilience in the human and nonhuman world are often underplayed in discourses on the Anthropocene. But you're living it! How do you envision the land of your grandchildren's children's children, and what traces of cultural heritage could you imagine they might carry on? How did your experience on the ground with the Winter Count group nourish your ideas about resilience and future? That work seems very different aesthetically from the clay work, in that it incorporates digital media and collaborative, participatory methods. Can you talk about some of the practices you've engaged with there? Do you have a message to people all over the world when they hear the latest reports on climate change and land degradation and feel like there is no hope?

Cannupa Hanska Luger: There is no hope! We're doomed as individuals so let's recognize that first. You are not going to live forever. You are going to die and become a part of the landscape. You are part of a continuum. So it's about recognizing that first. I am thankful that I will not live forever. I love that my life is finite because it makes apples delicious. It makes everything that I do special. What good does *hope* do for our grandchildren's children's children? Our hope does nothing for

them. However, we can work toward benefiting them. I don't want to hope that everything turns out OK. I want to die trying to make sure their lives are better.

I love the idea of engaging with our future generations and I think about that relationship in our work with Winter Count. Our project is really based in that idea of re-indigenizing. We are trying to use our current technologies to voice these conversations that I think will benefit future generations. We use objects and we make exhibitions in museums, but we also use digital media quite a bit because it is an incredible tool that is anchored in contemporary culture. Rather than posting selfies, we used it to share something stronger and more beneficial to us as a whole rather than us as individuals. Narcissism and lack of meaning have crippled us as a species. And our persistence on hoping and not actively working toward change has had a collateral effect on all. So, it's good to keep a light in your heart, but it's absolutely useless without practice and accountability.

Our mythical stories are bound to the geology and geography of place. I can take my sons to North Dakota and tell our creation stories by pointing to specific places. That's incredible. That's something that maybe Catholics or Jews can relate to by going to Israel and seeing those landscapes and seeing those ancient places. The lack of connection to place in the United States has affected the culture so much that there's a void that's trying to be filled through capitalist consumerism. So I want to touch people in that spot, wherever it is, so they can relate to land again in a more meaningful way rather than seeing land as a resource. Land should not be a resource. Water should not be a resource. *Oil* should not be a resource. Land and water should be recognized for what it is—and it's us. To dominate it is to dominate ourselves.

So, the source of our engagement at Standing Rock wasn't to stop the oil industry. It was to protect the landscape. And you know, the pipeline went through anyway. They dug under the lake and did all that stuff and I don't see it necessarily as a loss. It was incredible win that happened at that place because we came together. There was a level of solidarity and incredibly beauty under that umbrella of "Native American," that I have never seen before in my lifetime, that I got to witness firsthand. For instance, the Crow came to Standing Rock. The Crow and the Lakota have been enemies for long before the United States, long before Columbus. It's been an old grudge, one that was built around trying to control landscape, trying to control land and water and soil. And there was this broad apology and a burying of the hatchet, where they were like, "Our cultural differences are insignificant compared to us having to protect the land and protect our practices and our cultures." Once again, the protection of landscape *is* the protection of culture. And to make efforts toward that is to reestablish that incredibly beautiful connection to place that is our privilege as Native peoples in North America.

Alexandra R. Toland: In the purpose statement of Winter Count, which you coinitiated in the context of the NODAPL protests, you describe the reality of protecting both the tangible and intangible heritage of the land:

> Today we see natural cycles of life disrupted by the extraction and transportation of what we have come to call resources from the land … We bring together our minds as artists to cultivate gratitude and respect for water, land, and the interdependence of all things living in this world. Through our work, we bind together our diverse ancestry and cultures, to honor and protect water and land … We are listening, we are watching, we are holding up

reflectors, waving flags, singing the horizon and telling the story of how we are now … how we can be, what we can make for our children, and our grandchildren's children.[11]

While the tangible heritage of the soil is the subject of archaeological and paleopedological study, the idea of *intangibility* has gained legal footing in the United Nations Educational, Scientific and Cultural Organization (UNESCO).[12] Does the idea of intangible cultural heritage resonate with you? Do you see such declarations as a feasible attempt to protect landscapes for many of the reasons you've described already.

Cannupa Hanska Luger: Well, I think it's great to define it, but the question is, Does it work to protect the land? The UN is toothless when they come up against industries that lobby its programs. I like the idea of "intangible cultural heritage," as it is conceptualized by such institutions, but I just don't know if that's where we need to turn to protect it. I think the place to turn is the hearts and minds of our children—in education. It's in developing future generations that relate to places more than to resources. And we fail at that in our educational practices because that relationship to the land has been lost …

I think about when I was a kid I got to ride horses bareback down to the river and play with horses in the river. I don't see that for my kids, not where I used to go anyway. Where I used to go is not healthy enough to run horses and children in the river. And the threat of our extractive industries are making it worse. But the beautiful thing about water is that it can clean itself. And we can support that process. No matter how toxic the world has become in our lifetime, no matter how hot we make the world, the effort to change is a kind of hope. And I don't know if all the practices we've maintained to get us to this point are wrong or right. I just know that it is what it is and we can

Still from Winter Count's video *We Are In Crisis*, 2016, approx. three minutes.

Photo Credit: Winter Count, reprinted with permission from Winter Count.

make an effort to change it. I don't know if we would have made that effort, had we not gone down the road this far. So, it's a recognition of our own folly. And once again this goes back to the practice of working with clay. Failure. Failure is an incredible teacher, but you gotta recognize that you failed. You gotta be accountable for the fact that it's not working. And that's what I see right now.

There is an entire generation born into a world that recognizes that this is not working. When I talk to younger groups, one of the first things I say is, "I'm sorry." I'm sorry for where we're at right now. I'm sorry that you're born into this place, but I'm your ally. I'm your accomplice." So look at all the tiny condolences that we've made, all the things we've put aside to make things better, and recognize what it means to cooperate with one another. That's the real meaning of intangible cultural heritage—sharing. It is a lot easier than keeping away, than maintaining separation. You share and then you can move on to the next thing. It takes generations to keep things separate, to keep landscapes separate, to maintain borders and walls, so long that you forget what the original point of separation was. But if you share a thing, it's complete.

Alexandra R. Toland: Well, thank *you*, Cannupa, for sharing your ideas and vision with me today and for sharing your artwork with the greater community of artists and scientists and readers of this book. We have a lot of work cut out for ourselves, but as a shared endeavor we can all work towards re-indigenizing knowledge and culture, for the sake of future generations, and the heritage of the soil, water, and land that we are a part.

Endnotes

1. Cairns-Smith, A.G. 1985. *Seven Clues to the Origin of Life—A Scientific Detective Story*. Cambridge: Cambridge University Press.

2. See Chapter 10 in this volume by Roxanne Swentzell. For more on the art and life and work of Roxanne Swentzell, visit http://www.roxanneswentzell.net/.

3. For more on the Flowering Tree Permaculture Institute and the tradition of Pantes ovens, visit http://www.roxanneswentzell.net/roxanne_swentzell_permaculture.htm.

4. Wells, E.C. 2006. Cultural soilscapes. In E. Frossard, W.E. Blum, and B.P. Warkentin, *Function of Soils for Human Societies and the Environment*. London: Geological Society Special Publications.

5. For more information on the politics and aftermath of the Garrison Dam, see Ojibwa, March 10, 2010, Dam Indians: The Missouri River, Native American Netroots, http://nativeamericannetroots.net/diary/406; Rose, C., 2015, Echoes of Oak Flat: 4 Pick Sloan dams that submerged native lands, https://indiancountrymedianetwork.com/history/events/echoes-of-oak-flat-4-pick-sloan-dams-that-submerged-native-lands/; Berman, T., 1988, For the taking: The Garrison Dam and the tribal taking area, https://www.culturalsurvival.org/publications/cultural-survival-quarterly/taking-garrison-dam-and-tribal-taking-area.

6. Korody, N., March 11, 2016, Timothy Morton on haunted architecture, dark ecology, and other objects, http://archinect.com/features/article/149934079/timothy-morton-on-haunted-architecture-dark-ecology-and-other-objects

7. See, for example, Demos, T.J., 2016, *Decolonizing Nature: Contemporary Art and the Politics of Ecology*, Berlin: Sternberg Press.

8. For an in-depth discussion on the entangled discourse of decolonization, see Tuck, E., and Wayne Yang, K. 2012. Decolonization is not a metaphor. *Decolonization: Indigeneity. Education & Society* 1(1):1–40.

9. For more on soil deification, soil and religious symbolism and practice, see McNeill, J.R., and Winiwarter, V., 2006. *Soils and Societies*. Isle of Harris, UK: The White Horse Press; Patzel, N., 2015, *Symbole im Landbau*, München: Oekom Verlag; Landa, E.R., and Feller, C., 2010. *Soil and Culture*, New York: Springer Science and Business Media.

10. See, for example, Harvey, D., 2006, *Spaces of Global Capitalism: Toward a Theory of Uneven Development* (London: Verso Books) for discussion on the geographic and ethnic divisions of wealth and the ideologies of neoliberal geopolitics.

11. http://www.cannupahanska.com/wintercount/

12. UNESCO. 2003. *Text of the Convention on the Safeguarding of Intangible Cultural Heritage*. Retrieved February 14, 2014, from http://www.unesco.org/culture/ich/en/convention.

Stories from the Hills

Tales of the Lowland

Ruttikorn Vuttikorn, Myriel Milicevic, and Prasert Trakansuphakon

Play activist **Ruttikorn Vuttikorn** is an industrial design graduate and has been involved with toy design all of her professional life. She strongly believes that every child should have accessibility of quality play. For that reason, she has never limited herself to only designing toys. From 2007, Vuttikorn started collaborating with different organizations designing games that educate children to understand and solve different problems, such as disaster, environmental, social, and political problems. These experiences create interest in environmental and political issues. She travels around Asia conducting workshops on game design, helping local people translate local problems and solutions into fun and creative activities.

Myriel Milicevic is an artist and interaction designer. She explores the hidden connections between people and their natural, social, and technical environments, and presents these as practical-utopian models and stories. These explorations are mostly of a participatory nature,

Title image: This picture reflects old and new farming methodologies. The original rotational farm is situated above the terraced rice fields, which is a new method of planting rice for the Karen. These two methods of planting are free from toxic substances and produce a high yield of rice, corresponding to the increasing population in the community.

Photo: Chalit Saphaphak.

emerging from collaborations with artists, designers, activists, and scientists in the context of workshops, exhibitions, residencies, and field studies. She currently holds a professorship at the Department of Design at the University of Applied Sciences Potsdam. She has worked with Ruttikorn Vuttikorn on a number of other projects focusing on sustainability communication, including most recently a live role-playing game about biopiracy called "Crops and Robbers" that was featured at the International Garden Exhibition in Berlin (IGA).

Prasert Trakansuphakon is a specialist of indigenous studies in Thailand. Of Karen origin, he has a doctorate degree in sociology and has developed an expertise that he put to good use both in academic circles and in civil society as a researcher and senior nongovernmental organization (NGO) activist. He has authored many articles on local knowledge and traditional modes of subsistence practices among indigenous peoples, and has held numerous positions in various organizations at national and international levels since the 1990s. He was director of the Regional Indigenous Knowledge and Peoples in Mainland SEA (IKAP) for many years, and is currently the chairperson of Pgakenyaw (Karen) for Sustainable Development (PASD) and chairperson of the Inter Mountain Peoples Education and Culture in Thailand (IMPECT).

Known as a people of storytelling, the peoples in the hills have been handing down their genealogies mouth to ear over generations, without any form of written record or alphabet to keep the words, protected only by the high hills of Northern Thailand. How could these stories help us understand the close relationships of people to their land? What lessons can be learned from those who live in and with the forest, cultivating food as subsistence hill farmers, and how can their experience inform our urban ways of communal living? Traveling to the forests of indigenous people in highland communities Ruttikorn Vuttikorn and Myriel Milicevic stayed in two Akha villages. Later, Vuttikorn traveled to two Karen villages with Prasert Trakansuphakon. One of the villages is still following the ways of tradition based on rotational farming, while the other had changed its practice due to state policies and laws on mainstream cash crop agriculture. Although the indigenous elders still speak their own languages, traditional knowledge and practice among young people is dwindling because of the modern education system (based on assimilation policy). Many of the stories, poems, and songs have disappeared altogether with the tree and water spirits, ghost gates and swings, and traditional farming practices. Younger generations have replaced their handcrafted garments with regular clothes and rural animism with other religions, educations, and careers in the city. This dialogue chapter was conceived and written in conversation with many highland villagers and long dialogues between the authors.

The project was shown at the La Fin des Cartes exhibition in 2015 at Espace des Arts sans Frontières, Paris. An extensive documentation of the *Stories from the Hills* design research project can be found online at http://neighbourhoodsatellites.com/stories-from-the-hills/index.html.

Just like an elephant taking small steps, closing the circle after many years of wandering and returning to graze on the same land, we come back to the same fallow place and prepare it again for cultivation. We can tell by the trees the age of a forest. We cut the plants on a small patch of land, leave them to dry and burn the remains to ashes. Only in the first year we grow rice and a variety of vegetables there and then leave it for seven to eight years before we come back and start the cycle again. It is a form of natural subsistence economics. Some fields are steep. Some are far. Every year we decide anew with the permission of the spirits, who cultivates which piece of land. We share the water, the forest and the land. It is communal land, it doesn't belong to anyone.

What people of the Lowland call Slash and Burn and understand as a system of deforestation, we call Rotational Farming. It is an ancient agricultural system that coexists with nature.

This is what a boy was telling the visitors who came to the Karen village several decades ago.

The visitors had more questions, and so the boy continued:

Each one of us is connected to their own tree. Our umbilical cord is tied to a branch when we are born, and so we take care of this tree for all our life. The history of origin of Karen people comes from Mu Qa Khle (Grand Mom Bayan tree). Every life needs to come through the Grand Mom Bayan tree before birth and when people die they need to go back through the Grand Mom Bayan tree again for reporting. This means our Karen people are part of nature. We come from nature and go back to be part of nature.

My friends and I like to run around and play anywhere in the forest. My favorite place in the village is a small forested area in the location of the community well. Here everyone comes together. It is a source for both human and wild life. The young children like me and my friends have learned how to hunt and gather in this small forest.

One day, the Great Planner dreamed of a magnificent water system: a village that could store all the water in one body and use what is needed at the time for watering the fields. With a group of assistant planners he traveled to the village of the lively community forest and turned the well into a lake. Not only the well but also the traditional umbilical forest around the community was gone and had become part of land cultivation for cash crop farming as a symbol of modernization in a new system that people needed to follow. The Lowlanders gave orders that no forest should be cultivated in the traditional farming method anymore and they divided

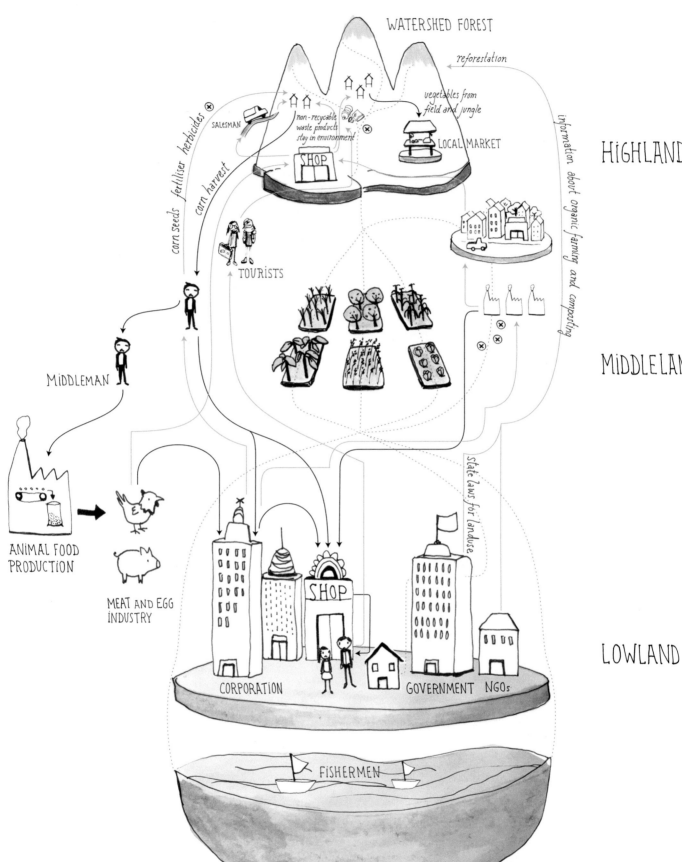

WATERSHED FOREST

reforestation

HIGHLAND

vegetables from
field and jungle

salesman

non-recyclable
waste products
stay in environment

LOCAL MARKET

SHOP

corn seeds fertiliser herbicides

corn harvest

information about organic farming and composting

TOURISTS

MIDDLELAND

Middleman

state laws for landuse

ANIMAL FOOD
PRODUCTION

MEAT AND EGG
INDUSTRY

SHOP

LOWLAND

CORPORATION

GOVERNMENT   NGOs

FISHERMEN

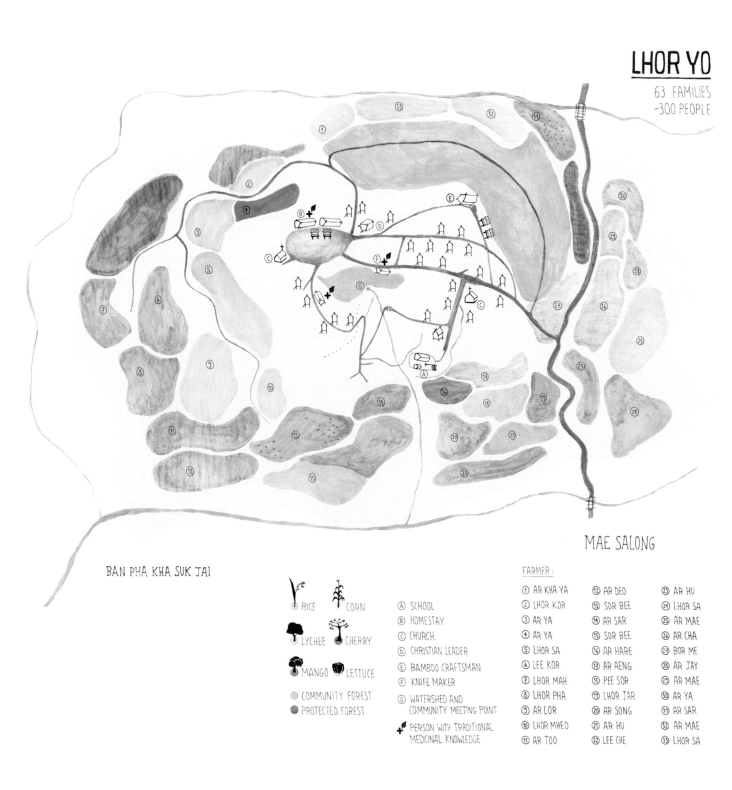

LHOR YO

63 FAMILIES
~300 PEOPLE

BAN PHA KHA SUK JAI

MAE SALONG

 Y RICE   🌽 CORN

 🌳 LYCHEE   🌲 CHERRY

 🌴 MANGO   🌿 LETTUCE

 ● COMMUNITY FOREST
 ● PROTECTED FOREST

Ⓐ SCHOOL
Ⓑ HOMESTAY
Ⓒ CHURCH
Ⓓ CHRISTIAN LEADER
Ⓔ BAMBOO CRAFTSMAN
Ⓕ KNIFE MAKER
Ⓖ WATERSHED AND
  COMMUNITY MEETING POINT
✚ PERSON WITH TRADITIONAL
  MEDICINAL KNOWLEDGE

FARMER :

① AR KHA YA	⑫ AR DEO	㉓ AR HU
② LHOR KOR	⑬ SOR BEE	㉔ LHOR SA
③ AR YA	⑭ AR SAR	㉕ AR MAE
④ AR YA	⑮ SOR BEE	㉖ AR CHA
⑤ LHOR SA	⑯ AR HARE	㉗ BOR ME
⑥ LEE KOR	⑰ AR AENG	㉘ AR JAY
⑦ LHOR MAH	⑱ PEE SOR	㉙ AR MAE
⑧ LHOR PHA	⑲ LHOR JAR	㉚ AR YA
⑨ AR LOR	⑳ AR SONG	㉛ AR SAR
⑩ LHOR MHEO	㉑ AR HU	㉜ AR MAE
⑪ AR TOO	㉒ LEE CHE	㉝ LHOR SA

the land among the villagers. Potions were sold to defeat pests and make the fields flourish year after year after year. But then, no more years after that. After some cycles the harvest started declining, the soil eroding, and insects dying. This new way of agricultural monoculture—a permanent land use without rotation—disrupted the intertwined relationships between nature, people, spirits, and the connections between the generations of the past, present, and future.

The ever-moving village now remained in one place. The land that provided food for the villagers soon produced corn to feed animals far away. The cows that used to fertilize the soil by walking around freely disappeared from the scenery. Chemical fertilizers were sprayed instead, or cow dung had to be transported long distances. The same company that provided agricultural chemicals also sold the seeds to the farmers, and sold meat and eggs through their system of discounters to the Lowlanders, as well as to the Highlanders. Instead of being self-reliant, the farmers became dependent on money to buy everything from the market. A seemingly easy income at first, these dependencies turned more and more farmers into debtors, unable to pay back their loans.

Where natural debris used to decay, now packaging of all kinds of products littered the forest. The water running down the hills to the Lowland became contaminated, passing many fields along the way toward the cities.

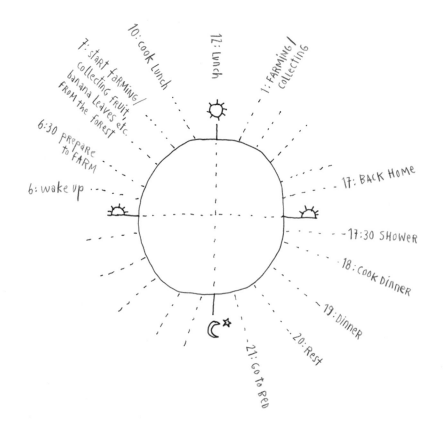

The boy, now the village elder: *The well is gone and with it the meeting place of our community. The water is too dirty to drink, because chemicals have flown down from the villages and their fields uphill.*

Another Karen village was able keep many of its traditions. Compared to the neighboring villages, the soil here is full of life. There is less harm by pests in mixed fields. Ants build their small soil chimneys. Flowers are pollinated by incredible butterflies and wild bees.

It's hard to believe that 30–40 years ago, this area had been commercially logged and devastated. Through knowledge, lore, and traditional practices, the village is today surrounded again by an amazing forest.

Village leader: *Since we have more bees, there are more flowers and the forest is full of fruit—far more than in other forests. We also have some income from the wild honey.*

Visitor: *This honey is very special; you could sell it at a higher price.*

Village leader: *We don't want to sell at a higher price. Everyone should have the possibility to buy it who cares about nature.*

There was a request from the U.S. to export 25,000 jars per year. We could do that, but we don't want to do that. Nature is not a manufacturer.

This village is very strong and can negotiate with government officers. But another one in another area can't. Their people have united their fields to keep doing rotational farming. They think, if the officers want to arrest rotational farmers, they have to arrest the whole village. However, burning one big piece of land is riskier. Four deer were found dead after a fire. The forest takes longer to regrow. The periods are shorter than they ought to be, due to limited space. The elephant has to speed up to return for grazing and close the circle.

Today, the stories of the hills tell about the low urban land: from there, healthcare, energy supply, schools, commercial products, cellphones, TV, and tourists have come, together with waste problems, contract farming, industrial corn seeds, fertilizers, and herbicides. The words that are used for these stories are Thai words, which have entered the local languages.

At the same time, people in the flat, urban land are unaware of how closely they are connected with the upland through the daily food and water they consume. They often hold prejudices against the hills and their people, seeing in them cultivators of opiates and drug traffickers—an image that is also supported by mainstream media. The highland communities are held responsible for deforestation and soil erosion and therefore a shortage of water supply, while the growing cash crop fields in the lowland demand ever more water.

Back in the Lowland, we met with product design students from Srinakharinwirot University in Bangkok and discussed what we had learned from the people in the hills. None of them had ever visited a village there before.

The students identified the problems they experience in the city. These ranged from air pollution, traffic jams, and other environmental issues to problems in families, unemployment, homelessness, corruption, and crime, which often lead to a system of mistrust among citizens.

Building on the knowledge we had learned from the hill people, teams of students came up with ideas to address their urban problems and find creative solutions for them.

Young villagers today try to establish their own forms of connections and economic systems, telling other stories to the Lowlanders. They open cafes with coffee from their hills, let people taste the wild honey and invite small groups of tourists to stay in their homes.

Karen people prepare the land for the next farming season. They cut down grass, weed, and trees before burning the land. They believe the ash will be natural fertilizer and help adjust soil condition. The preparation happens in winter where the tree trunk will be left (around 2 m high). If trees were to be cut during the rainy season, tree roots can rot. Trunks that are still alive will sprout in the next three to four months and will still anchor the soil. This eradicates problems like landslides that occur in agriculture areas.

Photo: Chalit Saphaphak.

On the way to a Karen village that completely depends on monoculture. Growing corn is so precious, that every piece of land, even cemeteries, are cultivated. That day the fields had just been sprayed and we were advised that it was better to not visit the village.

Photo: Chalit Saphaphak.

Fieldwork.

Photo: Chalit Saphaphak.

Other fieldwork: Terrace farming.

Photo: Chalit Saphaphak.

When the year of cultivation is ending, plants, vegetables, flowers, and seeds are collected. The land will be left for seven years for the soil to be naturally nourished. These fields that are left alone will be the source for food and medicine, and an area for wild animals.

Photo: Chalit Saphaphak.

Karen people harvest tea leaves in the community forest, as well as coffee and various kinds of fruits. Because the forest is free from chemicals like pesticides, tea and coffee that are harvested here have a very high price, and is in demand in the market.

Photo: Chalit Saphaphak.

Villagers conserve the forest also by giving space to wild bees. Honey has always been a big source of income for the community. However, these villagers only collect honey once a year to maintain the population of bees.

Photo: Chalit Saphaphak.

Rice farmer in an Akha village. Each hill farmer family grows mostly food for their own needs, with rice being a main crop. During harvest time, families help each other to cut the rice and thresh the paddy. Working together makes the labor easier for everyone.

Photo: Myriel Milicevic and Ruttikorn Vuttikorn.

In their community forests, hill people grow a big variety of plants that provide food or materials. To keep a balance between growing and taking, people need to know how and how much they can take from a plant, so that it will grow back. Bamboo provides people with material for construction and crafting objects, but also with food. Young shoots can be eaten fresh, and bamboo worms provide a rich source of protein on a daily basis.

Photo: Myriel Milicevic and Ruttikorn Vuttikorn.

Banana leaves are collected only from the wild species, because people don't eat those fruits. The house banana tree gets to keep the leaves so that the fruit will be sweet and full of nutrients.

Photo: Myriel Milicevic and Ruttikorn Vuttikorn.

When farmers go to work in the field, they only take a little food with them. They grow a "lunch garden" next to their rice fields, so they can go pick fresh vegetables for lunch and cook it right there. Once the field is left fallow, the huts themselves also turn to soil.

Photo: Myriel Milicevic and Ruttikorn Vuttikorn.

Woman from a Lisu village separating rice and husks.

Photo: Myriel Milicevic and Ruttikorn Vuttikorn.

Building an Akha village and keeping evil out

In the old days, if a community moved to another place, it was the shaman who found the right location for building a new village. To identify the right spot, he would drop a raw egg together with some rice. If the egg didn't break, it meant that it was a safe place to settle.

When the houses for all the families were built and ready, the shaman decided on the right day for setting up the ghost gate.

Still today, a ghost gate has to be built within one day together with the wooden statues of a man and a woman that will be placed next to it. Located on the path to the forest, the gate creates the border between the people and the world of the spirits, keeping evil and illness from entering the village.

Photo: Myriel Milicevic and Ruttikorn Vuttikorn

Together with younger and older people of Lhor Yo and Suan Pa, we made geographical and temporal maps, documenting the present and the past of the village.

Photo: Myriel Milicevic and Ruttikorn Vuttikorn.

Highland coming to Lowland: workshop with product design students from Srinakharinwirot University, Bangkok.

Photo: Myriel Milicevic and Ruttikorn Vuttikorn.

Sketches in the Sands of Time

Mandy Martin and Libby Robin with responses from Guy Fitzhardinge and Mike Smith

Mandy Martin is adjunct professor, Fenner School of Environment and Society, Australian National University. She is a practicing artist who has held major exhibitions in Australia, Mexico, and the United States, and has been exhibited in Asia and Europe. Her works are in public and private collections including the National Gallery of Australia, the South Australian Art Gallery, the Guggenheim Museum New York, the Los Angeles Museum of Contemporary Art, and the Nevada Museum of Art, Reno. Martin's works include painting and printmaking. Her environmental art workshops explore remote regions and include collaborative art workshops with Indigenous artists. She has published eight books on environmental art projects.

Title image: Mandy Martin, *Cool Burn*, 2015. MS 56 ochre, fluoro pigments, sepia ink, black paper, 28 × 28 cm.

Reprinted with permission from the artist.

Libby Robin, FAHA, is professor of environment and society, Australian National University. She is a historian of environmental sciences and an international leader in collaborative environmental humanities scholarship, including art and museum practice, natural sciences, and history. Her most recent books include *The Environment: A History* (Johns Hopkins, 2018), *Curating the Future: Museums, Communities and Climate Change* (Routledge, 2017) and *Desert Channels: The Impulse to Conserve* (with Chris Dickman and Mandy Martin; CSIRO, 2010).

Mike Smith, AM, FAHA, author of *The Archaeology of Australia's Deserts* (Cambridge, 2013), has worked for more than thirty years across the Australian arid zone. He has pieced together the human and environmental histories, including prehistory, cultural history, human ecology, and the history of ideas about Australian drylands, and the presentation of environmental history in museums, including *Extremes*, a major comparative exhibition of southern deserts in Australia, Africa, and South America.

Guy Fitzhardinge, AM, holds a PhD in rural sociology, and is a research affiliate of the Centre for the Inland, La Trobe University, Victoria. He is a pastoralist with major holdings in New South Wales, and a conservation leader in the new science of "co-management," a cultural and ecological endeavor that is important to the health of people and land in traditionally managed areas of Australia. He is director of the World Wildlife Fund Australia and director of the Northern Australia Indigenous Land and Sea Management Alliance Ltd.

Libby Robin, Mandy Martin, Guy Fitzhardinge, and Mike Smith have been working collaboratively for over a decade. This chapter offers a reflection on their collective scholarship, inspired by the invitation to explore partnerships between art and science in a larger interdisciplinary dialogue on soil protection in an age of accelerated environmental change. It reflects on multiple, longitudinal partnerships developed though environmental art projects led by the distinguished Australian painter Mandy Martin. This interdisciplinary conversation began in the 1990s, a time in Australia when environmental management and Indigenous ideas increasingly converged. In 1992, the High Court upheld Native Title for Eddie Mabo on the island of Mer ("The Mabo judgement"), and this led to the Commonwealth Native Title Act, passed in December 1993. Thus, the natural environment is now also cultural: more than just Western science is needed to understand the relationships between humans and the environment in this continent. Art works to break down the old nature–culture schism that had persisted from colonial times.

Sketches in the *Sands of Time* began with the sands. The ochres that are the basis for the suite of works that complement this essay come straight from the science of archaeology, the science of deep time. They are also specific to place. These sands mean different things in shifting contexts, as historian Libby Robin has explored. These particular sands gather together the changing momentum of environmental art–science projects, the partnerships between the artist Mandy Martin and conservationist Guy Fitzhardinge and archaeologist Mike Smith. These sketches portray landscapes at a time of Indigenous rights politics and land management, and they also reflect the rising awareness of accelerating biodiversity loss and land degradation through global climate change.

When Alexandra R. Toland invited Martin to submit an art-science dialogue for this volume, Martin focused on a collection of little vials of ochre Mike Smith had given her and the sketchbook she had started with those ochres. Sketching represents a bridge between *Strata*, an earlier project with Smith and Robin, and the 2016 *Arnhembrand* project with Guy Fitzhardinge and Hamish Gurrgurrku, underpinned by traditional understandings of country and conservation sciences, especially fire ecology and invasion biology. It is also tribute to the use of sketchbooks by field archaeologists. The powerful landscapes of Arnhem Land, of dramatic weather and complex history, are described and illustrated by the authors and artists with their unique disciplinary and cultural perspectives.

Grounded Painting: The Art of Mandy Martin

Mandy Martin has always worked with scientists. From an early age, her ecologist father, Peter Martin, drew her eye to the workings of nature. Later, she traveled with him on ecological field trips, sketching and painting, and drawing on his deep understanding of the relations between soils and plants. Much of her painting is physically "grounded," that is, the pigments and soils of the place she portrays are embodied in the artworks. Soil is the foundation of her art, as well as the ecological foundation of the landscapes she depicts. Her art projects always begin in place, on the ground. They often start with a "hot" political issue, such as gold mining in her home region of central western New South Wales; water management in the tributaries of Lake Eyre; or biodiversity management in the Desert Channels of South West Queensland. Increasingly, concerns about place have to be understood in partnership with Aboriginal custodians, such as the management of country in remote areas of Western Australia (as in the project *Desert Lake*) and Arnhem Land in the Northern Territory (as in the project *Arnhembrand*).

In the collaborative project *Arnhembrand. Living on Healthy Country* (2016), art and storytelling were used to raise international awareness about the work undertaken by the Indigenous communities living in the Djelk and Warddeken Indigenous Protected Areas (IPA) to preserve their unique cultural and ecological environments through the work of the Karrkad Kanjdji Trust (KKT). In this project, the "grounding" was about putting Western artistic practice in conversation with Indigenous understanding. All the artists involved in the *Arnhembrand* project worked on paper and canvas, prepared first with an ochre pigment or fluoro paint underpainting.

The groundedness of works in this and other environmental projects undertaken with Martin begins with local people and the soil on which they live. For this sketchbook Martin used ochres that connect local places and ideas about time—particularly the scientific ideas about deep time in Australia. In his work in the Northern Territory in the 1980s and in his subsequent work, Mike Smith amassed hundreds of soil samples for analysis and dating. Grains of sand are the crucial element for luminescence dating techniques, which Smith was among the first to use in Australia. Sands that have long been buried can yield signals upon first exposure that provide precise dates for the whole layer in the landscape, and therefore the artifacts that are touching that layer. The core of sand is sent to a laboratory to determine how long it has been since *that* grain of sand has been exposed to the sun. Archaeologists drill a core in the face of the trench, from which samples from underneath human artifacts, such as tools or ochres, may be extracted. Ochres can be identified as "human artifacts" when they are sourced from far away from the dig site (some special ochres associated with ceremonies were traded thousands of miles from their quarry to the occupation site where they were discovered). After his analyses were done, Smith kept matching samples from many different sites, some of them for up to three decades. Although color and texture can be important for testing ideas as they develop, there can be no further dating of grains of sand using luminescence once they are exposed to light.

The dialogue here is between place and time. The ochres or "sands of time" were a gift from Smith, Australia's most renowned desert archaeologist and author of *The Archaeology of Australia's Deserts* (Cambridge University Press, 2013). The geochemistry of the ochres enabled Smith to reconstruct and date many of the important trade routes of Aboriginal people living in the desert. His archaeology has been a major part of the discovery of the long human history (deep time) of Australia, which dates Aboriginal occupation at well over 55,000 years. In 2015, when Smith finally came to part with his tiny vials of ochre, all systematically numbered, he presented them to Martin. These were among the many soil samples from his archaeological digs right across Australia that synthesized the stories in his book. How should Martin use them to make art? She thought about this carefully:

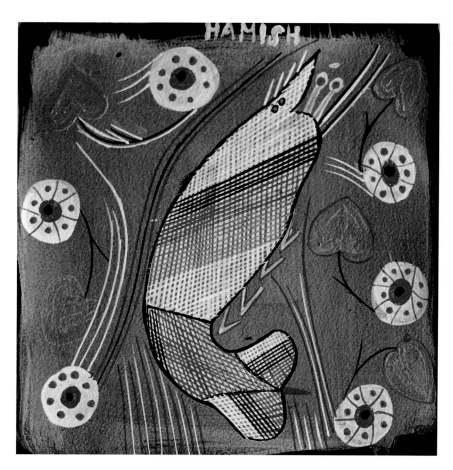

Hamish Garrgarrku, *Yabbie Dreaming*, 2015. (A yabbie is a freshwater crustacean—a prized food.) Ochre, fluoro pigments, acrylic on black paper, 28 × 28 cm.

Reprinted with permission from the artist.

Two Hills, Pamarr country, from the soil and pencil sketchbook of Mandy Martin, created on April 6, 2013.
Reprinted with permission from the artist.

Mike gave the ochres to me with the suggestion that I might like to use them to paint with. He was clear that he did not want the samples post-analyzed in the future. Most of the samples were in tiny glass vials. My paintings are usually encrusted with ochres and pigments, but these tiny soil samples—strongly colored and sandy from the Australian desert—would not work as canvas crusts. I decided that the best use for these colors was in a dedicated sketchbook.

Martin conceived the use of fluoro pigment combined with Smith's natural ochres as a way for the diverse group of Bininj artists to record their stories about healthy country, in a new way. We called it new way (or new wave) painting to distinguish it from the traditional Bininj way used by the Rarrk Masters on painted barks, funeral poles, and other wooden objects in Arnhem Land. Fluors are markers of the Anthropocene, reflecting the fact that many of the global drivers impacting on healthy country are connected to climate change, plastics, minerals, and other signs of the anthropogenic changes of our times. *Sketches in the Sands of Time* portray landscapes of the north, but use the materiality of deep-time archaeological samples from all over Australia. Together they celebrate the partnerships between arts and sciences, particularly archaeology

and conservation biology. The sketches embrace the long connections between Traditional Owners and their country, and explore the history of relationships between the peoples of the first and second settlements of the continent.

Deep Time and Mike Smith

The groundedness of the environmental works begins with local people and the soil on which they live. Literally. For *Sands of Time,* Martin has chosen works that use ochres that connect not just single places, but ideas, particularly the scientific ideas about deep time in Australia. The dialogue here is between place and time. The ochres or "sands of time" were a gift from Mike Smith, one of Australia's most renowned desert archaeologists. The geochemistry of the ochres enabled Smith to reconstruct and date many of the important trade routes of Aboriginal people living in the desert. His archaeology has been a major part of the discovery of the long human history (deep time) of Australia, which now dates Aboriginal occupation at 65,000 years. (Chris Clarkson, Zenobia Jacobs, Ben Marwick, et al. 2017. Human occupation of Northern Australia by 65,000 years ago. *Nature* 547:306–310. doi: 10/1038/nature22968.)

Mandy Martin, *Late Burning on Flood Plain from Djinkarr*, 2015. MS 16 ochre, sepia ink in sketchbook 28 × 28 cm.

Reprinted with permission from the artist.

Mandy Martin, *Fire Walk*, 2016. MS 4 ochre, fluoro pigments, sepia ink, black paper 28 × 28 cm.

Reprinted with permission from the artist.

The drawings and loose works on black paper slipped in between the leaves of the *Dirt Dialogues* sketchbook were drawn over four trips to central Arnhem Land, as I developed and worked on *Arnhembrand*. Living on Healthy Country, a collaborative art, science, and stories project with Bininj (Indigenous) and Balanda (white) participants. My time for drawing was limited; I usually managed one sketch or more a day early in the morning or often late in the day as we sat gazing out over the Liverpool flood plain at dusk from Djinkarr, an outstation about 35 km from Maningrida. The third trip in August was burning time, when the local people and Djelk Rangers were all out burning the landscape, practicing the many different kinds of *wurrk* tradition.

In his work in the Northern Territory in the 1980s and in his subsequent work, Smith amassed hundreds of soil samples for analysis and dating; grains of sand, the crucial element for luminescence dating techniques, which Smith was among the first to use in Australia. Sands that have long been buried can yield signals upon first exposure that provide precise dates for the whole layer in the landscape, and therefore the artifacts that are touching the layer. The core of sand is sent to a laboratory to determine how long it has been since *that* grain of sand has been exposed to the sun. The sands are extracted from the archaeological trench, using the precision archaeological *tradecraft* Smith describes later in a semifictionalized account of the technicalities and pleasures of his work. Tradecraft also demands drawing, as Smith explains. Since the writing is fictionalized, the (real) drawing included is just an example of his careful professional work. After his analyses were done, Smith kept matching samples from many different sites, some of them for up to three decades. Although color and texture can be important for testing ideas as they develop, there can be no further dating of grains of sand using luminescence once they are exposed to light.

In 2004–2005 historian of science Libby Robin arranged with Smith to make a trip to the west of the Northern Territory to his archaeological site at Puritjarra in the Cleland Hills. Puritjarra is the oldest dated site of desert settlement in Australia, a crucial cultural, historical, and scientific location. Robin and Smith used the archaeological science to trace a history of this remote place from the Ice Age to the present. Mandy Martin, in partnership with the Traditional Owners of the site, the people at Ikuntji (Haasts Bluff), explored this place through art. The resulting environmental art project, *Strata: Deserts past, present, and future,* was a transdisciplinary reflection on the cultural significance of how human time has deepened over the past few decades in Australia. The discoveries in this single place have proven how much older the Australian past has become in just our lifetimes. Australia's oldest dates have been found in the northern tropics,

Mandy Martin and Mike Smith,
Palimpsest, 2005. Found local
and sourced pigment, sand,
rock shelter floor matter, ochres
and acrylic on Arches paper,
30 × 200 cm.

Reprinted with permission from
the artist.

while in the shifting sands of the Aeolian desert it is has been extremely difficult to find places that have survived intact in ways that can be analyzed. Before this discovery at Puritjarra, archaeologists had speculated that no one could have lived in the dry and cold conditions of the last Ice Age in the inhospitable Australian desert. They figured that people had probably moved in as conditions got milder in the Holocene (over the last 10,000 years or so). Puritjarra changed all that. This was a remarkable rare place where people lived and left artifacts from the last Ice Age which could be found today. Because its entrance faced away from the prevailing winds, a six-foot trench here revealed a great depth of time: 35,000 years of human settlement. The time frames of Australian desert archaeology shifted by an order of magnitude because of the particular study of Puritjarra. These have since changed because of sand in another place, moving the accepted date of inhabitation in Australia to 65,000 years.

Tradecraft

Mike Smith

He sat in the trench, looking at the wall. He looked hard, with anticipation, taking in the earthy smell. If he stood up, his eyes were at ground level. This was the quintessence of an archaeological dig, he thought. In the wall of the trench the history of the site would be recorded, summarized, laid out. All excavations, he knew, were initially local history. The trick was to learn to read a section. It was like reading a language, where each deposit was written in a different dialect. Don't make a mess of it he told himself. Not at this site.

He enjoyed the sense of depth. Of being enveloped by the anatomy of a site. It tended to unsettle the perception of place. Walking about, he

became acutely aware of the layers beneath his feet, like peering into the depths from a glass-bottomed boat. He began to clean the wall of the trench with his trowel. Lightly scraping, but not digging, he exposed the fresh earth. Allowing the details to stand out.

The early morning humidity helped. The light wasn't optimal but the faint moisture made details and color stand out, before the dryness of the day bleached everything to an undifferentiated red-brown. People thought that archaeological sites were "layer cakes." One layer above another. So, they were—but in the desert the layers were cryptic, hard to see. He knew they were there, but he would have to work for it. At this site, subtle changes in texture, color, and fabric worked in his favor. Anything dug from a higher level would contrast with the underlying sediments. That is, if he looked closely enough. He went back to his work, cleaning the face, taking in the smell of earth, his eyes searching keenly.

Each trench, he reminded himself, would provide a different window on the internal structure of a site. The same layer could look quite different depending where he dug. Sediments, he knew, could vary in character across a layer. To understand the anatomy of this deposit, he needed a number of cuts.

A movement in the pit caught his eye. Looking around the trench, he saw a wasp trying to bury itself in some loose earth. In the corner was the limp body of a small snake. "I must burn an arc of spinifex in front of the cave, to give us some clear ground," he reminded himself. A cold night in the desert is crystal clear and quiet. The warm weather and rain a couple of nights back had brought the desert to life. People were often surprised by the difference. In the morning, he'd find half-eaten bodies of hopping mice on the site. Lizards and snakes were out and about. Scorpions. And centipedes as big as 10 cm in size. He made sure to clean up the inevitable pile of loose earth that had accumulated overnight in the pit. No matter how scrupulous he was, there was always this trickle of loose earth, the fallout of any excavation. He had trained himself to always clean this away before starting any serious work.

He turned back to the face, working systematically from the top to bottom so his spoil did not obscure features on the fresh surfaces. He put nails in the face to mark elevations, to key his section drawing into the trench plan. Each level, each horizontal time-slice, each spit of his excavation was already drawn, so that a detail seen in cross-section in the face could also be explored to some extent in plan.

Rocks and pebbles showed the bedding of the deposit. Subhorizontal. Good. It avoided the complications of a site with sloping strata. Feeling down the section, he felt for subtle changes in texture. Color alone was not much of a guide in these sites. He loved the physicality of the task: feeling, smelling, and inquiring. Here, the top layers were gritty, with lots of small spalls of sandstone, from weathering of the shelter walls. But the matrix of sand was a very fine sand, unmistakably dune sand. Aeolian sand he knew was 100–150 microns. The desert landscape was made up almost entirely of this 150-micron quartz sand, cycled and recycled, transported across the landscape and blown into sites like this. Almost always with a rind of rusty red iron oxide. The brighter the red, the older the sediment. And here was the deep red of antiquity.

The upper layer was brown, gritty sand, but halfway down the face it changed to a silty red oxidized sediment. He looked for a boundary. He knew a sharp boundary would mean an interface, or a line of erosion. Or perhaps a stillstand, where nothing had been laid down for hundreds perhaps thousands of years, creating a surface on which a composite of things had happened—a palimpsest. Almost as if a camera had stopped rolling and everything was now recorded on one frame. Overprinted. Superimposed.

He looked closer with his magnifier. He hadn't detected a surface when he'd dug down to this level, just a change in sediment. The ants had given the first clue that he was coming down onto a new layer. When he had scraped the surface with his trowel it was speckled with pinpricks of bright red sand. And these invertebrate excavators, with their tiny spoil heaps, had heralded the approach of a new layer. He was alert to this. But he'd not found a paleo-surface here.

The section confirmed this. There was no boundary here. Just a gradual change, over a few thousand years, in the character of sediments reaching the site. A change in sediment flux rather than a gap in time. He looked up, looking out onto the sand plain, beyond the shelter. The site was intrinsically connected to the landscape. He'd need to walk around, eyes tuned to the flows of sediments to get a better idea of what this change in sediment influx meant. How changes in the landscape had affected what entered the rock shelter. He'd do this later, quietly, almost surreptitiously. He liked to work things out in his head before worrying his crew of diggers. They were a really good crew—disciplined and hard working—but living conditions were inevitably hard after a few weeks in a remote bush camp: a rock shelf for a bed, a bit of fried damper, and a cup of sweet tea for breakfast. The crew was showing the strain. He wouldn't fuss about it now, and risk breaking the rhythm of their day, the trudging of exhausted men to a finish line. Not just yet. Soon enough they would all face the back breaking work of backfilling the excavations. One couldn't just leave the trenches open. Slumping trench walls would damage the archive that a site represented, the pages of a book destroyed before they could be read.

In the trench, he could see that the brown layer was a promiscuous jumble of stuff. There was so much here. The red layer was more austere: Zen rather than baroque. He could see several stone tools in the trench wall. His instinct was to dig them out, but his professional training forbade this. Excavation was like a surgical dissection. Sections had to be kept intact. Surfaces had to be kept scrupulously clean. It was so easy to lose any sense of the structure, if things were clogged with spoil, like an operation flooded by blood. He'd seen a dig degenerate into chaos, "a charnel-house of murdered evidence." Not here, he thought.

So much for the major strata. Now he needed to fill in the detail. In the red layer, he could see some pieces of red ochre, and a line of sharp-edged chips, just the edges projecting from the face of the trench. He knew these were flakes of silcrete. Stone tools as big as one's thumb. Each with "a memory of other hands." He could see that this band of tools marked an individual event, perhaps a faint trace of a hunter's camp 20,000 years ago. This abyss of time made him dizzy. With a measure of relief, his eye fastened on a lens of ash and charcoal toward the top of the face. Only a few millimeters thick but with lots of internal detail. Details within details. A line of yellow ash here. A thicker line of white ash there, interdigitating with some finely comminuted charcoal and a few larger lumps of charcoal at one end. This he knew was a sleeping fire. He'd exposed it in plan when digging the trench. There were half a dozen features like this, where people had slept between small fires to ward off the cold of a desert night. Later when he'd drawn it, he'd take the larger lumps of charcoal as a c14 sample. He'd collected many such samples during the dig, but a sample from a known feature—identified on a section drawing—was always valuable. Much better than detrital charcoal because it was not a float of occupation debris averaging out the everyday life of a century or more. It was fixed to an event, to a feature, to a moment in time.

He never tired of the magic of radiocarbon dating. A carefully labeled and sealed plastic bag of black charcoal, unassuming as it was, would be sent off to the laboratory for assaying. And back came an age, 950 ± 50 years BP or something. He liked to test himself. He had done enough work in the desert to have a good feel for age, but sometimes he was surprised. Once he'd been out by an order of magnitude. He'd thought 2000 years and the charcoal showed 20,000 years. He was unsettled then, and thrilled.

In the trench, he could sit and wallow in his craft, away from the hustling ambition of academic life, a tradesman archaeologist, concentrating on the immediacy of his task. He looked for other features. Toward the base of the red layer, he could see an old pit. Looking closely, he could just make out the sharp cut of the pit, accentuated by the fill. The pit introduced the coarsely textured sediments of the brown layer into the red silt. He could see the fill lines of grit and charcoal, each marking the slumping of earth into a pit. He found that he could make out the base of the old pit quite clearly, but where was the top? What level had it been dug from? He traced it with his trowel, carefully following the lines of grit, and could see where the pit had thrown out silty, red sand at the ancient ground surface from which it had been dug. People forget that while a pit brings younger sediments into older levels, it also does the reverse. The material dug out had to go somewhere and was just as much of a problem. He could see now why some of his radiocarbon dates seemed out of order in this part of the trench. If he found a pit during excavation, it was his practice to hold an inquest to determine what level it was dug from. But it wasn't always possible to tell. A few hundred years of feet scuffing up sediment would be enough to erase this sort of detail.

He set up his zero line, leveled it in, and began to draw the section. He felt he had a grip on things now. Until he understood the section, he could not begin to draw. He knew it was no good just following the color and drawing lines, then fixing them later when radiocarbon results came in. That was just "coloring by numbers." Good archaeology begins with understanding sediments *in situ*.

He relished the quiet task of drawing a section. It was a deliciously quiet period, a sort of dialogue with a site, a culmination of a dig and a summary of a site's history. He gave the crew other tasks—or the morning off—so that he could have it to himself. He worked with confident clean strokes keeping his drawing clean and precise. And working with a fine lead pencil. The old hands were adamant on this point. Ink would run when it got wet or covered with sweat. Pencil lines survived the travails of fieldwork better. A section drawing said a lot about an archaeologist; their neatness and sense of order and inquiry. He took pride in his tradecraft. It was better not to interrupt the flow before the drawing was finished, so he kept one person nearby to fetch

Mike Smith, *Malakunanja Section* drawing, 1989.

Reprinted with permission from the author.

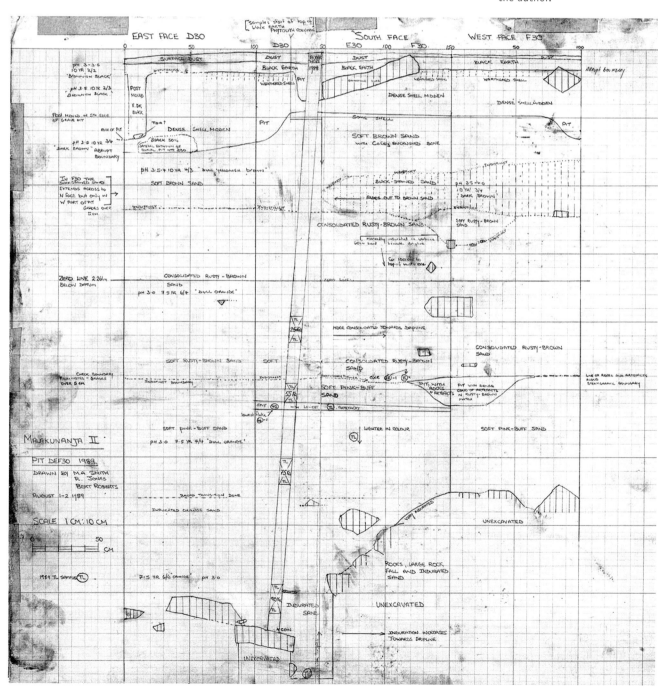

things, as climbing out of the pit was awkward. And it was important to have an apprentice involved: this was how he'd first learned his own craft.

When he'd finished drawing, he took his samples. He began at the bottom of the wall and worked up, so that his spoil didn't compromise subsequent samples. A neat line of sample bags, sealed and labeled, soon grew on the ground around the pit. The pollen or phytoliths in these—those bizarrely shaped plant microfossils—might give him a record of vegetation in the past. He reckoned it could be vibrant with change. After all, the desert had swung between wetter and drier climates for as long as people had made it their home. Combined with plots of stone tools and his growing series of radiocarbon dates, he loved the emerging complexity of a site's story. He sat back with a sense of satisfaction. A stratigraphic section, he decided, was a running commentary on the history of a place. And these layers, these sediments, intrinsically connected a site to a landscape.

He carefully dated and initialed his drawing.

Comanaging Country in Australia's Indigenous Protected Areas (IPAs)

Guy Fitzhardinge

As the wet moves in across the tropical Arnhem Land plateau, billowing clouds, flashes of lightning and darkening skies herald the imminent rain. Beginning with fat drops the size of peas, the cycle of wet and dry returns. Soon rivulets of water cascade down the face of sandstone rock that has offered shelter and home to people in the wet for tens of thousands of years.

On the walls and ceilings of the rock shelter, the art of long-departed inhabitants reminds us of the importance of these natural shelters to the people who lived in and were part of this landscape. For here, and across so much of the Australian continent, there was no separation in thinking or practice between social systems and natural systems. People belong to country, not the other way around.

Although this country has been occupied by people for thousands of years, its original inhabitants now call it "empty country." Government policy from colonial times to the present and the practical needs of a new economy have emptied it. There is no one there to care for it—to do the "housework" of hunting and burning, maintaining the sites and renewing the cultural links with the physical topography and its natural inhabitants. The rock shelters and their archaeological stories remain silent—mute testaments to generations of lives for which they played a vital role.

As the rain drips down from the shelter roof, so the dust of the floodplain slowly turns to mud and rivulets form beginning anew the cycle of wet and dry—of boom and bust and of the replenishment of nature's capital. However, with the emptying out of the country's people, a range of other introduced species has taken their place. Descendants of water buffaloes, brought to the Northern Territory from Indonesia between 1824 and 1849, now ravage the floodplain in increasing numbers. Feral pigs, introduced by Captain Cook in the 1700s, multiply and indiscriminately feed in waterholes where families source valuable bush foods. Introduced grasses, escaped from their original locations, now outcompete endemic species and take over, thereby destroying natural habitat and burning with an intensity that endemic species do not.

The end of the wet season marks the beginning of the season of work. With the landscape replenished and another cycle of growth over for flora and fauna, it's now time for the rangers to get out and carry on traditional practices of caring for country. This time, it's far more complicated. There are new feral animals to deal with—buffaloes and pigs—that are trashing the waterholes. Feral cats are seldom seen, but their effects are evident as they prey on small mammals and reptiles, now critically endangered across the plateau. With few people living out bush, traditional cool fires are no longer lit. People are needed for the precision burning that will restore the natural functioning of ecosystems.

Country is central to the cultural aspirations of the people here, but cultural traditions must work with other practical needs. For example, reclaiming the landscape through cool season burning may be at odds with requirements for hunting. Buffalo are an important food source in some communities, so how are numbers to be controlled in such a way that the food source is preserved while their negative impact on the landscape is minimized—all this in a rugged landscape that has few roads, few people, and extends for many thousands of square kilometers?

As the rain drips off the rock shelter overhead and nature renews itself as it has done for thousands of years, so is the need to support and renew the connection to this landscape with its people—not only spiritually and culturally, but also in very practical ways. This is important for the sake of the landscape and the values that it provides to its original inhabitants but also to Balanda, as white Australians are called here. If we want this incredibly beautiful and environmentally valuable natural asset to be preserved, we must empower its owners. Like the weeds and the feral animals, we as newcomers with our Eurocentric ideas and values have to learn from the experience of Traditional Owners and from ecological science, how to care for this country together.

A Visual Ethnography of Soils in Space and Time

Ekkeland Götze and Winfried E.H. Blum in conversation with
Alexandra R. Toland

Ekkeland Götze was born in Dresden, Germany, in 1948. He has been working as a painter, silk screen printer, engineer, and manager. In 1988, he moved to Munich. He has been occupied with *Earth* since 1989 and has been working at a conceptional *Image of the Earth* since then. Works by Götze are in public collections in Berlin, Dresden, Munich, and Wolfenbüttel, Germany; New York; Rotorua and Wellington, New Zealand; Ondini, South Africa; Dharamsala and Dehradun, India; and Luxembourg. He has realized projects all over the world i.e.: *TERRA DI SIENA* (Italy), *Berlin Wall—Death Strip* (Germany), *Icefire* (Iceland), *Europa* (Crete), *Ruaumoko, Kokowai,* (New Zealand), *Songlines* (Australia), *Zulu, Lapalala,* (South Africa), *Sinai* (Egypt), *Kailas* (Tibet), *Rice* (Japan), *Amazonas* (Brazil, Venezuela), *Go West* (United States), *Rarámuri* (Mexico), and *Maka Wakan* (Sioux, United States).

Title image: *Menabe—Das Große Rot (Madagaskar)*, *Die Baobabs*, Befasy, 2014, Erdbild-N° 830, Terragrafie auf BFK Rives 300 g auf Nessel auf Keilrahmen, 71 × 71 cm, Unikat. *Menabe—The Big Red (Madagascar)*, *The Baobabs*, Befasy, 2014. Earth Art Work-N° 830 on BFK Rives 300 g on canvas on frame, 71 × 71 cm.

Unique earth image reprinted with permission from the artist.

Winfried E.H. Blum is professor emeritus of soil science at the University of Natural Resources and Life Sciences (BOKU), Vienna, Austria (since 2009). He was previously professor at the University of Freiburg, Germany (1972–1975), at the State University of Paraná in Curitiba, Brazil (1975–1979); and professor and director of the Institute of Soil Research at BOKU University (1979–2009). His research and teaching activities are based across Europe, Africa, Latin and North America, Asia, and Australasia. He is a member of editorial boards of 27 scientific journals, with more than 800 publications since 1965 in 15 languages in the fields of soil science, agricultural and forest land use, and environmental protection. Blum is a member and honorary member of many academies and national or international learned societies, with numerous distinctions and awards.

> The colour and structure of my soil artworks explain not only their physio-geographic origins but also let you imagine the people which live on these soils, how they treat their soils, how they adore their soils, and pray to them as the most important basis of life.

The artist behind this statement has been traveling the world in order to collect soil materials with different colours and textures for his artwork. His selection and collection of soils is based on the observation of the local people living on these soils, and their personal relations to the soils and their soil use. During his travels, he always spends a considerable amount of time with the local people in order to understand their relationship with soils and nature before he collects soil samples for his artwork. This chapter explores the ethnographic approach to the soil-printing practice of artist Ekkeland Götze, with insight and reflection from the prominent soil scientist and former secretary general of the International Union of Soil Sciences (IUSS) Winfried E.H. Blum.

Alexandra R. Toland: Ekkeland, you've worked at so many different sites around the world and with so many different types of soils. Have you developed any expectations about the ways different soil samples will react to your "terragraphic" process?

Ekkeland Götze: I have one main technique that I use, that I invented in 1989 for all my works. All the earth pictures are made using the same technology, regardless of soil type. So, basically, the soil makes the image itself. As a printing process, the technique is absolutely, methodologically the same each time, and the only variable is the medium—the individual soil sample. So, the expectation from my experience is that the process results in a wide range of different colors and structures. I can't foresee what will happen beforehand. It happens in the moment of the printing process, which can be very exciting. Soil is an element like air and fire, and its release onto the paper is unique. There is an energy and spirit left on the paper that is palpable, compared to normal paint.

Alexandra R. Toland: Winfried, you've similarly worked in many landscapes around the world. Are there certain expectations you have when you enter a new landscape? What features do you look at to read the landscape? And would you agree that a certain methodological consistency (in scientific as well as artistic practice) is necessary in demonstrating the uniqueness of a particular landscape—its beauty, local meaning, and potential threats?

Winfried Blum: As a soil scientist, I first try to understand the origin of the different colors and structures based on the processes of soil development in different regions of the world, with their different geologies, climates, ecologies, and land uses. Colors very much depend on the elements contained in soils and their chemical behavior influenced by oxidation, reduction, the turnover of organic matter and other biological and physicochemical processes. This allows to estimate the possible functions of these soils, especially for satisfying the demands of local inhabitants for food, water, and energy. The uniqueness of a particular landscape is of course not only based on soil features, but also on the topography, the plant cover and its diversity, the animals living in this environment, and especially the action of humans, which have often intensively changed the overall natural conditions.

Alexandra R. Toland: Ekkeland, your terragraphic images are so stunningly different, which mirrors the uniqueness of the sites you choose to work in. Could you talk a bit about your relationship to site? I'm interested in where you choose to work, and if there are certain criteria you use to select sites …

Ekkeland Götze: Okay, it always depends, you know, but I work mostly in this manner: First I have an idea for a project, which means for example I want to go to a certain region, and then I contact the local people living there, which is not always easy. For example, in 2000, the Goethe Institute Sao Paolo invited me to do a project in Brazil called "Sinne der Erde" (Senses of the earth). I originally wanted to interview people from all the living Indian Tribes for the project. I thought there were maybe about 50, and then I learned there were over 350. Talking to all of them would have been impossible in the context of that art project. Additionally, I always ask permission to take soil materials from other places. And even this is complicated, because there are conceptual and legal differences of ownership of the soil around the world. Permission from land-owners is not the same as permission from land-guardians.

So, through several contacts in France and Brazil I met a very old Indian priest in Santarém and took the first soil samples. On the way from the delta of Rio Amazonas to the BazoCasiquare in Venezuela

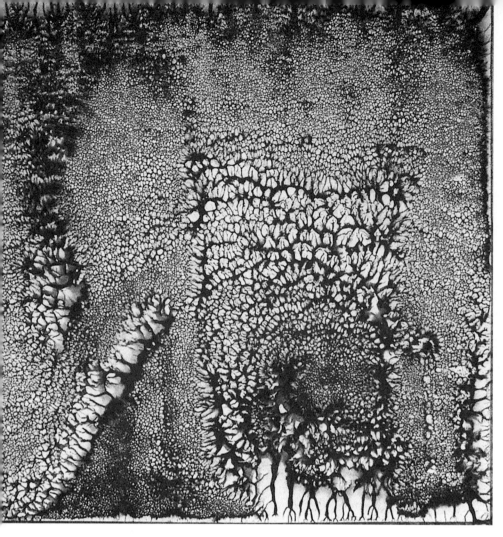

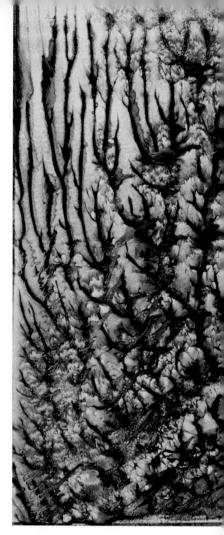

Eisfeuer (Island), Sprengisandur, 2003, Erdbild-N° 518 auf BFK Rives 300 g auf Nessel auf Keilrahmen, 71 × 71 cm, Unikat. *Icefire* (Iceland), Sprengisandur, 2003, Earth Art Work-N° 518 on BFK Rives 300 g on canvas on frame, 71 × 71 cm.

Unique image reprinted with permission from the artist.

Eisfeuer (Island), Herdubreidalindir, 2003, Erdbil 529, Terragrafie auf Hahnemühlen Bütten 350 g auf Nessel auf Keilrahmen, 100 × 100 cm, Unika *Icefire* (Iceland), Herdubreidalindir, 2003, Earth Work-N° 529 on Hahnemühlen paperstock 350 g canvas on frame, 100 × 100 cm.

Unique earth image reprinted with permission f the artist.

we took several soil samples. In Santarem, I rented a small boat, and we went up the Rio Tapajos. We met various people along the way who live in the local rainforest. One community elder told us that we couldn't take the soil away until they knew more about the project.

There was a meeting in the afternoon, and the whole community came to the school, and I had to explain what I was doing and what I wanted. The whole community was involved in the discussion—"Should we give this man our soil?" Some children were laughing, and after a short moment one person said yes. They invited me to come back in the evening and celebrate the gift of soil with a small ceremony of goodwill. The next morning, we went to the top of the hill above the village and gathered some samples. So, I always ask people, what is the most significant site in your community and in your home, in this way the people determine the place. The site selection is a process of conversation and exploration.

Alexandra R. Toland: Winfried, could you explain the site selection process from a scientific perspective? If you are pursuing a certain research question, how do you narrow down the exact

Die Azteken, Teotihuacan, (Mexico), Die Stadt der Götter, 2010, Erdbild-N° 717, Terragrafie auf Hahnemühlen Bütten 350 g auf Nessel auf Keilrahmen, 100 × 100 cm, Unikat. *The Aztecs*, Teotihuacan, (Mexico), The City of Gods, 2010, Earth Art Work-N° 717 on Hahnemühlen paperstock 350 g on canvas on frame, 100 × 100 cm.

Unique earth image reprinted with permission from the artist.

places to take measurements? I imagine maps and prior data sets play a significant role, but do local people sometimes point you in new directions? And have there ever been surprises in your fieldwork, especially, as Ekkeland mentions, regarding issues of land ownership and property rights?

Winfried Blum: The reasons for selecting a specific site or soil for scientific research can be extremely diverse, depending on the main goals of interest, mainly defined by the type of land or soil use. One goal can be to understand the type and the impact of local land use, for example, agriculture or the possibility to dig for groundwater resources. Other reasons can be the use of soil material for human ingestion (geophagy) as a kind of medical treatment or specific memorial reasons, for example, for places of ceremonial events.

As a scientist, I always select soils and sites according to a clearly defined research concept, which can be the analysis of specific solid, liquid, or gaseous soil components, the taxonomic classification of a soil, or the specific capacity of a soil for distinct kinds of uses like agriculture or forestry.

In many regions of the globe, no soil or site maps or any prior information about soil data exist. Here field observations, based on experience from other areas can help as well as information, derived from indigenous knowledge of the local population.

In many areas of Africa, Asia, and Latin America no land cadastre exists and the indigenous population believes that their land was given by god. However, in such areas very old and strict rules about soil protection and their uses exist, which were the basis of sustainable use of these soils and sites during many centuries. These soils are nowadays severely endangered by "land grabbing," leading to a loss of livelihoods for many local populations.

Alexandra R. Toland: Ekkeland, I'm fascinated by your ethnographic approach. In addition to your terragraphic work, you also make portraits and collect stories of the people you meet along the way. Thus, your artwork is not only about the soil, it's about the human connection to the soil. You're retelling stories in your art.

Ekkeland Götze: Yes, I collect many stories, for example, in Amazonia I collected nineteen stories. Some of them have never been published, and some are in my book. My aim is to make authentic projects in the locations I work, and this can only develop in discussion with the local people. So, I ask everyone I meet about what they think and how they feel about their landscape, their soil. And doing so, I collect ideas at the same time as I collect the soil.

Alexandra R. Toland: Winfried, has your work as a scientist ever taken this ethnographic approach?

How do you depend on local expertise in your work and do you consider your own work as a scientist a form of storytelling?

Winfried Blum: In many regions of the world I have used an ethnographic approach in order to understand the direct or indirect impact of the local population and of their surroundings on land and soil, using this knowledge for the selection of research sites or the sampling of soil and plant material. Some of the results were telling me specific stories about the life or the living conditions of the local people, sometimes also about their myths and religions, for example, in East and West Africa and in the Andine region of South America, in the Amazon basin of Brazil and in Mexico.

Alexandra R. Toland: How do you think artistic and scientific approaches complement each other in soil protection efforts?

Ekkeland Götze: I think my art speaks for itself. It gives people a wide range of ways to think about soil and offers an idea about the beauty of the soil and how it is a part of cultural landscapes.

Winfried Blum: In the case of the artist, Ekkeland's approach looks at the beauty of visible soil characteristics like color or structure in connection with the local people. In contrast, the scientist tries to measure specific soil characteristics, such as color or structure, to determine explanations to various phenomena. Both approaches indicate how differently persons with different disciplinary backgrounds can identify and explain soils as objects of natural beauty and as an important natural resource, thus sustaining humans and their environment.

Lessons from Emma Lake

A Metamorphosis of Science and Art in Landscape and Local Color

Ken Van Rees and Symeon van Donkelaar

In his art, **Symeon van Donkelaar** uses the land as pigment color. In using such local colors, his work rejects the dichotomy of a visible hue from a tangible material. By keeping his colors whole and unrefined, all the senses are present when he works with a pigment—yellow can have a sweet smell and red can squeak when handled. By working with the land in this unique way, he also allows the pigments to remain connected to the places where they originate. After foraging for soil, minerals, plants, and bones, he creates local color pigments and uses them to make artwork that reveals the culture, history, and spirit

Title image: Students painting during the evening at the Emma Lake Kenderdine Campus.

Image reprinted with courtesy of the artist.

present in the land visible through its local color. His use of such pigments in art is represented in a broad range of works—iconographic, diagrammatic, and hieroglyphic. All of these are informed by local color pigments that reveal the heritage of the land.

Ken Van Rees is a professor of forest soils at the University of Saskatchewan Canada and has researched the effects of natural forests and tree plantations on soil carbon in the prairies of Saskatchewan. He is passionate about experiential learning and has received many awards for his creative use of visual arts in his soil science field courses. He paints plein air with a group who call themselves the "Men Who Paint" and has painted landscapes across Canada and the high Arctic. As a consequence of his research site burning in a wildfire, he can also be found meandering through boreal forests after forest fires where his sense of play and experimentation become the "living canvas" while the burnt trees transcribe their carbon marks to his paper and canvas.

The two authors have known each other the past several years because of their interests in soils as pigments. They also have conducted several courses together on creating pigments from soils, and from these shared experiences have each written their respective dialogues and then sat down together while on a pigment-collecting trip to finalize the document. The mixing of the soil science and visual arts disciplines provides a unique story for highlighting the use of soil materials in cultivating creative experiences that can be historical, cultural, and spiritual in nature. Our stories represent journeys from either science to art or from art to science, where we both end up understanding that soil is connected to our heritage and that utilizing the earth as colored pigments helps facilitate a deeper connection to the landscape for both scientists, artists, and students.

In the Boreal Forest: A Journey from Science to Art

I (Ken Van Rees) have spent my career as a scientist investigating the role of soil in boreal forest regeneration and the production of short rotation woody crops, such as hybrid poplar and willow. My fascination with the relationships between vegetation and soils is an extension of my childhood, where as a kid in Ontario I spent a lot of time amongst trees, rocks and water. As an undergraduate forestry student, I spent the first three to four weeks of each university year out in the field. It was this experiential learning and sense of place, of being in the boreal forest, that I also wanted my students to experience when I began teaching. Once I became a university professor, I took my students on a three-day field trip to the boreal forest to understand how forest management and natural disturbance events impacted soils in my forest soils course. We would dig and scrape, feel and taste, measure and count the various characteristics shaping a soil and its vegetation.

In 2004, I was preparing to teach a new, weeklong experiential field course in northern Saskatchewan that would investigate the relationships between soils and vegetation. Several months before the course was offered I visited my mother in Ontario who wanted to go to a *Group of Seven* exhibition at the McMichael Gallery in Kleinburg. As a scientist fully engaged with my work, I had had very little time for arts and culture up to that point but ended up going to the exhibition with my mom. It turned out to be life changing. As I stood in the gallery observing the works of these renown Canadian landscape painters I had an extraordinary epiphany: "Why can't we paint landscapes in my soil science field course?" My next immediate thought was "Well *you* can't—you don't know anything about painting!" I wrestled with this idea, however, for a few days trying to figure out how to include art into my science field course. Is it possible to enhance student learning for science students by using artistic methods? I turned to my sister-in-law who was a high school art teacher in Ontario and began a dialogue of how I might tackle the subject. Her suggestion was to start simple and try oil pastels—they are not messy, could handle cold or rainy weather—they'd be simple and clean. She provided me with a few drawing rules to give the class, and off I went. You must understand that I was stepping way out of my comfort zone, because when I teach I like to know my material intimately to feel comfortable in delivering it. Art, however, was outside the realm of what I was trained to do, but I really felt compelled to take a risk and push forward with the experiment. In 1964, Koestler wrote *The Act of Creation*, which discussed how creative ideas happen when different intellectual disciplines collide. This theory resonated with me, giving me the courage as "a science guy" to engage with artistic practices.

Ken Van Rees, Mark making on watercolor paper with burnt trees in northern Saskatchewan, August 2017.

Image reprinted with courtesy of the artist.

From that experience I started taking painting lessons at Emma Lake, eventually joining a prominent Saskatchewan group of plein air painters called the Men Who Paint. We continue to exhibit regularly together and paint landscapes around the country—including remote locations such as Ivvavik National Park. My research program in the jack pine forests of northern Saskatchewan has also lead me to its own artistic endeavors. My research plots were burnt in a wildfire in 2010 and when I went to examine my melted equipment after the fire, I noticed that the burnt trees were creating interesting charcoal markings on my clothes. Seven years later, I'm still experimenting at this site and other recent forest fire locations with paper and canvas—exploring how these burnt forests can "draw" nonobjective artworks with their black carbon if I provide them with a canvas.

In September of 2004, I offered the Soils and Boreal Landscape course for the first time with twelve students at Emma Lake Kenderdine Campus (ELKC) in northern Saskatchewan. This facility started as an art camp in 1935 and has a rich art history with artists such as Barnett Newman, Kenneth Nolan, and Anthony Caro leading workshops there. ELKC was also instrumental as a place that ignited and fueled other innovative ideas, including many of the interactions I had with other scientists, artists, and staff. ELKC has been a "magical place" (as one artist described it to me) for learning, and it is where I took my first painting class that started me on my present trajectory as scientist-painter. For the past 12 years I have been incorporating artistic methods into my undergraduate soil science course, encouraging students to experiment with various techniques from drawing with charcoal to painting with acrylic paints.

A number of years ago I began to explore how I could engage my own disciplinary expertise of soil science more directly with the visual art world. I had read an article in the *Canadian Geographic* about an artist in Ontario, Symeon van Donkelaar, who used local materials—soil, rocks, and plants—to make pigments for his iconography paintings. Here was an opportunity to explore the use of the very material of my discipline to make art. Up until that point we had only used acrylics, oil pastels, charcoal, and other conventional painting and drawing media in my undergraduate field courses. So I sought out Symeon on a trip to Ontario in 2010 to discuss the idea of coming to Saskatchewan to help share his knowledge and skills of making pigments from natural materials for a graduate course …

Finding Local Colors: A Voyage from Art to Science

The earth, and its minerals and soils, first entered my (Symeon van Donkelaar) art through the practice of pilgrimage and the world of spirits. One spring evening while reading an account of the life of the Voyagers, those French Canadians who transported furs in canoes across Upper Canada in the eighteenth century, I happened upon a reference to the Porte de l'Enfer (Hell's Gate). This was a cave along the Mattawa River where these men believed that a monstrous spirit lived, which deeply frightened them. But, it was also a rare place where a sacred red ocher pigment had been collected by the Algonquin First Nations.

As I thought about this account, I found both the idea of a frightening sacredness and holy earth intriguing. But, the idea of a place having its own, unique earth pigment was practical enough for me to act upon. Without waiting for a more opportune time, my family and I set off the following week in search of this color.

What I knew as we headed out was:

- The cave is only accessible by the Mattawa River.
- Getting there would entail a couple-hour canoe trip in order to reach it.

What I didn't know as we headed out was:

- The river was swollen and raging from the spring thaw.
- We'd be paddling upstream the whole way.
- The mosquitoes at that time of year are insane.
- And, the mouth of the cave is halfway up a sheer cliff face.

Despite the raging rivers and blanket of mosquitoes typical in central Ontario every spring, we did manage to reach the cave. Red flowed out of its mouth, like a gaping wound in the rock, the color of a very special place. Not having any background with minerals or soils, I realized I didn't really know what I was looking for. But, I collected rocks at the cave's mouth and when I got back to my studio they produced a complex reddish-brown pigment. It wasn't bright, but it did shimmer when I painted with it.

This first experience with what I would later call, *Local Colours*, raised a new question for me: How many places are there in Canada that could offer earth pigments? I began to explore the village where my family and I live, Conestogo, and discovered it had actually had a pigment mill back in the 1890s, which produced paint that was shipped across the British Commonwealth. From an artist's perspective, I was beginning to suspect that every place might have its own local colors. When I was named the Artist in Residence (AiR) for the City of Cambridge, Ontario, the following year, I got my chance to test this idea.

My work as AiR centered on the 100-mile ART project, in which I proposed to only use the local earth from around 100 miles of Cambridge City Hall to make a traditional orthodox icon. I knew from the outset that the only possible way of doing this would be to collaborate with the community, and so I began talking with historians, farmers, and even scientists. In doing so I met the most amazing and generous people, who loved where they lived and were excited to share it with me. Reiner and Maggie, a couple of seasoned rock-hounders, began tagging along on my pilgrimages, sharing with me their eyes and experience. A few local mines opened their gates to us, and we got to explore from deep in the earth. The earth sciences department at the University of Waterloo even supplied a little Mastodon tusk from the last ice age in Ontario, which created the first ivory black from the region in perhaps 10,000 years. In the end, I ended up with a bright collection of pigments, made from this region's dirt, rocks, plants, and animals. On the opening night of the exhibition for this project I gave back to the community its place as art—and, I had created my first local color palette!

The adventure of foraging for local colors that defined the 100-mile ART project deepend into a pre-scientific, alchemistic exploration of the processes during my Masters of Fine Art (MFA) studies at the Transart Institute in Berlin, Germany. My concept of local color was stretched and deepened during this time in the conceptual art world, and the artistic result was the *Atlas of Canada's Local Colour*, an ongoing publication that takes the rocks, soils, plants, and even some bones, from different communities across the provinces and territories of Canada and documents their potential as pigment. Using processes such as crushing, grinding, washing, and firing (which do not have the aim of purifying but rather fulfilling each pigment), the resulting color changes during these processes is recorded onto the pages of the Atlas, with the resulting spidergram maps revealing the local colors of a community.

Subsequent to successfully defining my thesis on the concept of local color, I began experimenting with a third strand of how we can know the land and its soils—through culture and stories. These visual artworks are written with the local colors of the place and use the soil as pigment to reveal its story. It's in these works that I feel connected with a deeper way of knowing the soil through its heritage. In using the soil to embody stories of culture, history, identity, and spirit, people experience the revelation of knowing the land in a new way. My belief is that it's this embodiment that reveals the nature of soil in a way that has the potential to invoke a love for the land in people that can positively change their interactions with the earth during this time of environmental crisis.

My ongoing work has enjoyed some broader public attention in the media. Articles have appeared in a few educational journals and art blogs, but has also caught the attention of scientists—appearing in *YES! Magazine* and on the Canadian Broadcasting Corporation's *Science News*. One of my highlights was when it appeared in the publication, *Canadian Geographic*. It was after reading that particular article that Ken van Rees, a professor of the Department of Soil Science at the University of Saskatchewan, wrote me to ask if we could meet the next time he was in Ontario. This meeting with Ken began a conversation about how we could offer the experience of knowing soil from a scientific and artistic perspective and led to us collaborating the following year in a graduate-level course for both MFA and soil science students …

Next pages:
Symeon Donkelaar, *The Local Colours of Conestoga, Ontario*, local soil in egg tempera on paper, 76 cm × 112 cm, 2014.

Image reprinted with courtesy of the artist.

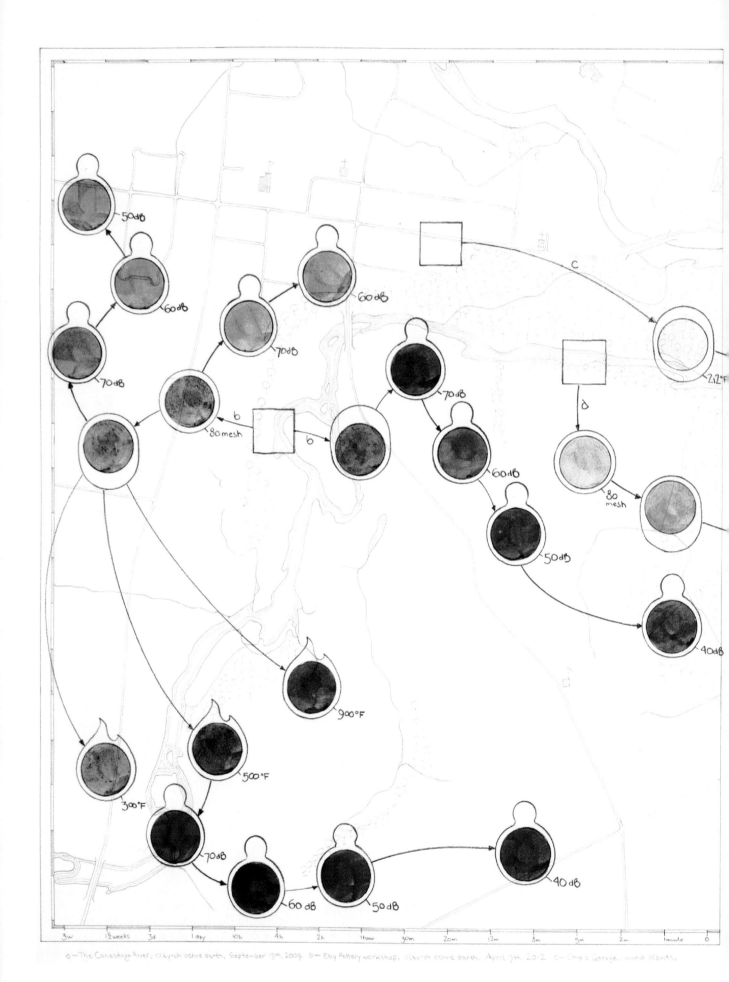

a — The Conestoga River, clayish ochre earth, September 19th, 2009. b — Eby Pottery workshop, clayish ochre earth, April 9th, 2012. c — Ship's garage, wood plants,

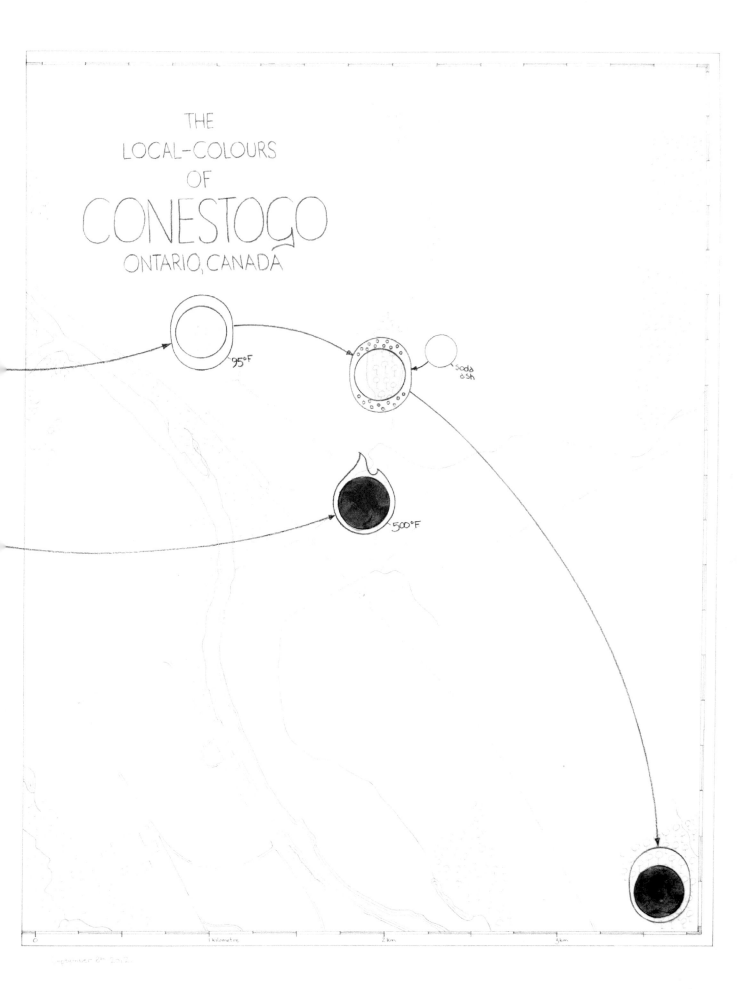

THE
LOCAL-COLOURS
OF
CONESTOGO
ONTARIO, CANADA

95°F

Soda
Ash

500°F

(Left) A thread of ochre along the bank of the Conestoga River. (Right) Symeon Donkelaar, *St. Peter of Conestogo, Ontario*, local soil in egg tempera on paper, 51 cm×183 cm, 2017.

Image reprinted with courtesy of the artist.

Art and Science at Kenderdine

In 2012 we offered our first pigment course at the Emma Lake Kenderdine Campus (ELKC) north of Saskatoon on the southern fringe of the Boreal forest. Allyson Glenn, a painter from the Art and Art History Department at the University of Saskatoon, and her art students joined us for the course. We wanted to go one step further in terms of who we taught. What would happen if you put together soil science graduate students with Master of Fine Art students in the same course? Would they interact? Would they learn from each other? How different would their perspectives be when interpreting landscapes with the pigments they made?

By the first session sitting around the fire, it didn't take long for students to let their guard down and begin sharing stories of how they ended up in their respective art or science disciplines. The week was filled with exciting discoveries of how to make pigments with colorful soils and minerals, and even a few animal bones (the burning of which produces a fine black), which helped us understand the role that fire, water, and grinding can have on color. We worked hard to create opportunities for the students to

Symeon Donkelaar, *Adam Naming the Animals*, resulting work of the 100 Mile ART Project using local color in iconography painting. Cambridge Centre for the Arts, Cambridge, Ontario, 2008.

Image reprinted with courtesy of the artist.

experiment, and were both surprised by some of the outcomes. Many of the perspectives shared by soil science students made for strong art and, at least in one case, pigment work we were privately convinced would fail produced an amazing color.

We also invited a soil chemist one morning to talk about the chemistry of color and what specifically happens when you apply heat to drive out the water in a yellow ocher pigment to turn it a deep red. In the afternoons we traveled to various boreal ecosystems so that the students could use their pigments to interpret their surroundings. By the end of the week there was a real sense of community among the whole group. We were united by hours of creating and experimenting with pigments and painting the mosquito-infested landscapes.

In times past, from the renaissance to the Bauhaus, students were trained in both the sciences and arts, but this trend is not as evident at many colleges and universities today. At a time when there's so much drive for encouraging creativity in education, we think our experience at Emma Lake demonstrates how rich the outcome can be when the disciplinary boarders of art and science are taken out of the equation. We also believe that the Emma Lake Kenderdine Campus was instrumental in helping to remove these barriers between arts and science because of the long history of cultivating a place where scientists, visual artists, musicians, and writers could come together in one place. Sadly, it was closed in 2012 because of budgetary constraints by the University of Saskatchewan and we hope that one day it may reopen to continue this cultural heritage of providing a space for the mixing of the disciplines and resulting creativity. For us, both as scientist and artist, our experience is that the crossover between disciplines continues to be a deep well of inspiration.

Emma Lake Kenderdine Campus.

Image reprinted with courtesy of the artist.

Soil Connoisseurship

Linda Weintraub

Linda Weintraub is an artist, curator, educator, museum director, homesteader, and author of several popular books about contemporary art including the first college eco art textbook titled *To Life! Eco Art in Pursuit of a Sustainable Planet* (UC Press). Others include *In the Making: Creative Options for Contemporary Art* (DAP and Thames & Hudson), and the forthcoming *What's Next? Eco Materialism and Contemporary Art* (Intellect Books). In this chapter, Weintraub proposes three strategies that answer the following question: How can the cumbersome entity known as contemporary civilization reverse its aversion to soil? First, the essay identifies a few of the astonishing qualities of soil that cancel soil's association with the mundane. Second, it recommends direct, multisensory interactions with soil as a means to cultivate soil delight. Third, it proposes that it is by cultivating soil connoisseurship that such revered standards of beauty as proportion and harmony can be applied to the crumbly brown substance that ensures fertility. Discerning such soil excellence is the job of connoisseurs.

Title image: Cultivating Compost.

Photo: Linda Weintraub, 2018.

Each day, vast swaths of soils are crushed beneath multiton vehicles and sealed beneath multistory skyscrapers. Soils that once were protected by meadows and renewed by trees are now exposed to winds and washed away by rain. Nasty concoctions of sewage, pesticides, herbicides, industrial effluents, medical waste, spilled oil, and chemical substances seep into its crevices. A common perception of soil is that it is so abundant it can be squandered without consequence. This perception is false. While fertile soils are becoming increasingly rare, this shrinking resource has *never* been abundant on our planet:

> 3/4 of Earth is water and 1/4 of Earth is land.
> 1/2 of land is available for human use.
> 3/4 of land available for human use is too rocky, too wet or too dry, or too hot or too cold for food production.
> Only 1/32 of the earth's surface is available for food production.[1]

Fewer than half the people alive today have access to functional interactions with soil. This is because 54% of the world's population lives in urban areas, a proportion that is expected to increase to 66% by 2050.[2] For these populations, soil makes feeble appearances in indoor planters and outdoor plazas. Otherwise it is obliterated by buildings, roads, sidewalks, and parking areas. The cardinal substance of planet Earth is thereby sealed from sight, beyond reach, inaccessible to physical interactions. Asphalt and concrete obstruct the "hospitality" that soil offers by cushioning our falls, sequestering our toxins, growing our food, and mediating between minerals and organic matter to secure our sustenance and well-being.[3] Urban populations sacrifice physical interactions with the infinitely varied qualities of soil that not only defines our location on Earth; it defines our position in the cosmos.

How can a culture of neglectful soil desecrators be converted into care-giving stewards? This volume lays the foundation for answering this crucial question. Its contributors offer resourceful and imaginative responses to the fundamental functions of soil, its properties, and its evolution, as well as its existential concerns and documented threats. They examine how soil factors into the creative practices of contemporary artists and scientists, and they explore the role of soil resources in determining social and environmental sovereignty. The collection of interviews and essays in this volume exceeds simply conveying scientific information and artistic interpretations of soil. It aspires to elevate soil's status on the hierarchy of social values, bestowing the esteem that is required to ensure its protection. While no "operating manual for restructuring social priorities" currently exists, one strategy for rescuing the planet's soils is suggested by the notion of connoisseurship, recognizing that we humans lavish care upon things we perceive as beautiful, pleasurable, rare, or valuable.

The word *connoisseur* is commonly defined as "to be acquainted with." This definition does not specify the character of such "acquaintance." It is both broad and intense. Connoisseurship is a skill that is cultivated through an extensive process of study and direct experience. This skill is practiced by evaluating and rendering critical judgments about a given cultural product. Thus, individuals are connoisseurs if they are fortified with the knowledge gained through experience and exposure that enables them to pass expert judgment regarding quality.

While art connoisseurs pinpoint the masterpiece qualities of a work of art, wine connoisseurs can detect the region where grapes were grown from a sip and a whiff of wine, and cigar connoisseurs not only appreciate the nuances of cigar taste and texture, they recognize the boxes, bands, and rings that encase them. Connoisseurs continually refine their perceptions, starting with generalities, then dividing these generalities into historic periods and geographical locations, then discerning the multiple styles within these periods and locations, then recognizing individual manifestations of these styles, and ultimately appreciating the nuanced differences within these individualized expressions. Regarding soil, connoisseurs aspire to discern between thirty-two reference soil groups (RSGs) and hundreds of subclasses.[4] They can differentiate soil profiles (from deserts, tundra, swamps, coasts, basins, temperate forests, and jungles); detect the subtle gradations of a soil's texture (sand, silt, loam, clay); color (tones of red, orange, gray, white, green, yellow, black); and aroma (sweet, pungent, acrid, metallic, musty, yeasty). Soil connoisseurs can reconstruct a soil's history by detecting signs of weathered mineral deposits and decomposed organic materials in a given space and time. Besides visual and somatosensory clues, soil connoisseurs discern the place of a soil's origin from the flavor of the crops it produces. In addition, their attentiveness translates into respect for the diversity of life-forms that make their homes in soil.

In all these ways, connoisseurs savor the finest soils, bestowing upon them the respect and dignity of masterworks. This book supports their efforts. Its editors and authors share the conviction that in an era beset with ecosystem failures, caring about soil is not merely an individual pursuit. Soil conservation and sustainable land-use depends upon establishing a culture of soil connoisseurship. This would entail convincing the public to care about soil as much as they currently care about electronic upgrades, sports playoffs, and computer games. By making connoisseurship a mass movement, fertile soil would become a source of regional pride and identity.

Soil tournaments are already being conducted by the National Collegiate Soils Contest and during the World Congress of Soil Science. These prestigious organizations sponsor soil judging competitions in which teams of students, trained by coaches, participate in competitions that test the contestants' abilities to describe soil profiles, practice soil taxonomy, and propose land use based upon soil and site characteristics.

Fluency

Connoisseurs of soil honor the essential role of "earth" (soil) to support life on "Earth" (planet). This prodigious transformation of inert substances into living matter occurs within the narrow zone where the bottom layer of sky and the top layer of our planet intersect. Connoisseurs focus on this territory because it offers the precise quantity and proportion of four essential ingredients that support life: water, air, organic matter, and mineral nutrients. This elemental assemblage accounts for every living entity that has ever existed on our special planet.

Connoisseurs also recognize that as much as soil provides a habitat for elemental forms of life, it also serves as the sepulchre for life's partner, death. Loved ones around the world are laid to rest within this dark, moist zone. In actuality, the deceased's tranquility is short-lived. Burials stimulate the dismantling actions by microorganisms and scavengers. Their feeding frenzy persists until the last organic molecule has been liberated to contribute to the production of fertile soil. This process is essential for the perpetuation of life. Unlike the aboveground food web that derives its energy from the sun, the energy that propels the underground food web emits from decaying organic matter called detritus.[5] When these energy sources are synchronized, the bacteria and fungi underground conduct the heavy work of nourishing aboveground populations. Thus, soil connoisseurs sharpen their perceptions and refine their assessments to monitor peat, humus, and clay as well as the myriad microscopic life-forms that occupy the soil.

Disciplined analysis enables connoisseurs to transcend generalities and appreciate the significance of each soil's unique components and interactions, locating combinations of attributes precisely within the range of known possibilities. At the same time, detecting subtleties and complexities enables connoisseurs to discern the gradations of poor, good, and excellent. Each interaction with soil, therefore, is undertaken in anticipation of discovery and excellence.

How, then, can connoisseurship be learned? Gaining fluency with soil, as with language, involves acquiring a "vocabulary" to describe the minerals, organic matter, gases, liquids, and countless organisms of the soil, and "syntax" of mineral and organic formations. Fluency involves distinguishing the soil norms from the anomalies and recognizing the conditions that produce these variations. The following strategies are designed to cultivate soil connoisseurship. Readers are encouraged to conduct this self-directed training and to hone their skills in sensing soil excellence so that soil is treated to the privilege, care, and attention enjoyed by other forms of connoisseurship.

Observation

Let us commence connoisseurship training with the simple act of observing, which is the skill set upon which art connoisseurship is founded. Within the European art tradition, the self-conscious practice of connoisseurship was not codified until the seventeenth and eighteenth centuries when merchants acquired the means to purchase fine art, but they lacked the knowledge to discern its qualities. In response to the merchants' desires to develop their own critical faculties, connoisseurship acquired a name, a definition, and a skill set.

Perceptivity, the keenness of observation and insight, is as fundamental for developing soil connoisseurship as it is for art. In both contexts, connoisseurship depends upon refining perceptions to discern technique (how something was created), composition (what it consists of), and form (how it is structured). When these attributes are compiled, connoisseurs are able to assign authorship (who/what created it) and appraise quality (in terms of aesthetics and fertility). In all these ways, connoisseurship, engages the intrinsic qualities of the object itself, in addition to published information about the object. Its conclusions, therefore, are a matter of judgment in combination with fact. As such, connoisseurship is invested with intuition and opinion, appreciation and disappointment, emotion and desire.

Since many qualities of soil are not observable, let us explore the simple act of touch as a means to develop a deep and abiding rapport with soil. The hand is exquisitely tuned to detect the innumerable textures that soils provide: sandy, gravelly, granular, gritty, slimy, smooth, and uncountable combinations of these qualities. Even simple tools like shovels, rakes, and spades interfere with the detailed knowledge that connoisseurs seek through bodily explorations. While some agronomists utilize such sophisticated soil testing tools as sonic sieve shakers that ensure speed and accuracy of their assessments, others engage direct physical interactions. For example, a manual published by the Food and Agriculture

Organization of the United Nations refers to the "feel method" of soil analysis.[6] It describes moistening a small sample of soil taken from just below the surface, and molding it to determine its structure. This technique confirms the ability of the human body to assess soil structure, a capacity that is often overshadowed by standardized particle analysis conducted in the lab. Connoisseurs are likely to appreciate visceral connections.

In addition to discerning the textural attributes, touching soil is reported to induce infusions of health and happiness. Credit for this advantage is assigned to a strain of bacteria called mycobacterium vaccae that inhabit soil. These microscopic organisms trigger the release of serotonin, which elevates mood, decreases anxiety, improves cognitive functioning, treats arthritis, and possibly inhibits cancer.[7] These pleasurable chemical suffusions associated with gardening are also available to connoisseurs of soil. Meanwhile, those who are reluctant to get their hands dirty, or who avoid contact with "unhygienic" substances, deny themselves this free, health-inducing opportunity.

Developing a connoisseur's rapport with dirt can expand from hands to feet, another physiologically complex appendage that offers abundant sensory inputs. Walking barefoot on unpaved surfaces activates neglected assemblies of muscles, tendons, and ligaments, by connecting elaborate networks of nerves that signal the brain about the texture, topography, compactness, moisture, and temperature of the soils they are contacting. Thus, going barefoot allows the foot to function in the manner evolution prescribed. Its proficient data-collecting capacities are blocked by shoes, particularly those that have insulating rubber or plastic soles.

In addition, shoes block the body from receiving the upward flow of Earth's energies. This is not some flimsy mystical concept. Soils are structurally magnetic. It is through the medium of soil that our bodies come in contact with the Earth's electrons, enabling us to resonate with the same electrical patterns as the earth. The *Journal of Environmental and Public Health*[8] reports that grounding our bodies in this manner induces physiological and electrophysiological changes that promote optimum health. This beneficial energy supply is invisible, free, reliable, and renewable. Shoes obstruct these generous offerings. Soil connoisseurship optimizes them.

Abandoning bedrooms offers an even more compelling way to access the mobile electrons within the global electron circuit. Although "sleeping under the stars" offers a romantic allure that may be missing from "sleeping on the ground," the practice of lying directly on the ground has recently earned its own name; it is called "earthing." Throughout history, humans have slept on the ground or on skins. Through this direct contact, the ground's abundant free electrons were able to enter the body, which is electrically conductive. Proponents assert that the Earth's subtle electrical fields are essential for proper functioning of the human immune systems, circulation, synchronization of biorhythms and other physiological processes, and may actually be the most effective, essential, least expensive, and easiest way to attain antioxidants.[9] While skepticism is likely to greet such extravagant claims,[10] the prestigious *Journal of Environmental and Public Health* reports, "Omnipresent throughout the environment is a surprisingly beneficial, yet overlooked global resource for health maintenance, disease prevention, and clinical therapy: the surface of the Earth itself. … Mounting evidence suggests that the Earth's negative (electron) potential can create a stable internal bioelectrical environment for the normal functioning of all body systems."[11]

The progression of physiological opportunities to connect with the soil culminates with the mouth. "Eat dirt" is not simply urban slang for accepting blame. It describes the practice of "geophagy," the ingestion of soil by humans and other animals. Dr. Sera Young, a nutritional scientist at Cornell University, reveals that medicinal dirt eating probably started long before there were humans, since parrots, baboons, and other animals are known to eat dirt too. Young explains that this powerful yearning to ingest soil, particularly among pregnant women, is widespread, archaic, and global. It is both medicinal and religious.[12]

Connoisseurs can include soil tasting within their zones of sensory experience. Bill Wolf, who has earned credentials as a soil connoisseur by developing organic foods, fertilizers, and pest controls, explains, "A very acid soil would crackle like those sour candies that kids eat, and it had the sharp taste of a citrus drink. A neutral soil didn't fizz and it had the odor and flavor of the soil's humus, caused by little creatures called "actinomycetes." An alkaline soil tasted chalky and coated the tongue."[13]

From Communication to Action

When soil connoisseurs communicate the discoveries they make through direct, multisensory contacts and discriminating discernments, they celebrate the very values that are typically derided as repulsive or dangerous. A widespread cultural perception of dirt locates it down low where excrement, bad behavior, malice, insults, and contamination are located. It exists alongside depression, despair, gloom, and pessimism. The opposite direction is summoned when people "lift" their spirits by "rising" up with "high" hopes and "lofty" ambitions. Dirt has been collecting derogatory connotations since the fifteenth century when the word *dirty*, meaning "smutty, morally unclean," first appeared. The phrase "dirty words" was first recorded in 1599; "dirty trick" in the 1670s; "dirty work" in 1764; "dirty linen" in the 1860s; "dirty look" in 1928; "dirty old man" in 1932.[14] These phrases link the Earth's principal life-sustaining substance to squalor instead of fertility. Likewise, the word *soil* can be traced to the Latin and Middle English, Old French, and Old German words for "pigsty": *solum, soile, souil, souiller, sol*.

The ramifications of the symbolic discrepancy between high and low may determine the prospects for the future of soil on Earth. As much as we humans pamper things that are beautiful, we tend to abuse what we dislike, avoid what we mistrust, and destroy what we fear. Uplifting terminologies and metaphors are needed to associate soil with its life-giving functions, and thereby invest it with protocols of honor and dignity. These efforts are well represented by the artists and scientists included in this book.

Soil connoisseurship involves more than simply acquiring knowledge and cultivating sensitivity. It is also a matter of heeding an impending environmental crisis. As this book reveals, securing soils for future generations is a joint effort of many disciplines. Within their diverse creative practices, soil is preserved, conserved, beautified, politicized, and revered. Some contributors are forging the ecological consciousness that features dirt's role at the frontier of cultural change. Others are in the trenches remediating and restoring soils. Those who address civilization's role in determining soil's fate explore interactions that are harmful as well as those that are beneficial. In all, connoisseurship is not an end; it is a means for preserving and creating fertile soil. It is hoped that readers will join the contributors to this volume by envisioning perfect soil and striving for improvement. This book is less an offering than an appeal to readers to minister, instruct, mediate, legislate, lobby, and write on behalf of soil security.

Endnotes

1. "How Much is Dirt Worth? Hitting Pay Dirt." Agriculture in the Classroom. Utah State University Cooperative Extension, page 2. Pdf.

2. "World's Population Increasingly Urban with More Than Half Living in Urban Areas." United Nations, July 10, 2014, http://www.un.org/en/development/desa/news/population/world-urbanization-prospects-2014.html.

3. William Bryant Logan, *Dirt, The Ecstatic Skin of the Earth* (W.W. Norton & Company; Reprint edition, January 17, 2007), 19.

4. See the Food and Agriculture Organization's World Reference Base for Soil Resources for more on international soil classification: http://www.fao.org/3/a-i3794e.pdf.

5. See Georg Dietzler and Ellen Kandeler, "A Public Chemistry of the Detritusphere," Chapter 34, this volume.

6. J.A. Adepetu, H. Habhan, A. Osinubi, eds. *Simple Soil, Water and Plant Testing Techniques for Soil Resource Management.* Food and Agriculture Organization of the United Nations. AGL/MSIC/28/2000, http://atl.org.mx/files/DirectricesAgricultores/8.pdf. Accessed June 2016.

7. D.M. Matthews and S.M. Jenks. "Ingestion of Mycobacterium Vaccae Decreases Anxiety-Related Behavior and Improves Learning in Mice." US National Library of Medicine, National Institute of Health. Electronic publication February 27, 2013 https://www.ncbi.nlm.nih.gov/pubmed/23454729. Accessed October 2016.

8. Gaetan Chevalier, Stephen T. Sinatra, James L. Oschman, Karol, and Pawl Sokal. "Earthing: Health Implications of Reconnecting the Human Body to the Earth's Surface Electrons." *Journal of Environmental and Public Health* vol. 2012, Article IID 291541. 2012. http://www.hindawi.com/journals/jeph/2012/291541/. Accessed August 2017.

9. Luke Sumpter, "Earthing: Studies Show Walking Barefoot Might Be An Essential Element of Optimal Health," May 11, 2015. http://reset.me/story/earthing-studies-show-walking-barefoot-might-be-an-essential-element-of-optimal-health/. Accessed April 26, 2018.

10. Ibid.

11. Gaetan Chevalier, Stephen T. Sinatra, James L. Oschman, Karol Sokal, and Pawel Sokal. "Earthing: Health Implications of Reconnecting the Human Body to the Earth's Surface Electrons." *Journal of Environmental and Public Health*, PMC3265077. January 12, 2012. https://www.ncbi.nlm.nih.gov/pmc/articles/PMC3265077/. Accessed August 2017.

12. Sera L. Young. *Craving Earth: Understanding Pica--the Urge to Eat Clay, Starch, Ice, and Chalk* (Columbia University Press, August 14, 2012).

13. Nicola. "Sweet and Sour Soils." *Edible Geography: Thinking Through Food.* December 9, 2009. http://www.ediblegeography.com/sweet-and-sour-soils/. Accessed October 2016.

14. The American Heritage Idioms Dictionary (Houghton Mifflin Company, 2002). http://dictionary.reference.com/browse/dirty. Accessed November 2016.

Underground Roots

Akilah Martin in conversation with Nance Klehm

Akilah Martin earned her doctorate from Purdue University investigating environmental quality as it relates to urban environmental modeling, social identity, and environmental science education. She is an environmental scientist rooted in the practice of enhancing soil, water, and overall environmental quality that will lead to a more sustainable future. Many of her research projects involve collaboration between community partners and students in a continued effort to bridge the gap between university research and community involvement. She considers herself to be a civically engaged scholar in that she crafts impactful learning experiences that in turn offer opportunities to create a more just and equitable society. She has envisioned and implemented a classroom that is not only in a building but outside in the

Title image: Nance Klehm, impressions from *The Ground Rules*, Chicago 2012–present. Social Ecologies' active process for this project includes the identification and negotiation of organic wastes generated by local businesses, the dedicated performance of collecting these wastes, the establishment of multiples community Soil Centers, the dialogue generated by town hall meetings and trainings, and the publication *The Ground Rules*. By trialing these community-based, soil-making research models in different cities, Social Ecologies hopes to reinvigorate the dialogue of the urbanscape as habitat, healthy soil as an element of healthy infrastructure, and ourselves as contributors and instigators toward more inspired action.

world seeing and actively solving problems through strategic thinking. Through this lens, she facilitates learning from a paradigm of empowerment, helping learners to understand the world from significant scholarly voices as well as from their own perspectives. In collaboration with different communities of learners, she has enlivened real-world applications in the classroom space to illustrate that you do not have to be a scientist to solve real-world scientific problems. She strives to shift learners from being passive to active learners by challenging them with authentic problems to solve.

Nance Klehm is a process-oriented ecologist, landscape designer, and artist/activist whose work has been reported on in national and international media outlets including radio, print, and many books. She is coauthor of *The Ground Rules: A Manual to Reconnect Soil and Soul* and is currently working on a second publication called *Dirt Work: Recreating Coherence in Urban Soils* as well as *The Soil Keepers* a book of interviews with people who work with soil. She was named a 2012 Utne Visionary and has been a two-time finalist for the Curry-Stone Prize, where she now is part of their Social Design Circle. Klehm is the director of Social Ecologies, which creates durational projects that aim to build healthy habitat and interaction through direct engagement of place with those who dwell there. Social Ecologies seeks to encourage holistic, systematic thinking through varying levels and degrees of project participation.

In a series of conversations and emails, Nance Klehm and Akilah Martin discussed their unconventional methods of reawakening citizens and students to reconnect and positively engage with the living soil heritage beneath their feet. Coming from two very different backgrounds and disciplines, they agree that soil is the fabric of our heritage. Soil heritage threads together a community's past, present, and future. What is evident in both women's work is that cultivation of the land and the people in an urban center requires unimaginable creativity and strategizing. Connecting people to the Earth and disconnecting them from past forces that kept them away, can lead to an ethos that provides opportunities for growth and leadership of our Earth's resources.

Nance Klehm: How do you find your learning audience outside the classroom?

Akilah Martin: I do a lot of volunteer work around the city and I've been in connection with other people who've put me in connection with others. I've done things with Citizens for Science, Project Exploration, IP Boeing Scholars. I'm really excited about working with the youth. Whenever the opportunity is presented to me, I'm ready to go and talk to them, to do any type of experiment with them, whether it's outside or in the lab, whether it's community gardeners who want to know about soil or neighbors because they have a backyard garden. I am on the board of directors at the Forest Preserve. I got on that because of a student in my class. It's connections like that that I live for, that help me get the word out. My main goal is to spread soil awareness—what it does, what it is, how important it is for our lives. I started this course for that specific reason, and so it's taken off lately.

Nance Klehm: You're talking about your "Urban Dirt" course offered at the School for New Learning, DePaul University for Adult Learners, which I had the opportunity to visit!

Akilah Martin: Exactly! If I just called it "Intro to Soil Science," nobody would register for that. Some people are still insistent on calling it dirt because they're not ready to accept that soil is actually a living, breathing system.

Nance Klehm: Well, they seemed super engaged and they are gonna graduate from "Urban Dirt" to your next class "Urban Soil"!

Akilah Martin: Most of my students haven't had biology or chemistry or any type of math or courses that would be prerequisites for a science course. And so instead I start with "What is soil?" to what are the physical properties of soil.

"Urban Dirt" course offered at the School for New Learning, DePaul University.

Photo credit: Nance Klehm. Image courtesy of the author.

We talked about soil microbiology today. We'll talk about soil chemistry. They'll have a chance to actually test their soil fertility with kits. We're going to look at how different textures are impacted by erosion from water. We're also going to build a levy and a water filter. They have to create a medium that is akin to natural soil.

We experiment in a pseudo lab just about every week, just to give them a hands-on, experiential learning type of situation. We're actually visiting Jackson Park next month to look at how the Army Corps of Engineers and other entities are working to build habitats (i.e., water, soil, vegetation) for native wildlife species.

I want to give them different ways of thinking about their environment and for them to ask themselves, "What is my impact on soil? What is the impact of soil on me? What do I need to

Dr. Akilah Martin School in Nairobi, Kenya (2011).

Note: This school was started by one of Dr. Akilah's student in Kenya.

Image courtesy of Akilah Martin.

do to be a more beneficial component of that system?"

Nance Klehm: I've seen that you've also done some work in the Niger delta and then I've also seen some images of what you've done in Kenya.

Akilah Martin: Yeah, the Kenya work is not necessarily directly related to soil. It was basically a slum where I was working. And so, one of the things that they've focused on for the past eight years, has been the issue of accessible fresh water and waste disposal. There's just waste everywhere, so we are working to educate people about waste and its impact, and figure out entrepreneurial businesses that can emerge from that.

Nance Klehm: So, what made you fall in love with soil?

Akilah Martin: I believe, when I was an undergraduate at Alabama A&M and I randomly chose soil science as a major, because I …

Nance Klehm: Randomly???!

Akilah Martin: Yes! It was completely random. I said, "I'm doing this!" In retrospect I had attended an ag high school where we learned everything but soil. I had always worked on science projects related to plants and soil fertility, without really knowing about soils. I chose Alabama A&M because of my mentors during a summer internship. They had attended AAMU and said I'd have a great learning experience there. I applied and received a full scholarship, so it was on from there. The significance of this choice did not become apparent until I began attending national conferences related to soil and saw that there weren't many people who looked like me. It wasn't a good feeling nor did I feel welcomed. I pressed on however. I always like a challenge and I knew there was nowhere to go but up with this soil venture. Some of the professors at AAMU took me under their wing and provided rich research experiences that helped set the stage for my future. I got the chance to travel to Germany to present research on soil remote sensing and also had an opportunity to work with NASA. I have to say, it was a mind-boggling experience to go from an Historically Black College or University to a Predominantly White Institute in the field of soil. I sometimes cried on my way back to grad school (I'd visit home on the weekends) because of the feeling of not belonging and not being treated well. I was determined to show that the study of soil is not just for white males in khakis! It has also been a struggle here in Chicago to connect learners to soil. But it is slowly becoming more visible in different communities. I know that when learners walk away from my course, they are armed with knowledge that they never expected to gain.

I have experienced a disconnection with people that comes from the ideals and notions about farming and slavery. The older generations still have gardens but many in the younger generations are far removed, not just from soil but from the environment in general. The total lack of respect boggles my mind. We have to find ways to connect the youth with the Earth. Disclaimer: I shouldn't state that *all* youth are disconnected, because there are younger generations who have a vested interest in their food production.

Beyond teaching, my research has been based in low-impact development, environmental health, and civic engagement … but I'm from Chicago where the main issues are soil availability and sustainability, and water policy. You can't ever forget water. One of the major functions of the soil is to filter water. And for Chicago, soil contamination is a real problem. People thinking about soil not as dirt and not just something that's just there, but something that's actually supporting *you and all of us.*

I just did a project on low-impact development integrating water and soil policy. These things should be at the top of the list, from elementary school classrooms to city hall. Let's talk about the environment, so that people start thinking, and start strategizing things that not only won't harm the environment, but will also be beneficial to the environment, and bring back ecosystems, and stabilize them.

Nance Klehm: Right on!

Akilah Martin: So, Nance, how did you get started on this "soil path"?

Nance Klehm: Well, I grew up on a 500-acre farm in rural Illinois, and my dad's idea of "church" was to walk around on our land for three or four hours and talk about the birds and the animals and the plants and the water and all the interrelationships that were happening. We had a large kitchen garden and orchard trees, goats and chickens, horses, and so many dogs and cats. We spent a lot of time outside working as well as playing. We identified with the land.

Strangely enough, I ended up studying in Washington, DC, got a degree in archaeology and then worked in Peru before I returned to Chicago for a job at The Field Museum of Natural History. When I moved to Chicago, I started growing food and I immediately noticed that city soils everywhere looked really bad to me, compared to the soils I grew up with. Even though I was growing food, I was much more interested in the soil that I was growing in than the produce, herbs, or ornamental gardens I was using in my landscape designs because soil is so subtle and complex and it supports all this growth.

I ran an ecological landscaping business for fifteen years. So, I have a very strong practical knowledge of plants and increasing soil health. Developing my business, Social Ecologies, of which The Ground Rules, the composting and remediation business is a part, was a way of designing regenerative systems and putting some pieces together for me.

I want to be helpful in translating my passion, in translating some basic scientific knowledge to people, whether they're literate or not, whether they're native English speakers or not. I really think these things are understandable and I want to help develop that curiosity and knowledge and pass on that love that I have.

The interest seems to be there now. Before, people thought I was just cuckoo for caring about something so silent as soil and plants. "You're so dirty all the time. What's up with that? I've never seen anyone with dirtier feet in the city," they'd tell me in my yoga class.

Akilah Martin: So, how do you get connected to the different places you've lived and worked in?

Nance Klehm: As a teacher, I've been invited to lecture or train others in community bioremediation, composting technologies, landscape interpretation, bioremediation, and soil biology. As an artist, I exhibit my work, which usually takes the form of an installation, performance, or public intervention … As an activist, I am often asked to develop ways of working with waste streams by reimagining their value through a social dynamic or to entirely design water and waste systems on a decentralized neighborhood or community scale … I've done projects in Rio, Doha, L.A., Portland, Warsaw, Aarhus, Port Au Prince.

Connecting to these places in the work that I do requires that I listen and observe well so I am able to enter the ethos and flow of the groups I work with, which have been so incredibly diverse in their resources, their beliefs, and their approaches to the same issues we all have—wanting and needing clean water and healthy soil and somehow both inheriting and actively creating the opposite. It is so connective and also so humbling.

I try to spend a little extra time volunteering on land-based projects as I travel. Some of those projects have been really inspiring, and some of those have been really horrifying. And I think it all gets me thinking or it gets me writing or doing different kinds of research or experimenting or exhibiting.

Akilah Martin: What is your impact on the different projects you're working on?

Nance Klehm: I have had extended engagement with projects locally in Chicago and a few in California. These are the projects that thrive—they have been absorbed and tweaked by the groups of people where they are. These projects have lives of their own and are mostly out of my hands. This is by my design. This is how I know they are working, when I am not needed anymore. But the other ones, where I pop in for a week or a month and do an intensive training and then pop out, those are a little harder to know what my impact is. I try to give it everything I got when I'm there. Sometimes I hear much later what my impact is and when I do, I am gratified. I want so deeply for people to feel better in their skins and in their hearts. I want this world to be a healthier place.

Akilah Martin: What is the most gratifying part of your composting and bioremediation business The Ground Rules?

Nance Klehm: I actually think it's my alone time outdoors! Where I get to sit and quietly observe or go for a really quiet walk outside and notice things. That's probably some of the deepest work I do. It's vulnerable and reflective and it feels really connective to me.

Nance Klehm, impressions from *The Ground Rules*, Chicago 2012–present. Social Ecologies' active process for this project includes the identification and negotiation of organic wastes generated by local businesses, the dedicated performance of collecting these wastes, the establishment of multiples community Soil Centers, the dialogue generated by town hall meetings and trainings, and the publication *The Ground Rules*. By trialing these community-based, soil-making research models in different cities, Social Ecologies hopes to reinvigorate the dialogue of the urbanscape as habitat, healthy soil as an element of healthy infrastructure, and ourselves as contributors and instigators toward more inspired action.

Photo credit: Nance Klehm.

And then secondarily when I work with others, I enjoy translating my appreciation and love of land, the metabolic, the biological, when I work with others through both practical and creative ways. The Ground Rules is a good vehicle to create a deeper awareness and increase peoples' perception and care for where they live.

What does soil heritage mean to you?

Akilah Martin: The many communities in the city of Chicago provide a compelling soil story. Soil is endlessly engaging and provides a home and comfort like none other. From the art, the gardens, the buildings, and the vegetation, to the culture of the diverse communities in Chicago. Currently, I am collaborating on a project related to soils and health in two diverse communities. One community you can barely see soil as most of it is covered by concrete, asphalt, and the other has beautiful park and green space. The patterns are similar in relation to their life expectancy and their environments rival one another in their propensity to provide unhealthy aspects. This project intends to provide a thread to the present soil and how it can be a powerhouse of health and evolution. These undertakings will create a heritage and legacy for community engagement with resources that sustain life—*soil*!

Nance Klehm: To me, soil heritage means the choices we make and actions we take, form the soils we live with and will become the soils we pass on.

The Ground Rules uses a process we have developed called "deep mapping." It is a process inspired by William Least-Heat Moon, an indigenous writer (William Least Heat-Moon, *PrairyErth: A Deep Map*, Houghton Mifflin, 1991.). Deep Mapping is a lengthy process of on-site ecological and biological study paired with anthropological and sociohistorical research to tell the environmental history of a site whether seen, cared for, and valued as such or one invisible, small or large, neglected or abused. Once we rebuild that deep map, we understand this site as a true place worthy of its inherent and emerging qualities, uses and relationships, and offer this forward.

The process of deep mapping rewrites a story of land and helps shift perceptions about where life starts and ends, about the boundaries and borders of our responsibilities and re-enlivens our connection to soil, land, the biotic community including ourselves. Our stories are embedded in the soil.

Soil Lovers Unite for a Down and Dirty Q&A

Beth Stephens and Annie Sprinkle in conversation
with Frederick L. Kirschenmann

Beth Stephens and **Annie Sprinkle** have been partners and collaborators for seventeen fertile years. Their Ecosex Manifesto launched a movement and officially added the E to GLBTQI. Their award-winning documentary film about mountain top removal coal mining in Stephens's home state West Virginia, *Goodbye Gauley Mountain: An Ecosexual Love Story*, is available on iTunes and from the distributor Kino Lorber (kinolorber.com). Stephens and Sprinkle just finished a new documentary, *Water Makes Us Wet: An Ecosexual Adventure*, an homage to water. Their new book, *Explorer's Guide to Planet Orgasm*, is inspired by ecosexual principles. Stephens is a professor and chair of the Art Department at University of California Santa Cruz. Sprinkle is a former sex worker, an artist, and college lecturer. In 2017, their work was exhibited in the prestigious international art exhibition, *documenta 14*. Together they run a center at UCSC, the E.A.R.T.H. Lab (Environmental Art, Research,

Title image: Annie Sprinkle and Beth Stephens, *Earthy*, Photo by Errick Petersen at Vortex Theater in Austin, courtesy of the artists.

Theory and Happenings; earthlab.ucsc.edu) and they welcome collaboration. For more information about their work, visit www.sexecology.org.

Frederick L. Kirschenmann, a longtime leader in national and international sustainable agriculture, is Distinguished Fellow for the Leopold Center for Sustainable Agriculture at Iowa State University and president of Stone Barns Center for Food and Agriculture in Pocantico Hills, New York. Kirschenmann oversees management of his family's certified organic farm in south central North Dakota and also has an appointment in the ISU Department of Religion and Philosophy.

In lieu of an abstract…

Q: What happens when two ecosexual artists and a farmer philosopher get together, via e-mail, and create a list of seventeen questions about what they think and feel about soil, then they each answer the questions individually?

A: They end up with a poetic, humorous, dirty, homage and experimental love poem to soil, of course.

Let's introduce ourselves by our past connections with soil in under fifty-nine words.

Beth Stephens: My grandparents were farmers, and as a kid, I worked in their garden, in the orchard, and with the farm animals. One of my favorite memories of my grandfather was that he made an earthworm farm underneath his beautiful grape arbor. He used an old refrigerator turned on its back and buried up to its door, to keep the soil cool for the worms. I loved digging in this fridge for fishing worms.

Fred L. Kirschenmann: After experiencing the dust bowl in the 1930s my father did not *ever* want *that* to happen to his farm again and so became a radical advocate for "taking care of the land" and by the time I was four years old he began instilling that value into me, so I grew up with a compelling value in my genes!

Annie Sprinkle: I had no interest in soil whatsoever until Beth and I did our *Dirty Ecosexual Wedding to Soil* in a performance festival in Krems, Austria, in 2014. We did a full-on wedding, where two hundred and fifty people came and could all marry the soil with us. So, when I learned more about soil and married it, I fell in love. I'm still creepy crawler phobic but am working on my bug and worm issues.

What are our current relationships to and with soil?

Beth: The *Wedding to Soil* vows ended with the line, "We vow to love, honor and cherish you soil until death brings us closer together forever." I look forward to becoming soil myself one day.

Fred: To me restoring the health of our soil is now *the* most important task before us—it is essential for sequestering carbon, reducing our water and energy consumption, and designing a resilient food system.

Annie: As I'm married to soil, I'm a very dirty woman.

Three things we love about soil are …

Annie: Soil is teeming with magical life giving properties, it holds me up, and its scent is divine.

Beth: It's complicated, dark, and mysterious.

Fred: It's an incredible community of living organisms, its self-renewing capacity, and the way it supports all life on earth.

How can we love soil better and more deeply?

Fred: We have to develop intimate relationships with it—stop thinking of it as "just dirt."

Annie: More gratitude! When I learned that one little handful of soil has more living creatures in it than there are humans on Earth, that was a mind-fuck that touched me deeply. To think that we walk on the backs of gazillions of creatures every day and we don't appreciate them is tragic.

Also by eroticizing soil, we could attract more interest and care.

Beth: We can compost, prevent erosion, and educate others about the importance of soil.

What is our favorite soil related word?

Beth: Earthy!

Fred: Life.

Annie: Fecund!

Are we lovers of soil, or are we lovers with soil?

Fred: Both. As Aldo Leopold reminded us, we are simply "plain members and citizens" of the land, so we are "with" soil, but we also have incredible responsibilities to care for the soil so we also need to be lovers "of" the soil.

Annie: Both. I'm definitely a lover with and to the soil, through my desire and my senses; through smelling, touching, tasting, listening to, and kissing soil. I love you Soil! Thank you for everything! You're a wonderful lover!

Beth: Both. I love it in a separate-from-myself way as well as consider myself part soil and therefore entangled with it.

Does soil care about us?

Annie: I sincerely hope so. But I wouldn't blame it if it hates us. Maybe it's half and half.

Beth: No. I don't think soil has the kind of consciousness that "cares about" people in the way that a person probably thinks that "caring about" would be. People and soil have a very symbiotic relationship, which most people have forgotten. So, people aren't caring for soil as they should but the soil never stops doing what it does, or nurturing as it can. It simply does so without thinking. Perhaps that is "caring for" or "taking care of" but different than "caring about." It is more simply "being soil."

Fred: Probably not in the way that conscientious organisms "care for" other organisms, but soil does care "for" us in that it is the foundation for all of life.

Does soil heal us? How?

Beth: Mud baths are healing. Some people eat dirt, notably pregnant women, for its healing nutritional properties. For me gardening, and especially putting my hands in soil is healing. It is the perfect antidote to typing endlessly on a computer.

Annie: Making cool mud pies as a child was deeply satisfying! Many important medicines are derived from and with soil—including many antibiotics.

Fred: Sir Albert Howard reminded us that healthy soil, healthy plants, healthy animals, and healthy humans were all part of "one topic." I grew up spending a lot of time playing in soil, and I cannot "prove" that that contributed to my relatively healthy life, but there is increasing evidence that it may have. It may be that in our modern culture we are just "too clean."

How do humans disrespect the soil?

Annie: By ignoring it. By poisoning it. I'm guessing that it probably doesn't particularly like being stepped on either.

Beth: They scrape it, push it, pave it, they just don't value it as being the life-giving substance it is. If they did, something like mountain top removal coal mining could never exist, nor would the mass paving of this planet.

Fred: Because we (as Leopold reminded us) treat it as a "commodity belonging to us" rather than "a community to which we belong."

How can people cultivate what Fred calls "ecological conscience"?

Beth: An ecological consciousness is one in which everything is connected to and dependent on everything else. It is systems thinking and being. One can cultivate it by observing nature and learning how the world works ecologically and then trying to be part of the ecological system while trying to do as little harm as possible and sustain the web of life.

Fred: As Leopold ultimately reminded us, this will have to be "a product of social evolution" so we still have to cultivate it!

Annie: Education is key. Learn/teach. I have so much more to learn! I'm also an "ecosex educator." After four decades of doing sex research I'm now exactly to where I need to be: bridging sexuality and environmentalism. Thankfully I have Beth as my muse and collaborator and we are on the unpaved path together. Ecosex has been a way for me to open up into "ecological conscience." It's working for many other people too. I also have to say that my experiences ingesting psychotropic plants helped me experience how everything is connected and one ecosystem!

Many people take pleasure from soil. How?

Beth: I love to compost. Composting is so *hot*! I also love to squish mud between my toes, and digging in soil always gives me pleasure.

Annie: I've noticed that lot of artists are working with soil these days and really digging it.

Fred: As Deborah Koons Garcia points out so beautifully in her documentary *Symphony of Soil*, we "can" relate to soil in a loving and caring way. And when we do, especially beginning as children, we take great pleasure from it.

Does soil feel our love? Can we give pleasure to soil? Does it matter?

Fred: We will likely never know whether that incredibly diverse community of life in soil can take pleasure from our loving care, but we can certainly relate to that community of life as if it could.

Annie: I do believe that soil can feel our love. But I am not sure. I hope someone out there has studied this question.

Beth: I love to get buried in soil during our performance art pieces such as our *Dirt Bed* series.

I think of soil as an extension of ourselves and so when physically interacting with soil, I imagine that the soil gets some kind of sensation out of the movements we make. I also think that organic ways of enriching soil such as adding in compost might pleasure the soil.

How can we be better lovers to and with soil?

Annie: Sniff it, lick it, mud wrestle in it! Do clay masks. Talk dirty to soil. Be gentle and sensual with soil. Give soil a massage with your feet. Lay naked on top of soil, or let it get on top of you. Sing to soil. Protect it from further abuse.

Beth: By empathizing with it. If we change the "i" to a "u" then soil becomes soul.

Fred: By relating to it as a "community to which we belong."

Dirt has a bad reputation. Does it need cleaning up?

Fred: Yes. In our culture dirt, like the microbes that live in it, is considered to be "bad" and must be removed from us. If we relate to it properly, dirt, like microbes, can help to build our natural immunities and make us healthier.

Beth: No, it needs to become dirtier. Society needs to realize the value of dirt instead of always trying to clean it up. The saying "cleanliness is next to Godliness" is one of the worst, most controlling sayings ever. It is used to judge people who are deemed closer to soil than others and therefore dirtier than those who have the income, wealth, and privilege to stay clean. I say dirty or soiled is where it's at!

Annie: Soil's bad, and its dirty reputation is part of its charm. What's wrong with dirty? I'm dirty and proud!

Can soil speak to us? If so, how, and what does it say?

Fred: When I take up hands full of soil in the spring before planting my crops, it definitely speaks to me—it lets me know when it is ready for planting, if it is sufficiently "healthy" to sustain the plants and/or animals I want it to nourish.

Annie: As an ecosexual, I use fantasy, my imagination to communicate with soil. If I really listen I will get a message, loud and clear. Maybe the soil is actually talking to me, or maybe it is my imagination. But who is to say? Soil says different things at different times depending on what soil I'm with, where, what time of day, and how I feel at the moment, or the phase of the moon. Soil often says things that mirror how I feel in the moment, but it is also full of surprises. I would love to develop better communication with soil.

Beth: I imagine the soil saying, "Life can be rich and full of surprises."

If ecosexual means imagining the Earth as our lover, are we ecosexuals?

Annie: I'm definitely an ecosexual. I didn't feel as connected to the Earth as a mother. I did feel gratitude. But when I imagined Earth as lover that gave me the paradigm shift I needed to connect more deeply with the nonhuman worlds. Finally! What a relief. I wanted so badly to connect. I love you Soil! Thank you for everything! You're a wonderful lover!

Beth: I've always been an ecosexual and in love with the Earth, even as a child.

Fred: I have been in love with soil as far back as I can remember—it was a relationship my father instilled in me.

What fantasy soil project would we love to do if we had a few million dollars to spend?

Fred: Work with artists to develop imaginative ways to engage the general public in a loving relationship with living soil.

Annie: I'd create a fantasmagorical soil amusement park with all kinds of soil experiences: soil tasting, soil sniffing, soil viewing, soils from around the world, soil under microscopes, soil rides, dirt beds to lay in … It would be outdoors and totally free! Everyone who went would fall in love with soil. I'd also like to finish the ecosexual documentary film about soil which Beth and I are working on.

Beth: I would resoil every mountain top removal coal mining site that I could to their original richly soiled condition so that the forests and plants could grow there again. Our last documentary was about the devastating effects of mountain top removal coal mining, including the removal of all the topsoil of over five hundred mountains. It breaks my heart. I'd love to be able to regenerate those mountains.

Do we have any last words, favorite sayings, or poems for soil?

Beth: My grandmother used to say, "Your ears are so dirty you could grow corn in them." I savored the knowledge that I was a garden and that I could grow things and thought it would be very exciting to grow ears of corn in my own human ears.

Fred: "I have seen so many delicate shapes, forms, and colors in soil profiles that, to me, soils are beautiful." Hans Jenny.

Annie: I love the poem that rock star artist Peaches wrote and performed for our *Dirty Ecosexual Wedding to Soil*:

Dirt

dirt dirt dirty dirt dirt
Stay away from the dirt
You'll get dirty
It's a mess
yes! yes!

We wanna get down and dirty
hit the dirt
dig in the dirt

dirt is a wonder
dirt is real
dirt is precious
dirt gives us breath
dirt will sustain us
dirt makes life

dirt is life
we need to be dirty
we need dirt

break it down earthworm
break it down
break it down

fungi love me
humble humus
all hail bacteria
the criteria

release the nutrients
release us
fertilize us
treat us like dirt
give us dirt

We wanna be soiled
Richly soiled
In sand, silt and clay
we will lay, we will lay

let's get dirt on our hands
cover us in dirt
make us dirty
dirt is the shit
dirt is the shit

Peaches

Documentation of Beth Stephens and Annie Sprinkle's *Dirt Bed* performance at Grace Exhibition Space in Brooklyn, New York, 2012.

Digital photo by Leslie Barany. Copyright Stephens & Sprinkle. Image courtesy of the artists.

Documentation of *Dirty Wedding to the Soil*, Krems, Austria, 2014.

Photo by Iris Ranziner. Image courtesy of the artists.

Documentation of *Dirty Wedding to the Soil*, Krems, Austria, 2014.

Photo by Iris Ranziner. Image courtesy of the artists.

Documentation of Peaches performing her poem *Dirt*, in collaboration with dancer, Keith Hennessy, at *Dirty Wedding to the Soil*, Krems, Austria, 2014.

Photo by Iris Ranziner. Image courtesy of the artists.

Documentation of *Dirty Wedding to the Soil*, Krems, Austria, 2014.

Photo by Iris Ranziner. Image courtesy of the artists.

Function 6

STABILIZER: Soil as platform for structures, infrastructures, and socioeconomic systems

Stabilizer

The sixth and final section of the book explores the soil as social stabilizer, literally the rock on which all human civilization stands. This function of soil conceptually overlaps with the cultural heritage function of the soil described earlier (are not cathedrals, pyramids and hydropower dams also a form of cultural heritage?). Of all the functions of the soil, the function as platform and stabilizer for human structures is perhaps the most riddled in soil conservation contexts. Unlike the cultural heritage functions of the soil, which are often intangible, this function appears at first as utterly counterintuitive to the very goals of soil stewardship. Disciplines such as soil mechanics and civil engineering calculate soil properties to plan a human universe of concrete and steel upon the porous architecture of the ground below. Buildings, bridges, highways, plazas, retaining walls, dams, dumps, subways, sewers, and power and communication lines could not exist if the soil failed to provide space and stability for their construction. But these structures are usually erected at the detriment of habitat, filtering, and food production functions of the soil. An interminable game of material displacement, sealing, and mitigation ensues. Partly inspired by ongoing debates within the international SUITMA ([Soils of Urban, Industrial, Traffic, and Mining Areas] is a working group of the International Union of Soil Sciences that meets regularly and works on an interdisciplinary basis to confront problems of the Anthropocene as they affect soils.) working group, the following contributions illustrate how visual artists, architects, landscape architects, and performance artists question, critique, and envision hybrid alternatives of this conflicted but necessary soil function.

Amy Green sets the stage by asking what the arts can offer to an already onerous discussion on the soil as a platform for human development. For a second time in the book, nuclear disposal on the surface of the Earth is discussed as an irreversible mark on the pedosphere. Can the properties of the soil actually bear this responsibility? The Center for Land Use Interpretation documents its response in a visual essay on the "Perpetual Architecture" of the American Southwest. Meanwhile, Lara Almarcegui talks rubble with Gerd Wessolek—building debris comes to symbolize the constant forces of urban change as sandy fill becomes a placeholder for soil in the ceaseless cycles of construction and demolition in the city, the focus of Almarcegui's work. In another interview by Gerd Wessolek, Betty Beier discusses her square-meter earth prints. These are hyperrealistic demonstration models of preconstruction spaces that serve as forensic evidence of soil lost to urban and industrial development. In chapters by Debra Solomon and by Ellie Irons and Jean Louis Morel, the biological potentials and sociopolitical scenarios of such urban spaces are explored. In chapters by Seth Denizen and Qiu Rongliang and interviews with Paolo Tavares, Antonio Guerra, and Sean Connelly by Alexandra R. Toland, mapping, speculative taxonomies, and indigenous land-use practices are discussed as alternative approaches to designing a more sustainable soil platform.

Seeing the Soil Platform

Amy Green

The arts, science, and education are the three main interests that have manifested in the weave of **Amy Green**'s life path. As an urban ecologist and environmental scientist, she has studied on two continents and researched the soil–water interface in cities with respect to emerging contaminants. As a Berlin-based singer, actor, and artist, she has worked professionally for over twenty years in diverse contexts such as in the theater installations of Freidrich Lichtenstein, medieval and new vocal music with various international ensembles, and with country western bands. Working as part of the Global Soil Forum, the soil and land team at the Institute of Advanced Sustainability Studies (IASS) in Potsdam, Germany, she was responsible for the conception and realization of the art and participatory elements of the *ONE HECTARE* exhibition in 2015.

Title image: The *ONE HECTARE* installation, Gleisdreieck Park, Berlin, 2015. In a 20-minute performance, one hectare in the public park is symbolically sealed as a visual protest to unabated land take in Germany.

Photo credit: IASS Potsdam.

Introduction

Humans relate naturally to land and soil as the space upon which to live their lives and build their homes. Most contemporary societies do not have effective mechanisms in place to regulate this activity such as to prevent the deterioration of soil resources upon which they depend. Despite ample evidence of the negative consequences of soil sealing and land conversion, land take rates remain high. In Europe, urbanization is the major cause of desertification and soil depletion in semiarid places like Spain and its Mediterranean neighbors.[1] Sealed or compacted soils can no longer absorb water and buffer their environments from flooding. Fertile soils once converted into urban space can no longer be used for the production of food. In our increasingly affluent world, in which the demand for land-intensive meat and energy production is on the rise, this loss becomes a global issue. The Global North imports large proportions of their products from developing countries inhabited by small-scale farmers who struggle to produce food for their families on degraded land.

The use of the soil platform is thus at the very heart of sustainability discourse. What does it mean to live a good life? What does sustainable resource use in the case of the platform function mean? What responsibility do we have to other living species on the Earth and to each other? Is less actually more? Will change come from individuals in terms of lifestyle choices, or governments making effective regulation? What is the role of innovation? The regulation of the platform function of the soil is complex in its social, economic, and ecological aspects, especially under the new demands for adaptation driven by climate change.

The creation of regulatory instruments to combat soil sealing and the spread of urbanization faces complex opposition. It is not possible for decision makers in a democracy to legislate the norms and aspirations of home building and happiness. The dream of a family home with a garden where the children can play and grow is a choice people should be able to make in free societies. But as societies, especially in the Global North, become increasingly individually minded, the value of soil resources is perceived through the lens of the neoliberal economic system. As our standard of living increases, the percentage of income spent on food decreases, and with it the value of agricultural land in relation to commercial development.

The arts can play a role in the search for solutions by providing a different way of seeing the soil platform. Art does not necessarily solve problems but can make us aware of the existence and extent of problems. It opens our eyes to see and our heart to feel and imagine. Art can communicate to the nonconceptual mind, protest the status quo, or provide a vision of the way forward. A narrative of sustainable land use will resonate if it is centered on a connection to quality of life. Car-free, walkable cities with affordable public transport, for instance, could mean more freedom—from congestion, air pollution, an so forth. Artistic approaches with a specific focus on the issues of soil as platform are discussed in the following in the context of a public exhibition realized on the occasion of the 2015 International Year of Soils.

ONE HECTARE Art and Participatory Exhibition

The *ONE HECTARE* art and participatory exhibition,[2] part of a broader effort to raise awareness and to foster a fruitful dialogue on the local and worldwide significance of soil, was organized by the Institute for Advanced Sustainability Studies (IASS). It took place from April 22–25, 2015, in the well-frequented public park, Park am Gleisdreieck near Potsdamer Platz in Berlin, and from October 15–27, 2015, at the Dresden Altmarkt, one of the oldest public squares in the city center. The exhibition included several new site-specific art works, performances, sculpture, sound installations, videos, and science communication materials in the form of a large installation, information boards, and animations. Parallel to the exhibition, there was series of free public workshops and events. These communication channels were designed to display the scientific and artistic input on equal footing.

Visual Protest: Have a Look from Above

One hectare of land (10,000 m²) was marked in white sports demarcation chalk on the park lawn. While the hectare is the international standard unit for measuring larger land areas, only specialized professionals like farmers, spatial planners, and real estate developers have a real sense of how much space a hectare really occupies. From the viewing platform, visitors could experience the size of the hectare, which was further divided into proportional use of land worldwide. Debunking the myth of a planet of limitless land resources, the installation drew attention to the scarcity of the resource soil, especially in regard to fertile soil available for the production of food, energy, and raw materials. The amount of land available worldwide for cultivation makes up a mere 10% of all land.[3] Fertile land is even more scarce, as 38% of all agricultural land worldwide is already degraded.[4] And while urban areas, including large and small cities, roads, and other infrastructure cover 3% of global land area, the current rapid expansion is largely at the expense of fertile cropland. Forest and grasslands converted to croplands come at a considerable ecological and climatic cost.

To demonstrate the rate of land take in Germany in real time, a symbolic paving of the hectare took place on April 22, 2015. The action was part

Hola Móra VI, Egill Saebjörnson. *ONE HECTARE* Exhibition, Berlin, Germany. 2015.

Photo credit: Shahram Entekhabi.

of a program of the third Global Soil Week, an international event and knowledge platform for high-level decision makers, social and natural scientists, civil society representatives and artists from a wide range of professional and cultural backgrounds.

In Germany alone, one hectare of land is built over every 20 minutes, translating to 77 hectares of land partially sealed beneath asphalt and concrete on average each day of the year. A large digital clock presided over the act of covering the park space with black material. The park itself was borne out of citizen activism to create public open space in the midst of the postunification land rush in Berlin. The aerial image is a visual protest of toothless regulatory instruments in Germany, which have set a goal to reduce land take to a national goal of 30 hectares per day by 2030. This visual protest was intended to be a media spectacle to spark public discussion of the issues surrounding soil conservation in the media.

Soil Platform As Home: A Look from Below

One solution to soil sealing would be to integrate our residential infrastructures in the earth, as cultures all over the world have done for thousands of years. Building directly into the ground better allows soils to carry out ecosystem services—to store water and carbon, and provide habitat to more than only human dwellers. The work of Islandic artist Egill Saebjörnson delves into this underground world.

A large mound of fresh dark brown soil, which at first glance appears to be an oversized gopher mound, is the temporary home of Móri, a globetrotting free spirit. A high-pitched voice can be heard, in the form of melodic fragments and contented babbling in the earthy Nordic Islandic, with snippets of German. The sound emerges from the entrance to Móri's hole as if the viewer has happened upon the inhabitant in the midst of a private conversation he is having with himself. Móri lives in the soil, just as many animals do, but through his vocal antics we easily relate to him as a being, without ever seeing him. It is also the whimsical atmosphere created through the sound, which also makes Móri's home seem so cozy and inviting to the viewer. The viewer remembers the innate connection to "nature" and to soil. *Hola Móra* reminds us that we have a very deep connection to the earth in and on which we build our homes.

Hola Móra is part of a series Saebjörnson has been erecting in different locations around the world. He describes the work as a collaboration with an inner voice, or a ghost. Saebjörnson writes, "We should ask Móri what

Matsogo, Lerato Shadi (Berlin/South Africa), 2013, video.

Photo credit: Erik Dettweiler.

his art means. He is not educated in art, and is shy. Please try anyway. Have a conversation with him from inside your head. He might answer. Maybe he can tell you a lot about soil, and about the significance it has for the Earth. And, perhaps in a conversation with him you can get ideas about how to organize our lives in cooperation with the Earth."

Soil and Speculation

A triangular slice of moist dark chocolate cake rests on a newspaper page printed with stock market information. A hand picks up a piece of cake and begins to knead it. We hear a voice singing. Her song weaves African Setswana folk song motifs into a dialogue among multiple characters on the subjects of belief, trust, and betrayal. She kneads the cake over and over again before reforming the mass into its original shape. Although she has neither added nor taken anything away from the cake, the cake undergoes a transformation from a mouthwatering delicacy into an unappetizing mass.

The film speaks to the connections among soil, food security, and speculation. Fertile humus is evoked through the dark color and moist texture of the cake. Fertile soil, needed to produce commodities such as grain, feedstock for the meat industry, biomass as fuel, and sugarcane needed for worldwide food industry, has thus become an attractive investment and object of speculation, and is increasingly scarce in countries with fast growing populations. The values steering the free market increasingly collide with the global goals of sustainable development such as in the case of food sovereignty. The instrument used by the elite class to make profits out of selling commodities, the stock market is not interested in sustaining soil fertility, but in extracting short-term profit. Consumers in rich countries such as Germany effectively import fertile land when they buy products grown in other countries. The European Union, the continent most dependent on land outside its borders, imports each year products that have been produced on 640 million hectares, or 1.5 times the size of its 28 member countries.[5] About 25% of global cropland area now produces commodities that are exported to land-poor but cash-rich countries.[6] The land resources used to produce export commodities have been transformed through the globalized economic system in such a way that sustenance can no longer be provided to local communities of small-scale farmers, similar to the transformation of the cake in the video from delicious to unappetizing and no longer edible.

Soil degradation due to speculation is an issue within the European Union as well. Because urbanized land for commercial and residential use can often generate large profits for both the speculator as well as the municipality, agricultural land that does not fall under conservation status is difficult to protect. Fertile soil is of higher value when it is used as a platform, paved, built on, and transformed like in Shadi's video.

Bird Cage, Shahram Entekhabi and Amy Green, 2015, site-specific installation of bird calls, field recordings, and caution tape. *ONE HECTARE* Exhibition, Berlin, Germany.

Photo credit: Shahram Entekhabi.

Platform As Cage: A Reflection on Urban Biodiversity and Succession

A structure of red-and-white caution cape is erected within a 200 m^2 area of a miniature forested nature conservation area in the center of Berlin. The caution tape functions as a reminder of the critical decline of biodiversity due to urbanization. The mini-woodland is merely a remnant of a once sprawling successional urban forest that developed on the abandoned site of a former train yard. The train yard turned public park at Berlin's Gleisdreieck was part of a green infrastructure plan designed as compensation for the reconstruction of Potsdamer Platz. The structure of caution tape is temporary and unfinished, akin to the transience of processes such as migration and ecological succession against the backdrop of a seemingly static urban landscape.

Inside this structure, a composition of bird calls and urban field recordings can be heard from ten speakers that are hidden in birdhouses mounted on trees. The composition deals with a soundscape of ecological succession of bird species in relation to the process of urbanization and is informed by urban ecological data, more specifically a study of the urban–rural gradient of bird species in Berlin[7] as well as historical surveys of bird populations in Berlin. What begins as a complex polyphony of endemic bird song is infiltrated and replaced with a sonically more homogeneous blend of mechanical and technical sounds associated with humans, as well as urban adapter species.

The process of urban development of land has pushed many of our historical avian companions and locally indigenous species out of the cities. Urbanites may recognize the few species that we encounter on a daily basis, such as the pigeon or the sparrow. However, the call of the grey partridge (*Perdix perdix* or *Rebhuhn* in German), a once common companion endemic to Berlin, featuring prominently in folk literature such as the tales of the brothers Grimm, is now completely unknown to urban dwellers today. The gray partridge, extinct in the city-state of Berlin, is found on the Red List in Germany as an endangered species as it can neither find suitable habitat in our cities, nor in a countryside where intensification of agriculture has led to the simplification of the landscape.

Habitat loss due to processes or urbanization and the intensification of agriculture have caused alarming trends in bird populations worldwide. In 2015, one in eight species of birds worldwide are at risk of extinction.[8] While in highly urbanized Germany around half of all birds are at risk of extinction.[9]

While the sounds of the city are now predominately caused by the people living in it, the biotic soundscape is homogenized. Caution tape also evokes in this installation the immediate association of a dangerous or no-go situation. Walls of caution tape contain a public woodland left to follow natural succession. Despite the high ecological value of urban

Tamara Rettenmund, *one-hectare/one day*, long durational performance, *ONE HECATRE* Exhibition, Berlin, Germany.

Photo credit: Shahram Entekhabi.

land left to natural succession, the seeming chaos of this natural process is viewed by many urban people as unkempt and even dangerous. As McKinney, an ecologist studying biotic homogenation notes, "Urbanites of all income levels become increasingly disconnected from local indigenous species and their natural ecosystems. Urban conservation should therefore focus on promoting preservation and restoration of local indigenous species."[10] The process of urbanization including soil sealing has a great impact on biodiversity above and below ground. Whether we will be able to reverse this trend, significant to the larger collapse of biodiversity known as the sixth extinction, has much to do with the management of the soil platform function.

Platform and Shelter

To explore the idea of soil as a platform providing the very basic need of shelter from the elements, the artist Tamara Rettenmund uses the medium of film and long durational performance. In the film *one hectare/one night*, Rettenmund spends a night sleeping in the Gleisdreieck Park and is seen wandering in stop motion in a green tunic, like a clay figurine onto the park space, the grassy park surface providing an uneasy shelter from the cold and the darkness of night. The viewer witnesses the series of sleeping positions from above, as one leads to the next. At sunrise, she rises and leaves the park. In the long durational performance, *one hectare—one day*, she spends several hours rolling from one sleeping position to the next. A team of assistants documents the process by tracing the forms of the various sleeping positions on the grass with chalk paint. Over time, the hectare is transformed into a mosaic of rounded body shapes. Normally the white outline of a body painted on the ground evokes the immediate association of a crime scene. In the context of the performance, the body shapes take on a playful quality with which the park visitors can interact. The soil platform appears soft and comforting inside these cloudlike shapes.

Soil itself cannot directly provide humans with shelter from the elements or the comforts of home. In fact we often do not even think of soil at all, when we consider our home, rather the structure and the possessions within. Our homes are not merely meant to provide shelter during sleep, but are kingdoms unto themselves. As we continually amass more things and larger objects, so grows our perception of the adequate size of our living spaces. Modern living rooms in 2017 must be as spacious as our sofa landscapes and our widescreen entertainment systems.

The platform function of the soil is such a conflicted issue in places like Germany due to what Günther Bachman, general secretary of the German

Council for Sustainability calls a "societal obsession with housing." Here, the amount of living space per capita has more than doubled from 20 m² in 1960 to over 40 m² in 2014.[11] Urbanized space, as well as the size of newly constructed buildings, continues to expand. Rettenmund reminds us with her performance that the essence of the platform function of soil, providing shelter and a home, is an inalienable human right. This being the case, the value of soil and land that provide for all of humanity must be viewed also in terms of the common good, and not only as something that can only be governed through private ownership.

The No-Waste picnic, cooked on site from foodstuff overflow, celebrates soil as community. *ONE HECTARE* exhibition. Berlin, Germany. 2015.

Photo credit: Shahram Entekhabi.

Conclusion

In developed countries such as Germany, the driver of land take is not due to population growth but economic growth. The German population has remained relatively stable since 1980 with roughly 80 million residents,[12] yet the economy has continuously grown up to rates of 5%, with only short phases of depression following financial crashes.[13] Land take continues unabated, while the federal government seems unable to take action to slow down the development of land. In 2011, it set targets to limit land take to 30 hectares per day in 2020. Obviously it is not going so well, considering current land take rates of 73 hectares per day.[14]

Despite all of this expansion, housing has become increasingly scare for the low and fixed income population, especially in cities. Gentrification due to real-estate speculation has caused the socioeconomic structure of entire city districts to change. Even in Germany, where the social safety nets provided by the government are relatively intact, the population of the homeless and impoverished seniors is increasing.

Urban land take is a global threat to sustainability. Societies need to protect their resource base of soil for climate mitigation and adaption, to provide habitats for the other species of the earth, as well as to produce enough food for their populations. In Europe and North America, soil sealing is a main cause of the deterioration of the soil/land resource base, yet the bitter reality of food insecurity does not affect its inhabitants. The myth of an unlimited land resource base must be debunked. In our present economic systems, we experience that everything is available for those who can pay. We do not worry about our supermarkets running out of food. The food that we eat is produced on land across the globe. Through the products we buy to support our lifestyles, we virtually import large areas of land in other countries.

Trends such as worldwide increase in meat consumption, land degradation, and climate change, put pressure on land resources. Fertile soil has become increasingly scarce. Cities consume agricultural land, and forests and wetlands are converted into agricultural land. There is a need for a societal discourse on the value of soil as a common resource of sustainable development. Action, both political and private, is necessary to change these trajectories.

Endnotes

1. Barbero-Sierra, C., Marques, M.J., and Ruiz-Perez, M. The case of urban sprawl in Spain as an active and irreversible driving force for desertification. *Journal of Arid Environments* 2013, 90:95–102. https://doi.org/10.1016/j.jaridenv.2012.10.014.

2. For more information about the *ONE HECTARE* exhibition, please see www.ein-hektar.de.

3. The State of the World's Land and Water Resources for Food and Agriculture. Managing Systems at Risk, 2011. The United Nations Food and Agriculture Organisation. http://www.fao.org/docrep/017/i1688e/i1688e.pdf. Accessed December 3, 2017.

4. Bai, Z.G., Dent, D.L., Olsson, L., and Schaepman, M.E. Proxy global assessment of land degradation, 2008.

5. Fader, M., Gerten, D., Krause, M., Lucht, W., and Cramer, W. Spatial decoupling of agricultural production and consumption. *Environmental Research Letters* 2013, 8(1).

6. Global Land Outlook, first edition. 2017. United Nations Convention to Combat Desertification. https://static1.squarespace.com/static/5694c48bd82d5e9597570999/t/59e9f992a9db090e9f51bdaa/1508506042149/GLO_Full_Report_low_res_English.pdf. Accessed on December 1, 2017.

7. Simon, U., Kübler, S., and Böhner. Analysis of breeding bird communities along an urban-rural gradient in Berlin, Germany, by Hasse Diagram Technique. *Journal of Urban Ecosystems* 2007, 10:17–28. https://doi.org/10.1007/s11252-006-0004-5.

8. The IUCN Red List of Endangered Species. 2015. https://cmsdocs.s3.amazonaws.com/keydocuments/IUCN_Red_List_Brochure_2015_LOW.pdf accessed on December 1, 2017.

9. Rote Liste der Brutvögel. Fünfte gesamtdeutsche Fassung, veröffentlicht im August 2016. https://www.nabu.de/tiere-und-pflanzen/voegel/artenschutz/rote-listen/10221.html#3. Accessed on December 1, 2017.

10. McKinney, M.L. Urbanization as a major cause of biotic homogenization. 2006, 127(3):247–260. https://doi.org/10.1016/j.biocon.2005.09.005.

11. Statistisches Bundesamt, Statistisches Jahrbuch. https://www.destatis.de/DE/Publikationen/StatistischesJahrbuch/StatistischesJahrbuch2017.html.

12. https://de.statista.com/statistik/daten/studie/1358/umfrage/entwicklung-der-gesamtbevoelkerung-deutschlands/. Accessed on December 5, 2017.

13. https://data.worldbank.org/indicator/NY.GDP.MKTP.KD.ZG?locations=DE. Accessed on December 5, 2017.

14. https://www.umweltbundesamt.de/en/topics/soil-agriculture/land-use-reduction#textpart-1. Accessed on December 5, 2017.

Perpetual Architecture

Uranium Disposal Cells of the Southwest

Center for Land Use Interpretation

The **Center for Land Use Interpretation** (CLUI) is an American research and education organization interested in understanding the nature and extent of human interaction with the earth's surface, and in finding new meanings in the intentional and incidental forms that we individually and collectively create. The organization was founded in 1994, and since that time it has produced dozens of exhibits on land-use themes and regions for public institutions all over the United States as well as overseas. The center publishes books, conducts public tours, and offers information and research resources through its library, archive, and website. The CLUI exists to stimulate discussion, thought, and general interest in the contemporary landscape. Neither an environmental group nor

Title image: A 71-acre radioactive tailings disposal mound built by the Department of Energy for the interment of 3.5 million cubic yards of material from two former Union Carbide uranium mill operations in the nearby town of Rifle, Colorado.

CLUI image with flight support from Lighthawk, from CLUI photo archive.

an industry affiliated organization, the work of the center integrates the many approaches to land use—the many perspectives of the landscape—into a single vision that illustrates the common ground in "land-use" debates. At the very least, the center attempts to emphasize the multiplicity of points of view regarding the utilization of terrestrial and geographic resources.

In 2012, the Center for Land Use Interpretation (CLUI) presented an exhibition in its research center in Los Angeles that explored the extent of uranium disposal cells in the United States. The exhibition included aerial photographs taken by the CLUI, interactive touchscreen maps, and detailed information pertaining to these unusual fortress-like structures in the landscape. In 2014 Alexandra R. Toland and Gerd Wessolek invited CLUI to present images from its Perpetual Architecture research project in a scientific session on "critical issues of radionuclides and their remediation" at the 20th World Congress of Soil Science in Jeju, South Korea. In this final section of the book, the CLUI presents images of eerie, windowless structures, sometimes up to half-mile in length, which contain the relics of uranium mill buildings and tailings. These unintended "land art" objects represent some of the most menacing surface features on the soil platform. Built to minimize erosion and other threats to their stability, the structures are at some level open to the atmospheric forces that erode all monumental surfaces. In this visual essay, CLUI presents the paradox of the platform function of the soil as the expectation of perpetual stasis of a natural body, whose very nature is to develop and change over time.

The radioactive mounds of America's nuclear programs exist today, with more to come. These disposal cells are tombs for the remains of uranium mill buildings and tailings, bulldozed into engineered enclosures constructed to limit contact with their surroundings for a thousand years. There are dozens of them—across the country, from Pennsylvania to Arizona—built within the past 25 years or so, mostly by the Department of Energy and maintained by their Legacy Management office.

The contents are not considered high-level radioactive waste, like spent fuel from nuclear reactors. That material has yet to find a permanent home. What these cells contain is radioactive tailings from uranium processing sites, as well as the demolished buildings and apparatus from the mills themselves. The amount of radioactivity in these cells varies, but is generally considered harmful to people if exposure takes place over sustained periods. Most of the radiation comes from uranium 238, which has a half-life of 4.47 billion years, nearly the age of the Earth itself.

Sometimes the disposal cell is built at the site of a closed-down uranium mill, but in most cases the cell is constructed a few miles from the mill site, in a location that is more distant from communities and waterways, and the material is trucked to the site for burial.

Many of the mills from the uranium boom years sat abandoned for decades, with the radioactive tailings and salvaged parts from the mill used as construction materials in the surrounding communities. In some places, the mounds of sandy-textured tailings were used to make concrete for sidewalks, patios, parking lots, houses, and even schools. These structures later had to be identified and torn down as part of the cleanup process, and the remains moved into these mounds.

In many cases, the containment structures are low polygonal piles with flat sloping tops lined with an outer coating design to be a tough skin, minimizing erosion and other potential threats to the stability and form of the radioactive debris inside.

Though anomalous and distinct in their form from their surroundings, the cells are meant to blend in geomorphologically, to integrate with the forces of drainage and erosion of the landscape. In arid environments, the outer shell is a layer of coarse riprap rock, golf ball– to softball-sized stones a foot or two thick. Beneath this is a clayey soil layer, a few feet thick, which covers the radioactive material below. When rain falls, the water passes through the riprap, then flows over the top of the less permeable clay layer, and down the sides of the pile, where troughs and channels take it away from the structure. The riprap is a carapace, holding the clay beneath it in place, and it also reduces the collection of organic material on top of the mound and the development of soil that would lead to growth of plants whose roots could eventually penetrate the clay layer. The low angle of the sides of the mounds, less than the angle of repose, keep the rock in place, and the form of the mound intact. In places with higher rainfall, the tops of the mounds are sometimes covered in soil and planted with grass. The soil and plants act as a sponge, soaking up the rain, and slowing down the runoff, which would otherwise form channels and eventually erode the clay barrier beneath. The soil and grass also reduce runoff by evaporating and transpiring moisture into the air. The shallow roots from the planted grass help keep the soil in place, while other plants that might sprout unintentionally, with potentially deeper roots, are extracted from the soil through regular maintenance of the pile.

The cells tend to be in arid regions in the southwest, as this is where the uranium was mined and milled: northwest New Mexico, western Colorado, and southeastern Utah, especially. But they were built elsewhere, too, as uranium mining and milling occurred in other states such as Texas, Washington, Oregon, and Wyoming. Uranium metal processing and engineering took place in dozens of states, including Massachusetts, New York, Pennsylvania, Missouri, West Virginia, and Ohio. In some cases, factories involved in the process of milling uranium have been razed and transported across the country for disposal in arid land disposal cells. In other cases, the factory site was bulldozed into a mound, capped, and left for the future.

Up to half a square mile in size, they resemble pyramids, ziggurats, or relics from a geometrical mound-building culture. Like the ancient tombs of Egypt, they are meant to be disconnected from the contemporary world, kept inert and intact for as much of forever as possible. They are nonplaces—isolated from the present, designed, and destined for the future.

A network of mounds in Clive, Utah, 80 miles west of Salt Lake City, is one of the major locations for the disposal of radioactively contaminated structures and equipment from across the United States.

CLUI image with flight support from Lighthawk, from CLUI photo archive.

A new disposal cell in Crescent Junction, Utah, scheduled to be completed after 2019, built to contain uranium mill tailings and debris from a mill site in Moab, 30 miles south.

CLUI image with flight support from Lighthawk, from CLUI photo archive.

A six-acre disposal mound for radioactive tailings, built by the Department of Energy, and located at the site of a former uranium mill in the town of Green River, Utah.

CLUI image with flight support from Lighthawk, from CLUI photo archive.

A 29-acre disposal cell created by the Department of Energy in 1995 to contain the contaminated tailings and mill buildings from a uranium mill in Gunnison, Colorado, six miles away.

CLUI image with flight support from Lighthawk, from CLUI photo archive.

A 68-acre disposal cell in Mexican Hat, Utah, contains radioactive material from two uranium mills located nearby, which were cleaned up as part of the Department of Energy's Uranium Mill Tailings Remedial Action Project.

CLUI image with flight support from Lighthawk, from CLUI photo archive.

Wastelands

Lara Almarcegui in conversation with Gerd Wessolek

The work of Spanish artist **Lara Almarcegui** often explores neglected or overlooked sites, carefully cataloging and highlighting each location's tendency towards entropy. In 2013 Almarcegui blew minds as Spain's representative to the 55th Venice Biennial, where she filled the interior of the pavilion with massive piles of building rubble similar to those used by workers during its construction. Working at a time of widespread urban renewal in Europe, she has remained a champion of overlooked, forgotten sites—creating guides for the cities' wastelands and even instigating their legal protection.

Lara Almarcegui lives and works in Rotterdam. She communicated with Gerd Wessolek in the fall of 2017 about her approaches to the materiality of urban sites in perpetual transformation.

Title image: Construction materials of the main hall, Secession, Vienna 2010.

Photo: Wolfgang Thaler.

Image reprinted with permission of the artist.

Gerd Wessolek: Lara, how did you become interested in working with anthropogenic soil substrates and the spaces they occupy—brownfields, landfills, construction, and demolition sites—and the residual corners of city life?

Lara Almarcegui: In the '90s when I started working, I was completely surrounded by building projects. There was a large excitement about constructing and designing, with almost no critical voices against the consequences of so much construction. I started my work willing to disagree with this state of things I was living in Amsterdam at the time. As a young artist, I wanted to do whatever was critical of the pace of construction around me. My projects were a means to defend wastelands or nonbuilt areas from development. These were very direct actions that were trying to do the opposite of construction. My first project, when I was 22, consisted on renovating a market hall scheduled for demolition to be replaced by a shopping mall. Standing against unrestrained urban development and design has always been my starting point.

Renovating the Gros Market a few days before its demolition, San Sebastian, 1995.

Image reprinted with permission of the artist.

Gerd Wessolek: Common to all places everywhere is that they are subject to change in a seemingly infinite cycle of demolition and development. Question: Is the topic of "transition" your real thematic center, because in life nothing on earth remains the same? Or are there other, underlying messages that you want to convey?

Lara Almarcegui: I have been focusing on places about to be transformed because they are in an "in between" moment, for a short time meanwhile a new development is started. These are places that do not correspond to any design. I have a problem with the fact that in the contemporary city, every space must fulfill a program or a function. Each space has been designed for a purpose. I can only feel free in spaces that do not fit someone's program or design, therefore I defend wastelands, even if they offer freedom for only a short period.

On the other hand, I also find it extremely interesting to get to know places that are about

Construction materials of the Spanish Pavilion, Venice Biennale, 2013.

Photo: Ugo Carmeni.

Image reprinted with permission of the artist.

Machefer, Biennale de Lyon, 2017.

Photo: Blaise Adilon.

Image reprinted with permission of the artist.

Construction materials of the water tower, Phalsburg, 2000.

Image reprinted with permission of the artist.

Mineral Rights, Tveitvangen iron deposit, Norway, 2015.

Biography adapted from Creative Time Summit 2013.

Image reprinted with permission of the artist.

to change—not just because of the urgency of the opportunity to observe a place that no longer will exist, but because those spaces often contain narratives that are not exactly what the city managers want to tell. They do not usually correspond with "the image" the city wants to sell of itself. But they tell a lot of interesting stories that relate to hidden processes of entropy and decay. They tell of strange and temporary uses, or maybe just wild nature.

Gerd Wessolek: You've worked with stones and sand, rubble, glass, bricks, and landfill soil. Which of the materials is most attractive to you aesthetically? What challenges or insights have you discovered with different materials?

In this book, we are interested in the various forms of knowledge artists incorporate in their practices. What kind of material and technical insight have you gained in working with these materials?

Lara Almarcegui: I love waste construction materials and rubble of all sorts. They speak of a construction that has failed and has been demolished, but they also present the materiality of the raw material. Rubble has the capacity to address the source of the construction material itself—for example, the sand or gravel that has been extracted to produce concrete. In this way, it makes me think on the true origins of a building, which is underground; it is the

extraction of natural resources, the geology that lies below our feet.

Gerd Wessolek: I'm of course also quite interested in rubble. Besides bricks, rubble is a complex mixture of different materials such as gypsum, ash, coal, iron, and slag, which all have different chemical and physical properties. These are subject to weathering processes over time, which can have an impact on groundwater quality. A good example is the sulfate leaching from WWII rubble, which we have been researching for many years. I wonder what significance you assign to the qualities of rubble.

For instance, in some of your work, you characteristically sort materials and then heap them in large quantities into exhibition spaces. In other works, you explore brownfields *in situ* and create elaborate guidebooks to these marginal spaces. What is the significance of the authenticity of particular materials and places in your work? For example, does it matter if people can conceptually connect the materials to the places from which they are sourced, or is it enough for people to simply enjoy the aesthetic experience of being confronted with the materials themselves?

Lara Almarcegui: I do want the materials to be identifiable to a certain location, and I work as site-specific as possible. I just finalized a work in Lyon that speaks to the changes in the city. My ambition is to present the changes of

the neighborhood "Confluences" (where the show happened), which is going under a big process of urban transformation. We tried to facilitate the identification with the demolished building and mainly with the exhibition space itself. The work calls for awareness of the construction materials of the space. So, my aim is to go as close to the subject as possible. But actually, I've found that close enough is never enough. Lately I've been approaching the subject so much that I go underneath. The basic question I never seem to answer is, What is below my feet?

Gerd Wessolek: Sometimes you point out the different materials of a building in terms of weight and volume. In other works, these quantitative analyses fade to the background. What is your intention of including the numbers behind construction (or demolition) and alternatively in leaving the numbers out?

Lara Almarcegui: I like to present the construction materials used to make a building so the public can have the experience of the large volume of resources. This means moving around hundreds of tons of materials in trucks. Another similar project is just an identification and calculation of the materials themselves. I want to provide a tool to understand what a building is made of, and it becomes even more exciting when we do those identifications not just with a building but a whole city. Lately I've been identifying the materials of islands. Presenting materials in this way, it is possible to read the geological layers as a resource for exploitation.

Gerd Wessolek: Soil is more than its mineral components, even rubble soil. I was looking for the living components of the soil in your work and didn't find much. Why?

Lara Almarcegui: My reference is very much grounded in construction materials and architecture, and not the living components of the soil. Therefore, I have been focusing on the physical elements of a building such as construction materials, as well as conceptual issues of land ownership and real-estate investments.

Gerd Wessolek: Your work makes close ties to landscape and urban development issues. Do you ever work with urban or regional planners, developers, or landscape architects? If so, what were some of your experiences like? What were the highlights or challenges of working together, the similarities or differences of opinion?

Lara Almarcegui: From the start of a project, I usually get information from architects and city planners, which helps me to understand what is going on in terms of city development. Further on, when I am producing a work, I need a lot more detailed research and architects have helped me as well. Though I get their help, my approach tends to be much more critical of development. I'm usually defending the vacant land and the nature and vegetation against their plans, so my approach is much more that of defending the environment. My view of the city is less abstract and distanced than their view. I really do not like observatories or lookouts; I do not want to distance myself enough to see systems and structures. On the contrary, I try to get as close to the city as much as possible.

The idea of building observatories, which architects often like and seem to find poetic, is for me a very insulting one. I do not need architects to build towering lookouts to instruct me from which point of view I should look. I want to look from wherever I want; I want to feel free to move around, to observe at the street level.

Gerd Wessolek: It occurred to me that the residents, planners, builders, and local citizens are not really visually present in your work. Why is that? Are the materials and sites themselves more interesting to work with as an artist than the people who live there, work there, or are possibly affected by building changes there?

Lara Almarcegui: When I started working, I often saw the public being documented in the artwork so often that they seemed to represent a sort of social participation. I often felt like the public was being "used" by the artist. I prefer to leave them on their own. They might or might not enter the wastelands of my project. But I do not want to justify the quality of the work by documenting them. One of the constants of my work is that it produces discussion on the existence, present and future, of land and construction. I aim to initiate this discussion but I do not aim to control or to document it. I like it to happen freely. If I despise architecture because it controls the city habitants too much, I try to do the opposite and embrace no control at all.

Gerd Wessolek: Artists, like scientists, can wear many hats. How do you see yourself? As observer and interpreter/commentator, or as designer and maker? Do you feel like your work can contribute to philosophies and practices of urban change, or do you see yourself as more of an archivist or modern archaeologist? Are there any new roles or new directions you'd like to develop in your work?

Lara Almarcegui: Because of my rejection to moralistic behavior and because of a very deep rejection of any superiority position as an artist, I really do not like to create models. I do not think my work should be to consider a model to follow, because I really do not think I can tell anyone what to do. I have deep problems with art and architecture placing itself in such a superior position. My aim is just to encourage a certain questioning of the constructed environment.

But I do think some of my projects could be called a form of activism because they have provoked change. For example, the protected wastelands are like a dream to me. I failed a lot when trying to protect wastelands, but at the same time, I respect activists who work more locally than me. I often try hard, but if, finally, I am not allowed to do what I want in a place I just try somewhere else. For example, after failing to get mineral rights to protect an iron deposit in the north of Germany, I tried in Spain near my home town. I tried hard with different strategies, but when it was proved that in no way I would be allowed to protect the iron ore there, I move the project to an iron deposit near Oslo in Norway. A better activist, would have tried harder for years in the same place. As I mentioned, I am not a good model.

The Earth Print Archive

A Forensic Documentation of Land Take

Betty Beier in conversation with Gerd Wessolek, with input from
Frank Glante and Jörg Katerndahl

The German mixed media artist **Betty Beier** conserves landscape and soil conditions before a transformation through human intervention starts. In her project, *Earth Print Archive*, the artist documents different forms of land change and land take in marginalized landscapes in different parts of the world. Over the last 20 years, the *Earth Print Archive* has grown to include more than 80 completed "earth prints" that document the ecological and social conflicts associated with the platform function of the soil. This chapter presents a series of reflections and images by the artist as well as a short exchange with Gerd Wessolek from the Technische Universität Berlin and Frank Glante, Head of the Section "Soil quality, Soil Monitoring" in the German Environment Agency. In addition to exhibiting her work in prominent galleries and museums, Beier has been using a handmade cargo bike to take her work to the streets of Stuttgart, Munich, Freiburg, and Mannheim. She uses extensive photo documentation and an earth print from Kivalina (Alaska) to talk to passersby in Germany about climate change and land take in other parts of the world. She also exhibited her work at the German Environment Agency in 2013 and the German Federal Ministry for Environment, Nature Conservation and Nuclear Safety in 2016. "From Turf to Picture," an essay on Beier's work that appeared in the 2008 catalog, Erdschollenarchiv (*Earth Print Archive*) by the art historian Jörg Katerndahl has been translated from German and edited for use in this chapter.

Gerd Wessolek: The functional values of soil usually dominate soil protection schemes. Soil's heritage, beauty, reverence, mystery, mythology, and cultural associations are rarely addressed by lawmakers. What are the advantages of factoring these nonfunctional components of the soil into soil conservation?

Frank Glante: The "missing link" is an apt concept here. We've lost our bond and our contact to the soil. We "experience" everything in our engineered world in real time, but we're no longer affected. We see things like hunger in other countries, landslides, floods, or the effects of climate change like drought and storms on the news, but we don't connect it with our own behavior. We don't know (or we ignore the possibility) that our consumer activity and our hunger for energy directly influences soil degradation on other parts of the planet. Our resource and energy requirements dissolve the land (e.g., as seen in dam projects like in Karanjukar, or coal mining areas). Our belief in progress includes an acceptance of this destruction. The same way that we are only marginally aware (i.e., via the media, but not because we feel directly affected) that bee and butterfly populations are declining, we can't imagine the ground falling out from underneath us. The problems of the third world are long off and far away from us until waves of migrants knock at the door. But for the most part, the problems are our own doing when we declare the soil a commodity, investment object, or income guarantee.

Can we succeed in making the soil *tangible* again?

Awakening consciousness for new values is one of the steps towards that goal. But the idyllic scenes in magazines like *Landlust* are deceptive! The presentation of beauty isn't the goal, but to draw from it a demand for the protection of the soil necessary for effective preservation of that beauty. This approach has been partially successful in nature and wildlife conservation; the soil conservation movement can follow suit.

Betty Beier: I have trouble with this question. I find the formulation of the question too romantic, and the basic underlying problems too complex. This is a cultural question. Everything has to pay off today, to be worth something. That includes our awareness of the soil. That means confronting land with respect. The soil's story is our story. Our dealing with the soil is a reflection of our culture and cultural achievement. Is it possible to see soil not as a resource or natural capital but as living ecosystem that needs to be taken care of in its own right? Land take or loss of soil is a threat to future.

Gerd Wessolek: Describe your personal relationship to the soil and its conflicted function as platform for human settlement and economic development.

Frank Glante: As a plant physiologist, soil for me was the substrate plants grow on; it was the basis for metabolism and energy exchange. While I was writing my thesis on the subject of symbiosis between plants and fungi (mycorrhiza) my view of the soil changed only slightly. Lab and field tests gave me a deeper insight into the functions of the soil, but my understanding remained a scientific one. My work at the German Environment Agency widened my view from the technical to the political dimension and at the same time into numerous adjacent technical and political fields.

Away from work, ecologically conscious gardening, environmental protection, and using my experiences in environmental sciences in my community has become part of my daily life.

But I was first truly able to appreciate the soil with a sense of wonder and humility during two journeys "to the soil" itself. In 2003 I took part in a pedological-ecological expedition to western Siberia. Within the space of 3½ weeks, we examined seven climate zones with diverse soils and vegetation. These were, in most cases, natural locations such as moors, taiga, steppes, alpine tundra, and desert. I was able to process the dissimilarity and richness of the soils and the power of nature on location and with all my senses.

I became conscious of how fragile the earth can be and of what kind of damage man can cause the soil in a short time span on a trip to Iceland in 2012. In the space of less than 200 years, the first settlers to Iceland cleared a large portion of the forests, converting the land to meadow, and Iceland is still battling the consequences today.

These on-site experiences thoroughly changed my view of the soil and motivated me to action more than reading scientific literature ever did. My new understanding of soil—these practical experiences affected both by negative and positive experiences has changed my work. I've become less patient— seeing, in part, ignorance and denial in the face of negative changes to the soil caused by our own behavior.

The cost of maintaining good soil and rehabilitating degraded soil isn't the issue; it's how long we can afford the destructive exploitation of this precious resource. Entire cultures have been destroyed by excessive exploitation of resources. In the past, the effects were regional. In the globalized world we live in today, the entire planet is affected.

Betty Beier: I was interested in the soil from a young age. I collected animals and plants and studied them. Later, nature and landscape became the focus of my art and I began experimenting with soil and synthetic resin. But I always felt

strange about it—taking earth from the field. Once I took a wrong turn with my car and ended up on a large construction site. Gray cement, desert landscape, building noise, movement, chaos. It was the Rieselfeld construction site (urban expansion) in Freiburg. There wasn't much left to see of the former humid biotope—the soil had been completely displaced. The earth there was considered a waste product and I was welcome to take it. From then on I visited the construction site on a weekly basis. My focus moved more and more to the soil. I took surface soil and conserved it before the digger rolled up. Then other construction sites began to interest me. Altenwerder (harbor expansion), China (dam project), the Ministergärten in Berlin, and so forth. From 2005 to 2007 I visited the Kárahnjúkar dam project in Iceland, taking different soil states from the landscape, and the changes to them caused by construction and excavation using plaster casts.

I not only took samples, I also asked questions because I didn't understand what I was seeing. I spoke to geologists and scientists. I learned how sensitive the soil is, how soil erosion comes to pass, and how quickly the soil can be destroyed.

The samples of earth I took (casted surface sculptures I call "earth prints") directly reflect the influence of construction on the landscape. They also testify to the strength of the wind and the damage caused by soil erosion.

Gerd Wessolek: What does the future look like for the soil? What needs to happen to reverse the process of soil loss?

Frank Glante: Our basic knowledge of the state of the soil, globally, is insufficient. This lack of information has already been recognized and will be remedied through activities like the Global Soil Partnership, the Global Soil Map, and others in coming years. But the information garnered can't be exclusive. It has to be available to soil users on site. Land rights (tenure rights), access to resources (like fertilizers) and markets, and the evolution of agricultural space by and for small farms has to be more vigorously developed. Investments in agricultural space are important, but they can't be made at the cost of indigenous peoples. Land grabbing is an example of the most negative kind.

Traditional agricultural methods have to be examined using scientific considerations and some reintroduced. Some of these methods, despite having proven their value over long periods of time, were replaced by putative technological improvements, or lost due to conflict or war.

Soil conservation also needs activists. Scientists must leave the comfort of their ivory towers and put their energies not only into scientifically interesting studies, but also into improving our soil and conserving its quality.

Activists also have to come from other fields. Scientific "tunnel vision" and jargon aren't appropriate communication modes for raising public awareness. A variety of subjects and information dissemination methods are necessary in order to foster consciousness for the soil.

We also need something like soil tutoring. Some initiatives have already taken root:

- Classes for preschool and early education teachers (e.g., the "soil window" project)
- Soil nature trails and soil gardens
- Soil museums

- Films on soil problems (e.g., *Dirt the Movie*)
- Art activities and installations
- Summer schools and field trips

But there is still plenty of room for growth.

Betty Beier: I agree with all points here. The soil is neither seen nor appreciated by many people, regardless of their background or origin. Why is that? I often ask myself that question. In his 1955 thesis, Tristes Tropiques, Claude Lévi-Strauss (translated by John Russel) describes it this way: "The relationship between Man and the soil has never been marked by reciprocity of attentions" (p. 98).

Events like the "International Year of Soil 2015," "World Soil Day," and the "Global Soil Week" were mainly happening in cities so far, not in rural areas. A broader dialogue must be initiated— between citizens, farmers, artists, and scientists.

I created the *Earth Print Archive* with the intent to increase nature appreciation, in general, and soil appreciation, specifically, and to show solidarity with people who are literally having the ground pulled out from under them. Climate change is wiping out indigenous peoples— including ourselves. Our landscape is changing, for the worse, and on an hourly basis. Dams swallow up land. Greed for resources destroys land. Single-minded use of the soil homogenizes the landscape. The irreparable damage is displacement and relocation, homelessness and hunger. And all this is happening at a breathtaking pace. If we don't make an about face right now, our ability to have any influence at all will be lost forever.

From Turf to Picture

The process of creating an earth print consists of a complex disentanglement of original and copy. Seen individually, the works are life-sized, square-meter reliefs mounted on an "invisible" foundation made of artificial resin and fiberglass that realistically imitate the land surface as it was encountered on site. What is essential for the approach is on the one hand the systematic selection of the sites and on the other hand their highly detailed documentation over time. The artist usually selects areas within and outside Germany that are accompanied by controversial discussions: villages that must be abandoned to make way for major hydropower dams; island communities pushed to the brink of existence by rising sea levels; city blocks in a perpetual state of demolition and reconstruction. Beier visits these sites regularly in order to document the changes of the landscape and the land surface, offering before and after glimpses of a place. It is often tedious to obtain the permits for entering the sites. The time-consuming work in often impassable terrain and the transport of heavy materials require good logistics and thorough planning. Thus, the actual work on a print does not begin with making the plaster cast and the photographic and graphic documentation on site, but long before—by following political discussions, making official requests, and planning travel for the necessary field work.

When the artist finally arrives on site, she selects the specific point of documentation following both her intuition and aesthetic judgment. Her next step is making the plaster casts, generally in four square blocks of 50 cm each in order to facilitate the transport of the plaster elements to the artist's studio. The plaster will take on the form of the land surface before it hardens. Afterward, the area will be cut out and its original context will thus be destroyed because elements of the natural surface will remain stuck on the plaster. Back at the studio, these plaster casts will ultimately be used as a matrix for the actual image carrier made of resin and fiberglass. During the transfer process the original elements like soil and plants embedded in the upper plaster layer will partly be transferred at a scale of 1:1 to the actual image carrier. During this process, the plaster form will be destroyed, rendering each earth print unique. The process of reconstruction and artistic follow-up modeling requires accurate photographic and graphic documentation at the original location. In the studio, the artist will then model a detailed land relief based on the original condition found on site. Original finds will be conserved and inserted into the reproduction. This reworking process is often complex and can be carried out for weeks or months after the original frame has been cast in plaster. The lost one square meter of land is transferred onto the relief print, creating a deceptively realistic copy of an already lost reality.

Beier's method is characterized by extreme naturalism almost reminiscent of an educational scientific demonstration model. Because the panels mirror the original state of the land and use original materials, they are characterized by a high degree of authenticity. The land formations, geological characteristics, and plant details will be reproduced as they occur in nature and permanently conserved. Because the territory of the reproduced surface areas has meanwhile been changed considerably, resulting in a destruction of their original context, the confined details of the *earth print archive* play a decisive documentary role. The resulting earth print is not only a representation of the site but also a relic. Although the surrounding areas of where the prints were taken meanwhile no longer exist in the form in which they were found, the earth prints gain the status of lingering artifact. Art and science, lifelike representation, forensic reproduction, and awareness-raising of the respective historical, political, and ecological background of land-use changes and land take form a symbiosis in the *Earth Print Archive*. The process and result of human intervention and development triggered by technological and economic human interests are at the center of Beier's artistic practice.

Jörg Katerndahl, 2008

Ministergärten History: The area between the Brandenburg Gate and Potsdamer Platz—where once the Ministerial Gardens, later Hitler's Reich Chancellery, and last the so-called death strip were located—now accommodates the Holocaust Memorial (Memorial to the Murdered Jews of Europe), as well as several representative buildings of German federal states. This property, called the Ministerial Gardens, and its immediate surroundings are burdened with history; from Bismarck to Hitler—all literally stood in the sand here. The excavation of the Nazi bunkers during the construction phase for the Holocaust Memorial and the government buildings between 1997 and 1999 triggered a debate on whether Hitler's shelter and the drivers' bunker ought to be protected as historic monuments and opened to the public. Between 1997 and 1999, I followed this debate as I accompanied the construction phase of underground demolition ("Tiefenenttrümmerung"), and secured traces. In 1997, between the Brandenburg Gate and Potsdamer Platz, I found a green space that attracted my interest. The first attempts to obtain an access permit failed. In February 1998, I was allowed access to the construction site with a permit that had meanwhile been granted. The area was demolished and the debris cleared. At this point, I started to visit the site on a monthly basis. The underground garage disappeared in 1998–99 bit by bit. What remained was the fine sand characteristic of the Brandenburg-Mark region.

Earth print: "Ministergärten 1997," sand from the construction site, acrylic, artificial resin on fiberglass, 100 × 100 × 7 cm, 1999
Betty Beier private ownership Karlsruhe.

Top left:

In 1997, between the Brandenburg Gate and Potsdamer Platz, I found a green space that attracted my interest.

The first attempts to obtain an access permit failed.

Ministergärten, Betty Beier, 1997.

Top middle:

In February 1998, I was allowed access to the construction site with a permit that had meanwhile been granted. The area was demolished and the debris cleared. At this point in time, I started to visit the site on a monthly basis.

Ministergärten, Betty Beier 1998.

Top right:

The underground garage disappeared in 1998/99 bit by bit. What remained was the fine sand characteristic of the Mark region.

Ministergärten, Betty Beier 1998–99.

Kárahnjúkar History: In 2005, the Icelandic dam project "Kárahnjúkar" attracted the interest of the international media. The planned reservoir and power plant on the highlands of Iceland will be the energy providers of an aluminum factory that is being built in the small town of Reydarfjördur on the island's east coast by the American Alcoa Corporation. The power plant will produce 4.5 billion kilowatt hours of electricity per year—more than half of the amount of electricity that is being produced in Iceland today. For the project, 57 square kilometers of a wilderness area were to be flooded. Iceland's glaciers, one of them being Vatnajökull, the largest glacier in Europe, supplies water to rivers and numerous waterfalls that flow through deep gorges and uninhabited wilderness and form a grandiose landscape. Since September 2006, the canyon—the Kárahnjúkar gorge—is blocked by a 193-meter-high dam. Waterfalls, fens, valleys, and small gorges have disappeared underwater.

I accompanied the dam project from 2005 to 2007, first visiting the area in 2005 on a bicycle with a special trailer I designed myself to transport the heavy plasterboards across the difficult terrain.

The Kárahnjúkardam August 2006, Betty Beier.

The wind molds and characterizes the landscape, that's why I chose this place. On a small hill, the wind has already carried off the top surface soil. This part is only sparsely covered with heath (Beitilyng), mountain avens (Holtasóley) and mosses. In June and July, the mountain avens are in bloom and bear a white flower. I made a plaster cast of the surface.

Earth print: "Kárahnjúkar 1" soil, items found in the area, (plant material/partially acrylic replica) loess, acrylic on GRP, 100 × 100 × 15 cm 2008, Betty Beier.

Kárahnjúkar, 10.7.2005, 64°54'21.5"N, 15°50'51.5"W.

Belo Monte History: Belo Monte is the third largest and one of the most controversial dams in the world. It is known to scientists as the "monster dam" because of the scale of deforestation of the rainforest and associated negative influences on the world climate. The operator consortium has also been accused of violating human rights and conducting ecocide, especially toward indigenous tribes. A number of environmentalists have been murdered in the wake of the project, most recently Luís Alberto Araújo in November 2016. The first turbines were put into operation on May 5, 2016. Large areas of the rainforest were lost in the floods. Villages with thousands of people had to be resettled. The size of the flooded area corresponds approximately to that of Lake Constance. The dammed-up river, Xingu, was once one of the last intact river systems in Brazil. Its biodiversity was as breathtaking and unique. From 2014 to 2016 I documented the upheaval in the summer months (eight earth prints, numerous photographs, drawings, and a video). A project report accompanying the first earth print is in progress and has already been shown in the Kunstraum Loft, Ansbach (2016), as well as the group exhibition *Naturliebe—Erneuerbare Haltung (For the Love of Nature—Renewable Attitudes)*, Artists Association Walkmühle, Wiesbaden, Germany.

Belo-Monte Dam, Betty Beier 2015.

Antonia Melo on her destroyed premises.
Forensic securing of evidence.

Altamira, Betty Beier 2015.

Earth Print "Rainforest."

Plant material, soil, finds, acrylic on GRP, 100 × 100 × 10–20 cm, Betty Beier (2016).

Find spot and date: Rainforest near Altamira (Para), Brazil, 23.06.2014 S 03°17.608, W 052°13.495.

Soil in the City

The Socio-Environmental Substrate

Debra Solomon and Caroline Nevejan

Debra Solomon is an artist and independent researcher, and food forest and urban soil-building expert. Her vision of the urban public space as an ecologically coherent "foodscape" posits a productive, radically greened and socio-natural city. As an artist working in the public space, Solomon has collaborated with local communities since 2009 to produce food-bearing ecosystems in the form of park-like food forests at public locations in Amsterdam and The Hague. In 2010 she founded Urbaniahoeve Design Lab for Urban Agriculture, which means "the city as our farmyard." From 2004 to 2012 she wrote Culiblog.org, a blog about "food, food culture, and the culture that grows our food." Solomon is currently pursuing a PhD on multispecies urbanism with professor Caroline Nevejan and professor Maria Kaika at the Amsterdam Institute of Social Science Research at the University of Amsterdam.

Title image: *En Necromasse, Mycelium,* Debra Solomon. 2015. Screen print on paper, 100 × 70 cm. Mycelium assertively colonizes a dead leaf, metabolizing its nutrients and distributing them to the benefit of the soil community at large.

Dr. Caroline Nevejan is Chief Science Officer of the City of Amsterdam and professor Designing Urban Experience at the Amsterdam Institute for Social Science Research at the University of Amsterdam. Previously she was associate professor in the Participatory Systems Initiative at Delft University of Technology. Specializing in artistic and design research, Nevejan's research takes place in interdisciplinary contexts, focusing on the design of trust in today's globalized and mediated societies. Nevejan was founder/director of Waag Society and program director at Paradiso and lab director at Performing Arts Labs in London, UK. Between 2006 and 2014 Nevejan was a crown member of the Dutch Council for Culture and the Arts. Between 2014 and 2016 she was a member of the supervisory board of *Het Nieuwe Instituut*, which aims to illuminate contemporary subjects related to architecture and design.

Considering that 75% of Europeans and 50% of the population worldwide lives in cities, it is important to ask, Why is urban soil largely absent from soil protection policy recommendations? This question deals with, on the one hand, issues of appraisal, appreciation, and value, and on the other hand, practical, innovative approaches to urban land-use as a countermeasure to land take. This chapter argues that soil is a fundamental element in future ecologies that needs to be addressed in urban theory, methodology, and practice. First, the art installation *Entropical* is discussed, which poses the question of the discrepancy in value creation by soil production versus value creation by mining cryptocurrency. Having identified the lack of soil awareness in today's globalizing world, the chapter describes in detail how future ecologies need to take urban soil as a starting point and how soil is actually part of a social system. The concept of soil ecosystem services seems to offer a foundation for integrating soil into future ecologies and as a basis for policymaking but is unfortunately entangled in economic metaphors and furthermore lacks participatory roles for humans. Exploration of a new paradigm for urban soil is discussed in the presentation of the art project of Urbaniahoeve, an Amsterdam-based foundation.

In cities, large swathes of soil are situated within public space landscaping and "green zones," places more regulated than the soils of conventional farmland and mainstream agriculture. With more than 50% of the world's population living in urban areas and 75% in the European Union,[1] urban soils are a logical starting point to initiate a change in perception of humanity's role in the natural world at large. New systems of valuing are necessary to rethink urban soils, their cultivation, and protection. This chapter explores a new paradigm for thinking about urban soil. Inherent in its argumentation is the notion that art and artistic research has the potential to offer radical realism and contingency[2] and is as such complimentary to scientific research.[3] Both scientific and artistic research are positioned in relation to one another in putting forward a new paradigm in which soil is considered an actor in its own right and which is engaged with human society in a reciprocal manner.

Soil as Value System: *Entropical*

The case study *Entropical*[4] is an artistic research project by Debra Solomon and Jaromil initiated in the International Year of Soils. The exhibition was situated in a glass pavilion in Amsterdam's Amstel Park, as part of the Zone2Source[5] program curated by Alice Smits. *Entropical* reflects Solomon's conceptual roots in the Land Art movement of the 1960s and '70s, in particular artist Robert Smithson's "Non-site" installations from 1969. Smithson, who wrote extensively on the topic of entropy in reference to urban development, displayed rubble from building sites as artistic material.

Metaphorically, the *Entropical* installation and resulting land-art work represent a reconciliation between the prevailing economic, ecological, and agricultural value systems, proposing an alternative "ecology" herein, in which at first human hubris, followed by human nurturing of soil processes position human activities within the soil-producing community underground.

Entropical consists of four art works in which the value and dynamics of the exchange of materials in the biological world is set against the abstract value of algorithms and computer calculations. It questions whether economic value systems can be brought into a direct productive relationship with ecosystem producers such as fungi in a time in which intensive computation is valued more than ecological regeneration. How, for example, could Bitcoin positively affect the rhizosphere, the layer of soil/soil life around the roots of plants?

The collective artworks of *Entropical* play with the concept of "entropy," the second law of thermodynamics, a condition of constant change in which materials and energy are transformed. But the term is also used in cryptography, where it refers to algorithmic processes and abstract information. Entropical therefore inquires into the incentive to produce ecological regeneration and value in an age in which running intensive computation (e.g., "mining Bitcoin") yields far more value than soil production and requisite ecological regeneration.[6]

The exhibition in 2015–2016 comprised the following four works: *En Necromasse*[7] (five screen prints), *Seven Layers*[8] (one screen print), *Resist Exist*[9] (typography on the windows), and in a separate darkened room at the back of the gallery space, the installation *REALBOTANIK*.[10]

En Necromasse depicts one of the soil's most valuable resources: the dead materials known as necromass. The title of the work plays with the notion of soil organisms working as a single community together, en masse. Without ever showing a speck of soil, the screen prints show dead, organic materials in the process of becoming soil, pointing to the economy of topsoil metabolism and production.

En Necromasse, Forest root, Debra Solomon. 2015. Screen print on paper, 100 × 70 cm. The chaotic episode of a fallen tree reveals its root structure, now a habitat for uncountable billions of organisms as it is transformed into an abundant feast for fungi.

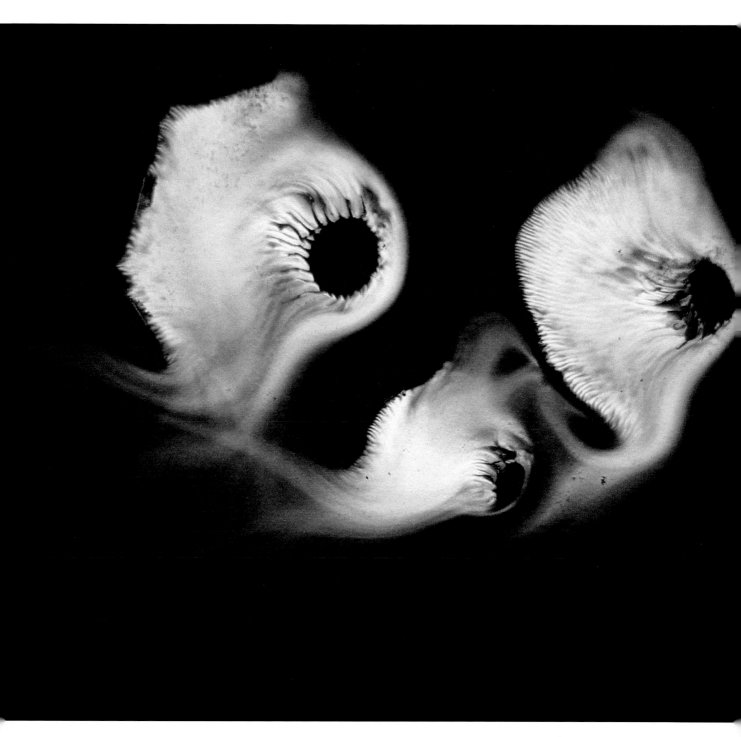

En Necromasse, Sporeprint, Debra Solomon. 2015. Screen print on paper, 100 × 70 cm. The spores of the parasitic oyster mushroom leave ghostly fungal drawings.

REALBOTANIK, Debra Solomon and Jaromil. 2015. Installation view with Bitcoin miner, oyster mushroom mycelium growing on screenprinted urban waste cardboard, dimensions variable.

Photo: Daniela Paes Leao.

Next page top:

REALBOTANIK, Debra Solomon and Jaromil. 2015. Special edition Monarch Bitcoin miner warms bags of soybeans and black beans inoculated with *rhizopus oligosporus*, during a workshop in which "tempeh" is a metaphor for soil aggregate. The heat provided by the Monarch miner, allows the *rhizopus oligosporus* mycelium to metabolize the beans, resulting in tempeh.

Photo: Debra Solomon.

In the REALBOTANIK installation, screen printed waste cardboard inoculated with oyster mushroom mycelia develops into thick mats warmed by the "waste" heat released by a computer mining blockchains, the technology behind the mining of cryptographic currencies such as Bitcoin. Heat as a by-product of the information and financialization industry is thus recycled in this installation in order to grow nutrients (for humans, for soil organisms, for plants) on cardboard, an abundant, noncontested urban waste material. After the exhibition, the mycelium mats that slowly take shape in the installation are used to restore poor urban soils by inoculating them with fungi as an act of nurturing.

During the two-month exhibition period two workshops were given in which the process of making the traditional Indonesian foodstuff tempeh becomes a metaphor for soil formation. The heat provided by the Bitcoin miner allows the tempeh fungus *rhizopus oligosporus* to grow and bind the (soy)beans together. Just as *Rhizopus* "mines" the beans for nutrients, so do soil fungi mine soil aggregates for nutrients.

REALBOTANIK elaborates on the almost poetical impossibility of a comparison between the abstract processes of value creation in finance and the material value creation of living processes.[11] The title refers to the term "Realpolitik,"[12] reflecting value attributions and technoscience essentialism used to describe resource exchanges within the soil organism and within computer/financial networks and the notable differential between "use value" and "exchange value" in market evaluations.

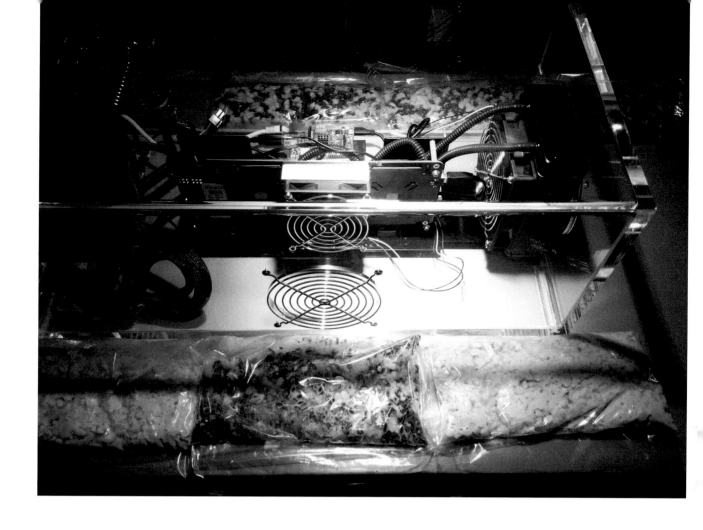

Soil as Social System: The Socio-Environmental Substrate[13]

To understand how urban soil is a social system, we need to look at the patterns and drivers of land take and soil sealing, as well as examine the status and socio-ecological potential of soils in the city. Urban and periurban space is often built upon what is or was *arable land,* that is, land suitable for food production. Urban sprawl, known in the literature as "urban land take" is the conversion of agricultural land into development and infrastructure. It reduces habitat and other ecosystem services. City soils are frequently paved over, "sealed" under roads, sidewalks, and public squares.[14] "Soil sealing" is a term used extensively in soil reports of the Food and Agriculture Organization of the United Nations, referring to the *disabling* of ecosystem functions so as to impair the soil's ability to perform essential ecosystem services such as water storage or carbon sequestration. Soil sealing is closely connected with "land take."[15] Though *land take* refers to a process that most often occurs outside the city when forests, wetlands, and arable fields are urbanized, it should seem obvious that it is a problem to be taken up in urban planning practice and policy as well as environmental science.

Land take is on the rise. "Urban land use deserves special attention as most human activities are concentrated in cities, and demand for the urban land-use patterns have a particular impact on the environment, for example, through soil sealing or whole sale change of landscapes."[16] But land take is also an indicator for urban growth, and in European Environmental Agency (EEA) assessments it is associated with economic development. Land take strongly indicates an economy that can afford to import its agricultural products, as well as requisite material resources, for example, water and soil nutrients, but also labor, and more abstractly "time spent" from elsewhere.

Soil sealing as a process of urbanization is not only associated with economic growth but also negatively with contemporary forms of colonialism. Soil sealing, in other words, represents not only a loss of "our" soils, but greatly impacts the land use, connected ecosystems, and eco-social functions of soils in other countries. Simply put, when we pave over our farmlands in a spate of "economic development," we are literally implying that we will "be doing our farming elsewhere," by other people, under other conditions. Deforestation and land degradation are often the result. Indeed, the SOER Assessment is clear, "The associated intensification leads to increased environmental pressure on water and soil resources in intensive farmland areas. In addition, abandonment of extensive farmland leads to a loss of biodiversity in the affected areas."[17]

The conversion of forests and grasslands to agricultural land is one of the most important drivers for habitat loss and greenhouse gas emissions worldwide. There are clear links between the use of farmland in Europe and global agricultural trends.[18] Simply put, urban soil sealing is, to no small degree, the equivalent of *soil stealing*. If we are to take an approach that harnesses this rising dynamic to tackle the problem of land take, it behooves us to look at processes of urbanization and the "spoils" of urban resource flows as the resource that they are.[19]

In urban public space, soil is figuratively, but also literally, hidden from view. Urban soils provide the foundation for all of the green landscaping and ecological infrastructure of our cities. Urban landscaping is based on urban soils and the higher the quality of the urban soil, the better its ecosystems can function. Urban soils in urban public green space are still vastly overlooked in policy papers on carbon sequestration and soil fertility, implying that urban soils are not widely considered as potential zones for food production and ecosystem-based adaptation for climate crisis. One reason for this lies in the notion that the larger the surface area, the higher the effect will be, and indeed, the footprint of urban soils is smaller than that of the rural hinterland. But certainly it is not only the physical size of the technical infrastructure of cities and their soils that must change, but also political instruments, municipal maintenance structures and schedules, education, and culture itself.

"Good" management, such as maintaining and building upon the soil's capacity for carbon and water sequestration, preventing erosion, and fostering biodiversity, can be done at all scales and land zone types. The addition of organic materials to urban soils using a combination of mulching and planting techniques[20] requires access to resources readily available, and as yet relatively noncontested in urban settings. Important drivers that positively affect a soil's ability to deliver ecosystem services (e.g., sources of organic material, access to [waste] water) exist in vast and largely untapped quantities in cities. Access to these same resources is usually unavailable in rural and conservation landscapes.[21]

Urban green zones feature "permanent" landscaping that does not require soil disturbance techniques like plowing. Because of a lack of disturbance, urban soils have been documented to contain more carbon sequestration services and higher levels of carbon than under "natural" conditions.[22] Although urban soils and their ecosystems have a considerably smaller footprint than rural and conservation lands, their potential impact in societal and ecological terms may be greater than previously thought, possibly surpassing the impacts of the rural/conservational context.[23]

In fact, the smaller size and fragmented nature of urban soils may even increase efficacy when it comes to the intricacies of project planning, implementation, participation, and the documentation of results at different levels of scale, and the future integration and transfer of innovative practices.

Soil as Economic System: Ecosystem Services

Ecosystem services[24] can be described as the sum total of capabilities and benefits that an ecosystem can provide. To give some examples, a soil ecosystem can provide habitat for soil biodiversity, water-carrying capacity (water sequestration), or can provide plant nutrition and crop health, which directly translates to societal cohesion[25] and human health. Ecosystem services are typically described in terms of four types of services that significantly affect human well-being.[26] From the Millennium Assessment Report, these include, "provisioning services such as food, water, timber, fibre, and genetic resources; regulating services such as the regulation of climate, floods, disease, and water quality as well as waste treatment; cultural services such as recreation, aesthetic enjoyment, and spiritual fulfilment; and supporting services such as soil formation, pollination, and nutrient cycling." From this list one can clearly understand that specific soil ecosystem services have an impact upon other ecosystem services.

Concretely, ecosystem services mean that if European urbanites don't want to pay for and build water drainage treatment plants to process an ever-increasing amount of storm water runoff due to an increasing amount of technically severe rain events in urban areas, it is a priority to ensure that green spaces and their soils, are capable of absorbing that water.

Urban soils are not only capable of sequestering water but also play an extremely important role in *carbon sequestration.* In agricultural contexts, carbon is represented by among other things (a.o.t.) *crop residues,* the woody leftovers of harvested crops, or in the rhizosphere, the food-rich channels of organic material, or the dead matter called *necromass.* Carbon stored in soil (carbon sequestration) not only positively affects soil fertility but is directly linked to positive climate adaptation, because carbon taken up by soil processes and left undisturbed is not released into the atmosphere as CO_2. The good news is that just about any soil, even pitiful urban soils, under proper stewardship and maintenance can be "encouraged" to sequester more carbon.

As convincing as the (Soil) Ecosystem Services Framework first appears, many authors have criticized its philosophical and political weaknesses, especially with regard to its attempts to monetize the values of such services in the marketplace. Among other soil scientists, Phillippe Baveye[27] has pointed out its fraught relationship with economic theory. In scientific communities back in the 1950s, there was a "belief that environmental deterioration … stemmed from the fact that, in large measure, the services provided by natural systems had no readily identifiable monetary value and were therefore entirely overlooked in economic and financial transactions."[28] Despite a sharp increase in publications on soil ecosystem services, "researchers have manifested very little interest in monetary valuation, undoubtedly in part because it is not clear what economic and financial markets might do with prices of soil functions/services, even if we could somehow come up with such numbers, and because there is no assurance at all, based on neoclassical economic theory, that markets would manage soil resources optimally."[29]

A number of alternative frameworks for valuation have challenged the ESS paradigm by focusing on the interdependency between such services, their spatial fluidity which makes values difficult to isolate, their intrinsic worth, and the connectivity between the biotic and abiotic components of the soil. These alternative frameworks include the soil functions concept formulated by Winfried Blum and adopted by the European Union, the Soil Food Web concept developed by Elaine Ingham and cited by the U.S. Department of Agriculture,[30] and more recently, concepts of temporality and care proposed by M. Puig de la Bellacasa.

Particularly the latter proposal by Puig de la Bellacasa[31] hails in an entirely new paradigm, in which soil actors are emphasized rather than soil services. In this view, humans fully join the community of soil producers with mutuality and reciprocity characterized by new roles for all ecosystem actors. In *Making Time for Soil: Technoscientific Futurity and the Pace of Care*, Puig de la Bellacasa describes the nervousness of soil treatment within a future fixated on promissory solutions as so exhausting, that even the process of soil regeneration is situated within the endeavor of reaching the "moment of production." This creates the effect of never actually being in the "temporality" of soil; the time it takes soil beings to make (top)soil. Even Ingham's Soil Food Web[32] concept, with its quantified soil constituents, is criticized by Puig de la Bellacasa for its lack of human actants aside from in the role of extractor, and where there is no caring human hand to serve up organic materials to the Soil Food Web's diverse soil communities. And in fact, if one surveys diagrams offered by other soil conceptualizations such as the soil ecosystem services framework[33] or the soil function(ing)s[34] diagrams, all are bereft of human agency in reciprocal roles of care.

The temporality of care which de la Bellacasa describes and which already forms the practice of many soil building practitioners, agroecologists, permaculturists, and gardeners; of engaging with the living soil community, is what gives Making Time for Soil so much resonance with practitioners-soil builders, as is the case with the Amsterdam-based foundation Urbaniahoeve.

Urban Soil Building as a New Paradigm: The Urbaniahoeve Technosol

In the previous sections soil is explored as value system, as social system and as economic system. The question is raised as to whether a paradigm shift is needed in which soil, including urban soil, is considered to be foundational for ecologies; for value, social, and economic systems; and in a larger sense, for the survival of mankind. The exploration of this new paradigm is practiced by Urbaniahoeve, which is discussed next.

The topsoil of the Urbaniahoeve Food Forest DemoTuinNoord (DTN)[35] is an example of a highly-functioning, nurtured, urban technosol.[36] This ongoing land art piece demonstrates how a change in the perception and management of urban soil can directly and positively affect the city. This anthropedogenic topsoil was developed *in situ* from organic material from the urban waste stream, within a period of just three years and which serves as a model for soil treatment in the public space.[37]

In Urbaniahoeve foodscape projects,[38] like the Food Forest in Amsterdam *DemoTuinNoord*, making time for topsoil production begins simultaneously with planting; applying organic material to the soil, allowing this layer to rot down and suffocate invasive weeds, brambles, and roots *in situ*, and planting through this layer at the onset. The process of adding unrotted organic material is continually attended and sustained. This presoil organic layer is immediately planted in the style of the forest garden,[39] with varied layers and types of vegetation chosen as food for humans and nonhumans alike, and as an aid in the process of topsoil creation. These steps are carefully monitored and repeated, and as previous layers of underlying mulch break down, new ones are added to maintain a blanket cover. The DTN soil is never tilled, nor are the planted beds walked upon, leaving the soil's ever-developing structure beneath the various mulch layers undisturbed.

Because the Urbaniahoeve soil-building methodology does not wait for organic materials to compost before application, the mulch layers provide habitat and food for a much wider range of soil building beings than if, as in other techniques, a layer of rotted compost was first applied. Typical materials of noncontested local urban waste streams comprise the mulching actions: cardboard layers up to 15 cm thick; 30–40 cm of wood chips from the municipal pruning; old blocks of mycelium spawn from a local fungi grower; and generous amounts of coffee grounds from

Topsoil of the Urbaniahoeve Food Forest Ecosystem, DemoTuinNoord, Debra Solomon, and Urbaniahoeve in close collaboration with the soil-building community of the food forest ecosystem. 2011– ongoing. Land art, 1200 m² × 60 cm. View shows spring 2015.

Photo: Debra Solomon.

local restaurants, cafés, and architecture studios. The most important soil amendments are simply what is most at hand.

In 2014 Urbaniahoeve collaborated with University of Wageningen (WUR) soil scientists to measure nutrients and other soil constituents present in the DTN technosol.[40] The levels of available soil nutrients were predictably high but the amount of organic matter present in the soil nearly reached an impressive 15%.[41] This high percentage of organic matter, not typically found in natural soils outside the tropics, is extraordinary considering that this location had just ten years previously been an abandoned and later depaved parking lot, made with the sand used for Dutch urban terraforming projects as its only "parent material."[42] What is especially noteworthy in the WUR soil study of the Urbaniahoeve technosol is this soil's extreme capacity to sequester water. To date, the spongy, black soil (which was not long ago merely sandy fill) is capable of absorbing *virtually* all of the water of the recent year's rainfall events. Instead of impacting urban water infrastructure with surges, the rainwater is safely stored in the soil and food forest ecosystem. This phenomenon is another kind of "banking"; storing and later tapping sequestered water sources during lengthy spring droughts.

Soil Portraits[43]

The WUR soil study provided an initial insight into the development of the Urbaniahoeve topsoil, but the chemical analysis as a technique was not without limitations. Considering that it was merely a snapshot of what was going on in the soil, this form of soil analysis is at the very least financially prohibitive as an observation technique for developing and nurturing (top) soils, in that such processes require frequent monitoring. In the case of

#26 24 x '6 | BHW-a | s 10cm | 1.925g /50 ml NaOH 1% blended samples 03.vii & 13.ix | 14.x.16 | AgNO3 1% ETOH 80% | 24.x.16 | humid room

Soil Portrait #26, Debra Solomon, 2016–ongoing. Soil chromatography, soil building, 580 × 580 mm. *Soil Portrait #26* (*SP#26*) is a soil chromatogram sampled from the self-made topsoil produced *in situ* by Solomon, Urbaniahoeve, and in close collaboration with the nonhuman soil-building community of the Urbaniahoeve Food Forest Ecosystem in Amsterdam Noord. *SP#26* depicts a young, well-aerated soil, comprised of a high percentage of partially decomposed organic materials. This illustrates that organic material, even when not completely decomposed, is accessible to many species of soil life present in this highly composed, urban technosol ecosystem.

Soil Portrait #55, by Debra Solomon, 2016–ongoing. Soil chromatography, soil building, 580 × 580 mm. *Soil Portrait #55* (*SP#55*) is a chromatogram of worm humus originating from the carefully "curated" vermicomposter at the Urbaniahoeve Food Forest in Amsterdam Noord. *SP#55* illustrates a young, well-aerated "soil," bursting with soil life, that has an abundance of partially decomposed organic material, and to which sand was recently added as a means of providing roughage for the worms' guts.

the Urbaniahoeve soil ecosystems and the (earth)worm ecosystem of the vermicomposter, both receive an intentionally composed "diet," replete with occasional and voluminous coffee shots and summer treats of comfrey leaf or nettles. Especially in the worm bin, one can bear witness to worms "joyfully" tucking into seasonal sources of nutrition. The ability to monitor the results of this enthusiasm in terms of soil/humus quality is therefore an imperative. A method for monitoring "worm enthusiasm" that is both affordable and technically easy to carry out on a regular basis is soil chromatography.

Soil chromatography[44] is commonly used by farm collectives in the Global South as a means to easily, quickly, and repeatedly check developing soils for changing indicators of soil health such as the presence of organic material, bioavailable nutrients, and soil life. As a data generator, soil chromatography does not quantify nutrient makeup like a chemical analysis, but provides a visual overview with which one can gain perspective into the state of the nutrient contents of any given sample. Most important, soil chromatography illustrates the interaction of soil life within the sample's materials in a visually evocative manner. Lay folk, normally uninterested in the subject of soils and their care, suddenly become fascinated when shown the mandala-like soil chromatograms, particularly the large-scale versions made of the Urbaniahoeve topsoils.

As an artwork, *Soil Portraits* is an ongoing series of large format (289 × 289 and 580 × 580 mm) soil chromatograms first produced during the Mondriaan Foundation's AGALab-Waag Society artist residency in 2016. The series consists primarily of chromatograms sampling the self-made topsoil produced by Solomon and Urbaniahoeve in close collaboration with the nonhuman soil-building community at the Urbaniahoeve Food Forest ecosystem. Solomon worked with Ruben Borge[45] to innovate a technique to make soil chromatograms up to sixteen times larger than the "normal" size used in the field, creating a magnification, primarily of the lightest sampled materials, the soil's organic matter. Produced all over the world by farmers and farm collectives, soil chromatography provides a way of analyzing soils under development in order to establish if the applied methods are effective.

Conclusion and Further Research

This chapter has argued that soil is fundamental to future urban ecologies and illustrated this with two case studies. The art installation *Entropical/ REALBOTANIK* shows the discrepancy of value creation between soil, which is fundamental to survival and new financial systems like Bitcoin.

Second, the case study of Urbaniahoeve's Food Forest ecosystem topsoil is discussed, which shows that it is possible to create high-quality topsoil in just a few years' time. New conceptual frameworks of soil ecosystems, consider both the soil's development *in situ* as well as the temporality of reciprocity and care. Adopting methodologies of observation that focus on the (soil) ecosystem as a whole bring into focus the activities and metabolisms of nonhuman members of the soil ecosystem community and produce the necessary feedback to summon sustained observance and nurturing.

Finally the effects of urbanization itself, such as a high level of availability of carbon resources, a lack of soil disturbance in urban greens, and a culture of increased knowledge-sharing practices, offer a unique opportunity that could allow urban soils to flourish and perform necessary ecosystem services. Urban soils are the low-hanging fruit of climate adaptation and food commons production when they are included in spatial planning policy and land-use legislation. Further research needs to investigate how intentional formats and typologies such as those shown here may be scaled-up to provide a means of water and carbon sequestration, food provision, and biodiversity, so as to make urban soil more resilient in order to significantly contribute to climate adaptation needs.

Endnotes

1. *Urban Europe: Statistics on Cities, Towns and Suburbs.* 2016 edition. Luxembourg: European Union, 2016.

2. Borgdorff, Henk. *The Conflict of the Faculties.* Leiden University Press, 2012.

3. Foreword by Manuel Castells in *Witnessing You: On Trust and Truth in a Networked World* by Caroline Nevejan. Participatory Systems Initiative, Delft University of Technology, 2012. www.being-here.net. p. 10.

4. The *Entropical* artistic research project was generously supported by Amsterdam Borough South Art and Culture, the Amsterdam Foundation for the Arts (AFK), and the Stokroos Foundation.

5. Zone2Source is an international platform that invites artists to develop projects inside and outside of the glass pavilions of the Amstelpark, in which alternative practices and experiences of our "natural" environment are being proposed. http://zone2source.net/en/home-2/, last accessed November 24, 2017.

6. Roio, D. Algorithmic Sovereignty (Doctoral Thesis). University of Plymouth, UK, 2018.

7. Solomon, Debra. *En Necromasse.* 2015. 5× screen prints on paper, 100 × 70 cm. *En Necromasse* is the result of the Centre for Contemporary Art and the Natural World's Soil Culture residency at Schumacher College in Dartington, United Kingdom, in 2015.

8. Solomon, Debra, and Jaromil. *Seven Layers.* 2015. Screen print on paper, 100 × 70.

9. Solomon, Debra. *Resist Exist.* Resisteresisteresistere 2015. Typography on windows, dimensions variable.

10. Solomon, Debra, and Jaromil. *REALBOTANIK.* Dec 2015–Feb 2016. Installation with Bitcoin miner, mycelium growing on screen printed waste cardboard, dimensions variable.

11. Roio, Algorithmic Sovereignty.

12. The term "Realpolitik" was coined by Ludwig von Rochau, a German writer and politician in the nineteenth century. Realpolitik appears and is described in his *Grundsätze der Realpolitik angewendet auf die staatlichen Zustände Deutschlands* (1853) to mean: "The study of the forces that shape, maintain and alter the state is the basis of all political insight and leads to the understanding that the law of power governs the world of states just as the law of gravity governs the physical world."

13. Swyngedouw, Erik. "Circulations and Metabolisms: (Hybrid) Natures and (Cyborg) Cities." *Science as Culture*, June 2006. https://doi.org/10.1080/09505430600707970.

14. European Environment Agency, *SOER 2010: The European Environment: State and Outlook 2010*. Soil, p. 14–15. Soil sealing happens when agricultural, forest, or other rural land is taken into the built environment. Sealing also occurs within existing urban areas through construction on residual inner-city green zones.

15. Ibid., Landuse, pp. 16, 24, and Soil, pp. 14–15, 23.

16. Ibid., Landuse, p. 16. "In 2000–2006 about 1000 km² of land was covered every year by artificial surfaces. Land take for urban area and infrastructure use increased between 1990 and 2000 by 5.7% across Europe, but with unequal distribution. This trend accelerated during 2000–2006—annual land take increased from 0.57% for 1990–2000 to 0.61% for 2000–2006

17. Ibid., Landuse. p. 28.

18. Ibid., pp. 16, 18, 20, 21, 24, 25.

19. Vasenev, Viacheslav, and Yakov Kuzyakov. "Urban Soils as Hotspots of Anthropogenic Carbon Accumulation," 2017. http://www.fao.org/3/a-bs019e.pdf, accessed September 15, 2017.

20. Lijing, Liu, and Paul Römkens. "Impact of Mulching in a No-Till System on Soil Quality from DemoGarden developed and maintained by Urbaniahoeve." Internship report, soil analysis commissioned by Urbaniahoeve. Wageningen University, June 2015.

21. Groeningen, Jan Willem van, Chris van Kessel, Bruce A. Hungate, Oene Oenema, David A. Powlson, and Kees Jan van Groeningen. "Sequestering Soil Organic Carbon: A Nitrogen Dilemma." *Environmental Science and Technologies: Viewpoint* 51 (April 20, 2017): 4738–4739. https://doi.org/DOI:10.1021/acs.est.7b01427.

22. FAO, Soil, Status of the World's Soil Resources.

23. Groeningen, "Sequestering Soil Organic Carbon."

24. World Resources Institute, "Ecosystems and Human Well-Being: Millennium Ecosystem Assessment." 2005.

25. Lal, Rattan. "Societal Value of Soil Carbon." *Journal of Soil and Water Conservation* 69, no. 6 (November/December 2014): 186A–192A. https://doi.org/doi:10.2489/jswc.69.6.186A. "Indeed, major concerns of the modern civilization, especially peace and tranquility (Lal 2014), are intricately connected with soil and its quality, sustainable intensification of agriculture, and climate-resilient farming through recarbonization of soil and the terrestrial biosphere."

26. Ibid., p. vi.

27. UMR ECOSYS, AgroParisTech-Institut National de la Recherche Agronomique.

28. Baveye, Phillippe, Jacques Baveye, and John Gowdy. "Soil 'Ecosystem' Services and Natural Capital: Critical Appraisal of Research on Uncertain Ground." *Frontiers in Environmental Science / Soil Processes* 4 (June 7, 2016):41. https://doi.org/doi:10.3389/fenvs.2016.00041.

29. Ibid.

30. Ingham, Elaine, Andrew Moldenke R., and Clive Edwards. "Soil Biology Primer [Online]." USDA, 2000. http://www.nrcs.usda.gov/wps/portal/nrcs/main/soils/health/biology/.

31. Puig de la Bellacasa, Maria. "Making Time for Soil: Technoscientific Futurity and the Pace of Care." *Social Studies of Science* 45, no. 5 (2015): 691–716. https://doi.org/10.1177/0306312715599851.

32. Ingham et al., "Soil Biology Primer."

33. Baveye et al., "Soil 'Ecosystem' Services and Natural Capital."

34. Forsyth, Tim. "Ecological Functions and Functionings: Towards a Senian Analysis of Ecosystem Services." *Development and Change* 46, no. 2 (2015): 225–246. https://doi.org/10.1111/dech.12154.

35. Solomon, Debra, and URBANIAHOEVE. *Topsoil of the Urbaniahoeve Food Forest, DemoTuinNoord.* 2011–ongoing. Land art, 1200 m^2 × 60 cm.

36. FAO, Soil, Status of the World's Soil Resources.

37. Römkens, Paul, and Debra Solomon. "To Mulch or not to mulch: betekenis van stadslandbouw voor bodemkwaliteit /the significance of urban agriculture for soil quality." *Eetbaar Rotterdam* (Dutch language blog). www.eetbaarrotterdam.nl/2016/10/to-mulch-or-not-to-mulch-betekenis-van-stadslandbouw-voor-bodemkwaliteit/. Accessed January 1, 2017.

38. Urbaniahoeve Foodscpe Schilderswijk (2010) the Hague, DemoTuinNoord FoodForest (2011) Amsterdam Noord, Foodscape Wildeman (2013) Amsterdam Nieuw West.

39. Jacke, Dave, and Eric Toensmeier. *Edible Forest Gardens: Ecological Vision and Theory for Temperate Climate Permaculture.* Vol. 1. Chelsea Green Publishing, 2005.

40. Römkens and Solomon, "To Mulch or not to mulch."

41. Ibid.

42. A term used in soil science to describe the mineral nature of the "original" material.

43. Solomon, Debra. *Soil Portraits|#26.* 2016–ongoing. Soil chromatogram of self-grown soil 580 × 580 mm.

44. Pfeiffer, Ehrenfried. "Chromatography Applied to Quality Testing." Bio-Dynamic Literature, 1984.

45. Ruben Borge from RockinSoils.com taught Solomon how to do soil chromatography and was invaluable in helping her innovate a technique to increase the size of the chromatogram, affording a 16 × magnification.

SOIL ASSEMBLY & DISSEMINATION AUTHORITY
2115–16 Employee Handbook

The Soil Assembly and Dissemination Authority (SADA)

A Thought Experiment in Building Tomorrow's Soils Today

Ellie Irons, in collaboration with Jean Louis Morel

Born in rural Northern California, **Ellie Irons** went to Scripps College in Los Angeles, where she studied art and environmental science. After falling in love with biology fieldwork, she began combining ecology with art. She relocated to New York City in 2005, and completed her MFA at Hunter College, City University of New York, in 2009. She joined the electronic arts PhD program at Rensselaer Polytechnic Institute in September 2017. Irons works in a variety of media, from video to workshops to gardening, to reveal how human and nonhuman lives intertwine with other earth systems. She is cofounder of the Next Epoch Seed Library and the Environmental Performance Agency, and a contributor to the Chance Ecologies project.

Irons regularly exhibits her work at nationally and internationally renowned institutions, including the Queens Museum, Flushing, Queens; Pratt Manhattan Gallery, Manhattan, New York; Goldfinch Gallery, Chicago, Illinois; Rixc Center for New Media Culture, Riga, Latvia; Bamboo Curtain Studios, Taipei, Taiwan; and Flora Ars + Natura Cabinet Program, Bogotá, Colombia.

Jean Louis Morel is professor of environmental biology at the University of Lorraine, where he teaches soil and environmental sciences. His research interests are dynamics of pollutants (metals and hydrocarbons) in soil-plant systems, evolution of soils strongly affected by human activities (urban soils), and applications for soil remediation (phytoremediation, agromining, soil restoration). He created, and led until 2012, the Laboratoire Sols et Environnement, a research group of forty persons of INRA and the University of Lorraine. He leads the GISFI, a scientific consortium of ten research groups devoted to the understanding of the functioning of brownfields and the development of processes for soil remediation. Morel chaired the international SUITMA group (Soils in Urban, Industrial, Traffic, Mining and Military Areas) of the International Union for Soil Science.

The collaboration, which was realized over e-mail and telephone, began with the realization that both of Ellie Irons and Jean Louis Morel were already dedicated to thinking critically about urban and anthropogenically altered soils. We discussed Morel's research (he shared several of his slide presentations) and his interest in storytelling as a form for conveying scientific knowledge. Irons was struck by Morel's assertion of the singular importance of soil as a resource, and mused on what it might be like if soil was given more prominence as a public good. We discussed several possible forms to explore this scenario (like a comic book, short story, drawing series) but decided to expand the psuedo-institutional form Irons had already employed in the Urban Soil Appreciation Initiative. This seemed to provide an opportunity to incorporate and build on the work Morel was already doing with SUITMA and provided continuity with the rest of Irons' artistic practice. After looking at the methodologies and terminology employed in Morel's research, Irons did a lot of sketching and drafting and shared various possibilities with Morel, who helped select the final terminology and made suggestions as to how to connect these imagined future scenarios to his current research. As a thought experiment in the potential role of soil in human population centers of the future, Irons and Morel present The Soil Assembly and Dissemination Authority (SADA). This hypothetical city agency, established in the second half of twenty-first century, is responsible for research, production, distribution, and outreach related to an essential (and no longer "naturally" available) resource: soil.

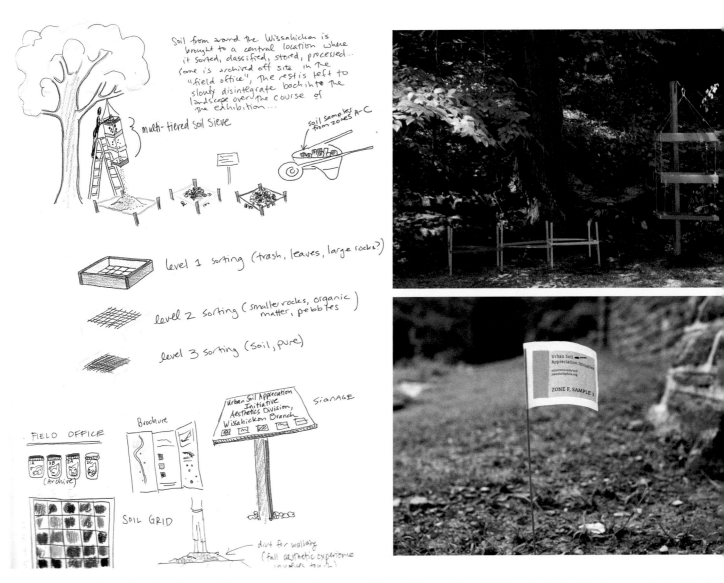

Soil from around the Wissahickon is brought to a central location where it sorted, classified, stored, processed... Some is archived off site in the "field office", the rest is left to slowly disintegrate back into the landscape over the course of the exhibition...

multi-tiered soil sieve

soil samples from zones A-C

level 1 sorting (trash, leaves, large rocks?)

level 2 sorting (smaller rocks, organic matter, pebbles)

level 3 sorting (soil, pure)

FIELD OFFICE

A 3B 2A 2C (Archive)

SOIL GRID

Brochure

Urban Soil Appreciation Initiative Aesthetics Division, Wissahickon Branch

SIGNAGE

dirt for walking (full aesthetic experience involves touch)

Proposal sketch and resulting public sculptural installation for *Urban Soil Appreciation Initiative (Aesthetics Division)*, a project realized as part of *New Trails*, an art festival held in Wissahickon Valley Park in Philadelphia in fall 2011. This iteration of the project consisted of a soil-sorting sculpture installed in the park and a temporary field office sited in empty retail space along Germantown Avenue.

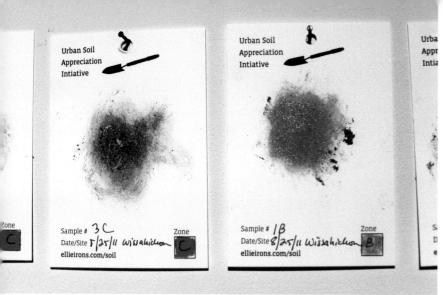

Sample # 3C
Date/Site 5/25/11 Wissahickon
ellieirons.com/soil
Zone C

Sample # 1β
Date/Site 8/25/11 Wissahickon
ellieirons.com/soil
Zone B

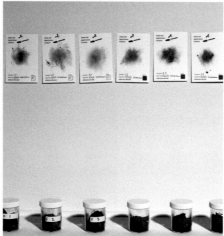

The field office portion of *Urban Soil Appreciation Initiative (Aesthetics Division)* included soil samples taken from throughout the park, demonstrating the diversity of urban soils in the area. In many cities in the United States at this time, soils found in cities were only categorized as "urban" without further, more detailed classifications.

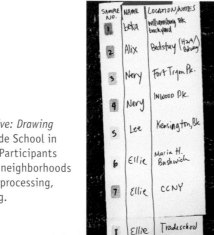

Urban Soil Appreciation Initiative: Drawing with Dirt workshop held at Trade School in New York City in spring 2012. Participants brought "wild" soil from their neighborhoods to a Manhattan storefront for processing, contemplation, and art-making.

Created initially to avert a biosphere-threatening crisis in soil availability in a postindustrial city in the American Midwest, the Soil Assembly and Dissemination Authority (SADA) has grown from a mayor's special initiative to a monolithic state institution. It manages many aspects of urban life, from real estate development to infrastructure to food production to waste management. Documents, memos, fliers, and other ephemera associated with SADA's 2115–2116 handbook are presented here, delivering a glimpse of a possible future scenario for urban soils, set in a landscape of humanly wrought environmental instability countered by continuing technological innovation. These documents, which build on contemporary research to speculate about the future of soil management, also reflect humanity's ongoing struggle to balance human consumption with the functioning of other earth systems.

As the result of a series of exchanges between a visual artist and a soil scientist, the SADA documents convey the melding of two our specialties into a new form: *a speculative bureaucracy for a hypothetical future.* While the results run the gamut from the playful to the mundane, the premise is dead serious. Through our treatment of soils today we create the anthropogenic paleosols of the future, and from many vantage points, those soils don't look too promising. The ability of future communities to feed their human populations will be based in part on how societies interact with soils today, and there is more and more evidence that human interventions are affecting them in dramatic ways.[1]

Jean Louis Morel's long-running research into Soils of Urban, Industrial, Traffic and Mining Areas (SUITMA) has provided an excellent foundation for this collaboration.[2] His insights into anthropogenically disturbed soils have been combined with my (Ellie Irons) work under the auspices of *The Urban Soil Appreciation Initiative*, a public-facing art project that involves seeking out and interacting with urban soils as a form of community building and consciousness raising. In constructing SADA, we have imagined "urban soil appreciation" in an extreme form: city-mandated engagement of all citizens in producing, protecting. and interacting with soils, both "natural" fossil-soils and human-produced artificial soils.

For both Morel and myself, the local habitat we experience on a daily basis has been an integral aspect in the development of our professional interests. Morel describes the evolution of his work with anthropogenic soils as closely tied to landscape around his research base in the Lorraine region of Northeastern France. This area has undergone rapid deindustrialization in recent decades, including the closing of mines

The journey from forest to parking lot, Broadway and Dekalb Avenue, Brooklyn, New York, April 2015–May 2016. This triptych is drawn from Ellie Irons' ongoing *Feral Landscape Typologies* series, which traces the structures and forms of open earth and feral land in Brooklyn, New York. Here, a so-called "vacant" lot with exposed soil and plant matter transitions to a concrete-covered surface.

and steel factories, revealing a highly degraded landscape full of poorly functioning soils.

Like any science, soil science is influenced by the environment where it is conducted. In areas which have been deeply affected by major economic and industrial changes (like in my home, the Lorraine region) research has been oriented towards brownfields (land upon which expansion, redevelopment, or reuse is complicated by the presence of hazardous substances, pollutants, or contaminants, often related to industrial history). Organizations like the French Scientific Interest Group on Industrial Wastelands (GISFI) have developed processes for working with these sites.

Polluted and degraded land requires remediation and restoration, which must be developed through appropriate soil research. To gain a new value from restored land, new ecosystems need to be built from the ground up. Effective soil functioning is an indispensable aspect of this process. Thus, the future is in soil engineering: the ability of human beings to create new soils with artificial materials. These soils must have the same functions and supply the same services as natural soils, but do so whenever and wherever people ask. The future of soils is in our minds and hands, and in that soil lies the future of our cities.[3]

Just as the deindustrialization of Lorraine has impacted Morel's research, the construction boom I've witnessed in New York City over the past ten years has influenced my art practice.[4] As vacant lots and

abandoned shorelines have become condominium towers and parking lots, I've become highly attuned to the bits of wilderness that emerge in the gaps. My artistic practice now focuses on how ecosystems function in highly urbanized landscapes, from managed waterways to feral animal populations. Most recently I've been exploring the spread of certain wild urban plant species in tandem with dense human populations. This of course involves contemplating the urban soil these plants grow in as well.

When I learned of Morel's commitment to seeing degraded soil revived to full health through the interventions of humans, my interest was piqued. Interrogating human entanglement with ecological systems drives much of my work, and I'm also drawn to research that occurs around boundaries where multiple disciplines blend into one another. Using the SUITMA concept to look at my city through the lens of urban soil science gives me license to think both about deep time and day-to-day choices, about the paleopedological record we've left in the soil over millennia and the challenges that highly anthropogenic soil will leave for those who encounter it in the future, human or otherwise.

From the average New Yorker's perspective, the connection between soil science and city life may not be immediately obvious. As part of The Urban Soil Appreciation Initiative, I've asked people participating in a workshop to bring me soil from their city block, preferably not from a planter or street tree pit. This is not always the simplest task. Finding soil in some New York City neighborhoods involves hunting for it, and reaching under fences or turning up paving stones can yield strange results. Soil is found, but with it comes a host of novel parent materials, from shreds of plastic and bits of safety glass to construction rubble and Styrofoam.

As Morel's research underscores, soil exists in a variety of situations in highly urbanized spaces, but much of it is compacted, coarse, full of anthropogenic artifacts, and it is often sealed beneath layers of impermeable surfaces.[5] The recently completed New York City Reconnaissance Soil Survey demonstrates as much: Beneath the sidewalks, parking lots, and buildings of my Brooklyn neighborhood there is a rich variety of soil types and structures, but all of this diversity is often out of view and inaccessible to the average citizen.[6] The places soil is revealed are generally designated as places of leisure (parks, ball fields) or blighted eyesores (abandoned infrastructure, vacant lots). Our SADA thought experiment recasts soil as an essential player in urban life and envisions a paradigm shift in which infrastructure, housing, transportation and access to food all rely on anthropogenically produced soils, while precious fossil-soils are interred in museum and university collections. This situation comes with its own bevy of problems, and although life is certainly not perfect under SADA's reign, the production and dissemination of soil is thoroughly embedded in daily life.

While I don't necessarily hope for the fulfillment of the future we've imagined, I am hopeful that research around anthropogenic soils will continue to expand and diversify. As more and more of us live in cities, we will need to find effective ways to restore our abused soils. This may well require large-scale production of soil through increasingly technological means. Combining long-running traditions of soil enrichment and production, from conventional composting to bokashi and biochar, with new methods for remediating and building soils at the state level provides a plausible starting point. For now, I'm rooting for the spread of the practice of cultivating healthy, lively soil in the nooks and crannies of our cities at the individual and community level, with some government support thrown in. Doing this now will not only improve the quality of life in the current moment, but may also allow for lower-impact, more ecologically secure consumption, and production patterns in the future. In short, let's do this work today, instead of waiting for SADA to force us into it tomorrow!

Endnotes

1. Lucas Reusser, Paul Bierman, and Dylan Rood. "Quantifying human impacts on rates of erosion and sediment transport at a landscape scale," *Geology*, G36272.1, first published on January 7, 2015.

2. See, for example, Nicholas Dickinson, Jean-Louis Morel, Richard K. Shaw, Gerd Wessolek, "IUSS SUITMA 6 International Symposium 2011," *Journal of Soils and Sediments* 13, no. 3 (2013): 489–490.

3. Jean Louis Morel, in an e-mail to Ellie Irons, February 24, 2015.

4. Irina Vinnitskaya, Bloomberg's Legacy: "The construction boom of NYC," July 27, 2013. *ArchDaily*. Accessed Mar 01, 2015. <http://www.archdaily.com/?p=405955>

5. Jean Louis Morel, Claire Chenu, Klaus Lorenz. "Ecosystem services provided by soils of urban, industrial, traffic, mining and military areas (SUITMAs)," presentation, annual meeting ASA-CSAA-SSSA, Long Beach, California, November 2014.

6. See New York City Soil Survey Staff. 2005. *New York City Reconnaissance Soil Survey*. United States Department of Agriculture, Natural Resources Conservation Service, Staten Island, New York.

2115 SADA Residential Pickup Schedule, ZONES 100-115

Pickup Schedule	Jan				Feb				March				April				May				June				July				Aug				Sept				Oct				Nov				Dec			
	5	15	22	30	5	15	20	28	5	14	19	29	7	16	21	28	5	15	22	30	5	15	25	30	5	15	20	27	6	16	22	31	5	14	21	29	5	15	23	31	5	15	22	30	5	15	22	31
A																																																
B																																																
C																																																
D																																																
E																																																

SCHEDULE A	SCHEDULE B	SCHEDULE C	SCHEDULE D	SCHEDULE E
Noble materials (safety category 1*)	*Anthropogenic materials (safety category 2*)*	*Hazardous materials (safety category 3*)*	*Inert secondary resources (safety category 1A*)*	*Conglomerates (safety categories 1A, 2)*
including organic food waste, residential agricultural by-products, uncontaminated living fabrics, surfaces and substrates	synthetic food waste, mechanized fabrics, surfaces and substrates, inert nonhuman animal byproducts (urban/domestic), phytoplastics	vintage plastics, neo-plastics**, silicone derivatives, operational genetic products, perscription medications	metals, hybrid metals, rare earths, glass and glass hybrids***	unsorted assemblages of anthropogenic and secondary resources

*See SADA 2115-2116 residential guidelines for a complete list of materials accepted for residential pickup in each safety category
**Zones 110-112 may contribute neo-plastics under a new pilot program. Information will be distributed. Visit your Residential Soil Management Office for details.
***10% increase in compensation for properly deconstructed and sorted inert secondaries as of 2/15/15

SADA
Do your part for the SADA Civilian Soil Effort

Excerpts from the 2115–16 SADA Employee Handbook. Further pages are available at ellieirons.com/projects/SADA.

SADA Standard Soil Profiles 2115

SADA Outreach Teams: Know Your Soils!

This pocket guide provides a breakdown of SADA's 2115 Soil Products, including technical information on parent materials, assembly procedures, practical applications, safety concerns, likely economic and cultural impacts, and exemplary projects. For a guide to terminology and acronyms, see the recently updated ICOMANTH glossary (International Committee for Anthropogenic Soils).

BASIC SOIL CATEGORIES

Corporate and Commercial Grades
- Infrastructure
- Built Environment
- Agricultural

Municipal Grades 1, 2, and 3
- Government and state projects
- State-sponsored defense and land reclamation bids

Domestic Grades 1, 2, and 3
- Consumer use
- Small scale commericial use

Artisinal
- PureSoil™ collaboration pilot project
- small batch vintage and synthetic fossil soils

Corporate & Commercial

1. INFRASTRUCTURE

Quick Summary Points:

- Derived largely from SUITMA Legacy Soils (urban, industrial, traffic, mining and military areas)
- Heavy duty, dense (aeration mechanisms must be installed in situ for optimal biotic activity)
- High water flow and regulation capacity
- Ideal base layer for shallow growth schemes
- Please note: *Variant A includes technosoils derived from decantation ponds and mine tailings of 21st Century origin (Andic, Calcaric, Hydric, Laxic, Thixotropic, Toxic) and must be installed at depths of at least 2 meters and permanently capped with Municipal and/or Domestic Grade soils*

Parent Material Breakdown:

Percentage	Material Type
10–15%	21st C Indust. Artifact Class A*
40–55%	SADA Synthetic Technosoil 1 or 2
25–35%	20th C Infrastructure Artifact Class B*
5%	SADA Aerating Compound

*Only 21st Century Industrial and Infrastructure Artifacts mined and processed by SADA and approved under the Joint ICOMANTH protocol for enduring use of non-fossil soil substrates are incorporated.

Corporate & Commercial

KEY EXAMPLES: INFRASTRUCTURE

Pont de la Terre, Zone 6

Designed by Jesula Alphonse of City Arc for the annual Water Projects Design Competition, this spectacular earthen bridge will span the Zone 6 reservoir, connecting the zone's residential bunkers more directly with the commercial district in Zone 7 and replacing the defunct water taxi service.

Currently under construction, it is made primarily of second generation SADA Infrastructure Soils with a skin of Grade A Agricultural facing which will be farmed by the Greenthumb Growers Association after a sun and windscreen for garden plots are in place (planned as part of phase 2). The common patios in Zone 6 residential bunkers on the north and east side of the reservoir provide excellent viewing opportunities. The bridge is expected to open to pedestrian traffic in August 2116.

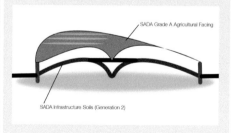

SADA Grade A Agricultural Facing

SADA Infrastructure Soils (Generation 2)

Excerpts from the 2115–16 SADA Employee Handbook. Further pages are available at ellieirons.com/projects/SADA.

Hybrid Landscapes

Ideas about Soil Chemistry and Urban Design from the United States and South China

Seth Denizen and Qiu Rongliang

Seth Denizen is a researcher and design practitioner trained in landscape architecture and evolutionary biology. He has received a number of design awards, including first prize in the ARCH + Bauhaus Dessau Foundation Out of Balance Competition 2013, and has published texts on art and design with the Asia Art Archive, LEAP International Art Magazine of Contemporary China, Volume, and Fulcrum, among others. He continues to sit on the editorial board of *Scapegoat Journal: Architecture/Landscape/Political Economy*. After three years of teaching architecture at the University of Hong Kong, he has moved to the University

Title image: Demonstration base for remediation of heavy metal–contaminated soil. Guangdong Provincial Key Laboratory of Environmental Pollution Control and Remediation Technology and SUN YAT-SEN University.

Reprinted with permission from Qiu Rongliang.

of California Berkeley where he is currently completing a PhD in geography. His doctoral research investigates the vertical geopolitics of urban soil in Mexico City, where he is working with geologists and systems ecologists to characterize the material complexities and political forces that shape the distribution of geological risk in the urban periphery.

Qiu Rongliang is distinguished scholar and dean of the School of Environmental Science and Engineering, at Sun Yat-sen (Zhongshan) University in Guangzhou. Since 1992, he was responsible for the teaching of "Soil Environmental Science," the course that was highly evaluated and was named as "Excellent Important Course" in 2006 by the university. Also, he has been involved in teaching several other subjects and has played an ongoing role in planning and development of teaching in the environmental chemistry area. He is on the editorial boards of the *Chinese Journal of Applied Ecology, Chinese Journal of Environmental Science, Journal of Soils and Sediments*, and *Journal of Agro-Environment Science*. His research interests include phytoremediation of heavy metal–contaminated soil; organic pollutants in water and soil environments; and soils of urban, industrial, and mining areas.

In the following dialogue, Qiu Rongliang and Seth Denizen sat down in the School of Environmental Science and Engineering at Sun Yat-Sen University to discuss the relationship between urban planning and soil science in China's industrial south. The two discussed the present and future challenges of soil taxonomy in urban settings, from New York City to Guangzhou, including new methods of integrating biochar technologies and wastewater management into urban planning schemes, rice production in industrial areas, and the interface between local politics, economics, and geology. Qiu discussed his recent work on the soil chemistry of cadmium availability in several different rice species, and its relation to China's hybrid urban-rural landscapes. Denizen discussed this work in relation to his recent project, *The Eighth Approximation: Urban Soil in the Anthropocene*, in reference to unfinished versions of soil taxonomy in the past, which used to be called approximations. Denizen's vision includes digital renderings of "gentrified regolith/citified soils," textural classes for bricks and concrete, the geomorphology of construction debris, street soils, and dredged landfills. The dialogue concluded with a discussion of what a merged soil science and urban design curriculum might look like, and a consensus that more integrated approaches between the disciplines are needed.

Seth Denizen: When I started learning about the history of soil taxonomy in the United States one of the fascinating stories to emerge was the Cold War politics that shaped the taxonomic discussions of the postwar period. The United States had largely inherited its approach to soil taxonomy from the Russians in the early part of the twentieth century, and into the 1950s some of the most influential architects of soil taxonomy in the United States, like Curtis Marbut and Hans Jenny, were avid readers of Russian soil science. But at a certain point it became extremely unfashionable to refer to the work of Russian scientists in the United States, and ultimately a very different system of soil classification emerged. I've always been curious to know where China stood in relation to these two systems. How did soil taxonomy develop in China given this polarization?

Qiu Rongliang: The first taxonomic system we used in China was mostly the Russian system, but over the course of the twentieth century the Chinese system has incorporated more of the American approach. Now there is a very easy way to convert a Chinese classification into the analogous American classification. You know, the Russians were very influential in China and in the 1950s there was a serious attempt by the Russian soil scientists to establish large rubber plantations in China based on the soil genetic classification system. Since 1980, U.S. soil taxonomy has had a large effect on the Chinese Soil Taxonomy, which was finally published in 1999. Now these two systems are used simultaneously in China. But farmers in China have always used the soil names from the genetic system, which are easy to understand, such as red soil, Latosolic soil, Latosol, yellow soil. Even the soil taxonomy with diagnostic layers sounds more scientifically reasonable.

Seth Denizen: The soil knowledge of farmers is always pretty sophisticated in its specificity to the soils that they work. The major international systems of soil taxonomy are always trying to get to the level of specificity of farmers, even if this knowledge is not consolidated in the same way. It's funny, in the American system they started out calling their soil taxonomies "approximations," because there was this recognition that the project was unfinished, or maybe even that the project was unfinishable. But after the "7th Approximation," they dropped this terminology and now it's just called "Soil Taxonomy," which gives the impression that it's done, that we have finally arrived at something complete. Now that human-made, anthropogenic soils are being thrown into the mix, quite literally, it seems like these major international taxonomies are less finished than they've ever been. Something like a layer of rubble sediment in the subsoil can't just be appended to a preexisting soil series in the taxonomy, like an extra topping on an ice cream, because different chemicals will interact with different soils differently. So, for instance I don't think there is such a thing as a contaminated inceptisol; there is just some kind of new soil that we haven't studied before. In some sense, soil contamination can be thought of as the production of a soil that has never before been seen, whose properties are specific and need to be studied specifically.

Qiu Rongliang: Yes, my laboratory is doing some of this kind of work. For the last fifteen years, since 2002, we have been looking at the way in which cadmium contamination affects rice production differently depending on the type of soil and the species of rice. Cadmium contamination is a serious problem in South China, mainly due to mining activity, rapid urbanization processes, industry, and the use of fertilizers with relatively high metal contents. It turns out that some species of rice absorb cadmium from the soil and transfer it into the rice grains more than others, and the difference is very significant. The mechanism that prevents this transfer of cadmium to the rice grains has just recently been understood and it is very interesting. Let me describe it briefly. As you know, rice grows in water, and the roots of the rice actually excrete small amounts of oxygen, like the bubbles made by an underwater diver. In iron-rich soils, like the highly weathered red soils we have here in Southern China, this oxygen can combine with the iron. Generally, in water, iron is a reduction species, but with oxygen it can become oxidated. So, the oxygen made at the root surface oxidizes the iron in the soil and starts to form a kind of metal crust around the root. This metal crust can then exclude other heavy metals like cadmium.

Seth Denizen: So, the rice can grow metal roots that protect against cadmium? That sounds like science fiction.

Qiu Rongliang: It depends on the species of rice. Some species produce a lot of oxygen at their roots and some produce less. The species of rice that produce more oxygen at their roots will form this crust at their roots if they are in iron-rich soil. So, producing safe rice grains depends both on the soil and on the plant. It's not science fiction, just soil chemistry! One of the challenges we face in this area is that the really dangerous soil contamination is in agricultural soil where the contamination is not very severe. If the soil contamination is very high, then nothing grows and this prevents the heavy metals from entering into food. The really dangerous areas are where the contamination is not high enough to kill the rice, but high enough to be incorporated into the food.

Seth Denizen: That's a really fascinating example. In one soil the cadmium contaminates the food supply, and in another soil it stays in the ground as a kind of silent participant in the soil profile. If we care about human health, then iron-rich

cadmium-contaminated rice fields must be considered a distinct soil from non-iron-rich cadmium-contaminated rice fields, given the right species of rice. The cadmium in the soil that was deposited by industry can be seen as a kind of deposition like any other process of soil deposition. In this sense both the rice we breed and the metals we deposit in the soil are soil-forming factors that would be significant to a soil taxonomy of Southern China. I wonder to what extent we can think of all urban processes as soil-forming factors? Hans Jenny's famous soil forming factors are climate, organisms, relief, parent material, and time, but what about cities? Are cities the sixth factor? And if we can plan cities, why can't we plan soils? How could soils be planned in South China? Not only is it a very specific form of urbanization in Southern China, but it's a very specific kind of soil, so I think this would have to be a highly local process. Planning a city's soil would need to take into account the processes of urbanization that are specific to its economy, its politics, and also its geology. So we wouldn't be able to export one political or economic system into the wrong geology, and we can't apply one kind of geological knowledge to the wrong economy. So maybe we could imagine your project with rice as a very specific kind of project—a local project—to try to understand a postindustrial rice-soil, so essential to Guangzhou.

Qiu Rongliang: Well, I think soil science will be very important to this region in the coming decades, and we need to work together with scientists from different fields like urban planning. We need to consider not only urban policies and economies but we also need to combine technologies like biochar and wastewater management. Back to the role of soil functions, we will need to rebuild a lot of sealed soils. These are soils that have been covered with concrete by urban development and then abandoned.

Seth Denizen: Yes, maybe we need to be completely rethinking this. Maybe what we need to say is that the soil scientist of the future is also trained as an urban planner or landscape architect. If we could try to imagine what the ideal soil scientist in Guangzhou would be, maybe she is someone who is able to manage wastewater systems in a region with over 30 million people.

Qiu Rongliang: Unfortunately soil scientists have no voice in the city planning process now. I also think that soil scientists don't care a lot about urban planning. So, from both sides we need to make some changes. I think maybe the first step is to begin to change the way we educate our students, to increase the consciousness of these issues. It's only been a little over thirty years since the Open-Door Policy in China. Our soils have changed a lot in thirty years.

Seth Denizen: What a great idea! What could the curriculum look like for this soil science degree? Semester 1: Soil Chemistry of Urban Development in South China. Semester 2: Soil Forming Factors in South China since the Open-Door Policy. How would we even begin to teach the kind of specific knowledge necessary to act in such a complex region?

Qiu Rongliang: This is a very interesting question. Maybe we should merge several schools. Actually, the city-planning department of our school in Guangzhou is just one floor below the soil science department, but they have a different teaching system. They are just below us, in the same building, but our students, frankly, know nothing about that. It's not easy for our students to know each other in different departments because our specializations are so narrow. We should think about how to change this.

Seth Denizen: I think this region of China could be one of the most exciting places to do this.

What would a soil survey of South China look like? One of the reasons why soil taxonomy really matters, it seems to me, is that you can't make a map of soils unless you have names to put on the map. You have to name the soils to map the soils. If we look at South China, we are dealing with a kind of endless semiurban periphery between Hong Kong and Guangzhou that forms a kind of archipelago of urban development with different intensities of industry and agriculture mixed into the interstices of a dense transportation network with a lot of land and a lot of soil floating between boundaries … if you were going to try and map this region one of the problems that would immediately arise is that there are no names for these soils. What's the name of a periurban, postindustrial soil planted with cadmium-resistant rice at the bottom of an elevated road called?

Qiu Rongliang: Yes, you are right, we have names for the natural soils but not that many for the urban soils. We also must ask, Is it still soil? Maybe not. What kinds of soil count as soil? We don't really know. We still have to complete a soil survey of Southern China. In the past thirty or more years in China we've paid a lot attention to water and air. Now it is the time to focus on soil!

Seth Denizen: I recently did a small project trying to imagine what a system of naming urban soils might look like in New York, although after talking to you I am already regretting that the project didn't focus on Southern China. In New York, the project was about trying to reconnect empirical soil knowledge to the history and politics of the city. So, for instance there is a soil series named for Robert Moses, who demolished huge swaths of Brooklyn to make the Brooklyn-Queens Expressway. Most of the construction debris was just mixed back into the soil, and the new neighborhoods were built on top. So, the soils there have an architectural specificity, in that they are made from the specific materials that were used to build houses and apartment buildings in the early twentieth century in New York. I also consider roads to be soils, which is a little contentious amongst the urban soil taxonomists. During the late 1950s when the U.S. highway system was built there were government standards that specified a precise system of layers, or horizons: asphalt, then the asphalt binder course, the aggregate base and subbase, compacted subgrade, etc. We know that over time this profile will weather into a very specific soil, which I named for the politician responsible for the road network. Basically, I am going back to the Russian system of soil taxonomy that we began discussing, where the process by which the soil is formed, its genesis, is privileged over its specific morphology for taxonomic purposes. The difference is that the Russians, like Vasily Dokuchaev, did this out of a love of geology, whereas my taxonomy does this out of a political commitment to keep empirical knowledge in dialogue with urban history, and in dialogue with its consequences. I think any postclimate change soil science will have to begin to operate this way to be successful.

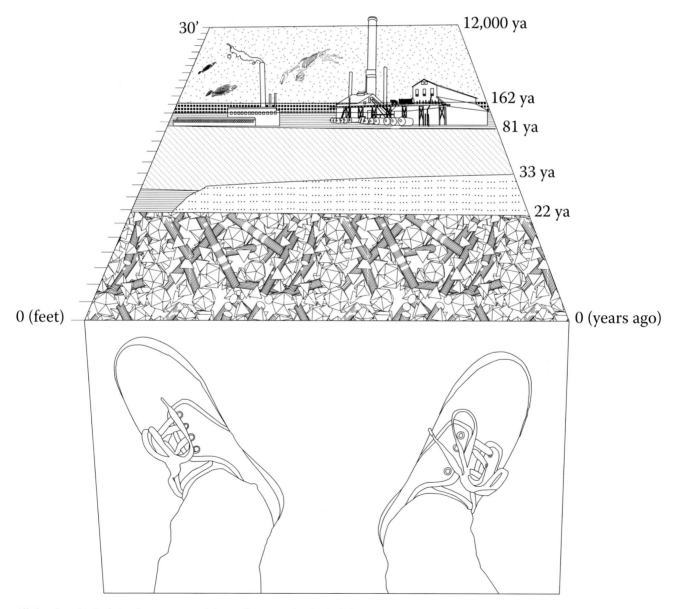

All drawings by Seth Denizen, excerpted from a larger work: The Eighth Approximation: Urban Soil Taxonomy in the Anthropocene.

GENTRIFIED SUBST—TO—GEOMORPHOLOGICALLY DREDGED EMBRY—TO—ANCIENT WONDER WHEEL

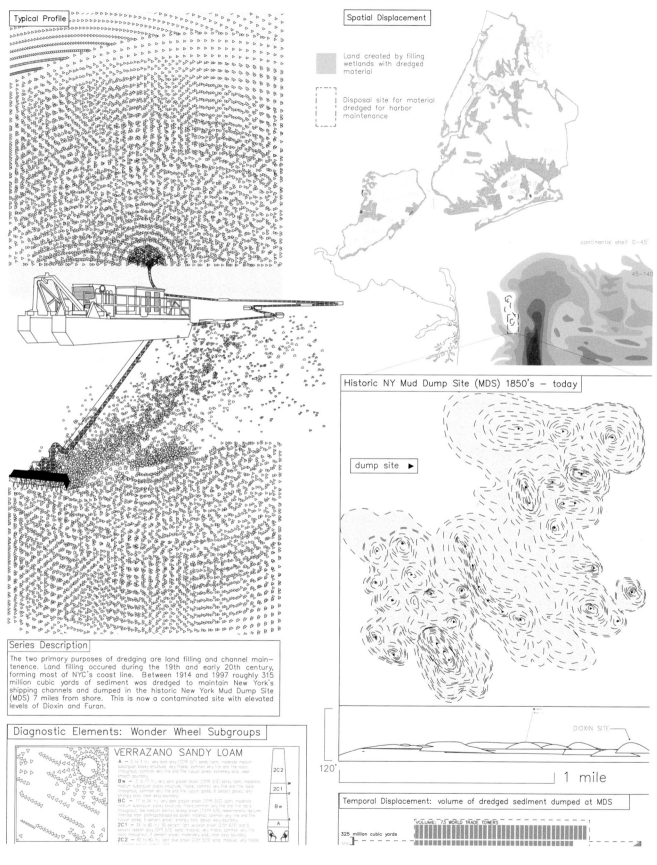

Typical Profile

Spatial Displacement

- Land created by filling wetlands with dredged material
- Disposal site for material dredged for harbor maintenance

continental shelf 0–45'

45–140

Historic NY Mud Dump Site (MDS) 1850's – today

dump site ▶

DIOXIN SITE

120'

1 mile

Series Description

The two primary purposes of dredging are land filling and channel maintenance. Land filling occured during the 19th and early 20th century,
forming most of NYC's coast line. Between 1914 and 1997 roughly 315
million cubic yards of sediment was dredged to maintain New York's
shipping channels and dumped in the historic New York Mud Dump Site
(MDS) 7 miles from shore. This is now a contaminated site with elevated
levels of Dioxin and Furan.

Diagnostic Elements: Wonder Wheel Subgroups

VERRAZANO SANDY LOAM

2C2
2C1
Bw
A

Temporal Displacement: volume of dredged sediment dumped at MDS

VOLUME: 73 WORLD TRADE TOWERS

325 million cubic yards

Profile Key

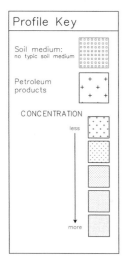

Soil medium:
no typic soil medium

Petroleum
products

CONCENTRATION

less

more

Typical Profile

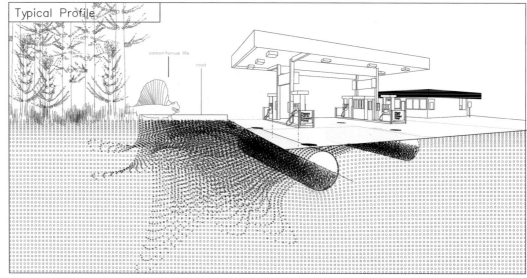

carboniferous life

road

Series Description:

According to the EPA, 35% of underground petroleum storage tanks (UST's) in the United States were leaking in 1986, totalling 1.4 million tanks. Since then the EPA has initiated a program of UST cleanup. The minimum number of petroleum brownfield sites today is 200,000, or roughly half of all brownfield sites recognized by the EPA. Each leaking tank typically results in between 23 and 38 cubic meters (30 to 50 cubic yards) of contaminated soil.

Temporal Displacement — 359-299 million years

359 —299 Million Years Ago   Jurassic   present

The vast majority of the worlds oil reserves were produced during the carbon-iferous. The temporary indigestability of lignin caused vast swamps, which under heat and pressure, became coal, oil, and natural gas.

Textural Polygon

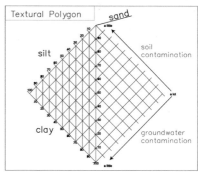

sand

silt

clay

soil contamination

groundwater contamination

Spatial Displacement — Oil production to US refineries

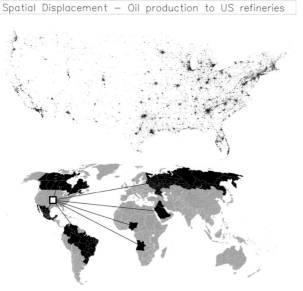

US GAS STATIONS
Exxon, Mobile, Gulf, Flying J

OIL EXTRACTION

Soil Survey:
Frank's Service Station, NJ

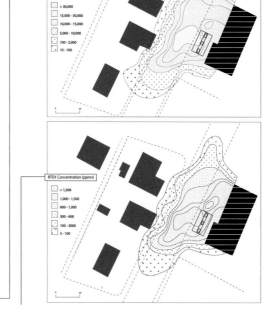

C5+ CONCENTRATIONS (ppmv)

> 30,000
15,000 - 30,000
10,000 - 15,000
5,000 - 10,000
100 - 5,000
10 - 100

BTEX Concentration (ppmv)

> 1,500
1,000 - 1,500
600 - 1,000
300 - 600
100 - 3000
5 - 100

Diagnostic Elements

Soil can be contaminated with petroleum hydrocarbons of three distinct classes:

CYCLOALKANES
Compounds such as cyclopentane, cyclobutane,

H   H

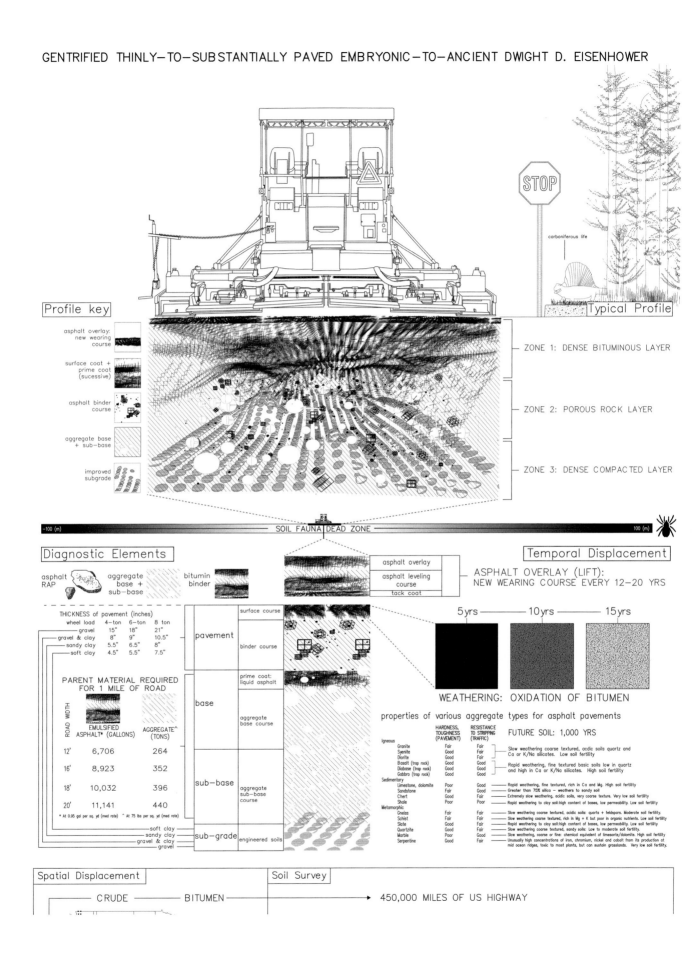

carboniferous life

Typical Profile

Profile key

asphalt overlay:
new wearing
course

surface coat +
prime coat
(sucessive)

asphalt binder
course

aggregate base
+ sub–base

improved
subgrade

— ZONE 1: DENSE BITUMINOUS LAYER

— ZONE 2: POROUS ROCK LAYER

— ZONE 3: DENSE COMPACTED LAYER

−100 (m)   SOIL FAUNA DEAD ZONE   100 (m)

Diagnostic Elements

asphalt RAP   aggregate base + sub–base   bitumin binder

asphalt overlay
asphalt leveling course
tack coat

Temporal Displacement

ASPHALT OVERLAY (LIFT):
NEW WEARING COURSE EVERY 12–20 YRS

5yrs ———— 10yrs ———— 15yrs

WEATHERING: OXIDATION OF BITUMEN

THICKNESS of pavement (inches)

wheel load	4–ton	6–ton	8 ton
gravel	15"	18"	21"
gravel & clay	8"	9"	10.5"
sandy clay	5.5"	6.5"	8"
soft clay	4.5"	5.5"	7.5"

surface course

pavement

binder course

prime coat: liquid asphalt

base

aggregate base course

PARENT MATERIAL REQUIRED
FOR 1 MILE OF ROAD

ROAD WIDTH	EMULSIFIED ASPHALT* (GALLONS)	AGGREGATE^ (TONS)
12'	6,706	264
16'	8,923	352
18'	10,032	396
20'	11,141	440

* At 0.95 gal per sq. yd (med rate)   ^ At 75 lbs per sq. yd (med rate)

sub–base

aggregate sub–base course

soft clay
sandy clay
gravel & clay
gravel

sub–grade   engineered soils

properties of various aggregate types for asphalt pavements

FUTURE SOIL: 1,000 YRS

	HARDNESS, TOUGHNESS (PAVEMENT)	RESISTANCE TO STRIPPING (TRAFFIC)	
Igneous			
Granite	Fair	Fair	Slow weathering coarse textured, acidic soils quartz and Ca or K/Na silicates. Low soil fertility.
Syenite	Fair	Fair	
Diorite	Good	Fair	
Basalt (trap rock)	Good	Good	Rapid weathering, fine textured basic soils low in quartz and high in Ca or K/Na silicates. High soil fertility
Diabase (trap rock)	Good	Good	
Gabbro (trap rock)	Good	Good	
Sedimentary			
Limestone, dolomite	Poor	Good	Rapid weathering, fine textured, rich in Ca and Mg. High soil fertility
Sandstone	Fair	Good	Greater than 75% silica — weathers to sandy soil
Chert	Good	Fair	Extremely slow weathering, acidic soils, very coarse texture. Very low soil fertility
Shale	Poor	Poor	Rapid weathering to clay soil:high content of bases, low permeability. Low soil fertility
Metamorphic			
Gneiss	Fair	Fair	Slow weathering coarse textured, acidic soils: quartz + feldspars. Moderate soil fertility.
Schist	Fair	Fair	Slow weathering coarse textured, rich in Mg + K but poor in organic nutrients. Low soil fertility
Slate	Good	Fair	Rapid weathering to clay soil:high content of bases, low permeability. Low soil fertility
Quartzite	Good	Fair	Slow weathering coarse textured, sandy soils: Low to moderate soil fertility.
Marble	Poor	Good	Slow weathering, coarse or fine: chemical equivalent of limesonte/dolomite. High soil fertility.
Serpentine	Good	Fair	Unusually high concentrations of iron, chromium, nickel and cobalt from its production at mid ocean ridges, toxic to most plants, but can sustain grasslands. Very low soil fertility.

Spatial Displacement

—— CRUDE —— BITUMEN ——

Soil Survey

→ 450,000 MILES OF US HIGHWAY

The City as Forest

Cont

Cartographic Reflections on Land Use in Brazil

Paulo Tavares and Antonio Guerra in conversation with Alexandra R. Toland

Antonio Guerra is a Brazilian geographer, with a PhD from King's College London and post-doc from University of Oxford and University of Wolverhampton. He is a professor in physical geography, at Federal University of Rio de Janeiro, where he coordinates LAGESOLOS–Laboratory of Environmental Geomorphology and Soils Degradation. He has written several books and articles on geomorphology and soils degradation, and now he is looking at geodiversity, geotourism, and geoconservation, relating art and science. Guerra is also a senior researcher with CNPq (Brazilian Research Council).

Paulo Tavares is a Brazilian architect and urbanist based in Quito and London. His work is concerned with the relations between conflict and space as they intersect within the

Title image: Page from a newspaper report about the "pacification" of the Xavante published in the magazine *O Cruzeiro* between 1946 and 1949.

Image courtesy of Paulo Tavares/autonoma and Bö'u Xavante Association.

multiscalar arrangements of cities, territories, and ecologies. Tavares teaches design studio and spatial theory at the School of Architecture, Design and Arts of the Pontificia Universidad Católica del Ecuador in Quito, and previously held teaching posts at the Centre for Research Architecture–Goldsmiths, and at the Visual Lab of the MA in Contemporary Art Theory, also at Goldsmiths, United Kingdom. His work has been exhibited at the CCA: Centre for Contemporary Arts, Glasgow; Haus der Kulturen der Welt, Berlin; Portikus, Frankfurt; ZKM Center for Art and Media; PROA, Buenos Aires; and the Taipei Biennial. The project presented in this chapter was conducted in collaboration with Bö'u Xavante Association research team, Dario Tserewhorã, Domingos Tsereõmorãté Hö'awari, Magno Silvestre, Marcelo Abaré, and Policarpo Waire Tserenhorã, with translation and consultancy by Caimi Waiassé and Cosme Rité.

The platform function of the soil has historically stood in contradiction with other soil functions such as food production, water filtration, and climate regulation.[1] As physical foundation for human infrastructure, buildings, roads, dumps, and city skylines, the soil platform stands as antithesis to the natural environment. However, recent research in the Amazon has proven this needn't be so.[2] As Paulo Tavares has argued, Amazonia is a cultural landscape that has been shaped by human intervention for millennia: "indigenous modes of inhabitation, both in the pre-colonial past and in the modern present, not only leave profound marks in the landscape but also play an essential role in shaping the forest ecology … The botanical structure and species composition of the Earth's largest biodiversity refuge is to a great extent a heritage of indigenous design."[3] By envisioning the forest as city, an inhabited space with a long history of infrastructural planning, we might conversely be able to envision the city as a multilayered forest rather than a platform for solely human needs. In a series of e-mails and a Skype call I (Toland) interviewed Tavares about his work to preserve parts of the Amazon forest as architectural heritage. In a parallel interview, Antonio Guerra shared his insight as a geographer and geomorphologist about the present state of soil degradation in Brazil, which is to a very large extent the result of colonial destruction of indigenous forest infrastructures. By investigating the cultural histories of spatial planning practice and governance, a completely different picture of the soil platform function emerges.

Alexandra R. Toland: Antonio, soil erosion in Brazil is currently estimated at around 800 million metric tons a year.[4] What percentage of that erosion is due to deforestation, and what lessons do you think planners and scientists could learn from indigenous design practices to meet the spatial needs of a growing population? Could Brazil's cities of today host food forests of tomorrow?

Antonio Guerra: Brazil is a large country with extreme pedogenetic diversity, which influences the potentialities and limitations of land use. This diversity is enhanced by regional cultural differences in terms of settlement and development. Although it is difficult to calculate exactly what percentage of erosion is due to deforestation, it is evident that vegetation clearance plays a major role in all sorts of land degradation in Brazil, and that soil erosion and land degradation occur both in urban and in rural areas.

Although soil erosion and mass movements are two forms of land degradation, they present different modes of occurrence and consequently different ways of being identified, monitored, and remediated. As humans play a huge role in these geomorphological processes, the best way to avoid both forms of land degradation is to act preventively, that means understanding the risks of soil erosion and/or mass movements and avoiding them. I would like to call attention here to land degradation in Rio de Janeiro State, where soil degradation is mainly due to soil erosion and mass movement, which causes river siltation, one of the main problems of the rivers in the state. Gully erosion can be found in several parts of Rio de Janeiro State, due to the sort of agriculture and cattle grazing without conservation practices that lead to severe land degradation in several parts of the rural areas of the state. These areas are naturally fragile and have low soil fertility. On the other hand, the urban soils, such as in Petrópolis Municipality especially in mountainous areas, are prone to mass movements, causing the deaths of thousands of people in the three last decades.

Planners and scientists could learn many lessons from indigenous design practices to meet the spatial needs of Brazil's growing population. They do not degrade the soil because they implement many soil conservation measures in their land-use practices, such as agroforestry, and preference of polycultures over monocultures. This type of land use is guided by the planting of trees, together with horticulture, maize, cassava, etc. They only clear the vegetation for their current needs,

Gully erosion on an oxisol in pasture lands in Silva Jardim Municipality, Rio de Janeiro State.

Photo by Antonio Soares da Silva.

Landslide scar with a condemned house in Quitandinha District, Petrópolis Municipality, Rio de Janeiro State.

Photo by Antonio Jose Teixeira Guerra.

whereas in commercial monoculture systems, large portions of land are cleared for profit, without considering the dangers of soil erosion, pollution, acidification, and salinization, which has ultimately resulted in the desertification of some parts of Northeastern Brazil.

Could Brazil's cities of today host food forests of tomorrow? Despite the wisdom of indigenous land-use practices, I do not believe Brazil's cities of today could host food forests of tomorrow. I do not think this is a possibility because most Brazilian cities are not planned for this and land degradation, such as mass movements, soil erosion, contamination, floods, river siltation, and other types of degradation can be seen in most Brazilian cities. In addition, there is no space within the city limits to host food production or forestry, which is a shame, because that could be a solution for many environmental problems that we see in many Brazilian cities.

Alexandra R. Toland: Tell me a little more about the environmental problems Brazil is facing today. What is the state of soil degradation and what needs to happen, even if reforesting whole cities is not an option?

Antonio Guerra: As a geomorphologist and soil scientist I see the soil as one of the main aspects for our survival on Earth. I also see the soil as one of the main elements of geomorphological processes acting to produce different land forms. Soil is the result of rock weathering, where climate, together with living creatures and topography, create several types of soils, with different textures and colors. They are essential for the growth of plants and animals. Soils also retain water and release it for the rivers, springs, and lakes. Its quality and amount depend on the way we treat the soils. Therefore, we must look after them with care and to protect the soils as an obligation of the whole society.

Brazil covers 8,547,403 km² and is divided into five regions (Northern, North Eastern, Central Western, South Eastern, and Southern). The diversity of climate, geology, topography, biota, and human activities has contributed to the considerable diversity of soil types and thus soil erosion problems. Erosion and land degradation is a global problem and poses a major problem in Brazil. This hazard affects both urban and rural areas within the extensive national territory. In turn, these problems have serious environmental impacts and pose socioeconomic problems. It is important that the soils are conserved for the present and future generations. Although erosion is a natural phenomenon, human activity accelerates erosion processes. Soil erosion on U.S. agricultural soils causes the loss of an average of 30 tonnes per hectare per year (t ha^{-1}−y^{-1}); some eight times quicker than rates of soil formation. A survey by EMBRAPA (the Brazilian Ministry of Agriculture) suggested the situation in Brazil is often worse. In some places erosion rates exceed 20 t ha^{-1} y^{-1} (EMBRAPA, 2002). For instance, erosion was estimated to average ∼40 t ha^{-1} y^{-1} over 6 million hectares of cultivated land in the State of Rio Grande do Sul. I find this very high, especially because soil formation is a very slow process, and it would take thousands of years for nature to form new soils and to recuperate these affected lands.

The main causes for soil degradation in Brazil differ in urban and rural areas. In the first case, urban growth without planning generates large impermeable surfaces in cities, causing different types of degradation, such as mass movements on hillsides and floods in the low lands. In rural areas, the lack of conservation measures in many parts of Brazil allows for unregulated surface sealing, leading to an increase in runoff,

which can generate high rates of soil erosion and impoverishment of the soils. Gullies and rills can be seen in different parts of Brazil.

If we do not use soils with care, they will become degraded in different and sometimes irrecoverable ways, including erosion, mass movements, salinization, loss of organic matter, sealing, acidification, eutrophication, loss of biodiversity, and desertification. Therefore, scientists have a great responsibility of researching these environments, and at the same time informing farmers and local authorities about how to use and manage the land without causing damage to it.

Alexandra R. Toland: Could you talk about the importance of maps and mapping in your work, in terms of understanding, evaluating, and communicating spatiocultural phenomena. Antonio, you've worked at various scales to aid conservation planning. How does scale affect social structures in terms of decision-making power and local governance?

Antonio Guerra: Cartographic methods are rather significant for environmental surveys. The use of aerial photographs and satellite images enhance the quality of our geomorphological and

Gully erosion on a rural area in Petrópolis Municipality, Rio de Janeiro State.

Photo by Antonio Jose Teixeira Guerra.

Rill bifurcation on a trail in a very compacted soil in Paraty Municipality, Rio de Janeiro State.

Photo by Luana de Almeida Rangel.

pedological studies, since they give more accuracy to what we are looking at. In addition, the use of geoprocessing makes it easier to map areas under different scales, to determine conservation units and spatial planning.

In Brazil, the problem is that for most areas of the North Eastern, Central Western, and Northern Regions the maps are usually at a very low detailed scale (1: 100,000). This lack of accuracy in cartographic detail affects the social structures of decision-making power and local governance. For example, in Ubatuba Municipality, São Paulo State, where the cartographic scale available is 1:50,000, which does not contain all the details needed for a precise map, it was possible to draw a map containing details related to contour lines, soil sample site collection, and the trails assessed for this research work. This work is part of a PhD thesis and gives an example of how maps can be fundamental for good guidance to the local authorities, which would be lost had we not produced this map and accompanying technical report. The civil engineers, geographers, architects, economists, urbanists, and ecologists, who work at the local government and do not have

the expertise to produce this type of work, would not know what to do regarding environmental risks and the rules regarding natural resources use and protection.

Alexandra R. Toland: Two important commonalties in the literature on transdisciplinarity is the positioning of research at the interface of "real-world problems" and the assumption of teamwork throughout the research process.[5] I wonder how contemporary examples of transdisciplinarity can offer new strategies of planning and governance for humans but also for nonhuman citizens such as animals, rivers, and maybe even built structures should they contribute positively to the greater ecology they inhabit. Antonio, talk about the importance of transdisciplinarity in your work and what you've learned from the art-science workshops you've organized in the past.

Antonio Guerra: This is a very good question. I have been working with different professionals and see how useful transdisciplinary experience is in terms of offering new strategies of planning and governance, for humans as well as nonhuman citizens such as animals, rivers, soils, and built

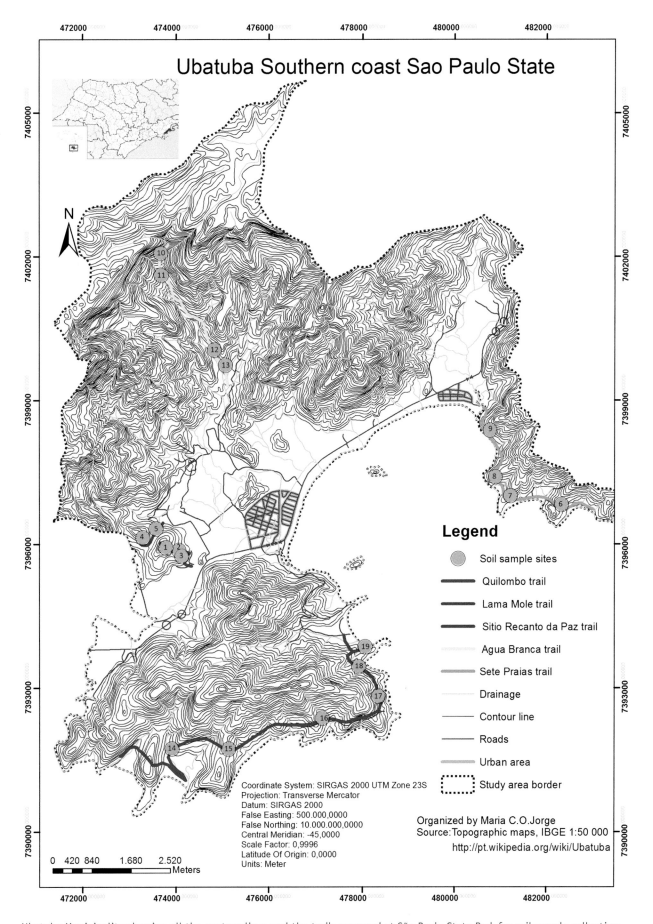

Ubatuba Southern coast Sao Paulo State

Legend

- ⬤ Soil sample sites
- ▬▬ Quilombo trail
- ▬▬ Lama Mole trail
- ▬▬ Sitio Recanto da Paz trail
- ▬▬ Agua Branca trail
- ▬▬ Sete Praias trail
- ─── Drainage
- ─── Contour line
- ─── Roads
- ▬▬ Urban area
- ┄┄┄ Study area border

Coordinate System: SIRGAS 2000 UTM Zone 23S
Projection: Transverse Mercator
Datum: SIRGAS 2000
False Easting: 500.000,0000
False Northing: 10.000.000,0000
Central Meridian: -45,0000
Scale Factor: 0,9996
Latitude Of Origin: 0,0000
Units: Meter

Organized by Maria C.O.Jorge
Source:Topographic maps, IBGE 1:50 000
http://pt.wikipedia.org/wiki/Ubatuba

0   420 840   1.680   2.520
                                    Meters

Ubatuba Municipality showing all the contour lines and the trails assessed at São Paulo State Park for soil sample collection and trails assessment.

Johann Moritz Rugendas, *"Forèt Vierge Pres Manqueritipa, Dans La Province De Rio De Janeiro"* (Mata Virgem Perto De Mangaratiba). In English, the title of the painting is *Tropical Forest, near Mangaratiba Municipality (this is in Rio de Janeiro State)*. In: Malerische Reise In Brasilien, 1835. This painting was used for the folder cover of the I Workshop of Art and Science, carried out at the Federal University of Rio de Janeiro in April 2017.

structures, to contribute positively to the greater ecology we all inhabit. In the past I have worked with ecologists, biologists, geologists, architects, soil scientists, artists and, of course geographers, like myself. Although each of these fields has different methodologies, our tasks are much easier because despite different approaches, we all are concerned about making a better, more sustainable world. Geography is a fundamentally very transdisciplinary science, since we must understand the physical as well as the socio-economic environment, and the interactions between them. This is what the other sciences, here cited, do but in a more compartmentalized way. Therefore, our experience in working together with these different scientists is very rewarding, since we may give our contribution to society, in a much more positive way.

My experience in organizing art-science workshops started in 2017, together with Professor Raphael David dos Santos Filho (architect), from the Faculty of Architecture, Federal University of Rio de Janeiro, where we organized the 1st Workshop of Art and Science. We managed to bring together 125 people, including professors, graduate, and undergraduate students, and people interested in sci-art collaboration from outside the university. Professor Michael Augustine Fullen from the University of Wolverhampton, UK, gave the inaugural lecture of the workshop, with the title "Nature Has no Concept of Waste." It was a two-day workshop, and from that event there has been a good integration between participating scientists and artists, coming from different areas of knowledge. We took a field trip to São Bento Open Museum, in Duque de Caxias Municipality, about 30 km from Rio de Janeiro City, where the participants had the opportunity to visit different types of mound shells, which are remnants of indigenous cultures from before 1500, when the Portuguese first came to Brazil. After all, although artists and scientists often speak different languages, so to speak, what unifies us is the concern about the landscape and the integrated reflection on the historical course of the landscape. Bearing this in mind, the workshop discussions were very fruitful, and inspired us to organize the 2nd International Workshop of Art and Science in 2018. We invited colleagues from Europe and the USA to come together from August 20–22, 2018. We will be expecting over 250 people from different Brazilian States and different countries for a three-day workshop, with time for roundtable discussions, posters, oral presentations, and field trips.

NUNCA UMA ALDEIA chavante fôra devassada como naquele vôo sensacional de nossos companheiros, em um avião pilotado pelo Major Antônio Basilio. Passearam durante 2 horas sôbre as habitações dos chavantes, que, desesperados, enviavam muitas flechas contra o bimotor de tela de pano, atingindo-o três vêzes.

CHAVANTES NA GUERRA

Fotografias de JEAN MANZON ★ Texto de DAVID NASSER

RENDERAM-SE os chavantes, segundo dizem, e aproveitando essa oportunidade, vossa revista deseja lembrar a nunca esquecida reportagem de David Nasser e Jean Manzon, "Os chavantes!" Evidentemente, não ficaria bem à gente dizer que êsse feito jornalístico não encontra paralelos nem antes, nem depois, em tôda a história da imprensa brasileira. Seria imodéstia alegarmos que a grande página escrita com o risco da própria vida por nossos dois companheiros foi lida em todos os países, reproduzida por "Images du Monde", na França, por "Life", nos Estados Unidos, e por dezenas de outras publicações. Não ficaria bem à própria revista que patrocinou o feito de Nasser e Manzon chamar a atenção dos senhores para certos detalhes, tais como os 6.000 quilômetros percorridos, desde o Rio até Kuluene, nas vizinhanças do Xingu. Fique bem ou não, seja imodéstia ou não, a verdade é que já não será possível escrever a história dos chavantes sem que a reportagem que hoje vos devolvemos ocupe um plano altíssimo, tão alto quanto aquêle em que colocou a própria imprensa brasileira, porque, como disse o famoso repórter norte-americano Richard Dyer, da I. N. S., "essa façanha jornalística está entre as dez maiores de todos os tempos e depois dela, seus autores deveriam ser aposentados com os honorários integrais."

ACONTECE que o mêdo de um repórter é diferente. Êle receia a morte, sente calafrios, jura que não mais se envolverá em assuntos dessa espécie, mas não vai embora. Os chavantes estavam lá em baixo, eu sei, o avião fôra atingido, também sei, uma vontade doida de voar para bem longe. E depois? Lá estavam os olhos investigadores dos companheiros procurando em mim a veia jugular do mêdo, a que revela as palpitações do coração. Mas, o essencial aí está. (CONTINUA NA PÁG. 16)

DAVID NASSER JEAN MANZON MAJOR ANTÔNIO BASILIO

Johann Moritz Rugendas, "Forèt Vierge Pres Manqueritipa, Dans La Province De Rio De Janeiro" (Mata Virgem Perto De Mangaratiba). In English, the title of the painting is Tropical Forest, near Mangaratiba Municipality (this is in Rio de Janeiro State). In: Malerische Reise In Brasilien, 1835. This painting was used for the folder cover of the I Workshop of Art and Science, carried out at the Federal University of Rio de Janeiro in April 2017.

Alexandra R. Toland: Paulo, your recent research in the Amazon envisions the forest as a city. Human settlement is designed within the spatial logic of the forest. Trees and rivers become part of the architectural infrastructure. Nature is not outside the city; it is the city. The forest is a cosmopolis for multiple human and nonhuman inhabitants who operate at multiple temporal and spatial scales.

Please explain this recent work with the Bö'u Xavante Association, which is featured in the curatorial research project by Anne-Sophie Springer and Etienne Turpin, *Reassembling the Natural*.

Paulo Tavares: There is a video that is a part of a larger work now on display in the exhibition, *Disappearing Legacies: The World as Forest (Verschwindende Vermächtnisse: Die Welt als Wald)*, curated by Sophie and Etienne at the Centrum für Naturkunde (CeNak), Zoological Museum Hamburg. In the video you see Xavante elders taking a ride out to the forest, and then navigating their way through a former village using their memories of the place and different maps and aerial photos. The piece is titled *Trees, Vines, Palms and Other Architectural Monuments*. The work specifically focuses on a group of trees that show evidence of habitation in the forest. From the aerial photos you can see they have grown in the shape of an arc. They grew this way because that was the shape of the Xavante village. Our goal is to consider those trees a kind of cultural heritage, as we would consider the dark earth soil (Terra Preta)

a kind of archeological heritage. The point is to understand those trees as *architectural* remains and not simply trees.

Alexandra R. Toland: How old is the arc structure? How old are the trees?

Paulo Tavares: We don't know how old the trees are. We imagine this village is from the nineteenth century, but we don't know exactly. What we proposed for the exhibition, in fact, was to commission a study to carbon date the soil, measure the trees, and analyze seeds. We wanted to conduct botanical and archeological research as part of the art work. But we realized that it's much more complicated than that. In the end we proposed a kind of draft petition to the United Nations and to the Brazilian Historic and Cultural National Heritage Institute to accept the trees as architectural monuments. Getting them protected would be the end of the project.

Alexandra R. Toland: The Amazon forest is a so-called biodiversity hotspot, which environmentalists are also trying to protect. But you've described these forests as architectural heritage. Do you make a distinction between natural heritage and cultural heritage, and between how these are represented? How can trees be read as architecture, and why is such a reading important in an age where most humans are born in or have moved to cities?

Paulo Tavares: The project at CeNak addresses this kind of liminal space between nature and culture and the way it relates to

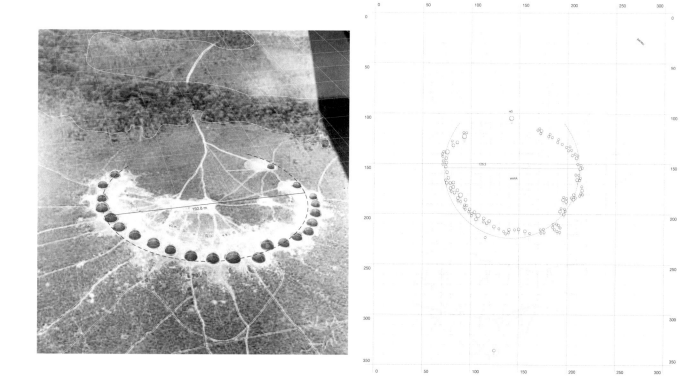

Archaeological site of Tsinõ. The botanic formation of trees and palms is the lasting evidence of this indigenous settlement. Different from the other settlements that were mapped, Tsinõ is a recent village, founded in the early 1960s after the communities of Marãiwatsédé were forced to settle next to the headquarters of the Suiá Missu farm.

Image courtesy of Paulo Tavares/autonoma and Bö'u Xavante Association.

the natural history museum, where the work is exhibited. We are proposing a kind of shift, or gesture of dislocation, from the natural history museum, which houses soil samples and botanical artifacts to be read as "Nature," to the museum of culture, and in this case architecture. We have approached the Center for Canadian Architecture (CCA), which is one of the most important architectural archives in the world and hosts the drawings of very celebrated architects. And in Brazil, we are working with the Institute of Artistic and Historical Heritage of Brazil, which is the main federal body that protects national heritage sites. The point of these collaborations is to present the forest not as nature, but in fact as architecture.

We think this is particularly important because Western knowledge about nature has been very much produced by erasing traces of indigenous settlement from the landscape. By removing the human mark, the Amazon forest could be colonized, categorized, and finally enclosed inside of the natural history museum. The cultural heritage of forest peoples was on the other hand read as ethnographic artifacts: objects, body paintings, rituals—what is known as immaterial heritage. So, the project opens dialogue with this colonial history of understanding of what heritage is and the way in which the museum as an institution has many different layers that represent that heritage.

Alexandra R. Toland: The village you speak of is at least 100 years old, but the spatial technology is probably much older. I wonder what you have learned about changes in indigenous land-use over time. Could you talk about the temporal trajectories of indigenous land-use in the Amazon?

Paulo Tavares: Well, first we need to understand that the process of change in these communities has been a process of violence with very severe impacts. In South America, these people are the last remains of a holocaust. In the case of the Xavante, we are dealing with a violence of the military dictatorship in the way that these populations have been displaced and removed. Before that, many suffered from epidemics brought in by the whites. So we need to acknowledge that what we might frame as tradition has

Identification of the vestiges of the village Sõrepré, the most ancient geopolitical and cultural center of the A'uwe-Xavante ancestral territory, probably founded in the early nineteenth century. In the contemporary satellite image we see the same site as it is today. Note the presence of an arc-shaped formation of trees that still preserves the old village's layout isolated in the middle of an area completely deforested. Grown out of a soil fertilized for more than a century by consecutive Xavante generations, these trees must be so old and productive that even the plantation owner decided to lose space in soy production to preserve them.

Images courtesy of Paulo Tavares/autonoma and Bö'u Xavante Association.

been violently damaged over time. Territories are much smaller. People have been settled. Missionaries and the government tried to modify the ways of life and transform people into rural laborers. So this is one thing that I think we should punctuate. Recuperating culture is a form of resistance. Maintaining traditions, myths, and so on of course cannot be the same as before, as if it were a static moment in time. Some elements endure and are preserved because there is a kind of social consciousness for them to be cultivated. In the present context this is important because we are all bombarded by media all the time. So, to care for the land and to maintain traditional ways is very future-oriented. This way of understanding the land manifests itself in some of the most contemporary ways of dealing both with climate change and food securities. That is to say the essential needs for the future of the human species is something that they have already produced. They have figured out carbon storage, for example. Indigenous lands are statistically the most preserved carbon-rich lands in the world. This means that to enforce indigenous rights is to combat climate change. They also employ and cultivate systems of agroforestry. We have been seeing how agroforestry has been studied as a way of getting out of the trap of a commodity food production system that is really killing the planet. So practices of planting with more biodiversity is something that is very ancient and at the same time very contemporary.

Alexandra R. Toland: The forest as city provides a contrasting image to the actuality of most urban metropolises, characterized by pavement and sprawling construction. Can you give a specific example of an enduring practice that urban planners and architects of today could adapt and modify? For example, could past designs such as the arced village you describe apply to cities of the future? Can we apply principles of indigenous land use to periurban spaces in danger of disappearing?

Paulo Tavares: That's a difficult one. As I said, some practices could be applied as they are very much future-oriented and relate to the survival of our species. So there are definitely lessons to be learned, but I think the question is less on the level of form and more about the politics of landscape. What is the relationship between humans and nonhumans? What is the relationship or the role of other species in the way we deal with the landscape? How do we intervene with the landscape? These are important questions.

I call this type of epistemic understanding of space multispecies urbanism. And I think that's definitely something that we can learn. There are many technologies being applied or developed in these lands that make us understand the contemporary condition of our active space, both urban and nonurban.

Alexandra R. Toland: Paulo, you've used historical maps and remote sensing images, mainly from government sources, to investigate the land use technologies of the Xavante people. These maps do not necessarily reflect the spatial understandings of the people who have occupied that territory for countless generations. In your

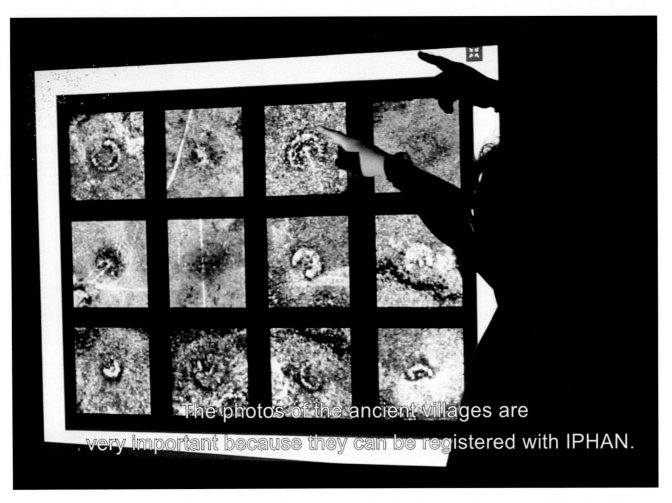

The photos of the ancient villages are very important because they can be registered with IPHAN.

Film still from *Trees, Vines, Palms and Other Architectural Monuments*, a project by Paulo Tavares/autonoma in collaboration with Bö'u Xavante Association. In this scene Caimi Waiassé describes the use of satellite images during a presentation of mappings in Etenhiritipá, August 2017:

> These photos are important to testify to the existence of the ancient villages in this region. When the indigenous claim their lands the governors always ask: "Where are the documents to prove?" Here are the photos that the government itself registered, and that will serve for the Xavante to prove the existence of the villages. They are very important because they can be registered with the National Institute of Artistic and Historic Heritage (IPHAN). This institution deals with vestiges of the past. Its mandate is to demarcate ancient sites, and also where the vestiges of the Xavante's ancestors are. There are cemeteries there, it is from these sites that the Xavante people spread to other regions. For this reason IPHAN needs to demarcate these areas, so they are respected even if they are located inside farms, to make sure the farmers won't enclose them with fences and turn these areas into plantations. Today we only have these photos, we don't have any other official document that guarantees the protection of these sites. We can negotiate with the farmer, we won't take his land, but only assure the delimitation of the territory of the ancient village, so these places can be respected.

Image courtesy of Paulo Tavares/autonoma and Bö'u Xavante Association.

video, *Memory of Earth*, a Xavante warrior elder, Dario Tserewhorã, assesses one of those maps with a critical eye. He says, "The limits of the territory make too many curves, forming a shape similar to a horn that ends in a stream that flows to the Xingu River Basin. But most places occupied by the Xavante are left outside the area, like the Black Stone Mountain Range … I don't like this map. It doesn't say anything. It's a lie, a fake. If I made this map, it would not be like this with so many curves. This looks like an anteater looking for food …"

Given this critique, I wonder how you think mapping, as a technology, can change people's

Identification of archaeological traces of ancient Xavante settlements in KH-9 satellite images from 1976.

Image courtesy of Paulo Tavares/autonoma and Bö'u Xavante Association.

relationships with the land. Does showing someone a map have a political effect?

Paulo Tavares: There's a very long answer to that, in which one could do a kind of genealogy of the map and examine the many possibilities of measuring, drawing, and representing space/territory as it has shifted throughout different human populations over history. But I guess the short answer is to say that maps have always been associated with some system of power, appropriation, and control over the land and its populations. The way in which power is procured is very much related to formations of nation states and centralized forms of power, and more specifically, with colonial practices of appropriating and expropriating land and people. So, cartography and cartographic knowledge are a form of colonial control. Maps are instruments of empire, essentially, in which territory is reduced to the realm of the visible. But a map is never simply a representation of a space, it's always the creation of that space by means of representation. So maps have a political effect

in as much as they allow you to represent a territory or to expose a situation.

In our work, we reappropriated some of the very maps that had been used to occupy the lands of the Xavante people. For instance, we recently used declassified imagery from a Cold War–era satellite surveillance program launched by the U.S. military in the early 1970s. We also used maps from a survey conducted by the Brazilian Air Force, which was one of the first mapping projects of the territory. It was done in 1966, two years after the military took power. These were very much framed both as a way of shaping the nation, or the image of what this nation is, but also as a kind of colonial project to map out resources and occupy territory. So, in our work, there is a way of trying to produce a different type of territory that shapes the sort of extended imaginary and frames what we call the modern colonial nation-state.

Alexandra R. Toland: How did the community participate in the research

Patches of trees identified as remains of ancient Xavante village Tsino, last village inhabited by the A'uwe-aXavante of Maraiwatsede before the deportation of 1966

Documentation of field expedition to the archaeological site of Bö'u, the old center of Marãiwatsédé. The site is characterized by a thick jungle-like formation that has preserved the original circular layout of the ancient village. Bö'u was probably founded in the late nineteenth century and existed until the forced removals of the 1950s. This region is known as Bö'umoahö, "place of production," in reference to the abundance of natural resources found in the area and the past of prosperity lived in the great village.

Images courtesy of Paulo Tavares/ autonoma and Bö'u Xavante Association.

process? And what will happen with these maps, with this knowledge you've gathered?

Paulo Tavares: Our work is very pragmatic and situationally grounded. We want these villages to be considered architectural heritage because we want to protect them. So it's a very pragmatic, political effect that we want to achieve, but also a way of reimagining what we are. What is this nation? What is the concept of the nation-state? And how is this connected to colonialism? There is a kind of imaginary map being drawn here, or a map that affects the imagination in a political way.

Regarding participation, the community participated in certain ways. We were guided by the elderly people, some of whom are 80 or 90 years old. They cannot draw a map as we would draw, but they led the field expeditions. They would say, "Well, you see

that hill? You see that forest? You see that river?" They have a really sharp cartographic knowledge of those landscapes, which have changed a lot. So, they fully participated in the project. I would download satellite images and bring them archive research from the city and present it to them and they would give me feedback.

And the way we did the footage was that I didn't translate anything while we were in the field. One of the guys spoke a little Portuguese, but we made the decision that they would speak and then we would translate and see what was happening later. So that's the way the work was conceived.

Now we're setting up a center together with other collaborators. After this project in this specific region, we felt we needed to set up what we call a Center for Geography and History where we would try to repatriate all the archives of this land, or at least have a copy of those maps and archives, and we would start to produce our own cartographic information on the ground. Indigenous knowledge is always under threat of being mined, so to say, by top-down mapping projects. So it's really a question of not appropriating but subverting the kind of maps of the state by producing our own autonomous types of maps and mapping.

Alexandra R. Toland: Paulo, you have written that Amazonia "demands deeply ethical engagements and transdisciplinary approaches on the part of designers." Why is transdisciplinarity important to your work and what advice would you give urbanists, architects, artists, and activists similarly striving toward more eco-ethical spatial design?

Paulo Tavares: In my work there's this kind of ambition, or feeling, or desire, or understanding that one cannot occupy just one space. Things have multiple lives in different forums, or different formats, or different frames, or in different spaces. So, every work that I have produced has had its own sort of political life. They have occupied spaces in the fields of architecture and art, but at the same time can be translated into other forms. The idea that we can

incorporate architecture both as medium for communication and as a way of producing knowledge is super important. So we can speak of transdisciplinarity, but also of transversality, which is how we understand that those spaces we operate in are not rigidly defined. Their borders are much more porous in this contemporary space, and it is important to think about how your work can have multiple lives in multiple medias and formats. The boundaries between different systems of knowledge are much more complicated than we think. And it's almost impossible, I think, to do this kind of research without relying on other disciplines.

Endnotes

1. For more on the soil functions paradigm, see Winfried E.H. Blum, 2005, Functions of soil for society and the environment, *Reviews in Environmental Science and Bio/Technology* 4(3):75–79.

2. See, for example, A.C. Roosevelt, 2013, The Amazon and the Anthropocene: 13,000 years of human influence in a tropical rainforest. *Anthropocene* 4:69–87.

3. Paulo Tavares. 2016. *In The Forest Ruins.* Superhumanity, December 9. http://www.e-flux.com/architecture/superhumanity/68688/in-the-forest-ruins/.

4. Merten, Gustavo H. and Minella, Jean P. G. 2013. The expansion of Brazilian agriculture: Soil erosion scenarios. *International Soil and Water Conservation Research* 1(3): 37–48. Open access: http://www.sciencedirect.com/science/article/pii/S2095633915300290

5. See definition on transdisciplinarity in, for example, T. Jahn, M. Bergmann, and F. Keil, 2012, Transdisciplinarity: Between mainstreaming and marginalization, *Ecological Economics* 79:1–10.

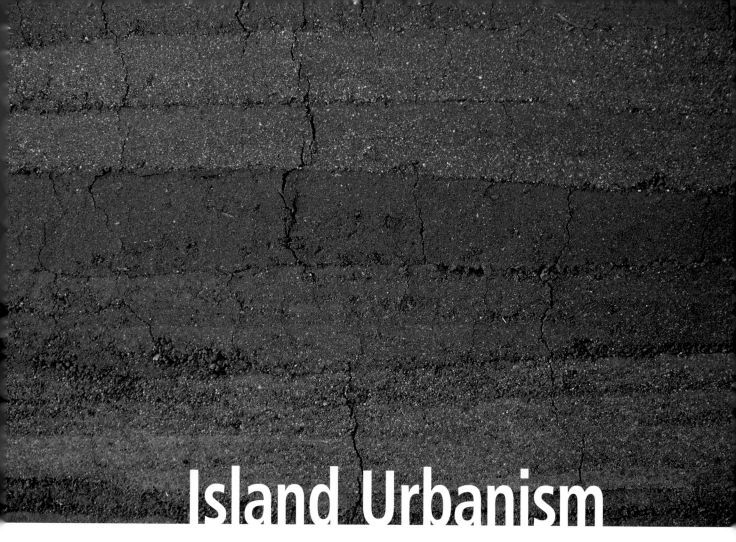

Island Urbanism
The City as Soil System

Sean Connelly in dialogue with Alexandra R. Toland

Sean Connelly focuses on systems and softness. (Rocks may be examples of softness, especially where land moves air, becomes wet, and feeds.) As director of After Oceanic (www.ao-projects.com), his focus on ecology and economy concerns the interactions of material, information, form, and time as technologic systems that humans design, and redesign. As a Pacific Island architect, landscape architect, and visual artist, Connelly's work strives to expand the boundaries of justice-advancing futures for our changing planet.

Title image: Detail, *A Small Area of Land (Kaka'ako Earth Room)*, ii Gallery, 2013.
Image courtesy Sean Connelly.

Most civilizations from the beginning of modern human history rose up along rivers and oceans. From the decline of the Sumerian culture in the Fertile Crescent to the worldwide sealing of thousands of square kilometers each year in the name of development, humans continue to overburden the watershed in a variety of ways. After first perishing from unsustainable slash-and-burn practices, Hawaiians figured out ways to construct their civilization within the flows of the watershed. They recognized connections between mountains and ocean, and carved out fish ponds and terraced gardens. They developed settlements, political structures, and cultural heritage sites based on their understanding of the watershed. After contact with European settlers and annexation to the United States, Hawaiians were forced to abandon these practices. In an age of climate change and rising sea levels, the question is, How can the lost technology of the watershed be restored and applied to contemporary urban planning strategies? I (Toland) interviewed Sean Connelly about the landscape and soils of Hawai'i, his practice as an architect and sculptor, and his visionary design manifesto, Hawai'i Futures. According to Connelly, indigenous Hawaiian concepts of design offer island cities in Hawai'i and elsewhere valuable lessons for the future. It is symbolic to end this section and this volume with Sean Connelly's vision of the city as complex soil system. Rethinking urbanism in terms of the watershed shifts the focus from buildings and streets to large-scale infrastructure, political organization, and indigenous knowledge. It is in this rethinking that the platform function of the soil becomes a key to protecting soil functionality as a whole.

Alexandra R. Toland: Sean, you utilize a range of visual approaches in your work, from large-scale minimalist sculpture to GIS-supported maps, to diagrams of economic recovery plans of Hawaiian watersheds. This broad methodology allows you to speak to multiple audiences that may have different needs. How do audiences respond to your work and where do you see your work having the most impact? What have you learned in your practice as an architect, say, that has helped you in your work as a sculptor? Conversely, how has your sculptural practice helped you think differently about the land and its waterways; the city and its infrastructures?

Sean Connelly: A common starting point for my work as an architect with a focus on research and art installation is the fact that over 90% of the population in Hawai'i is urban. Ninety-one percent of our energy is derived from imported fossil fuels. Nearly 90% of our food is imported from elsewhere. As an island in the middle of the world's largest ocean, 63% of our seafood is imported because the fisheries are either contaminated with microalgae that cause ciguatera poisoning in people who eat contaminated fish, or the reef itself is dead or dying. Nearly all freshwater streams on the island of O'ahu are channelized and contain a bacterium called *Leptospira* that causes an infectious disease in animals and humans. This is just the surface of Hawai'i's environmental issues and doesn't even begin to describe the urgency of our social inequity problems, although they are related.

Despite these negative facts, Hawai'i is still gorgeous. That gaze is part of what makes the work so difficult because for many people, they can't see the dangerous outcomes of our existing systems such as single-use land-use districts. On the other hand, many residents in Hawai'i do feel the effects in other forms, some of which include the high cost of living because everything is shipped in.

We know there are challenges for humans living on the world's most isolated landmass. Prior to European contact in 1778, Hawaiians supported a population nearing 1 million while continually mitigating environmental degradation as issues arose. We had oral histories and knowledge systems that developed here on the islands for several thousand years. This indigenous knowledge is embedded into language and practice through values, emotions, and relationships. Since these systems predate writing, a visual and perceptual approach is necessary. The lack of writing doesn't indicate a lack of sophistication but rather exposes the immense scale of time that indigenous knowledge contains—so vast that it surpasses the longevity of written form. Indigenous knowledge is old because it emerges directly from the interactions between a place and the human anatomy, cognition, and ecology that are essential to the capacities of which our species have evolved. This is generally consistent in core values around the world. Because of this continuity, projects like Hawai'i Futures may focus on Hawai'i, but the ideas behind it are universal and apply to multiple audiences in a way that moves beyond professions. It strikes at a level that is core to our species.

Hawai'i Futures is a virtual intervention for island urbanism and the future of the oceanic city that portrays a watershed framework for the recovery of the indigenous Hawaiian land-use system most commonly known as the *ahupua'a*. It's filled with small maps and diagrams that were both produced with ArcGIS and Rhino to visualize the past and future of ahupua'a.

As a retroactive manifesto, the website is very prescriptive. After I launched the site in 2010, I was also interested in creating ways for people to examine the same essential questions that led me to create Hawai'i Futures in the first place. One of those questions regards the relationships between people and land. This not something that can be drawn but must be experienced. A while later, Trisha Lagaso Goldberg presented an opportunity for me to produce an art installation. *A Small Area of Land (Kaka'ako Earth Room)* was the first sculpture that followed Hawai'i Futures. Art installation became a platform to explore research in a tangible way. While architecture presented a way of thinking conceptually about building and land-use as a system of interactions, sculpture provided an output to render those concepts into a pure material form free from the particular market constraints that often subdue the building.

To ask a question about land, I propose asking soil itself: What is our relationship to it? I brought 16 tons (32,000 lbs) of volcanic soil and sand into the gallery. But one of the conceptual strategies of the project that I felt was most effective was the ability for the form to take something as familiar as soil, and make it feel novel again, perhaps even divine. While soil is experienced across a horizontal plane beneath us, the sculpture essentially turned the ground onto its side. The soil was now face-to-face with the viewer. You could feel the weight and mass of the earth object in the room. It had a comforting aroma. It absorbed sound in the room, resulting in an almost flat feel of space contrary to its large presence in the room.

From a technical point of view, this project was great because it gave me tangible experience in dealing with soil types in a way that discussing it academically does not allow. I had to learn about soil types because I was endeavoring to use it structurally. With sixty volunteer assistants, we sifted and stacked soil and sand into layers to reveal an abstraction of itself, something similar to the culinary process of spherification where something solid undergoes liquefaction before reentering a semisolid version of itself. I harvested a loamy soil from the mountaintop, which was used for the base, and sourced an expansive clay soil from the valley, which was used for the upper half of the sculpture. Conceptually, this turned the watershed upside down, while also allowing the sculpture to expand and erode at eye level without completely falling apart. It was like chemistry, but rather mixing soil types to control a change in physical form.

This sculpture got a great response and eventually landed a feature on the news and publications like BLDGBLOG. People from all around came in to see it. Some folks sang to it, some prayed, others took naps in the room. Some people seemed scared of it; others wanted to kiss it. The only folks who seemed not to like the sculpture were, ironically, farmers—they see soil face-to-face every day. But for people who rarely get to see soil in their daily life– they wept. One thing I learned from this experience, I will say, is that while architecture taught me about time, energy, and scale, sculpture taught me about material, industry, and interaction.

Alexandra R. Toland: Could you give us a picture of the Hawaiian watersheds you grew up with on the island of O'ahu?

Sean Connelly: Very lovely. Growing up surrounded by water, the ocean was all I knew of the horizon. The ocean is so encompassing it seems to carry the sky while land penetrates in procession to carve the sky into freshwater streams. Some of that water returns to the sea, while the rest percolates through the pyroclastic rock into the aquifer within the island. While it may take only half a day for water from the mountain to enter the ocean via stream, it takes twenty-five years for water to seep into the aquifer.

The island I grew up on, O'ahu, gets its name in part because when approached from the sea the elevation of the mountain resembles an altar, or ahu. This altar is the sunken peak of an extinct

Detail, *A Small Area of Land (Kaka'ako Earth Room)*, ii Gallery, 2013.

Image courtesy Sean Connelly.

mass of two major shield volcanoes. Three million years ago these volcanoes coalesced to a height of some 3500 m that have since slumped, eroded, and in some areas completely collapsed into the sea. Today, the island peaks at 1227 m—two-thirds its original height—and encompasses an area at sea level that 2 million years ago was a high mountain covered seasonally with snow. The island continues to shrivel.

Hawai'i presents an interesting survey of soil types when mapping the volcanic lifecycle of the islands over time. For those who are unfamiliar, the Hawaiian Islands range in age from 1 million to 11 million years, with the rest of the old archipelago continuing from sea level into submerged lands 85 million years old. Within the array of the eight main islands, the most obvious change corresponding with the progression of age is hydrological. Younger islands are larger and voluptuous because streams are still in fetal formation, while older islands become eroded as streams reveal themselves into explicit watersheds—valleys, wetlands, and all.

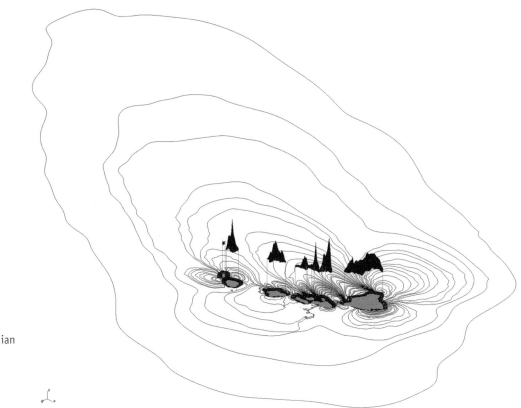

Digital model of Hawaiian Islands viewed from beneath the sea.

Image courtesy Sean Connelly.

A more overlooked aspect of volcanic island age and change are the implications of erosion on soil genesis. The most abundant soil type on the younger islands is andisol, which ironically is the rarest soil type on Earth, occupying only 1% of total world area. This soil is a black and fluffy soil resulting from volcanic ash. Yet, on the older islands, most of that original andisol has been transformed through biological and hydrological processes into a different range of soils today, or has sunken beneath sea level. To illustrate the volcanic soil cycle in a poetic way, an andisol is relatively nutrient poor because it is relatively young and undeveloped. But it's great for forests, and so this soil emerges into a forest, which overtime drives nutrients and organic compounds into the soil to essentially evolve the soil into a different order altogether. As the island erodes, the formation of ridges and valleys become more distinct. Soils that remain lodged at high elevation are exposed to bulk rains and become leached of nutrients, which then percolate down mountainsides into the valleys,

where soils become very rich and fertile, or so filtered that all that remains is an accumulation into clay. While sand on continents is silica, sand on volcanic islands is coral pulverized by the ocean, making beaches literally skeletons. On younger islands where corals have yet to grow, black sand is formed of pulverized basalt. Green sand is pulverized obsidian. I could go on about how fantastic and varied the soil systems in Hawai'i are.

From this soil and its corresponding flora, a system and ecology emerges; the island becomes a source of freshwater (wai). Islands are important because they yield life. The greatest form of wealth is waiwai. In the Hawaiian watershed, the water cycle is visible as a single unit of many units repeating continuously—sea sky land sky sea land. You can see clouds form at sea, and then watch them move as the island pulls the clouds over them like a magnet. As they approach the island you can see them either remain white and float higher into dissipation or darken and become

The Island of O'ahu (left) contains less andisol (black) than the younger island of Hawai'i.

Image courtesy Sean Connelly.

snagged on the mountain ridges to release rain. In some places you can see the cloud accumulating on one side of the ridge, which then extrudes the cloud on the other side in the shape of the ridgeline. Of course the dynamics at play are more complicated than the visual experience I portray, but the ability to see all these transactions happening between the different elements and features of the environment from mountain, soil, stream, ocean, and sky influenced Hawaiians with profound notions of system and state. The watershed itself is the greatest formation of sun energy because it stacks all the energy forms together at once. It's the original super power. Hawaiians were experts at tracking this system, and designing ways to respond to and mimic its nutrient flows with a real-time capacity that continental humans today are only now beginning to see.

Alexandra R. Toland: In the soil scientific conceptual framework of soil functionality, watershed functions are generally associated with the interface function of the soil—or the soil's ability to filter water, regulate nutrient fluxes, and buffer toxins in the environment. This vital function renders the soil a site of fluid interaction and transformation. The filter function is often cut off in cities by layers of asphalt and cement, giving way to another significant soil function—that as platform for human structures. Talk about these soil functions in your work. Is the watershed key to protecting soil multifunctionality? How can the soil platform function, through good design and planning practice, be harmonized with other soil functions such as food production, flood and climate regulation, and habitat for human well-being as well as nonhuman creatures?

Sean Connelly: This is a really fun question, actually, and leads me to talk about one of the maps I am currently working on for my Hi-Atlas project. Imagine the future soil types that will emerge from our era of human activity. Since we can think of soil in terms of functions, one of those functions is to serve as the platform for human structures. What would it change if we were to think of those human structures as a soil type in and of itself? What does that make a building, or a cell phone, or even a city? This fantasy has really got me excited, especially since, if we remake cities into soil systems—which is a species in and of itself—this would in effect make the places we live sacred again.

In Hawai'i, soil types were one of many considerations that informed the organization, allocation, and cultivation of resources of the island. These resources were traditionally managed as an ordered sequence of land divisions that organized continuous space between land, sea, and sky into accessible and productive units of wealth and information. Most commonly, these units are called ahupua'a, which references the altar used to mark the edge where ahupua'a meet. These boundaries were fluid and could expand or contract on a seasonal basis depending on rainfall or productivity based on shifts in climate. Divisions with an ahupua'a are called 'ili, which means surface or skin. The notion of 'ili is continuous between land and sea, as both are also understood to operate fluidly. Land is a fluid that moves much slower than water. The boundaries of an 'ili may pertain to soil types, water flow, wind patterns, as well as burials and other spiritual phenomena.

To talk about the soil platform function is a discussion about trade-offs. In discussing trade-offs, soil becomes a platform for more than human structures, but value systems, and then behaviors that perpetuate those values. The concept of the soil platform is important to consider with regard to value systems because not every soil type is meant to function as a platform for building. For instance, vertisol is a very expansive clay that expands and contracts daily according to weather and microclimate. The very liquid nature of vertisol would cause settling in any structure built upon it. While a structure can be engineered to withstand the constant settling of the vertisol, this soil type is also very fertile and great for cultivation. Technically the best and highest use of this particular soil is for food production because ultimately it would be cheaper to build a structure on more solid ground that also may be more

difficult to grow food upon. Fertile soil on a remote volcanic island is such an expensive resource that it's best used for farming. Buildings are placed on the periphery to occupy places where food will not grow. This is a very Polynesian approach to the idea of soil as platform and is contrary to the American system of land use, which drives from a different notion of wealth and value.

Today's modern approach to land division breaks the continuous space of 'ili and ahupua'a in many ways. Current land use is not fluid, but rigid because Euclidian zoning does not relate necessarily to annual shifts in weather or microclimate. Instead today's zoning is based on real estate and voting precincts. Current land use furthermore separates where people grow food, and fragments the watershed between mountain and ocean, failing to adequately protect the land areas around streams, beaches, and wetlands. So what you have today is a land-use system that is essentially disconnected from the land that is being used, such that areas zoned as urban districts are actually lands that would be better suited for prime agriculture, and agriculture districts that are not fertile or contain endangered plant species. In my opinion this is rather primitive compared to the ancient system of land use that was particularly calibrated to the specific nuances of each site. There was a precision involved, and that notion of precision was also very fluid and adaptive.

Alexandra R. Toland: You've made the point that it could take several generations to rebuild urban infrastructure. I wonder if we realistically have that much time. I'm wondering on the one hand, what can be done right now at a policy level to get the ball rolling, and on the other hand, how disaster can act as a catalyst for sustainable rebuilding and regrowth. The tragedies of recent hurricanes in the Caribbean or tsunamis in Southeast Asia seem to demand something better than business as usual. It is so hard for people to think beyond what is

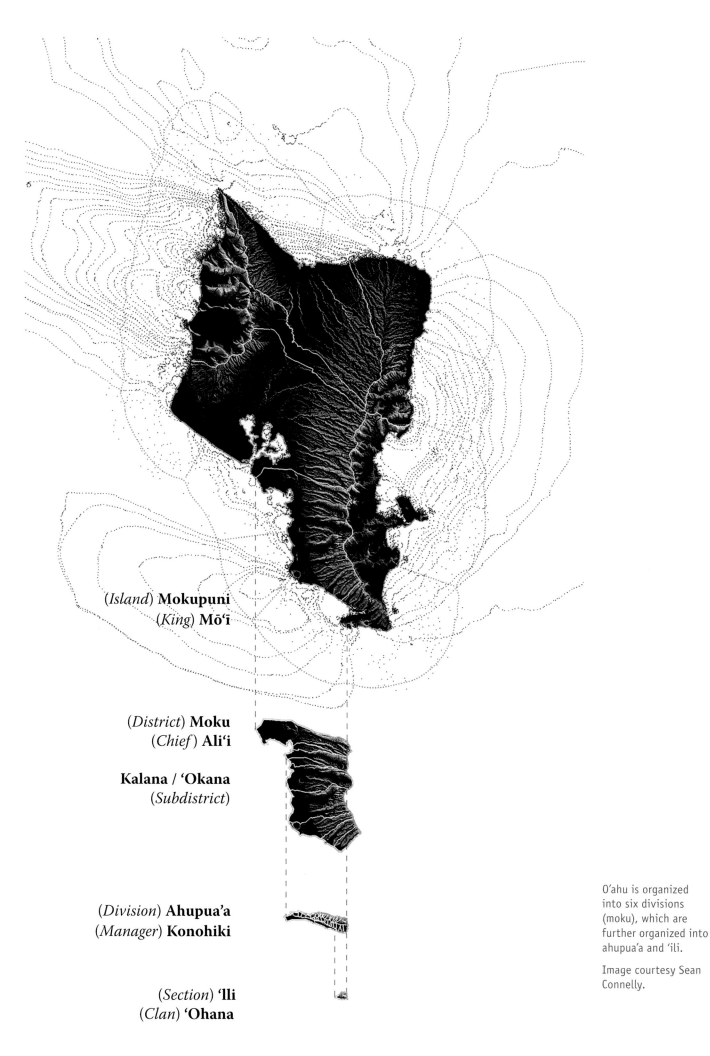

(*Island*) **Mokupuni**
(*King*) **Mōʻī**

(*District*) **Moku**
(*Chief*) **Aliʻi**

Kalana / ʻOkana
(*Subdistrict*)

(*Division*) **Ahupuaʻa**
(*Manager*) **Konohiki**

(*Section*) **ʻIli**
(*Clan*) **ʻOhana**

Oʻahu is organized into six divisions (moku), which are further organized into ahupuaʻa and ʻili.

Image courtesy Sean Connelly.

already in front of them: the streets and diverted streams, and cement-lined canals. What do the fields of risk planning and disaster management offer urban planners, lawmakers, and architects? How can the logic of the watershed be used to protect people instead of threaten them?

Sean Connelly: I'm very glad you asked this question because it describes the more general question of what's next. Oftentimes people will see the sense in these ideas and the focus of my work, but then they ask about practical solutions. Practical for some means cost effective, while for others it means that it takes less time. And then there are those people who want solutions that are both cheap and fast. On the contrary, I feel that part of the problem of the business as usual case is a notion that "practical" is cheap and fast. Soil and watersheds are not cheap and fast; they absorb a lot of energy and time.

Hawaiians understood tragedies and vulnerability, and understood that the most practical approach was not to work against the watershed but to plug into the system as the main platform of their infrastructure and cultivation technologies. In the cultivation system you'll see a range of applications for growing food in dryland fields, in streams, in wetlands, and in reefs. In time, the human system and natural system influence each other to become a single entity. There is no distinction between humans and nature in the Hawaiian worldview, just a continuum of interactions that are made culturally relatable in the notions of familial relationships.

The land-use system we have today represents a complete change from the cultivation systems that came before us, yet people can still be scared of change. The idea of reclaiming the watershed through legislative mandates and policies could be welcomed if it meant that we could eat fish from our reefs again. On the other hand, I am skeptical of top-down approaches because they are often linked to corporate developers and global financial resources that don't align with indigenous values. For instance, we could recover the areas of streams along watersheds to create space for recreation and agriculture, however, that would inadvertently lead to gentrification of the neighborhoods that now become more desirable because the stream is essentially like Central Park. Even in the poorest areas on O'ahu, homes along the beach sell for a million-plus dollars.

Ten years ago it was rare to hear a person in municipal government here say the term "sea level rise." Nearly everyone knows the term now and what's at risk, although not all of them necessarily believe it's true. Yet, regardless of climate change, volcanic islands sink. Regarding policy, it is important to overlay a future scenario plan of land uses based on the cultural data we have available and based on the parameters outlined in Hawai'i Futures. If the outlines of this future scenario plan were drawn today, when a catastrophe strikes, that plan would activate. For instance, if an area is currently developed, but in the future, because of its proximity to the shore, if there is a storm surge that damages those structures, they cannot rebuild. Instead, the insurance money goes toward relocation. Of course, there would have to be exceptions to the rule. For instance, if a structure is the home of a caretaker that manages a fishery, and that home is destroyed in a hurricane but the fishery recovers, then that home should be rebuilt. But in that case the home is essential to the operation of the fishery as a cohesive infrastructure, compared to a purely residential situation. Most likely, to compliment this future scenario for land use, there also needs to be future building typologies that perhaps don't exist yet. In terms of the future, the watershed is a means to protect people because we evolved to manage them. Nearly everything we understand about what the environment affords

1-mile

kua'iwi

mala
(*dryland frields*)

kuauna

lo'i kalo
pondfields

kuapā

loko i'a
fishponds

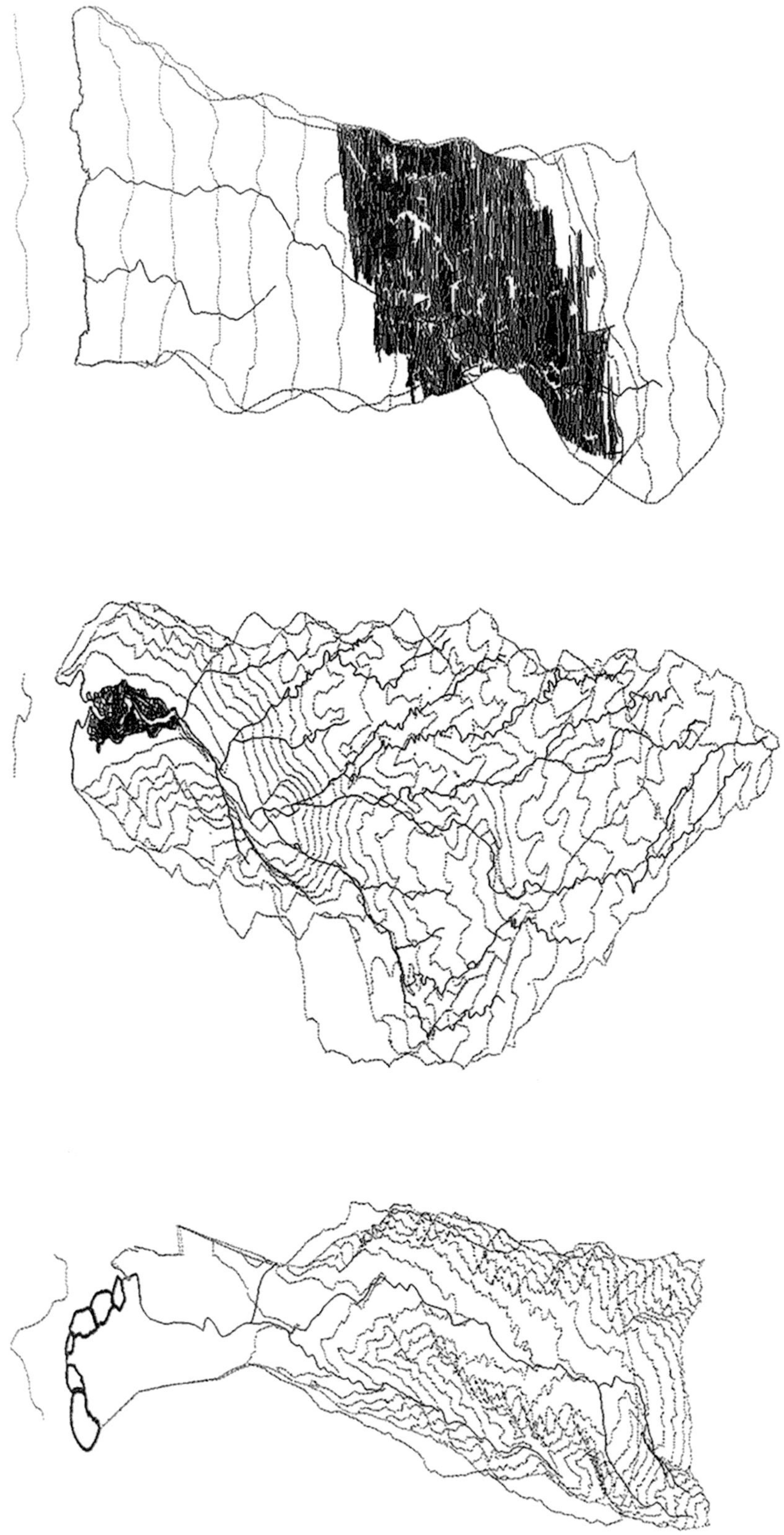

Three types of cultivation systems.

Image courtesy Sean Connelly.

us is perceived from the information a watershed contains. Reading the watershed in terms of slope, water flow, ground cover, wind, and sun together inform what areas are at risk for flooding or storm surge, what areas will feed us, what areas will grow materials to clothe and shelter us, and where to bury our loved ones.

Soil literally holds the physical matter of our ancestry and thus holds our link to the future. Time in Hawaii is reversed—the past is in front of vision not behind it. Soil is our link to the past, and the more we understand it, the closer we see our future.

Alexandra R. Toland: This chapter is the last chapter in a long book about new ways of thinking about the soil, its functions, meanings, services, inhabitants, and existential threats. It ends with a manifesto as it began. So I want to thank you for sharing your inspiring work and vision for the future. We will need it in our work ahead.

Recovering ways to repeat the success of Hawaiians thriving on Earth's most remote landmass is critical to the future of human survival. The environmental demise resulting from human survival today is attributed to choices thought to be logical and cost-effective. With O'ahu seen as a tipping point, this site offers moments to reconfigure what is considered typically infeasible or impractical, as reasonable interventions to evade the crime and catastrophes of climate change. The scale of unfolding crisis in this era of the capitalocene is so deep and vast, the coming generations will be faced with the inevitable need to recover ahupua'a as a major human technic of economy.

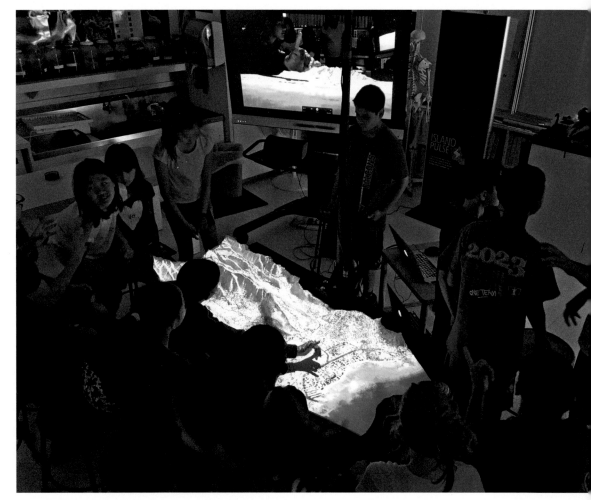

Hawai'i Futures light projection of GIS renderings on satellite-based 3D model of an ahupua'a carved out of foam.

Image courtesy Sean Connelly.

Hawai'i has a year-round growing season, every formation of clean energy, and diverse soils and microclimates. YET, over 90% of us live in costly and congested urban areas importing 90% of our food and energy...

RECOVER AHUPUA'A

HAWAI'I
FUTURES

AN INTERVENTION FOR ISLAND URBANISM

HAWAI'I FUTURES summons the habits of mind in which the cycles and surfaces of *wai (water)* organize the physical and emotional processes that craft city life. This virtual intervention creates a space to reclaim the notion of the city according to what it means to live on the volcanic islands of Hawai'i, versus a continent.

Recovering ways to repeat the success of Hawaiians thriving on Earth's most remote landmass is critical to the future of human survival. The environmental demise resulting from human survival today is attributed to choices thought to be logical and cost-effective. With O'ahu seen as a tipping point, this site offers moments to reconfigure what is considered typically infeasible or impractical, as reasonable interventions to evade the crime and catastrophes of climate change. The scale of unfolding crisis in this era of uncertainty is so deep and vast, the coming generations will be faced with the inevitable need to recover ahupua'a as a major human technic of economy.

The parameters of recovery illustrated by each of the following images may provide visual insights to redesign Hawai'i's broken land-use system back into a resource of sustenance, again. It is a way for Hawai'i to begin the discourse of the types of large-scale infrastructural changes the coming generations are bound to face. Utilizing the island itself as a technology of wealth can secure justice-advancing futures on our changing planet.

Screenshot of http://www.hawaii-futures.com

Image courtesy Sean Connelly.

Afterword and Acknowledgments

This book represents the fruits of over five years of fieldwork and an even longer engagement with the subject of soil in the arts. What started as an exhibition of artists' posters and a film screening program at the 20th World Congress of Soil Science in Jeju, Korea, grew into a long and multi-stranded conversation on soil functions and their meanings and values. Looking back, the pace of the book's development often mirrored the slow pace of soil formation itself. Weathered by setbacks, failed funding applications, stalled communication and even death, the project was continuously built up again by the generosity of its contributors and perseverance of its editors. We are deeply grateful to all authors involved in the creation of this book for their patience and belief in the project, and for their authentic contributions, insightful feedback, and commitment to dialogue.

The interdisciplinary nature of this book is perhaps most symbolically reflected in our publishing relationship with CRC Press, a division of Taylor and Francis. For an academic publisher known for their scientific journals and textbooks to take on a project of such size, thematic scope and aesthetic aspirations, represents for us an openness to pluralistic modes of knowledge creation and sharing, towards which this book strives. We would therefore like to thank the production team at Taylor and Francis for their keen eyes to detail and design, especially Teena Lawrence at NovaTechset, and Randy Brehm for taking on the project in the first place.

For their critical feedback and editing suggestions we would like to thank our colleagues, Myriel Milicevic, Winfried Blum and Björn Kluge. For opening up the concept of boundary objects in transdisciplinary stakeholder contexts, I (Toland) am indebted to the hours of conversations and co-teaching experience with Bettina König, Anett Kuntosch and Anne Dombrowski at the Humboldt University's IRITHEsys Integrative Research Institute. For their bibliographic expertise and wealth of knowledge and advice on publishing in the crossover fields of cultural and ecological studies and art, I (Toland) would like to thank Lorena Carràs and Jean-Marie Dhur of Zabriskie Books. For her impeccable transcription work on most of the interviews included in the book, we would like to thank Amanda Anderson. And on a personal level we would also like to thank our families for their patience, support and guidance, especially Ulf Kypke, Peter and Gerry Toland, and Bärbel Wessolek.

We would also like to recognize and express our appreciation and admiration of artists and cultrual reearchers who we have worked with and written about in the past and would have liked to include in this volume. Their ongoing work is an inspiration to diverse audiences worldwide and their ideas are very much alive in this book. These include the curators and art theorists, Clive Adams (director of the Centre for Contemporary Art and the Natural World and co-curator of the Soil Culture project, 2015), Beatrice Voigt (curator of the Boden Leben Conference, 2012, and longtime supporter of soil and arts), and Hildegard Kurt and Shelley Sacks (initiators of the Earth Forum and other methodologies of soil-related social sculpture). We also acknowledge the work of several artists, architects, filmmakers, and performance artists who have focused on soil properties in metaphor and material, including: Anneli Ketterer, Marianne Greve, Joel Tauber, Philip Topolovac, Philip Beesley, Smudge Studio, James Cassidy, Chris Fremantle, Deborah Koons Garcia, Natalie Jeremijenko, Mathias Kessler, Koichi Kurita, Helen Lessick, Perdita Phillips, Bonnie Ora Sherk, Touchstone Collaborations and Frollein Brehms.

We would similarly like to recognize our colleagues in the soil research community whose commitment to communication and raising soil awareness encouraged the creation of this book in the first place. We are particularly grateful to the ideas pursued by Daniel Hillel, and other "elders" of the soil science community for their insight on how soil touches every part of human culture. In many ways, *Field to Palette* can be seen as an intellectual result of many years of dialogue within the International Union of Soil Science's Division 4 (The role of soil in sustaining society and the environment) and related work in the German Soil Science Society's Commission 8 (Soils in Society and Education) and the Soil Science Society of America's Division on Soil Education and Outreach. The members of these groups aim to gather and report on research being conducted about soils in other disciplines, as well as to serve as a point of knowledge transfer between soil societies and educators, politicians, business leaders, and the general public. For their openness to interdisciplinary dialogue and interest in finding new ways of approaching soil understanding, we are grateful to: Christian Feller, Jock Churchmann, Nikola Pätzel, Cristine Carole Muggler, Jae Yang, John Kim, David Lindbo, Eric Brevik, Thomas Sauer, Richard Doyle, Damien Field, Sabine Grunwald, Ganga Hettiarachchi, Viktor Chude, Budiman Minasny, John Selker, Luca Montanerella, Birgit Wilhelm, and Klaus Müller.

Field to Palette appears at a time in which the IUSS' Division 4 has been increasingly reflecting on the cultural embeddedness of land use and soil protection measures, as well as its own connective role within the culture of the international soil science community. Following a series of books that document a diversity of approaches to understanding, managing, and valuing soil in ancient and contemporary societies (e.g. Footprints in the Soil – People and Ideas in Soil History, edited by B.P. Warkentin, 2006; Soil and Culture, edited by E.R. Landa and C. Feller, 2010; The Soil Underfoot – Infinite Possibilities for a Finite Resource, edited by G.J. Churchman and E.R. Landa, 2014), a new working group was formed by Christian Feller and Nikola Patzel in 2016 to address these concerns: Cultural Patterns of Soil Understanding (CPSU). This working group invites the expertise of artists, anthropologists, environmental historians, psychologists, and other humanities scholars, signaling a *qualitative turn* in the field of soil science. For these new perspectives we are grateful and hope that this book can serve as a first major contribution to that working group, and inspire further dialogue between the natural sciences, arts, and humanities.

Editors

Dr. Alexandra R. Toland is a visual artist and environmental planner with research interests in ecosystem services, urban ecology, soil and culture, and the Anthropocene. She is junior professor for arts and research at the Bauhaus University of Weimar and has previously lectured at the Technische Universität Berlin, University of Arts Berlin (UDK), Humboldt Universität zu Berlin, and Leuphana University. She co-chaired the German Soil Science Society's Commission on Soils in Education and Society from 2011 to 2015 with Gerd Wessolek and continues to write and make artwork about soil.

Dr. Jay Stratton Noller is professor of landscape pedology and head of the Department of Crop and Soil Science at Oregon State University. His research focuses on morphologistics and human interactions with soils in modern and ancient agricultural and forest landscapes of the Middle East, Europe, and the Americas. His experience crosses disciplines of soil science, geomorphology, art, and archaeology and his work as an artist at Soilscape Studio LLC is internationally recognized.

Dr. Gerd Wessolek is a soil physicist and painter who has pioneered efforts at giving soils and soil science a broader exposure to wider audiences through presentations, exhibitions, and soil art projects. Information on his research on urban soils in the vadose zone and an online gallery can be found at http://www.boden.tu-berlin.de. Since 1999 he has been chair of the Soil Protection Department at the Technische Universität Berlin.

Printed in the United States
by Baker & Taylor Publisher Services